Fast Liquid Chromatography–Mass Spectrometry Methods in Food and Environmental Analysis

Fast Liquid Chromatography–Mass Spectrometry Methods in Food and Environmental Analysis

Editors

Oscar Núñez
University of Barcelona, Spain

Héctor Gallart-Ayala
ONIRIS LABERCA, France

Claudia P B Martins
Thermo Fisher Scientific, France

Paolo Lucci
Pontificia Universidad Javeriana, Colombia

ICP

Imperial College Press

Published by

Imperial College Press
57 Shelton Street
Covent Garden
London WC2H 9HE

Distributed by

World Scientific Publishing Co. Pte. Ltd.
5 Toh Tuck Link, Singapore 596224
USA office: 27 Warren Street, Suite 401-402, Hackensack, NJ 07601
UK office: 57 Shelton Street, Covent Garden, London WC2H 9HE

Library of Congress Cataloging-in-Publication Data
Fast liquid chromatography-mass spectrometry methods in food and environmental analysis / edited by Oscar Núñez (University of Barcelona, Spain) [and three others].
pages cm
Includes bibliographical references and index.
ISBN 978-1-78326-493-3 (hardcover : alk. paper) -- ISBN 978-1-78326-494-0 (electronic)
1. High performance liquid chromatography. 2. Mass spectrometry. 3. Food--Analysis.
4. Environmental protection. I. Núñez Burcio, Oscar, editor.
TX548.2.L55F37 2014
664--dc23
2014026684

British Library Cataloguing-in-Publication Data
A catalogue record for this book is available from the British Library.

Typeset by Stallion Press
Email: enquiries@stallionpress.com

Printed in Singapore

Dedicated to Velhote
C. Martins

Dedicated to my parents,
Martina and Tiago
P. Lucci

Dedicated to my parents and sister
O. Núñez

Contents

"Learn from yesterday, live for today, hope for tomorrow. The important thing is not to stop questioning."

Albert Einstein

Preface

Liquid chromatography coupled to mass spectrometry is one of the most prominent techniques in analytical science. It has brought together a group of editors who, just like many of you, have been working (and struggling) in the development of liquid chromatography–mass spectrometry (LC–MS) methodology for the analysis of a variety of contaminants in food and environmental samples.

The story behind *Fast Liquid Chromatography–Mass Spectrometry Methods in Food and Environmental Analysis* goes back to 2012, when Oscar Núñez received an invitation to participate in the special issue *High-Performance Columns and Their Operations* published by Journal of Chromatography A and edited by Professor Nobuo Tanaka. The invitation to write a book on the same topic soon followed.

Our main objective was to address different methodologies concerning the use of fast liquid chromatography coupled to mass spectrometry in food and environmental analysis. In addition, some of the new trends such as on-line sample preparation, direct analysis and high resolution mass spectrometry are also discussed. We have tried to cover different aspects of the analytical workflow, including sample preparation, chromatographic separation and mass spectrometry analysis, as well as quantification and confirmatory strategies in three main sections:

(i) Novel approaches to achieve fast and ultrafast separations (ultrahigh-pressure liquid chromatography (UHPLC) with sub-2 µm and core-shell particles, monolithic columns and high temperature) and the use of complementary stationary phases, such as HILIC and perfluorinated reversed phases;
(ii) On-line sample treatment procedures — On-line Solid Phase Extraction, Molecular Imprinted Polymers and Turbulent Flow Chromatography — coupled to fast liquid chromatography, direct analysis (including desorption electrospray ionization (DESI) and direct analysis in real time (DART)) and the use of liquid chromatography high resolution mass spectrometry;
(iii) Relevant LC–MS applications focusing on the analysis of different groups of contaminants, including pesticides, mycotoxins, food packaging contaminants, perfluorinated and polyphenolic compounds.

This book compiles the work of many authors who are considered experts on many of the topics covered. We would like to acknowledge their work, time and brilliant contributions to this book.

Oscar Núñez
Héctor Gallart-Ayala
Cláudia P.B. Martins
Paolo Lucci

November 2014

Part 1

Fast Liquid Chromatography Advances

Chapter 1

UHPLC Separations Using Sub-2 µm Particle Size Columns

Julie Schappler, Jean-Luc Veuthey, and Davy Guillarme

School of Pharmaceutical Sciences, University of Geneva, University of Lausanne

1.1. General Introduction to Ultrahigh Performance Liquid Chromatography (UHPLC)

Since the early days of chromatography, various authors have demonstrated the utility of reducing particle size in liquid chromatography;[1] indeed, it is well known that the chromatographic efficiency (N) is proportional to particle diameter (d_p), according to the following equation:

$$N = \frac{L}{h \times d_p}, \qquad (1.1)$$

where L is the column length and h is the reduced plate height. The first particles (100–200 µm) were developed in the 1950s for liquid chromatography (LC), prior to smaller porous particles (in the range of 10 µm) in the early 1970s, although packing reproducibility was an issue at that time. Irregular micro-porous particles were used throughout the 1970s, until spherical material was obtained. In the 1980s, 5 µm became the standard particle diameter and in the early 1990s, 3–3.5 µm particle diameters became commercially available; the latter demonstrated 30–50% faster analysis times and higher efficiencies compared to 5 µm particles. In 2004, the breakthrough came with the introduction of porous silica of very small particle size (1.7 µm), which enabled better resolution compared to the current

5 µm or 3.5 µm materials.[2] Several column suppliers currently offer columns packed with particles in the range of 1.5–2 µm.[3]

The interest of decreasing particle size in HPLC is illustrated in Fig. 1.1, which shows the Van Deemter curves obtained for columns packed with 5 µm, 3.5 µm, and 1.7 µm particles. As the particle diameter decreases, the optimal linear velocity (u_{opt}) is shifted to higher values: u_{opt} = 0.7 mm/s for 5 µm, 0.9 mm/s for 3.5 µm, and 2.1 mm/s for 1.7 µm. It is then possible to work at high flow rates with small particles without any loss in efficiency. In agreement with Fig. 1.1, the corresponding H value decreases with the particle size: H_{opt} = 12.3 µm for 5 µm, H_{opt} = 8.8 µm for 3.5 µm, and H_{opt} = 3.9 µm for 1.7 µm. Thus, for a same column length, efficiency can be improved by a factor of three with a column packed with 1.7 µm particles instead of 5 µm. It is also possible to work at flow rates higher than the optimal value without major efficiency loss since the mass transfer is improved using 1.7 µm instead of 5 µm, as shown by the flatter Van Deemter curve.

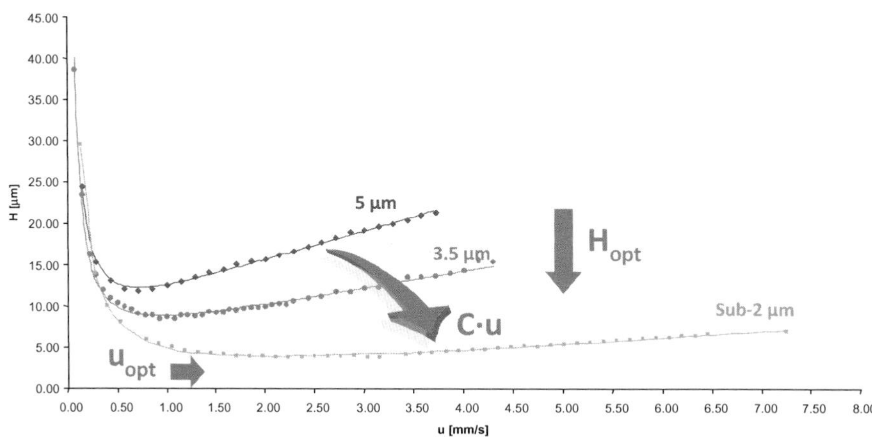

Figure 1.1. Impact of particle size reduction on the Van Deemter curves. Columns: XTerra, RP18, 4.6 mm × 150 mm, 5 µm; XTerra, RP18, 4.6 mm × 50 mm, 3.5 µm; Acquity BEH Shield, RP18, 2.1 mm × 50 mm, 1.7 µm. Adapted with permission from Nguyen, D.T.T., Guillarme, D., Rudaz, S., Veuthey, J.L. (2006). Chromatographic behaviour and comparison of column packed with sub-2 µm stationary phases in liquid chromatography, *J. Chromatogr. A*, **1128**, 105–113. Copyright (2006) Elsevier.

However, the reduction of particle size induces a significant increase in column pressure drop (ΔP), proportional to the square of the particle size, according to Darcy's law (and even to the cube of the particle size when working at the optimal linear velocity, the latter being also inversely proportional to particle size):

$$\Delta P = \Phi \frac{\eta L u}{d_p^2}, \tag{1.2}$$

where η is the mobile phase viscosity, L the column length, and Φ the flow resistance. Considering this constraint, it is required to employ columns packed with sub-2 μm particles exclusively on a new generation of instruments compatible with ultrahigh pressures, as stated by John Knox back in 1977.[4] He mentioned that ultrafast LC (i.e. short analysis time but low resolution) would require a new generation of particles and instrumentations. Particles of 1 μm or 2 μm and column lengths between 20 mm and 40 mm should be used to obtain $t_0 \approx 10$ s with reasonable pressures. Due to the strong reduction of the retention volume and the high applied-flow rate, the primary instrumental limitations would be the injector and detector performance (i.e. the injected quantity, the detector time constant, and the cell volume), as well as the system upper pressure limit. For this reason, 20 to 30 years have been spent to develop sub-2 μm particles and short columns. Today, such columns are available from several providers and instruments compatible with pressures in the range 1,000–1,300 bar are also accessible.

1.2. Proof-of-Concept of Ultrahigh-Pressure Liquid Chromatography (UHPLC) During the 1990s

Before the commercialization of UHPLC technology in 2004, the proof-of-concept was demonstrated by the groups of Jorgenson and Lee during the 1990s with a few impressive chromatographic separations.

The first outstanding separations performed at a ΔP_{max} of 4,100 bar were described in 1997 by Jorgenson *et al.* using fused-silica capillary (30 μm I.D., 52 cm length, packed with 1.5 μm non-porous particles). With this setup, a number of plates between 140,000 and 190,000 was reached for small molecules, together with analysis times lower than 10 min.[5] In 2003, Jorgenson's group increased the upper pressure limit and further reduced the particle size, to obtain exceptional chromatographic performance. A separation of five compounds under isocratic conditions was obtained (50 μm I.D., 43 cm length, packed with 1.0 μm non-porous particles) at a pressure of more than 7,000 bar. As illustrated in Fig. 1.2, analysis time was reduced to less than 4 min and plate count ranged between 190,000 and 300,000.[6]

In the meantime, Lee *et al.* also investigated UHPLC with capillary columns (29 μm I.D., 12.5 cm length, packed with 1.5 μm non-porous particles).[7] They constructed a homemade chromatographic apparatus compatible with pressure up to 3,600 bar. Separations of several benzodiazepines and herbicides were reached in less than 60 s and 100 s, respectively, with efficiency ranging from 20,000 to 30,000 plates.[8] Lee *et al.* also studied the combination of ultrahigh pressures with elevated temperatures.[9,10] Indeed, because high

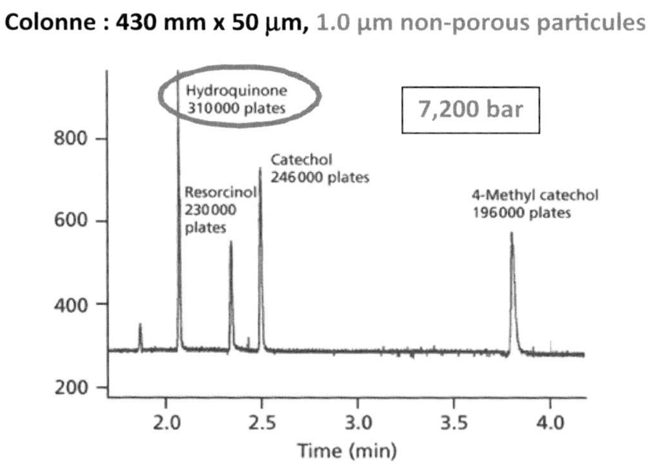

Figure 1.2. Chromatograms obtained with a column packed with 1.0 μm particles at run pressures of about 7,000 bar. Adapted with permission from Jerkovich, A.D., Mellors, J.S., Jorgenson, J.W. (2003). The use of micron-sized particles in ultrahigh-pressure liquid chromatography, *LCGC Eur.*, **16**, 20–30.

temperature induces a reduction of the mobile phase viscosity, system pressure drop remains reasonable. For this series of experiments, a capillary column (50 µm I.D., 14.5 cm length, packed with 1.0 µm non-porous particles) packed with a zirconia phase was employed due to its chemical stability at elevated temperatures. A separation of benzodiazepines was performed in less than 1.2 min at 100° C at a pressure of only 1,480 bar, while five herbicides were resolved with excellent efficiency in 60 s at 90° C and 1,800 bar.

As illustrated with these few examples, in the early days of UHPLC Jorgenson *et al.* demonstrated that UHPLC was a viable strategy for attaining high plate count ($N > 100{,}000$), ideal for the separation of complex mixtures of compounds. Conversely, Lee *et al.* mainly used UHPLC for performing rapid analysis of small molecules (in few minutes) with limited efficiency ($N < 30{,}000$).

1.3. Benefits of UHPLC Technique: Speed, Resolution, Solvent Consumption and Sensitivity

By adequately selecting the column length in UHPLC, it is possible, from a theoretical point of view, to increase the throughput by a factor of 9 compared to conventional HPLC while maintaining similar chromatographic efficiency. For example, if the original HPLC separation is carried out on a 150 mm, 5 µm column, then a 50 mm, 1.7 µm stationary phase should be selected in UHPLC, to achieve equivalent performance. The L/d_p ratio is similar between these two columns (equal to 30,000 and 29,400 respectively), generating equal plate number, in agreement with Eq. (1.1). The analysis time is reduced nine-fold, due to the three times shorter column and three times higher linear velocity. Such ultrafast separations have been experimentally obtained both in isocratic and gradient modes, and analysis times in the range 1–5 min can thus be expected.[11,12] This enhanced throughput is illustrated by the HPLC and UHPLC separations of a standardized extract of a widely used phytomedicine, *Ginkgo biloba* (Fig. 1.3). In agreement with the theory, a nine-fold reduction of the analysis time was obtained by transferring the 60 min HPLC gradient (on a 150 × 4.6 mm, 5 µm

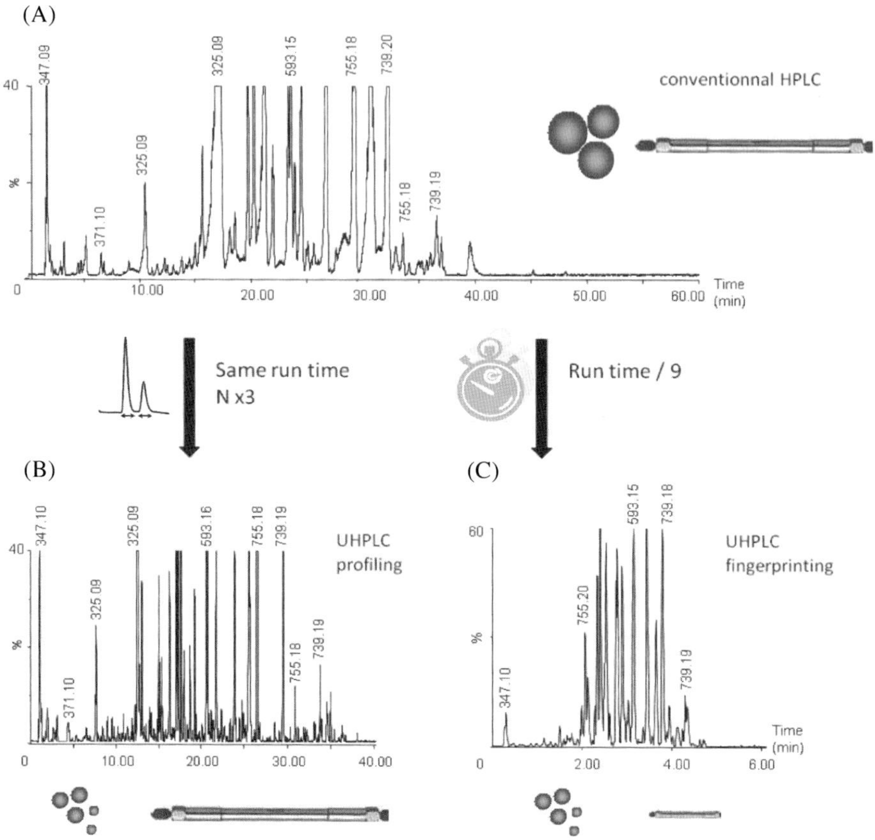

Figure 1.3. Comparison of chromatograms of a standardized *Gingko biloba* extract using method transfer. (A) Classical HPLC analysis carried out on a 150 mm × 4.6 mm, 5 μm column, with gradient of 5–40% ACN in 60 min at 1 mL/min. (B) HPLC method transferred on a 150 × 2.1 mm, 1.7 μm UHPLC column, with gradient of 5–40% ACN in 60 min at 0.350 mL/min. (C) Geometric method transfer to a 50 × 2.1 mm, 1.7 μm UHPLC column (with the same phase chemistry), with gradient of 5–40% ACN in 6.76 min (i.e. nine-fold reduction) at 0.600 mL/min. Adapted with permission from Eugster, P.J., Guillarme, D., Rudaz, S. *et al.* (2011). Ultrahigh-pressure liquid chromatography for crude plant extracts profiling, *J. AOAC Int.*, **94**, 51–70.

HPLC column, Fig. 1.3A) to a short UHPLC gradient (on a 50 mm × 2.1 mm, 1.7 μm UHPLC column, Fig. 1.3C).[3]

On the other hand, it is hypothetically possible to increase the efficiency by a factor of three in isocratic mode by keeping strictly identical column lengths in both HPLC and UHPLC, between columns packed with 5 μm and 1.7 μm. For instance, up to 35,000 plates can be reached with the latter using a 150 mm column, as demonstrated in Fig. 1.3B where a notable resolution improvement was obtained in the gradient mode, using the same gradient time as in regular HPLC. Some separations involving longer UHPLC columns have been widely reported in the literature and showed very high efficiency. However, optimal flow rate conditions are difficult to attain, due to the important backpressure generated by long columns packed with sub-2 μm particles. To further improve the compatibility of long UHPLC columns with existing instruments possessing an upper pressure limit of 1,000–1,300 bar, mobile phase temperature can be increased in the range 60–90°C. Under such conditions, the mobile phase viscosity is reduced and the pressure drops. The performance of UHPLC using long columns and elevated temperature (up to 90°C) was also demonstrated for large molecules, such as peptides.[13]

Another important benefit of UHPLC is the significant decrease in solvent consumption, which was particularly beneficial during the acetonitrile shortage in 2009. This is often a neglected argument for UHPLC since people focus their attention on the possibility of achieving high throughput and/or high-resolution separation. However, when comparing the solvent consumed by a regular HPLC method (involving a column of 150 × 4.6 mm, 5 μm) with that consumed by a UHPLC stationary phase (of 50 × 2.1 mm, 1.7 μm), the amount is decreased 14-fold, due to the important reduction of column length and internal diameter. Indeed, the solvent consumption is directly proportional to the column dead volume, which is reduced by a factor of 14 between the two sets of conditions.

Finally, from a theoretical point of view, a sensitivity increase can be expected with the UHPLC technology compared to HPLC, providing the same detector and the same column length are used. This increase is related to the higher efficiency produced by the UHPLC column, which results in narrower peaks. It should be noted however that if the injected volume is correctly scaled and an identical efficiency is maintained between HPLC and UHPLC (using similar L/d_p ratios), then there is no reason for better sensitivity in UHPLC, unless the detection technology has been improved. This is the case, for example, with modern UV detectors, including optofluidic waveguide technologies, or with the most recent mass spectrometry (MS) devices with enhanced ion transmission.

1.4. Method Transfer between HPLC and UHPLC

The term 'method transfer' is only valid as long as the stationary phase chemistry and mobile phase composition remain identical between HPLC and UHPLC. Numerous method transfer calculators are available from both commercial and academic sources[14] and integrate the equations that are described in this section. The user has to indicate the original and final column dimensions, as well as the mobile phase flow rate, injection volume and gradient profile. Then, the new conditions for UHPLC are automatically computed.

1.4.1. *Rules and examples in the isocratic mode*

In the isocratic mode, only two parameters have to be geometrically modified: the injection volume and the mobile phase flow rate.[11]

In order to avoid a detrimental band broadening related to the instrumentation and to maintain comparable sensitivities between HPLC and UHPLC, the injection volume has to be adapted in agreement with the column volume. As a rule of thumb, the injected volume should represent about 1–2% of the column volume in HPLC. The injected volume in UHPLC can be calculated by holding the ratio of the column dead volume to the injected volume constant between regular HPLC and UHPLC. The UHPLC injected volume (V_{inj2}) can be calculated using the following equation:

$$V_{inj_2} = V_{inj_1} \cdot \frac{d_{c2}^2}{d_{c_1}^2} \cdot \frac{L_2}{L_1}. \tag{1.3}$$

In Eq. (1.3), subscripts 1 and 2 are related to HPLC and UHPLC column dimensions, respectively. For example, between a HPLC column of 150 mm × 4.6 mm, 5 μm and a UHPLC column of 50 × 2.1 mm I.D., 1.7 μm, the injected volume has to be decreased

14-fold. Based on these calculations, an injected volume of 0.5–2 µL is conventional for a UHPLC column of 50 mm × 2.1 mm, 1.7 µm.

The mobile phase flow rate should also be adapted to remain as close as possible to the optimal linear velocity value of the Van Deemter curve (u_{opt}). In HPLC, the u_{opt} value is completely independent of the column length but is directly proportional to the square of the column diameter and inversely proportional to the particle size. For a successful method transfer, the product $u \cdot d_p$ must be held constant to account for simultaneous changes in the column diameter and particle size of the support. The UHPLC flow rate (F_2) can be calculated using the following equation:

$$F_2 = F_1 \cdot \frac{{d_{c2}}^2}{{d_{c1}}^2} \cdot \frac{d_{p1}}{d_{p2}} \tag{1.4}$$

In theory, between a regular 5 µm HPLC column and a 1.7 µm UHPLC column, the mobile phase linear velocity should be increased three-fold according to the Van Deemter representation (Fig. 1.1). However, because the internal diameter is also reduced between HPLC and UHPLC, the mobile-phase flow rate should be decreased by 1.6-fold between these two column dimensions. For instance, if the HPLC separation is carried out on a 150 mm × 4.6 mm, 5 µm column at a flow rate of 1 mL/min, the UHPLC column of 50 mm × 2.1 mm, 1.7 µm should be operated at 600 µL/min.

Many applications involving pharmaceutical substances can be found in the literature to highlight the interest of transferring an HPLC method to UHPLC. As reported in Fig. 1.4, a commercial pharmaceutical formulation composed of the active pharmaceutical ingredient, two antimicrobial preservatives and a possible degradation product was analysed for quality control purposes.[11] The original HPLC separation took approximately 8 min at a flow rate of 1 mL/min on a conventional RP18 column, whereas the analysis time was reduced to only 1.2 min under UHPLC conditions at optimal mobile phase flow rate (F = 613 µL/min), enabling similar resolution and selectivity between both methods. The analysis time could be further reduced to 45 s at a pressure of 1,000 bar and a flow rate of 1 mL/min working

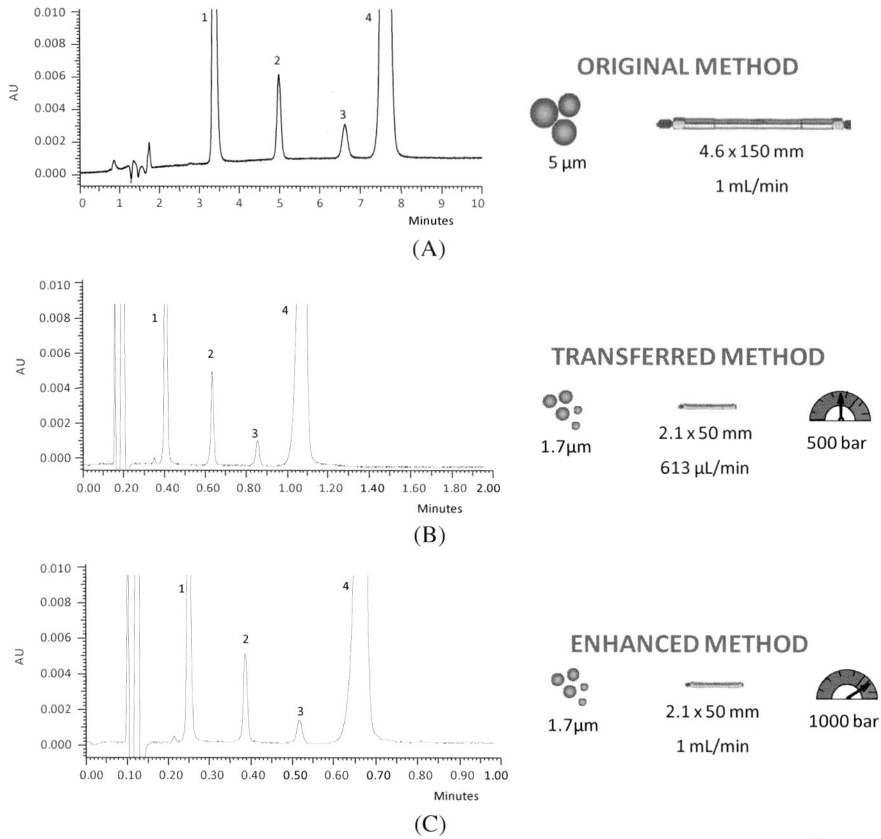

Figure 1.4. Isocratic method transfer from regular HPLC to UHPLC. Separation of a pharmaceutical formulation in isocratic mode. Elution order: (1) methylparaben, (2) 2,6-dimethylaniline, (3) propylparaben, (4) lidocaine.

beyond the optimal mobile phase velocity in UHPLC, without a significant loss in efficiency (due to the flat Van Deemter curve). In this last case, the resolution was reduced by approximately 15%, whereas the analysis time was decreased 11-fold.

1.4.2. *Rules and examples in the gradient mode*

The conversion of a gradient method involves rules that are more complex than those applied in the isocratic mode. The mobile phase flow rate and injection volume should be scaled in a similar way to the isocratic mode. In the case of linear or multi-linear gradient elution, the gradient profile can be decomposed into a combination of isocratic and gradient steps.

For any isocratic step (i.e. initial isocratic step, isocratic step during a multi-linear gradient, and final re-equilibration time), the ratio between the isocratic step time and the column dead time (which depends on the mobile phase flow rate, column I.D. and length) should be kept constant. The UHPLC isocratic step (t_{iso_2}) can be determined using the following equation:

$$t_{\mathrm{iso}_2} = t_{\mathrm{iso}_1} \cdot \frac{F_1}{F_2} \cdot \frac{d_{c_2}^2}{d_{c_1}^2} \cdot \frac{L_2}{L_1}. \tag{1.5}$$

For a method transfer between a regular 150 mm × 4.6 mm, 5 μm column and a 50 mm × 2.1 mm, 1.7 μm column, all the isocratic steps should be reduced nine-fold, at the optimal mobile phase linear velocity. The same reduction factor should be applied for scaling the re-equilibration time. Indeed, the usual 15–20 min required in HPLC experiments is reduced to approximately 2 min in UHPLC.

For slope segments, the rules originally described by Snyder and Dolan should be strictly followed.[15] To keep the relative position (i.e. the selectivity) in the chromatogram unchanged, it is important to scale the gradient volume in proportion to the number of column volumes, whereas the initial and final gradient composition (%B) should be kept constant. The new gradient time (t_{grad2}) can be expressed as:

$$t_{\mathrm{grad}_2} = \frac{(\%B_{\mathrm{final1}} - \%B_{\mathrm{initial}_1})}{\mathrm{slope}_2} \tag{1.6}$$

The gradient slope should be calculated to keep the product of the gradient slope and the column dead time constant. The gradient slope (slope_2) can be expressed as:

$$\mathrm{slope}_2 = \mathrm{slope}_1 \cdot \frac{d_{c_1}^2}{d_{c_2}^2} \cdot \frac{L_1}{L_2} \cdot \frac{F_2}{F_1}. \tag{1.7}$$

In the method transfer from a regular HPLC column of 150 mm × 4.6 mm, 5 μm column to a UHPLC 50 mm × 2.1 mm, 1.7 μm column, the gradient slope should be increased nine-fold.

Because the gradient elution mode is more popular than the isocratic method, an important number of applications dealing with the gradient method transfer between conventional and sub-2 μm column packings can be found in scientific literature. As reported in Fig. 1.5, an HPLC method was developed to separate an active pharmaceutical ingredient from 11 different impurities.[12] The separation was achieved using a 150 mm × 4.6 mm, 5 μm, RP18 column and was further transferred to UHPLC using a 50 mm × 2.1 mm, 1.7 μm RP18 column with a strictly similar chemistry. The original separation was performed in approximately 27 min prior to an efficient transfer to UHPLC, which enabled a separation in less than 3 min. Both separations were equivalent in terms of sensitivity, peak capacities, and resolution. In order to further increase the throughput, the UHPLC mobile phase flow rate was increased to 1200 μL/min (generating approximately 1,000 bar), leading to a complete separation of the complex mixture in approximately 1.6 min.

1.5. The Importance of Instrumentation in UHPLC

All the instrumental constraints involved in fast LC are summarized in Fig. 1.6 and described in detail in this section.[16]

1.5.1. *Extra-column band broadening*

The success of fast and ultrafast separations depends on the intrinsic column efficiency, but also on its conservation by minimizing the extra-column band spreading that occurs inside the instrument. Extra-column band broadening effects are particularly critical using short columns with reduced internal diameter (e.g. 50 mm × 2.1 mm I.D.). Various papers have discussed the extra-column effects as a major

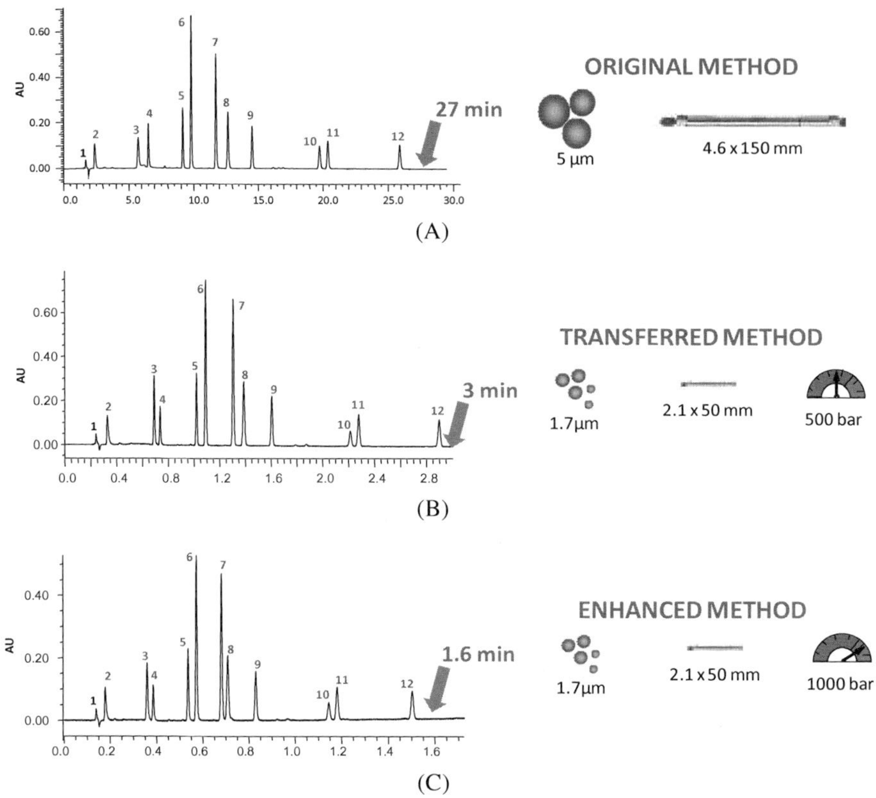

Figure 1.5. Gradient method transfer from regular HPLC to UHPLC. Separation of a pharmaceutical mixture containing the main product (6) and 11 impurities. Adapted with permission from Guillarme, D., Ruta, J., Rudaz, S. *et al.* (2010). New trends in fast and high-resolution liquid chromatography: a critical comparison of existing approaches, *Anal. Bioanal. Chem.*, **397**, 1069–1082. Copyright (2010) Springer.

concern that negatively impacts the apparent column efficiency under UHPLC conditions.[17–19]

The total band broadening is the sum of the column and the extra-column broadening, as shown by the following equation:

$$\sigma_{tot}^2 = \sigma_{col}^2 + \sigma_{ext}^2. \tag{1.8}$$

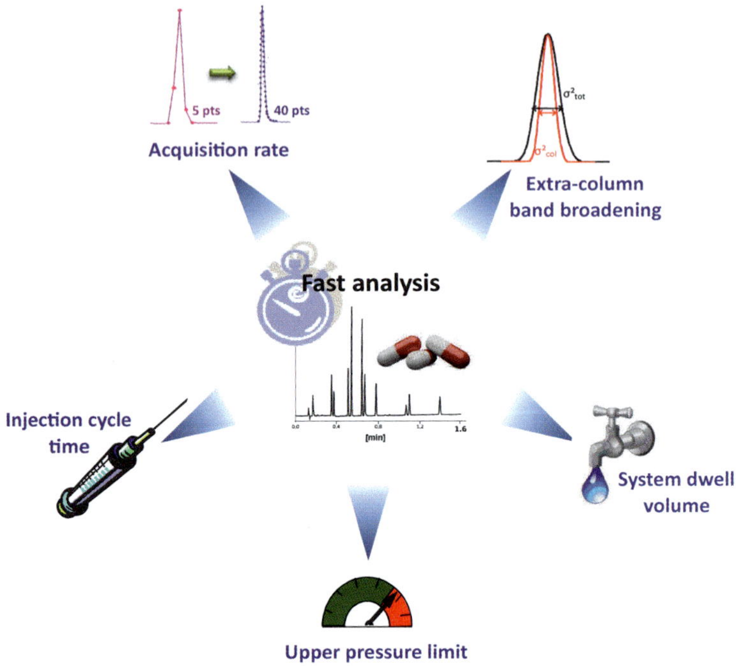

Figure 1.6. Instrumental constraints related to UHPLC experiments. Adapted with permission from Fekete, S., Kohler, I., Rudaz, S. *et al.* (2014). Importance of instrumentation for fast liquid chromatography in pharmaceutical analysis, *J. Pharm. Biom. Anal.*, 87, 105–119. Copyright (2014) Elsevier.

In general, the ratio between extra-column variance, σ^2_{ext}, and total variance, σ^2_{tot}, should be less than 10% to have an acceptable loss in efficiency.

The dispersion linked to the chromatographic column itself (σ^2_{col}) can be expressed by:

$$\sigma^2_{col} = \frac{V_R}{\sqrt{N}} = \frac{V_0.(1+k)}{\sqrt{N}}, \tag{1.9}$$

where σ^2_{col} is the column variance (in μL^2), and V_R is the retention volume, a function of column dead volume (V_0) and retention factor (k):

$$V_R = V_0 \cdot (1+k) \tag{1.10}$$

Extra-column band broadening can be expressed as the sum of three main dispersion sources:

$$\sigma^2_{\text{ext}} = \sigma^2_{\text{inj}} + \sigma^2_{\text{det}} + \sigma^2_{\text{tubing}}, \tag{1.11}$$

where σ^2_{ext} is the extra-column variance, and σ^2_{inj}, σ^2_{det}, and σ^2_{tubing} are the variances due to the injector, detector, and tubing, respectively. A more detailed expression of Eq. (1.11) can be expressed as:

$$\sigma^2_{\text{ext}} = K_{\text{inj}} \cdot \frac{V^2_{\text{inj}}}{12} + \frac{V^2_{\text{cell}}}{12} + \tau^2 \cdot F^2 + \frac{r_c^4 \cdot l_c \cdot F}{7.6 \cdot D_m}, \tag{1.12}$$

where V_{inj} is the injected volume, r_c and l_c are the tubing radius and length, respectively, V_{cell} is the flow-cell volume, τ is the detector time constant, and K_{inj} and K_{cell} are constants linked to the injection mode and the UV cell geometry, respectively. Considering these equations, the following criteria must be fulfilled to attain sharp peaks in UHPLC, when working with short columns of 50 mm.

The tubing length should be as short as possible between injector and column inlet and between column outlet and UV or MS detector,[20] and the diameter of the tubing should be selected as a compromise between a reasonable pressure and low volume. Thus, a system plumbed with 0.005″ I.D. tubing and zero dead-volume fittings seems optimal for UHPLC experiments. The latest generation of instruments is plumbed with 0.003″ I.D. tubing, but the residual system pressure is relatively high.

As previously discussed in the section dedicated to method transfer, the injection volume should be scaled in agreement with column geometry. Because most of the experiments conducted with UHPLC are performed with a 50 mm × 2.1 mm I.D. column (V_0 = 120 μL), the injected volume should be between 0.5 and 2 μL to limit band broadening, and up to 5 μL if the solvent used to dissolve the sample is less eluent than the mobile phase (band compression effect). In addition, a fast injection cycle time (less than 30 s) is mandatory for analysis times shorter than 1 or 2 min.

Finally, the detector cell volume and time constant should also be adequately selected. The volume of the UV cell should be <2 μL, while a pathlength of 10 mm should be maintained to avoid sensitivity decrease. The detector time constant has to be sufficiently fast ($10\ ms < t < 100\ ms$) due to the thin peak widths experienced in UHPLC (only a few s) but it should not be too low, otherwise the sensitivity can be reduced because of a high background noise induced by a weak time constant.

To evaluate the compatibility of a UHPLC instrument with a given column geometry, it is advised to experimentally measure the extra-column volume and/or variance. For optimal compatibility with fast separations, the extra-column volume should be less than 20 µL.

1.5.2. *System dwell volume*

The system dwell volume (V_d), also known as gradient delay volume of a chromatographic system, represents the volume from the mixing point of solvents to the head of the analytical column. In other words, it represents the delay that is observed until the selected proportion of solvent reaches the column inlet after the beginning of the gradient. The sample is thus subjected to an undesired additional isocratic migration in the initial mobile phase composition.[16] Two types of pumping systems are commercially available for HPLC operations: (i) high-pressure mixing systems, where the dwell volume comprises the mixing chamber, the connecting tubing, and the autosampler loop; and (ii) low-pressure mixing systems, combining the solvents upstream from the pump, where additional tubing as well as volume of the pump head is added to the components of the high-pressure mixing system.[21] In the case of conventional HPLC systems, typical dwell volumes are in the range of 0.5–2 mL and 1–5 mL for high-pressure and low-pressure mixing systems, respectively.

The dwell volume may differ from one instrument to another, but it can be easily measured according to several procedures described in the literature.[21] The dwell volume of all UHPLC systems ($\Delta P > 600$ bar) currently available on the market was recently reported.[16] Compared to conventional HPLC instruments, which possess V_d between 0.5 mL and 5 mL, UHPLC systems have average dwell volumes of 300–400 μL, with the best UHPLC systems having V_d of ~100 μL, and up to 1 mL for others. Two main concerns related to large system dwell volumes are involved when performing fast separations in LC: (i) unreliable gradient method transfer between columns of different geometries, which can be avoided by considering the dwell volume during gradient method transfer calculations; and (ii) analysis times longer than expected, due to the creation of an additional initial isocratic step.

In conclusion, it is straightforward to determine which type of column geometry can be employed on which instrument, after adequate system characterization. In theory, small V_d are highly recommended for fast and ultrafast analysis, but some issues have been reported with various UHPLC systems equipped with small mixers. A problem of excessive blending noise can occur, caused by inadequate mixing of mobile phases from the binary pumps.[22,23] This blending noise can depend on the pump design (i.e. piston column, mixer volume, and presence of a damper). As a result, cyclical perturbations synchronized to the pump strokes can be observed on the UV signal, leading to a sensitivity reduction. The phenomenon is particularly relevant when the UV detection is carried out at low wavelengths (<230 nm) with volatile mobile phases (e.g. formate or acetate salts). To avoid this problem, larger mixing volumes can be used, but at the expense of increased dwell time.

1.5.3. *System upper pressure limit*

In order to benefit from all features of columns packed with fully porous sub-2 μm particles, a chromatographic system possessing an extended upper pressure limit should be used. A large variety of chromatographic apparatus compatible with pressures between 600 bar and 1,300 bar have been developed in conjunction with new column technologies. Compared to the works of Jorgenson and Lee (ΔP_{max} up to 7,000 bar), the upper pressure limits of commercial UHPLC systems are more reasonable. The restricted pressure limitation can easily be explained by the construction design, robustness, and safety of UHPLC systems and columns. In addition, under very high-pressure conditions, negative effects can be observed on chromatographic efficiency, selectivity, retention, and repeatability. Regarding the instrumentation, there are many design trade-offs that are required to balance system characteristics (robustness, sensitivity, and safety) and intrinsic performance. Today, there is a wide choice of columns packed with fully porous sub-2 μm particles, compatible with pressures up to 1,300 bar. These stationary phases are (i) hybrid silica material prepared from two monomers (i.e., tetraethoxysilane and bis(triethoxysilyl)ethane), which incorporates ethylene bridges (the high degree of cross-linking ensures strong mechanical and hydrolytic stability);[24] and (ii) classical silica packed under very high pressure conditions.

In order to evaluate the achievable throughput and resolution on commercially available UHPLC instruments, the kinetic plot methodology can be implemented. This approach allows for visualizing the performance limits of a given stationary phase technology on a particular instrument at an upper pressure limit.[25,26] Using this approach, it has been demonstrated that a chromatographic system possessing an elevated upper pressure limit enables reduction of the analysis time for a given plate number. For example, for a column packed with fully porous 1.7 μm particles, an efficiency of 10,000 plates can be attained in 1.9 min at a pressure of 400 bar, while it requires only 0.7 min and 0.6 min at 1,000 bar and 1,200 bar, respectively.[16] In other words, analysis time is always inversely proportional to the $\Delta P_{max,}$ and elevated system pressure is always beneficial for maximizing kinetic performance. In the case of relatively low plate count (i.e. 10,000 plates), the three-fold reduction in analysis time between a system possessing a ΔP_{max} of 400 or 1,200 bar remains negligible, since the separation can be performed in less than 2 min in both cases. Conversely, for high-resolution separations (i.e., 100,000 plates), analysis time can be reduced by three-fold between systems possessing a ΔP_{max} of 400 and 1,200 bar. In this case, it corresponds to a significant analysis time reduction, from around 3 hr to only 1 hr using columns packed with 1.7 μm particles.[16]

These observations confirm that working with a chromatographic system possessing an elevated upper pressure limit is particularly relevant for high-resolution analysis (high plate count).

1.5.4. *Detector data acquisition rate for UHPLC operation*

Another technical limitation encountered in UHPLC is related to the detector acquisition rate. The narrow peaks produced by state-of-the-art stationary phases packed with sub-2 µm particles — down to only 1s in some cases — can be a real issue. For quantitative determination, 10 to 15 acquisition points per peak are recommended to correctly define the chromatographic peak and obtain suitable performance. This criterion can be easily fulfilled with spectroscopic detectors such as UV and UV diode array detection (UV–DAD), frequency-domain (FD) fluorescence, or even with evaporative light scattering detectors (ELSD) and corona aerosol discharge (CAD), as data acquisition rates up to 200 Hz have been reported with the last generation of detector devices. However, achieving high acquisition rate with MS devices is more difficult since the latter must be compatible with short dwell times and fast duty cycles.[27,28] Only the latest generation of instruments meets these requirements. In this context, the most powerful triple quadrupole (QqQ) devices are now able to perform multiple reaction monitoring (MRM) experiments in only 1 ms, while the best quadrupole–quadrupole time-of-flight mass spectrometry (QqTOF/MS) instruments on the market are able to acquire *m/z* ratios from 100 to 1,000 only in 10 ms with a resolution of ~40,000 FWHM (full width at half maximum).[27]

1.5.5. *Overview of UHPLC instruments and columns on the market*

Today, there is a wide choice of instruments compatible with pressures above 400 bar and columns packed with porous sub-2 µm particles, from various suppliers. An exhaustive list of existing UHPLC systems and columns can be found elsewhere.[3,29]

In order to select an appropriate UHPLC instrument, the cost is certainly a crucial argument but it is also important to analyse in details the technical specifications of all available instruments on the market. The most important features remain the maximal available pressure and corresponding flow rate. For commercial apparatus, ΔP_{max} varies between 600 bar and 1,300 bar. As previously discussed, for fast and ultrafast UHPLC separations of simple mixtures, the use of small particles is noticeable but the interest of extremely high pressure drop is much more limited. For high throughput experiments, the UHPLC instruments with pressure limits around 600–800 bar can provide a suitable solution at affordable price.[30] Conversely, for the characterization of complex samples, necessitating high resolution separation, long columns packed with sub-2 µm particles have to be employed, and thus, a system with maximal pressure of 1,000–1,300 bar is mandatory to work at acceptable flow rates.[30] In addition to the pressure capability of the UHPLC system, the instrument has to be adapted to operate in fast and ultrafast mode with reduced column volumes. For this purpose, systems presenting low dead volume (i.e. reduced injection volume, UV cell, tubing length, and I.D.), high acquisition rate (>40 Hz), and small gradient delay volume (<200 µL) have to be preferentially selected. A comparative study of various UHPLC systems has been performed lately, taking into account all instrument specifications.[31]

Another important aspect of UHPLC is the selection of adequate stationary phases that provide sufficient selectivity as well as reasonable performance and lifetime. Today, the number of columns packed with porous sub-2 µm particles is important, with more than 100 supports and 10 phase chemistries accessible from at least 15 different providers, demonstrating the opportunity to transfer almost all existing methods from HPLC to UHPLC. All these stationary phases are not equivalent in terms of pressure tolerance (from 600 bar to 1,300 bar) and particle size (from 1.5 µm to 2 µm), but also pH and temperature acceptance range. Some performance comparison between the different phases can be found in the literature[32,33] and data for column lifetime were also published.[34] Lifetimes of

UHPLC and regular HPLC columns were estimated and 1,000 to 2,000 injections can be performed on a column. However, this result can be strongly affected by the nature of the analysed samples, supplier, pressure, and mobile phase flow rate.

1.6. The Importance of Frictional Heating under Very High Pressure Conditions

When columns packed with very fine particles are operated at high mobile phase velocities, high pressure drops are generated.[35,36] Consequently, heat is induced by friction of the mobile phase percolating through the column bed at very high pressure (frictional heating phenomenon). The generated heat dissipates along and across the chromatographic column, inducing the formation of axial (longitudinal) and radial temperature gradients. The radial temperature gradients are mostly observed under isothermal conditions (e.g. the column is placed in a forced-air oven or in a thermostated water bath) and influence column efficiency. The longitudinal temperature gradients occur when the column wall temperature is not kept constant (adiabatic case: e.g. the column is placed in a still-air oven). A longitudinal temperature gradient may alter the retention and selectivity through changes in the average column temperature, whereas column efficiency is not seriously affected. Several approaches were employed to experimentally determine the temperature amplitude variation in UHPLC columns.[37–39] The temperature differences between the column inlet and outlet are in the range 13–20 °C and 3–16 °C for various lengths of 2.1 mm and 1 mm I.D. columns, respectively. The difference of temperature inside the UHPLC column induced by frictional heating is illustrated in Fig. 1.7. The first series of chromatograms was performed at a constant temperature of 30 °C and increased pressure between 100 bar and 1,000 bar. The second series was realized at a constant pressure of 100 bar and increased temperature between 30 °C and 50 °C. This study demonstrates that the pressure increase has the same effect on retention and selectivity as a temperature increase.

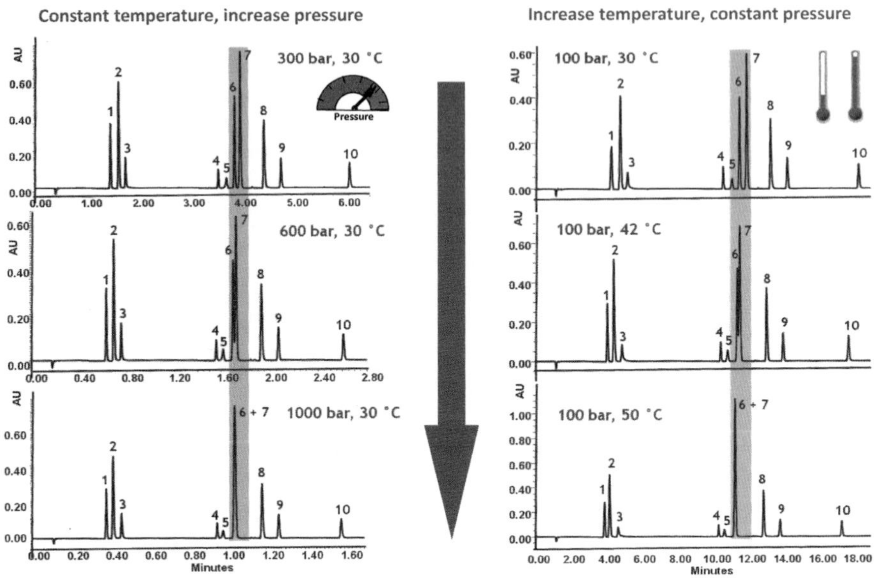

Figure 1.7. Influence of increased temperature at 100 bar and increased pressure at 30°C on an Acquity BEH Shield, RP18, 50 mm × 2.1 mm, 1.7 mm column. Elution order: (1) paracetamol, (2) salicylic acid, (3) catechin, (4) ethacrynic acid, (5) oxycodone, (6) dexamethasone, (7) indapamide, (8) nortriptyline, (9) gestrinone, and (10) thioridazine. Adapted with permission from Novakova, L., Veuthey, J.L. and Guillarme, D. (2011). Practical method transfer from highperformance liquid chromatography to ultrahigh-performance liquid chromatography: the importance of frictional heating, *J. Chromtogr. A*, **1218**, 7971–7981. Copyright (2011) Elsevier.

The dissipation power is directly proportional to the mobile phase flow rate (F) and column pressure drop (ΔP), according to:

$$\text{Power} = F{\cdot}\Delta P\,. \tag{1.13}$$

Therefore, two solutions can be applied to limit the frictional heating phenomenon: (i) reduction of the column inner diameter and (ii) reduction of the backpressure inside the column. Because current UHPLC instrumentations are hardly or even not compatible with columns of less than 1 mm I.D. (due to the strong contribution of extra-column variance to peak broadening), the best solution to alleviate the frictional heating is to set the upper pressure limit of instruments at a reasonable value (i.e. 1,000–1,200 bar).

1.7. Comparison of UHPLC Performance with Other Modern LC Strategies

To compare various chromatographic techniques, several approaches can be implemented.

The simplest one consists in constructing Van Deemter curves for various types of packing technology, as reported in Fig. 1.1, although such representation does not take into account the permeability of the support.

It is also possible to draw kinetic plots, which are based on the extrapolation of the Van Deemter (H, u) and permeability (K_{v0}) data obtained with a column of a given length, to shorter and longer columns, generating the upper pressure drop withstood by the chromatographic system. The final representation includes extrapolated N and t_0 values, calculated by simply reorganizing the Van Deemter and permeability data. Then, it becomes straightforward to determine (i) the minimum time required to reach a particular efficiency, and (ii) the maximal efficiency that can be achieved for a given analysis time. A simplified way to construct kinetic plots using only two simple equations was recently proposed by Desmet *et al.*[25,26]:

$$N = \frac{\Delta P_{\max}}{\eta}\left(\frac{K_{v0}}{u \times H}\right) \tag{1.14}$$

$$t_0 = \frac{\Delta P_{\max}}{\eta}\left(\frac{K_{v0}}{u^2}\right), \tag{1.15}$$

where $\Delta P_{\max}$ is the maximal system pressure (or column mechanical stability), η is the mobile phase viscosity, and H the height equivalent to a theoretical plate.

Because kinetic plots are not readily accessible for non-initiated people, a simplified approach was recently suggested to visualize data obtained from the kinetic plot methodology.[30] An illustration is provided in Fig. 1.8 for isocratic and gradient modes. In Fig. 1.8A, the lowest time required to attain a plate count of 10,000 is reported on the *y*-axis, while the maximal efficiency that can be reached with a column dead time of 30 min is reported on the *x*-axis. The performance of various analytical conditions, including silica-based monoliths, columns packed with fully porous 5 μm and 1.7 μm particles as well as core-shell 2.7 μm particles, using ambient and high temperature (90 °C), at various upper pressure limits, can then be compared. High-throughput separations (*y*-axis) require columns packed with small particles and should be carried out at increased temperatures. Conversely, high-resolution separations (*x*-axis) have to be performed with highly porous material, such as monoliths. The maximal temperature and pressure drop should be increased as much as possible, since both parameters are beneficial for increasing the plate count and throughput. Figure 1.8B reports similar types of data for gradient elution mode. In this representation, the lowest time required to attain a peak capacity of 100 is reported on the *y*-axis, while the maximal peak capacity that can be reached with a gradient time of 3 hours is reported on the *x*-axis. The performance obtained in the isocratic and gradient modes is similar, except for the UHPLC and high temperature (HT)-UHPLC strategies, which become more attractive for high-resolution separations in gradient versus isocratic mode. The fused-core (superficially porous particles) and UHPLC technologies are both very attractive for maximizing throughput and resolution in gradient mode. Whatever the selected strategy, the use of high temperatures is an additional parameter to improve gradient performance.

1.8. Conclusions and Perspectives

The use of columns packed with sub-2 μm particles under very high-pressure conditions offer obvious benefits over the traditional HPLC approach. The analysis time can be drastically reduced (on average by a factor 5 to 10) without compromising efficiency. It is also possible to drastically enhance resolution, with a limited impact on analysis time. In addition, the solvent consumption can be severely

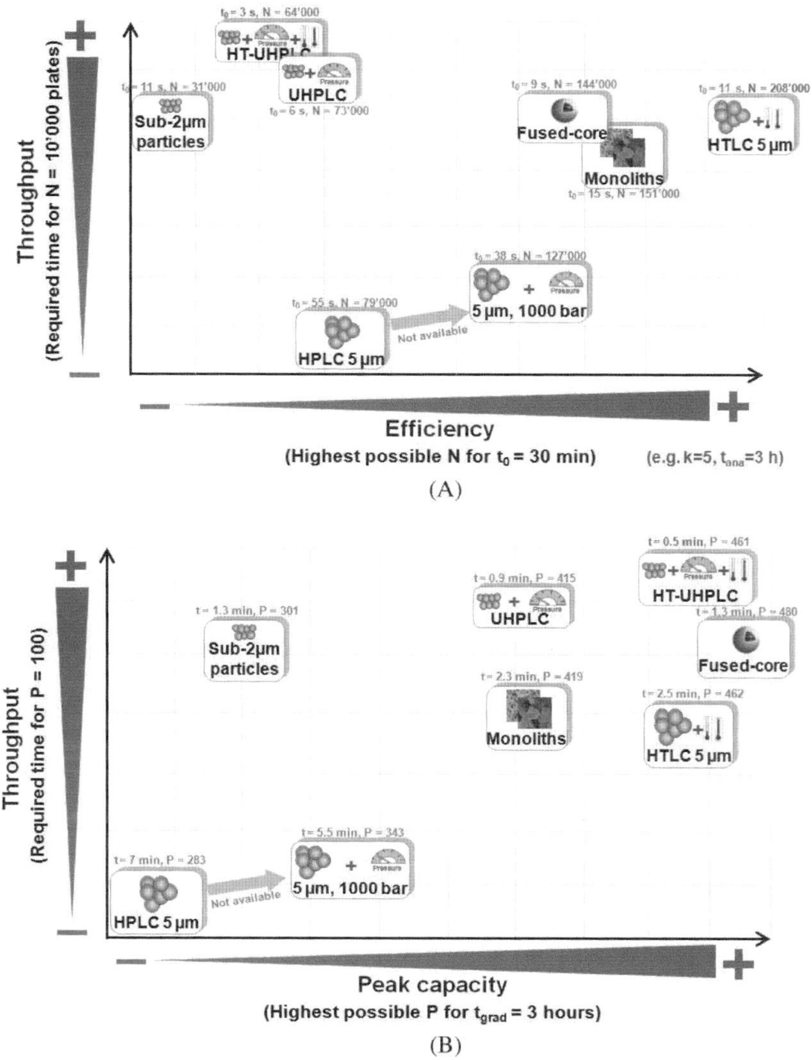

Figure 1.8. Performance of various LC strategies. (A) Isocratic mode: impact on throughput (t_0 for N = 10,000 plates) and maximal efficiency (N_{max} for t_0 = 30 min), considering a maximal pressure of 200 bar for monoliths, 400 bar for HPLC, HTLC and sub-2 µm, 600 bar for fused-core, and 1,000 bar for UHPLC, HT-UHPLC and HPLC 1,000 bar. (B) Gradient mode: impact on throughput (t_{grad} for P = 100) and maximal peak capacity (P_{max} for $_{tgrad}$ = 3 h). Adapted with permission from Guillarme, D., Ruta, J., Rudaz, S. *et al.* (2010). New trends in fast and high-resolution liquid chromatography: a critical comparison of existing approaches, *Anal. Bioanal. Chem.*, **397**, 1069–1082. Copyright (2010) Springer.

decreased thanks to the use of shorter and thinner columns. Last, the method transfer between HPLC and UHPLC is quite straightforward, due to the wide range of stationary phase chemistries and dimensions available in UHPLC mode. In order to take advantage of all benefits of UHPLC, the instrument specifications should be considered and a system possessing low extra-column volume, small dwell volume, fast injection cycle time and data acquisition rate, and high upper pressure limit is highly recommended.

As an alternative, the use of columns packed with superficially porous particles (also known as fused-core or core-shell) is expanding very quickly and appears as a serious competitor to the UHPLC technology, since very good chromatographic performance was achieved with sub-3 µm superficially porous particles.[40–42] This column technology could easily outperform UHPLC, thanks to a new generation of columns compatible with elevated temperatures (up to 90 °C) and high backpressure (1,000–1,200 bar).

References

1. Majors, R.E. (2005). Fast and ultrafast HPLC on sub-2 µm porous particles — where do we go from here?, *LCGC North Am.*, **23**, 1248, 1250–1255.
2. Mazzeo, J.R., Neue, U.D., Kele, M. *et al.* (2005). A new separation technique takes advantage of sub-2 µm porous particles, *Anal. Chem.*, **77**, 460A–467A.
3. Eugster, P.J., Guillarme, D., Rudaz, S. *et al.* (2011). Ultrahigh-pressure liquid chromatography for crude plant extracts profiling, *J. AOAC Int.*, **94**, 51–70.
4. Knox, J.H. (1977). Practical aspects of LC theory, *J. Chromatogr. Sci.*, **15**, 352–364.
5. MacNair, J.E., Lewis, K.C. and Jorgenson, J.W. (1997). Ultrahigh-pressure reversed-phase liquid chromatography in packed capillary columns, *Anal. Chem.*, **69**, 983–989.
6. Jerkovich, A.D., Mellors, J.S. and Jorgenson, J.W. (2003). The use of micron-sized particles in ultrahigh-pressure liquid chromatography, *LCGC Eur.*, **16**, 20–23.

7. Lippert, J.A., Xin, B., Wu, N. *et al.* (1999). Fast ultrahigh-pressure liquid chromatography: on-column UV and time-of-flight mass spectrometric detection, *J. Microcol. Sep.*, **11**, 631–643.
8. Wu, N., Lippert, J.A. and Lee, M.L. (2001). Practical aspects of ultrahigh pressure capillary liquid chromatography, *J. Chromatogr. A*, **911**, 1–12.
9. Xiang, Y., Yan, B., Yue, B. *et al.* (2003). Elevated-temperature ultrahigh-pressure liquid chromatography using very small polybutadiene-coated nonporous zirconia particles, *J. Chromatogr. A*, **983**, 83–89.
10. Xiang, Y., Yan, B., McNeff, C.V. *et al.* (2003). Synthesis of micron diameter polybutadiene-encapsulated non-porous zirconia particles for ultrahigh-pressure liquid chromatography, *J. Chromatogr. A*, **1002**, 71–78.
11. Guillarme, D., Nguyen, D.T.T., Rudaz, S. *et al.* (2007). Method transfer for fast liquid chromatography in pharmaceutical analysis: Application to short columns packed with small particle. Part I: Isocratic separation, *Eur. J. Pharm. Biopharm.*, **66**, 475–482.
12. Guillarme, D., Nguyen, D.T.T., Rudaz, S. *et al.* (2008). Method transfer for fast liquid chromatography in pharmaceutical analysis: Application to short columns packed with small particle. Part II: Gradient experiments, *Eur. J. Pharm. Biopharm.*, **68**, 430–440.
13. Eugster, P.J., Biass, D., Guillarme, D. *et al.* (2012). High-resolution comprehensive profiling of Conus venom by ultrahigh-pressure liquid chromatography coupled to time-of-flight mass spectrometry, *J. Chromatogr. A*, **1259**, 187–199.
14. HPLC calculator: software for chromatographic performance evaluation and HPLC method transfer. (2009). Available at: http://www.unige.ch/sciences/pharm/fanal/lcap/telechargement.htm [consulted in October 2013].
15. Dolan, J.W. and Snyder, L.R. (1998). Maintaining fixed band spacing when changing column dimensions in gradient elution, *J. Chromatogr. A*, **799**, 21–34.
16. Fekete, S., Kohler, I., Rudaz, S. *et al.* (2013). Importance of instrumentation for fast liquid chromatography in pharmaceutical analysis, *J. Pharm. Biomed. Anal.*, **87**, 105–119.

17. Fekete, S. and Fekete, J. (2011). The impact of extra-column band broadening on the chromatographic efficiency of 5 cm long narrow-bore very efficient columns, *J. Chromatogr. A*, **1218**, 5286–5291.
18. Gritti, F., Sanchez, C.A., Farkas, T. *et al.* (2010). Achieving the full performance of highly efficient columns by optimizing conventional benchmark high-performance liquid chromatography instruments, *J. Chromatogr. A*, **1217**, 3000–3012.
19. Fountain, K.J., Neue, U.D., Grumbach, E.S. *et al.* (2009). Effects of extra-column band spreading, liquid chromatography system operating pressure, and column temperature on the performance of sub-2-micron porous particles, *J. Chromatogr. A*, **1216**, 5979–5988.
20. Spaggiari, D., Fekete, S., Eugster, P.J. *et al.* (2013). Contribution of various types of liquid chromatography–mass spectrometry instruments to band broadening in fast analysis, *J. Chromatogr. A*, **1310**, 45–55.
21. Dolan, J.W. (2006). Dwell Volume Revisited, *LCGC North Am.*, **24**, 458–466.
22. Dong, M.W. (2007). Ultrahigh-pressure LC in pharmaceutical analysis: performance and practical issues, *LCGC North Am.*, **25**, 656–666.
23. Dong, M.W. (2010). 'Ultrahigh-pressure LC in pharmaceutical analysis: benefits, impacts and issues', in Wixom, R.L. and Gehrke, C.W. (eds), *Chromatography: A Science of Discovery*, John Wiley & Sons, Hoboken, New Jersey, pp. 328–332.
24. Wyndham, K.D., O'Gara, J.E., Walter, T.H. *et al.* (2003). Characterization and evaluation of C18 HPLC stationary phases based on ethyl-bridged hybrid organic/inorganic particles, *Anal. Chem.*, **75**, 6781–6788.
25. Desmet, G., Gzil, P. and Clicq, D. (2005). Kinetic plots to directly compare the performance of LC supports, *LCGC Eur.*, **18**, 403–409.
26. Desmet, G. (2008). Comparison techniques for HPLC column performance, *LCGC North Am.*, **21**, 310–320.
27. Rodriguez-Aller, M., Gurny, R., Veuthey, J.L. *et al.* (2013). Coupling ultrahigh-pressure liquid chromatography with mass spectrometry: constraints and possible applications, *J. Chromatogr. A*, **1292**, 2–18.
28. Guillarme, D., Schappler, J., Rudaz, S. *et al.* (2010). Coupling high-pressure liquid chromatography with mass spectrometry, *Trends Anal. Chem.*, **29**, 15–27.

29. Novakova, L. and Vlckova, H. (2009). A review of current trends and advances in modern bio-analytical methods: Chromatography and sample preparation, *Anal. Chim. Acta*, **656**, 8–35.
30. Guillarme, D., Ruta, J., Rudaz, S. *et al.* (2010). New trends in fast and high-resolution liquid chromatography: a critical comparison of existing approaches, *Anal. Bioanal. Chem.*, **397**, 1069–1082.
31. Cunliffe, J.M., Adams-Hall, S.B. and Maloney, T.D. (2007). Evaluation and comparison of very high-pressure liquid chromatography systems for the separation and validation of pharmaceutical compounds, *J. Sep. Sci.*, **30**, 1214–1223.
32. Nguyen, D.T.T., Guillarme, D., Rudaz, S. *et al.* (2006). Chromatographic behaviour and comparison of column packed with sub-2 µm stationary phases in liquid chromatography, *J. Chromatogr. A*, **1128**, 105–113.
33. Petersson, P. and Euerby, M.R. (2007). Characterisation of RPLC columns packed with porous sub-2 µm particles, *J. Sep. Sci.*, **30**, 2012–2024.
34. Nguyen, D.T.T., Guillarme, D., Heinisch, S. *et al.* (2007). High throughput liquid chromatography with sub-2 µm particles at high pressure and high temperature, *J. Chromatogr. A*, **1167**, 76–84.
35. de Villiers, A., Lauer, H., Szucs, R. *et al.* (2006). Influence of frictional heating on temperature gradients in ultrahigh-pressure liquid chromatography on 2.1 mm I.D. columns, *J. Chromatogr. A*, **1113**, 84–91.
36. Vanhoenacker, G., David, F. and Sandra, P. (2009). Increasing productivity in the analysis of impurities of metoclopramide hydrochloride formulations using the Agilent 1920 Infinity LC system, Agilent Technologies Application Note, 5990-3981EN.
37. Gritti, F. and Guiochon, G. (2008). Complete temperature profiles in ultrahigh-pressure liquid chromatography columns, *Anal. Chem.*, **80**, 5009–5020.
38. Gritti, F. and Guiochon, G. (2008). Ultrahigh-pressure liquid chromatography: column permeability and changes of the eluent properties, *J. Chromatogr. A*, **1187**, 165–179.
39. Novakova, L., Veuthey, J.L. and Guillarme, D. (2011). Practical method transfer from high-performance liquid chromatography to ultrahigh-performance liquid chromatography: the importance of frictional heating, *J. Chromatogr. A*, **1218**, 7971–7981.

40. Guiochon, G. and Gritti, F. (2011). Shell particles, trials, tribulations and triumphs, *J. Chromatogr. A*, **1218**, 1915–1938.
41. Fekete, S., Oláh, E. and Fekete, J. (2012). Fast liquid chromatography: The domination of core–shell and very fine particles, *J. Chromatogr. A*, **1228**, 57–71.
42. Ruta, J., Zurlino, D., Grivel, *et al.* (2012). Interest of columns packed with shell particles in the pharmaceutical field, *J. Chromatogr. A*, **1228**, 221–231.

Chapter 2

Core-Shell Column Technology in Fast Liquid Chromatography

Oscar Núñez[a] and Héctor Gallart-Ayala[b]

[a]Department of Analytical Chemistry, University of Barcelona, Spain
[b]ONIRIS, Laboratoire d'Etude des Résidus et Contaminants dans les Aliments (LABERCA), France

2.1. Introduction

Nowadays, there is a growing demand for high-throughput separations, and laboratories belonging to many different areas, such as toxicology, clinical chemistry, forensics, doping, and environmental and food analysis, are interested in cost-effective methodologies with reduced analysis time. High performance liquid chromatography (HPLC) is a common and well-established separation technique frequently used to solve multiple analytical problems, as it is able to separate quite complicated mixtures, of low and high molecular weight as well as different polarities and acid–base properties, in a variety of matrices. But lately, conventional HPLC alone has not been enough to solve all the analytical problems, especially related to the increasing number of analytes to be analysed in very complex matrices; when selecting this separation technique the compromise is related to either the analysis time or chromatographic resolution. Fast and ultrafast chromatographic methods can overcome the limitations experienced by HPLC when analysing such sample sets, by yielding high resolution within a reduced analysis time without a loss on separation efficiency.[1–6]

In general, in order to carry out such fast separations the column length must be decreased and the linear velocity of the mobile phase must be increased. The 3–5 cm long, narrow-bore (2.0–2.1 mm), columns have become then very popular in all those fields interested in using rapid and efficient procedures for qualitative and quantitative analysis, in order to cope with a large number of samples and to reduce the time required for delivery of results. Additionally, an easy transfer of well-established and validated conventional HPLC methods to ultrafast LC ones is of interest in many application fields, especially in pharmaceutical and food-control laboratories.

However, reducing analysis time and guaranteeing the quality of a separation in HPLC requires high kinetic efficiency. A general approach to increase the separation power, then, is to enhance the column efficiency. Among the several modern approaches in HPLC methods that enable the reduction of analysis time without compromising resolution and separation efficiency, ultrahigh-pressure liquid chromatography (UHPLC), either using sub-2 μm totally porous particle packed columns[7] or porous shell columns (with sub-3 μm superficially porous particles),[3,8] is one of the most frequently used. Recent research have reported both advantages and disadvantages of columns packed with core-shell and totally porous sub-2 μm particles. In this chapter, the possibilities of fast LC by applying new, very efficient columns packed with core-shell particles, especially focusing on the fields of environmental and food analysis, will be addressed.

2.2. Fused-Core Technology: Benefits and New Possibilities

The concept of pellicular particles was introduced in 1969 by Horvath *et al.*,[9] in order to reduce analyte diffusion distance to minimize mass transfer, and having as a consequence highly efficient and fast chromatographic separations with lower pressure. Later, Kirkland demonstrated that 30–40 μm diameter superficially porous packing particles provided faster separations when compared with the large porous particles.[10] Later still, the core diameter was reduced to 0.5 μm and used in fast peptide chromatographic separations. Figure 2.1a shows a typical fused-core particle. Today these columns

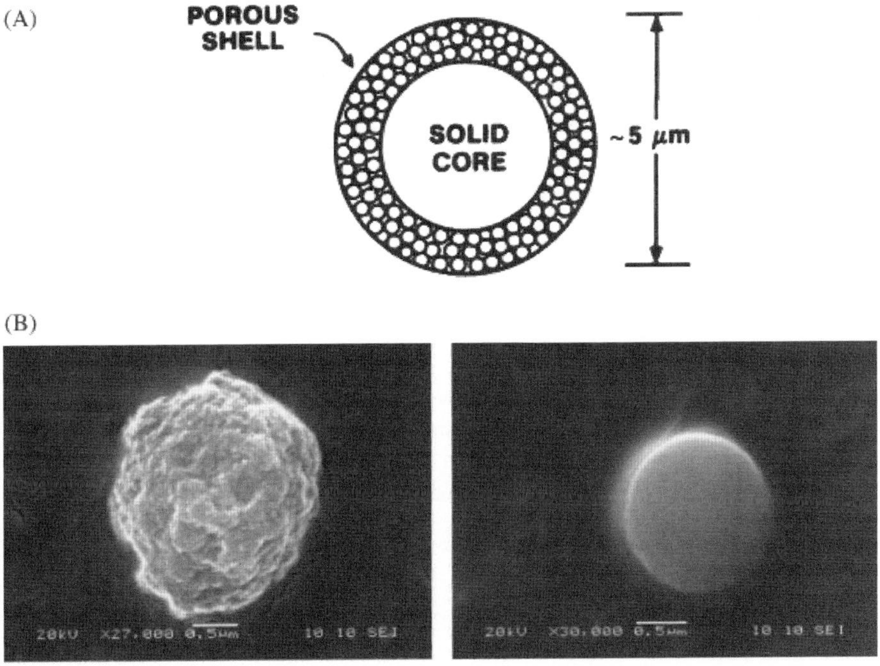

(a) fused core particle (b) sub 2 μm porous particle

Figure 2.1. (A) Scheme of a Fused Core particle. Reproduced with permission from Kirkland, J.J., Truszkowski, F.A., Dilks, Jr, C.H. *et al.* (2000), Superficially porous silica microspheres for fast high-performance liquid chromatography of macromolecules, *J. Chromatogr. A*, **890**, 3–13. Copyright (2000) Elsevier. (B) SEM picture of Ascentis Express shell (fused core) particle (right) and Waters UPLC BEH 1.7 μm porous particle (left). Reproduced with permission from Fekete, S., Fekete, J. and Ganzler, K. (2009). Shell and small particles: evaluation of new column technology. *J. Pharm. Biom. Anal.*, **49**, 64–71. Copyright (2009) Elsevier.

(namely porous-shell, fused-core and core shell) are commercialized in different diameters, the most commonly used in environmental and food analysis[4] being: Ascentis fused-core silica columns (Sigma–Aldrich), consisting of silica particles of a 1.7 μm fused core and 0.5 μm layer of porous silica coating, creating a total particle diameter of 2.7 μm; Kinetex (Phenomenex) with a 1.9 μm fused core and 0.35 μm layer of porous silica coating, obtaining a 2.6 μm particle; and Accucore (Thermo Fisher Scientific), which also has a total particle diameter of 2.6 μm. The main advantage of the core-shell particle columns is the diminution of both the longitudinal diffusion B coefficient[11] and the Eddy dispersion A term. In addition, this type of particle presents a shorter diffusion distance because of the presence of the solid fused core, which is impenetrable by analytes, decreasing the resistance to mass transfer (assigned the C term in the Van Deemter curve).[12] Fekete *et al.*[13] reported that the C term for shell particles is about two times higher than in sub-2 μm totally porous particles. This effect can be attributed to the rough surface of the particles in which the mass transfer is reduced through the outer stagnant liquid. Figure 2.1b illustrates scanning electron microscopy (SEM) pictures of a fused core and a totally porous particle. However, Guichon and Gritti report that the mass transfer improvements contributed by the limited shell thickness are negligible for small solutes, in which improved efficiency arises primarily due to a reduction on the eddy dispersion; the narrow particle size distribution is not responsible.[14]

Another benefit of this type of particle is that lower back-pressures are required, compared to totally porous particles providing the same number of chromatographic plate numbers. These advantages, in combination with the use of UHPLC systems, allow the use of smaller particle sizes down to 1.3 μm. However, when separating small molecules (molecular weight < 200) the 1.3 μm core-shell columns are useful only when relatively low peak capacities ($P < 300$) are required, corresponding to very fast gradients.[15] Sanchez *et al.*[16] tested some columns with very small core-shell particles (1.0–1.4 μm). As expected these columns showed excellent low minimal plate heights corresponding to a plate count of over 450,000 plates/m. However, this type of column produced a very high back-pressure, limiting their performance. An additional consideration proposed by the authors was that with these very small particle sizes, an UHPLC system with low extra-column variance (< 102 μL) is necessary in order to maintain the high efficient separation achieved by the column.

For readers interested in a better understanding, we recommend a list of scientific publications and reviews about the fundamental and practical aspects of porous shell particles technology[3,13–20] in liquid chromatography.

Although an important evolution in the development of new columns for fast liquid chromatography has occurred, some instrumental limitations are still present. Normally the use of fused-core particles for a fast chromatographic separation is justified because the lower back-pressure obtained with this type of column allows the use of conventional LC systems. However, as recently reviewed by Fekete *et al.*,[3] the LC instrument plays a very important role in obtaining and maintaining the efficiency achieved by the use of fused-core particles columns. In fact, extra-column volume (injector system, connector tubing) and variance are very critical when using small particle size columns with smaller internal diameter than the standard ones. Another important parameter to take into account, especially when a gradient elution mode is applied, is the system dwell volume (V_d); this parameter represents the volume from the mixing point of solvents to the head of the column. Indeed, in terms of system volume and variance, conventional HPLC systems (maximum pressure 400 bar) are definitively unable to maintain the high efficiency of those columns. Furthermore, if gradient elution is used then the dwell volume should be considered. Conventional HPLC systems present dwell volumes of 0.5–5 mL (Fig. 2.2), which are unacceptable for use in high-efficiency LC separations.

Another interesting benefit of using core-shell columns occurs when dealing with the direct hyphenation of on-line solid phase extraction (SPE) methods to ultrahigh-performance liquid chromatography. This direct hyphenation is challenging when using totally porous sub-2 μm particle size columns. Firstly, the high flow rates generally used in UHPLC (> 400 μL/min), in combination with the particle size, generate high column back-pressures, which is not directly compatible with the conventional on-line SPE systems that operate at low back-pressures. To overcome this problem Gallart-Ayala *et al.*[21] developed an on-line SPE UHPLC electrospray ionization tandem mass spectrometry (ESI–MS/MS) method using a core-shell column as an analytical column. The low back-pressure provided by this column enabled its direct hyphenation with a conventional on-line SPE system. This method allowed the direct analysis of bisphenol A (BPA) and its chlorinated derivatives in 1 mL of water

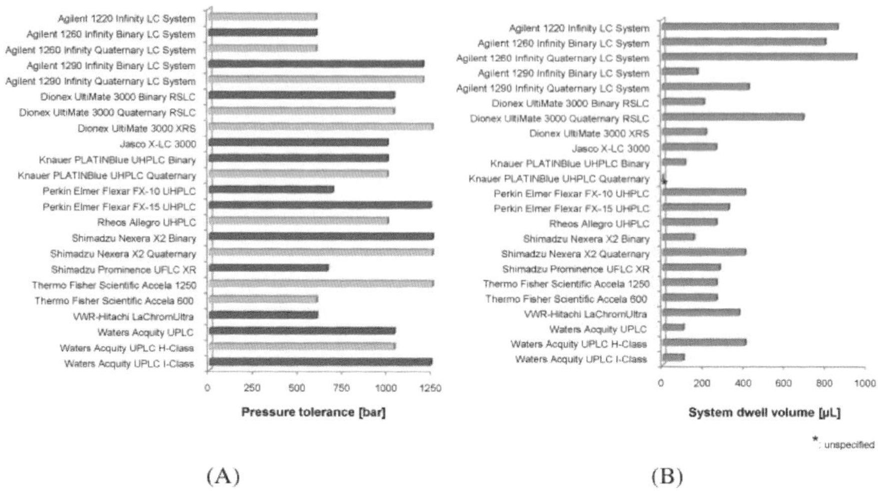

Figure 2.2. Graphical representation of (A) pressure tolerance and type of pumping system and (B) standard system dwell volume of all the UHPLC systems ($\Delta P > 600$ bar) commercially available. It is important to note that the dwell volume of some of these instruments can be modified by bypassing the damper and mixer, or by changing the mixing chamber. Reproduced with permission from Fekete, S., Kohler, I., Rudaz, S. *et al.* (2014). Importance of instrumentation for fast liquid chromatography in pharmaceutical analysis, *J. Pharm. Biomed. Anal.*, 87, 105–119. Copyright (2014) Elsevier.

samples at ng/L level in less than 10 min. Later on, this methodology was applied to the analysis of BPA and other bisphenols (such as bisphenol F, bisphenol E, bisphenol B and bisphenol S) in soft drinks by the direct injection of 1 mL of soft drink sample.[22] However, in this case an important matrix effect (80–95%) was observed due to the presence of matrix components that caused ion suppression in the ESI source, as can be seen in Fig. 2.3a. In this work several strategies to reduce the matrix effect were evaluated, and the authors concluded that only when the analytes were higher-retained in the analytical column and forced to elute in a cleaner chromatographic area was the matrix effect reduced, as shown in Fig. 2.3b. This fact shows that in some cases to obtain a good identification and quantitation of the target analytes it is necessary to sacrifice the analysis time, although by using core-shell columns the total analysis time will be lower than

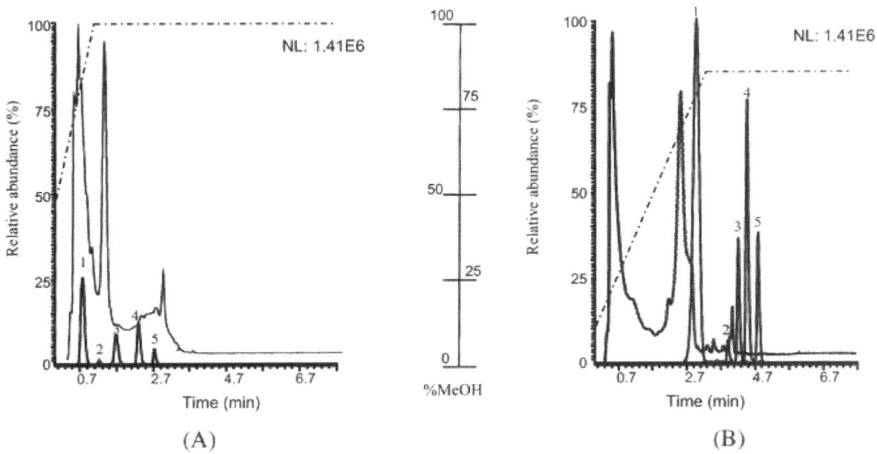

Figure 2.3. On-line SPE UHPLC-ESI-MS/MS and LC-UV at 228 nm chromatograms of a glass cola sample spiked at 10 µg/L. (A) ESI at ambient temperature, gradient elution 0 min, 50:50 MeOH:water; from 0 min to 1 min, linear gradient up to 100% MeOH, and (B) Heated-ESI at 300 °C, gradient elution 0 min, 15% MeOH; from 0 min to 2 min, a linear gradient elution up to 80% MeOH and then isocratic step (3.5 min). Compounds: 1, Bisphenol S; 2, Bisphenol F; 3, Bisphenol E; 4, Bisphenol A, and 5, Bisphenol B. Reproduced with permission from Gallart-Ayala, H., Moyano, E. and Galceran, M.T. (2011). Analysis of bisphenols in soft drinks by on-line solid phase extraction fast liquid chromatography-tandem mass spectrometry., *Anal. Chim. Acta*, **683**, 227–233. Copyright (2011) Elsevier.

one obtained with totally porous sub-2 µm columns, while keeping similar column efficiency.

2.3. Overview of Columns on the Market

The advancements in the manufacture of fully porous silica produced smaller and smaller particles, with the latest generation being sub-2 µm diameter particle columns, which were introduced in 2004. The performance achievable with these particles and the instrumentation capable of operating columns packed with them initially constituted what is collectively called UHPLC. However, the dominance of fully porous sub-2 µm particles was short-lived, with the reintroduction of core-shell particles in 2007. This new generation of core-shell particles took advantage of the many advances in silica sol-gel technology, providing increased surface area (typically 100–150 m^2/g) and a very narrow particle size distribution. The first commercially available sorbent in this new generation of core-shell particles was the Halo® from Advanced Material Technologies (Wilmington, DE). The Halo particle was 2.7 µm in diameter with a shell thickness of 0.5 µm. Columns packed with Halo core-shell particles give efficiency values close to those of sub-2 µm fully porous sorbents but at significantly lower operational pressure.

Because of the very high operational pressures required to operate with columns packed with sub-2 µm particles near their optimal linear velocity, it has been incorrectly assumed by many scientists that ultrahigh pressure was required to perform ultrahigh-performance liquid chromatography. The appearance on the market of core-shell sorbents demonstrated that the ultrahigh-efficiency characteristic of UHPLC could be easily achieved even at conventional HPLC pressures, as will be illustrated later in this chapter with relevant applications. Since the introduction of the Halo core-shell sorbents, several similar sorbents have been released onto the market by other manufacturers. A summary of the most relevant companies supplying core-shell technology columns, along with their physical properties and stationary phases available, is shown in Table 2.1. The number of stationary phases in columns with core-shell column technology is increasing every year. As can be seen in Table 2.1, a variety of reversed-phase (C18, C8, amide etc.), hydrophilic interaction liquid chromatography (HILIC), fluorinated (such as pentafluorophenyl (PFP) alkyl stationary phases), cyano, and phenyl-hexyl stationary phases, as well as some specifically designed for polar compounds and for peptides, are available on the market from different companies.

In 2009 Phenomenex (and later other companies) introduced core-shell columns under the trade name of Kinetex in two particle sizes (2.6 µm and 1.7 µm), the latter being the first sub-2 µm core-shell particle commercially available on the market; today, Phenomenex also produces new-generation C18 columns packed with 1.3 µm core-shell particles. The practical possibilities and limitations of this column technology, as well as its performance

Table 2.1. Summary of the most relevant core-shell columns available on the market.

Column	Supplier	Particle size	Stationary phases	Surface area (m^2/g)	Pore diameter (Å)
Halo	Advanced Materials Technology (Wilmington, DE)	2.7 μm	C8 C18 Peptide ES-C18 Phenyl-Hexyl HILIC Penta-HILIC PFP RP-Amide ES-CN	130	90
Kinetex	Phenomenex (Torrance, CA)	1.7 μm	C8 C18 XB-C18 Phenyl-Hexyl HILIC PFP	100	100
		2.6 μm	C8 C18 XB-C18 Phenyl-Hexyl HILIC PFP	100	100
		1.3 μm	C18	100	100
Accucore	Thermo Fisher Scientific (Waltham, MA)	2.6 μm	RP-MS C18 C8 aQ (polar endcapped C18) Polar Premium Phenyl-Hexyl PFP Phenyl-X C30 HILIC Urea-HILIC	130	80

(*Continued*)

Table 2.1. (*Continued*)

Column	Supplier	Particle size	Stationary phases	Surface area (m^2/g)	Pore diameter (Å)
Nucleoshell	Macherey-Nagel (Düren, Germany)	2.7 μm	C18 Phenyl-Hexyl PFP HILIC	130	90
Poroshell 120	Agilent Technology (Palo Alto, CA)	2.7 μm	C18 C8 Phenyl-Hexyl SB-Aq (polar compounds) Bonus –RP (alkyl amide) HILIC EC-Cyano	120	120
Ascentis Express	Supelco (Bellefonte, PA)	2.7 μm	C18 C8 Peptide ES-C18 RP-Amide Phenyl-Hexyl HILIC ES-Cyano F5 OH5 (polar compounds)	150	70
Sunshell	ChromaNik Technologies (Osaka, Japan)	2.6 μm	C18 C8 RP-AQUA PFP Phenyl HILIC-Amide	150	90

compared with other reference columns packed with 1.7 μm, 2.6 μm and 5 μm core-shell particles, were recently assessed by Fekete and Gillarme.[23] Using the Van Deemter representation, an H_{min} value of only 1.95 μm was achieved, corresponding to efficiency of more than

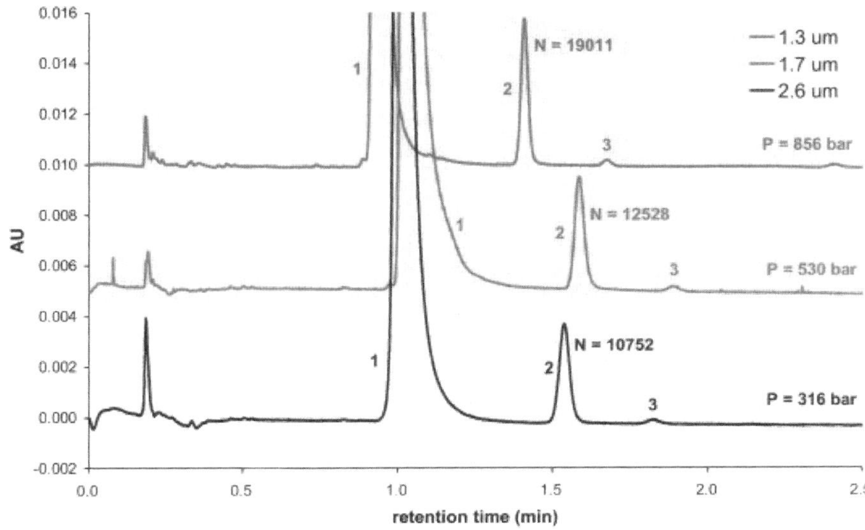

Figure 2.4. Impurity profiling of α-estradiol. Peaks: 1, α-estradiol; 2, main impurity; 3, minor impurity. Columns: Phenomenex Kinetex 50 mm × 2.1 mm, 1.3 µm, 1.7 µm and 2.6 µm. Mobile phase: water:acetonitrile 60:40 (*v/v*), flow rate: 0.5 mL/min, temperature: 25 °C, injection volume: 0.5 µL (25 µg/mL estradiol), $\lambda = 210$ nm. Reproduced with permission from Fekete, S and Guillarme, D. (2013). Kinetic evaluation of new generation of column packed with 1.3 µm core-shell particles., *J. Chromatogr. A*, **1308**, 104–113. Copyright (2013) Elsevier.

500,000 plates/m and about 25,000 plates for a 50 mm column length. For comparison purposes, a good column packed with fully porous sub-2 µm particles is able to produce “only” 300,000 plates/m. Fekete and Guillarme illustrated the gain in performance afforded by 1.3 µm core-shell particles with two real-life separations, as depicted in Figs 2.4 and 2.5.

The first example (Fig. 2.4) shows the chromatographic separation of a major and a minor impurity of α-estradiol obtained with three columns of 50 mm × 2.1 mm and packed with 1.3 µm, 1.7 µm and 2.6 µm core-shell particles. The major impurity (peak 2) eluted with a number of plates (*N*) of 19,011, 12,528, and 10,752 on the 1.3 µm, 1.7 µm and 2.6 µm materials, respectively (without correcting for extra-column band variance). Under the same experimental conditions, the column packed with 1.3 µm core-shell particles

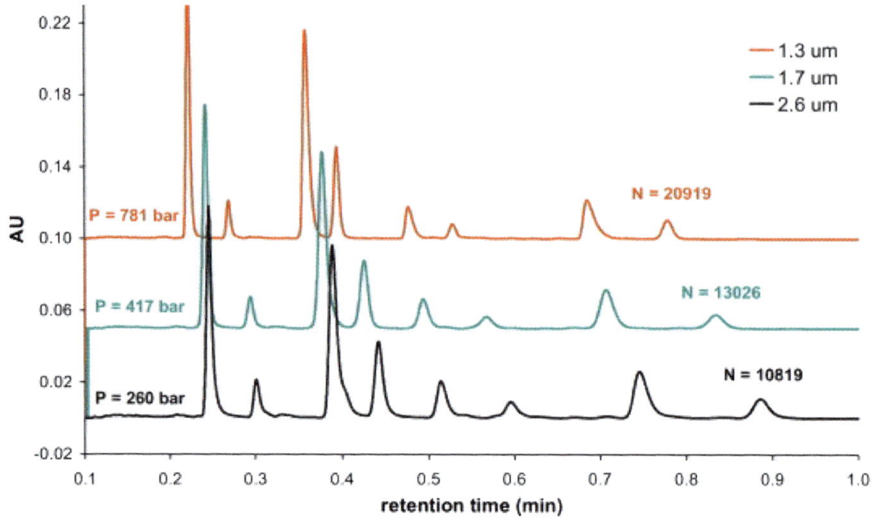

Figure 2.5. Fast separation of cashew nut extract. Columns: Phenomenex Kinetex 50 mm × 2.1 mm, 1.3 µm, 1.7 µm and 2.6 µm. Mobile phase: water:acetonitrile 14:86 (*v*/*v*), flow rate: 0.8 mL/min, temperature: 25 °C, injection volume: 0.3 µL (25 µg/mL estradiol), λ = 280 nm. Reproduced with permission from Fekete, S and Guillarme, D. (2013). Kinetic evaluation of new generation of column packed with 1.3 µm core-shell particles., *J. Chromatogr. A*, **1308**, 104–113. Copyright (2013) Elsevier.

generated a column back-pressure of 856 bar, while the 1.7 µm and 2.6 µm core-shell particle columns produced only 530 bar and 316 bar, respectively. Hence, this example confirms the possibilities offered by the new 1.3 µm core-shell packing and also confirms its low permeability. Moreover, the retention on the 1.3 µm packing was slightly reduced compared to the other two columns.

The second example (Fig. 2.5) illustrates a sub-1 min long separation of cashew nut extract, in contrast to the 25 min long separation of the same extract reported in the literature using a conventional 150 mm × 4.6 mm, 5 µm particle size column.[24] The last peak eluted with N = 20,919, 13,026 and 10,819 on the columns packed with 1.3 µm, 1.7 µm and 2.6 µm core-shell particles, respectively. This gain in analysis time and column efficiency clearly proves the advantage of very efficient core-shell particles.

In spite of the exceptional efficiency and success of sub-2 μm fully porous and sub-3 μm core-shell particles, it seems that the pharmaceutical industry still prefers to use columns with conventional 3–5 μm particles. In order to meet this necessity, Phenomenex also commercialized columns with larger core-shell particles (3.6 μm, Aeris columns) for the separation of proteins,[25,26] and very recently Supelco (Ascentis Express), Advanced materials Technology (HALO-5) and Phenomenex (Kinetex) introduced new 5 μm superficially porous particles with a ~3.3 μm core and 0.6 μm porous shell.[27,28]

2.4. Core-Shell Column Technology in Food and Environmental Analysis

The number of liquid chromatography (LC) and liquid chromatography–mass spectrometry (LC–MS) methods proposing the use of columns packed with core-shell particles has considerably increased in the last few years because of the high column efficiency achieved without excessive increasing of column back-pressure. Today, a number of important food applications,[4,22,29–42] as well as a small proportion of environmental applications,[4,43–46] using this column technology can be found in the literature, and several authors have compared the performance of core-shell column technology with that of sub-2 μm columns when dealing with this kind of applications.[4,31,35,45] The aim of this chapter is not to fully review the state-of-the-art in this topic, but in this section some relevant examples reported since 2011 will be addressed.

Shaaban and Górecki proposed an ultrahigh-performance LC method for the simultaneous determination of 25 emerging contaminants (such as veterinary antibiotics, non-steroidal anti-inflammatory drugs, steroids, central nervous system stimulants and preservatives) in surface water and wastewater samples using a 2.6 μm core-shell superficially porous particle column as an alternative to a 1.7 μm fully porous particle column.[45] The chromatograms obtained using both columns are depicted in Fig. 2.6.

While all analytes were fully resolved on the column packed with fused-core particles under the same conditions (resolution between

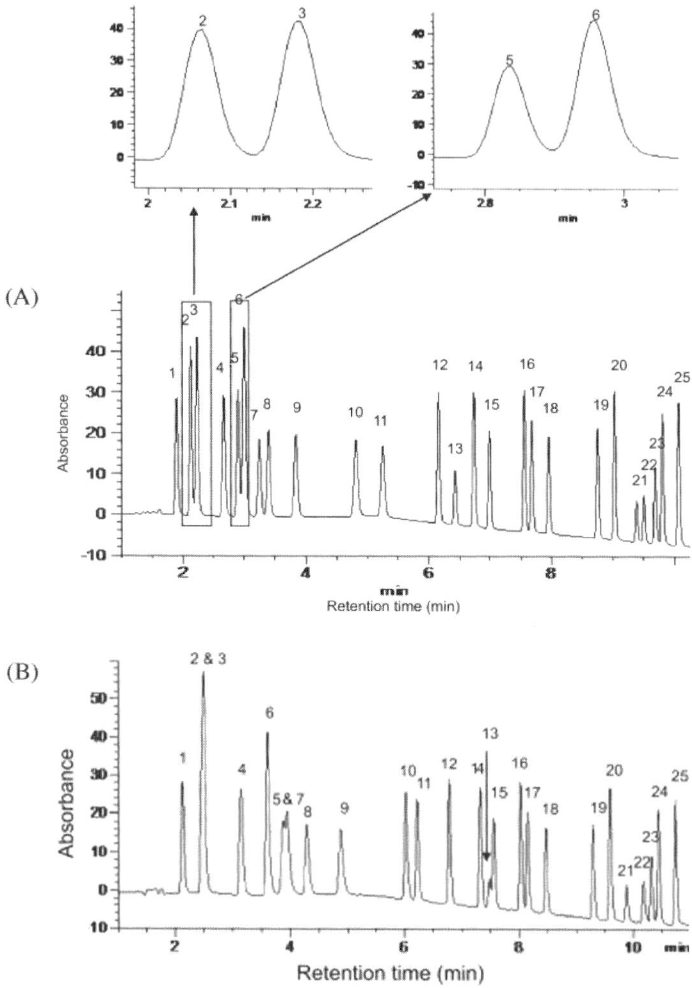

Figure 2.6. Chromatograms of a mixture of 25 emerging contaminants. (A) Using a core-shell Kinetex C18 (150 mm × 4.6 mm, 2.6 μm) column (Phenomenex). The inset shows baseline separation of peaks 2, 3 and 5, 6. (B) Using a fully porous sub-2 μm Zorbax Stable Bond C18 (150 mm × 4.6 mm, 1.8 μm) column (Agilent Technologies). Peak identification: 1, sulphanilamide; 2, theophylline; 3, acetaminophen; 4, sulfacetamide; 5, caffeine; 6, sulfadiazine; 7, sulfathiazole; 8, sulfapyridine; 9, sulfamerazine; 10, sulfamethazine; 11, sulfamethoxypyridazine; 12, sulfamonomethoxine; 12, acetyl salicylic acid; 14, sulfamethoxazole; 15, methylparaben; 16, sulfadimethoxine; 17, sulfaphenazole; 18, ethylparaben; 19, propylparaben; 20, ketoprofen; 21, 17α-ethinyl estradiol; 22, estrone; 23, fenoprofen; 24, flurbiprofen; and 25, diclofenac.

Figure 2.6. (caption continued) Analytes were separated by gradient elution at 30 °C using a mobile phase consisting of acetonitrile (solvent B) and water with 0.5% acetic acid (solvent A). Flow rate: 1–1.3 mL/min. Reproduced with permission from Shaaban, H. and Górecki, T. (2012). Fast ultrahigh-performance liquid chromatographic method for the simultaneous determination of 25 emerging contaminants in surface water and wastewater samples using superficially porous sub-3 μm particles as an alternative to fully porous sub-2 μm particles., *Talanta*, **100**, 80–89. Copyright (2012) Elsevier.

2.09 and 18.06, above the critical value of 1.5 in all cases) within an analysis time of 10 minutes, co-elutions of many analytes were observed on the column packed with fully porous particles and with longer analysis time. For example, the critical pair of theophylline and acetaminophen (peaks 2 and 3 in Fig. 2.6) showed a resolution of 2.25 in the core-shell column against complete co-elution in the sub-2 μm particle size column. Moreover, the system pressure observed when the separation was performed on the column packed with core-shell particles was 355 bar compared to 520 bar when the column packed with fully porous sub-2 μm particles was used under the same mobile phase composition and flow rate conditions. The low back-pressure obtained for the column packed with core-shell particles is advantageous and compatible to conventional HPLC systems, while the use of ultrahigh-pressure instrumentation (≥ 600 bar) is required when the separation has to be performed on columns packed with fully porous sub-2 μm particles. The changes in selectivity observed between columns could be attributed to differences in chemistry of both C18 stationary phases.

Gallart-Ayala *et al.*[31] compared the use of a totally porous sub-2 μm particle size column (Acquity BEH C18 50 mm × 2.1 mm, 1.7 μm) with a partially porous core-shell column (Ascentis Express C18 50 mm × 2.1 mm, 2.7 μm) for the fast LC–MS/MS analysis of bisphenol A-diglycidyl ether (BADGE), bisphenol F-diglycidyl ether (BFDGE) and their derivatives in canned food and beverages. Fig. 2.7 shows the structures of the analysed compounds and the LC–electrospray–MS/MS chromatograms obtained for BFDGE and BFDGE derivatives by using a triple quadrupole mass analyser.

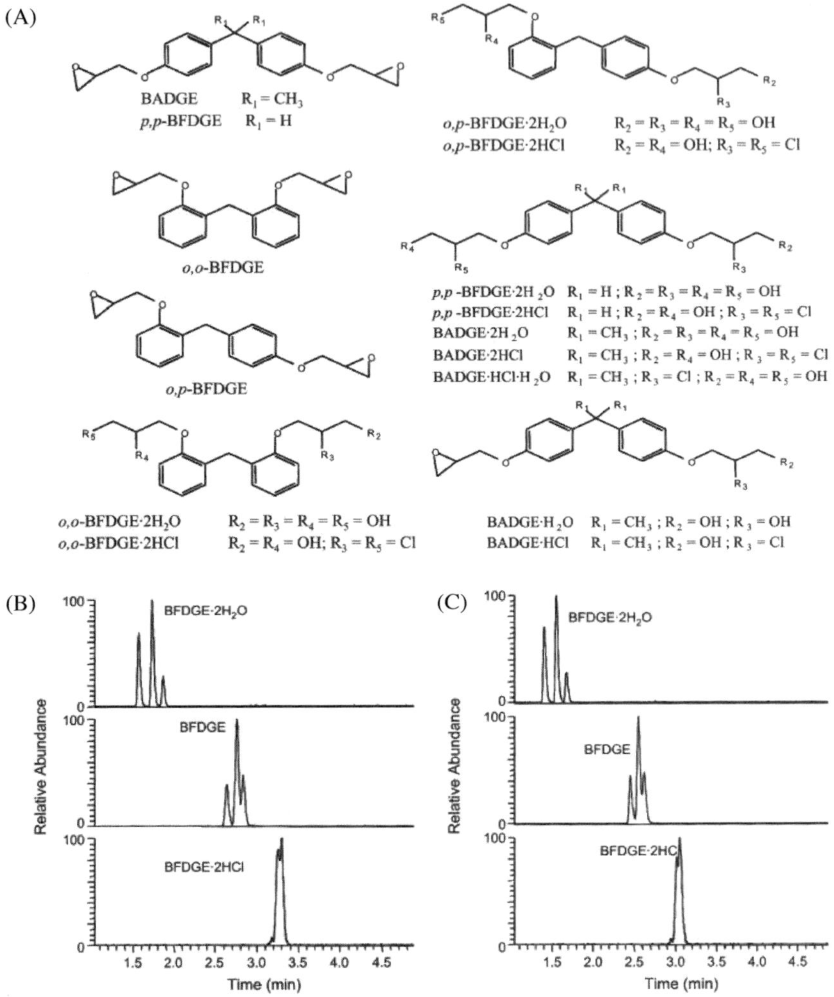

Figure 2.7. (A) Structure of bisphenol A-diglycidyl ether, bisphenol B-diglycidyl ether and their derivatives. (B) Chromatographic separation of BFDGE, and BFDGE·2H_2O and BFDGE·2HCl isomers using a totally porous Acquity BEH C18 column (50 mm × 2.1 mm, 1.7 µm). (C) Chromatographic separation of the same compounds using a core-shell Ascentis Express C18 column (50 mm × 2.1 mm, 2.7 µm). Methanol:ammonium formate/formic acid (25 mM, pH 3.75) gradient elution at 600 µL/min and 50 °C was used in both cases. Reproduced with permission from Gallart-Ayala, H., Moyano, E. and Galceran, M.T. (2011). Fast liquid chromatography–tandem mass spectrometry for the analysis of bisphenol A-diglycidyl ether, bisphenol F-diglycidyl ether and their derivatives in canned food and beverages., *J. Chromatogr. A*, **1218**, 1603–1610. Copyright (2011) Elsevier.

Both columns provided similar resolution and efficiencies for the separation of these isomers, although the core-shell column showed lower back-pressure of 200 bar against 513 bar for the sub-2 μm column. The authors proposed the use of this column for further studies because its low back-pressure permitted the increase of the column length up to 150 mm, which allowed the baseline separation of all BADGEs, BFDGEs and their hydrolyzed derivatives, including isomers.

The analysis of BPA and chlorinated-BPA and brominated-BPA compounds was also compared with a core-shell Ascentis Express C18 column against a totally porous sub-2 μm Acquity BEH C18 column,[4] and the chromatograms obtained are shown in Fig. 2.8. As can be seen, both columns provided similar column efficiency with the advantage that the core-shell column provided lower column back-pressure (300 bar against 725 bar), making it possible to

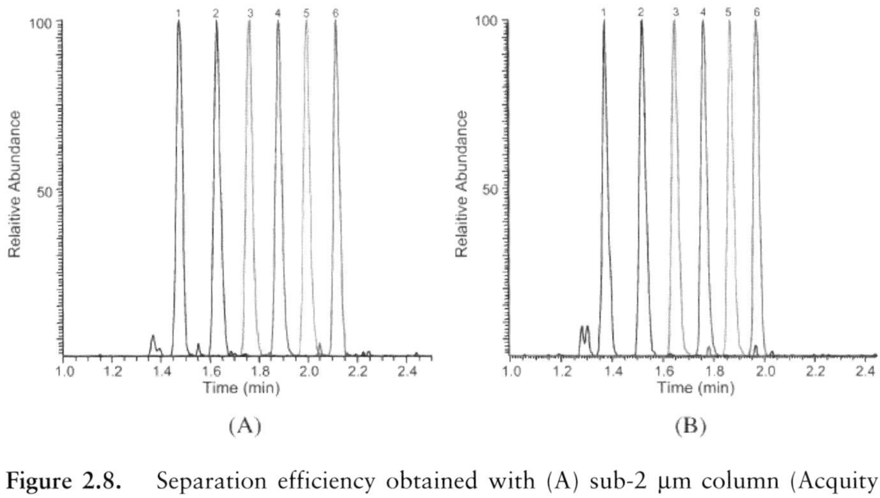

Figure 2.8. Separation efficiency obtained with (A) sub-2 μm column (Acquity BEH C18 50 mm × 2.1 mm, 1.7 μm particle size) and (B) core-shell column (Ascentis Express C18 50 mm × 2.1 mm, 2.7 μm particle size). Chromatographic conditions: gradient elution with 80:20 water (component A) and methanol (component B) at 600 μL/min. Peak identification: 1, BPA; 2, monochloro-BPA; 3, dichloro-BPA; 4, trichloro-BPA; 5, tetrachloro-BPA; and 6, tetrabromo-BPA. Reproduced with permission from Núñez, O., Gallart-Ayala, H., Martins, C.P.B. *et al.* (2012). New trends in fast liquid chromatography for food and environmental analysis, *J. Chromatogr. A*, **1228**, 298–323. Copyright (2012) Elsevier.

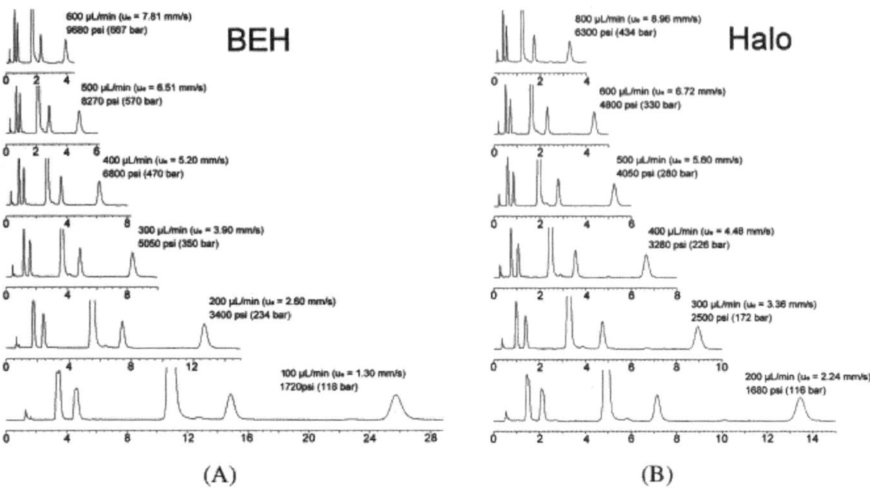

Figure 2.9. Chromatograms of artichoke extract obtained with (A) a totally porous Acquity BEH-C18 column (50 mm × 2.1 mm, 1.7μm) and (B) a Halo core-shell C18 column (50 mm × 2.1 mm, 2.7 μm) at different linear velocities. Mobile phase: Methanol:water 88:12 (*v*/*v*); injection volume: 2 μL; UV detection: 320 nm. Reproduced with permission from Wu, J., Qian, Y., Mao, P. *et al.* (2013). Separation and identification of phenolic compounds in canned artichoke by LC/DAD/ESI-MS using core-shell C18 column: a comparative study., *J. Chromatogr. B*, **927**, 173–180. Copyright (2013) Elsevier.

achieve a fast chromatographic separation using conventional HPLC systems.

Recently, Wu *et al.*[35] carried out a comparison between core-shell and sub-2 μm C18 columns for the separation and identification of phenolic compounds in canned artichoke by LC–diode array detection (DAD)/ESI-MS. The UV chromatograms obtained for the artichoke extracts with both columns at different linear velocities and under the same chromatographic conditions are depicted in Fig. 2.9.

It can be seen that when the back-pressure of the Halo-C18 column increased from 116 bar to 434 bar, the analysis time yielded a 10-fold decrease from 13.4 min to 3.2 min, while keeping a baseline separation. So, the improvement in peak efficiency does come at a cost, because pressure increases with the inverse square of the decreasing particle diameter.[47] For the 1.7 μm totally porous particle-packed column under similar separation conditions (Fig. 2.9A), the back-pressure increased from 118 bar to 667 bar, reflecting an increase in linear velocity from 1.3 mm/s to 7.8 mm/s, accompanied by a decrease in the separation time from 26 min to 4 min. The authors described that the core-shell column also showed similar Van Deemter kinetics, but lower reduced plate height and a flatter C-term in Knox plot than the sub-2 μm column. Similar efficiency separations could be achieved on conventional HPLC systems using this core-shell column, saving the expensive cost of ultrahigh-pressure instrumentation; thus the authors selected the core-shell column for rapid LC/DAD/ESI–MS analysis of the phenolic compounds in artichoke extract.

As mainly described in the three examples addressed in this section, one of the most important advantages of core-shell columns against totally porous sub-2 μm columns is that similar efficiencies could be achieved with lower back-pressures, making it cheaper to change from conventional LC separations to fast LC ones without the requirements of ultrahigh-pressure (UHPLC) instrumentation. This is a very interesting advantage when multiresidue analyses are intended. For instance, Wang *et al.*[32] compared the LC–ESI–MS/MS analysis of 151 pesticides in grains when using a conventional 3 μm porous particle column (Atlantis dC18, Waters, USA) with a 2.6 μm core-shell particle column (Kinetex C18). The multiresidue separation of these pesticides was reduced to 9.22 min (retention time of last retained pesticide Dodemorph) with a total analysis time of 12 min when the Kinetex C18 column was used, compared to 20.41 min (retention time for Dodemorph with a total analysis time of 35 min) when conventional Atlantis dC18 column was used, with an additional improvement in efficiency and sensitivity.

2.5. Concluding Remarks

Ultrahigh-performance liquid chromatography is extensively used in food and environmental applications because of the reduction in analysis time without the compromising of chromatographic resolution and efficiency. Actually, the introduction of core-shell particles into the market, and the good performance in terms of efficiency and chromatographic resolution in combination with lower back-pressures, make these columns a reliable competitor with sub-2 μm totally porous particle columns. Although it is accepted that the use of core-shell particles demonstrated that ultrahigh-efficiency chromatographic separations can be achieved at conventional HPLC pressures, the use of appropriated liquid chromatography systems is recommended to maintain the high efficiency obtained with those columns.

The excellent capabilities and robustness obtained with core-shell particles have allowed the employment of smaller particle sizes down to 1.3 μm. The use of these columns in combination with a UHPLC system with low extra-column variance will permit improved chromatographic resolution and efficiency.

References

1. Nguyen, D.T.T., Guillarme, D., Rudaz, S. *et al.* (2007). New trends in fast liquid chromatography, *Chimia*, **61**, 186–189.
2. Yamaguchi, T., Tanaka, K., Goto, T. *et al.* (2008). Application of ultra-fast liquid chromatography to food analysis, *Shimadzu Hyoron*, **65**, 93–108.
3. Fekete, S., Olah, E. and Fekete, J. (2012). Fast liquid chromatography: the domination of core-shell and very fine particles, *J. Chromatogr. A*, **1228**, 57–71.
4. Núñez, O., Gallart-Ayala, H., Martins, C.P.B. *et al.* (2012). New trends in fast liquid chromatography for food and environmental analysis, *J. Chromatogr. A*, **1228**, 298–323.
5. Zacharis, C.K. and Tzanavaras, P.D. (2013). Trends and applications of fast liquid chromatography in bioanalysis, *J. Chromatogr. B: Anal. Technol. Biomed. Life Sci.*, **927**, 1–2.
6. Núñez, O., Gallart-Ayala, H., Martins, C.P.B. *et al.* (2013). State-of-the-art in fast liquid chromatography–mass spectrometry for bio-analytical applications, *J. Chromatogr. B: Anal. Technol. Biomed. Life Sci.*, **927**, 3–21.
7. D'Orazio, G., Rocco, A. and Fanali, S. (2012). Fast liquid chromatography using columns of different internal diameters packed with sub-2 μm silica particles, *J. Chromatogr. A*, **1228**, 213–220.

8. Fekete, S. and Fekete, J. (2013). 'The potential of shell particles in fast liquid chromatography', in Xu, Q.A. (ed.), *Ultra-High Performance Liquid Chromatogaphy and its Applications*, John Wiley & Sons, Hoboken, New Jersey, pp. 133–167.
9. Horvath, C. and Lipsky, S.R. (1969). Column design in high-pressure liquid chromatography, *J. Chromatogr. Sci.*, 7, 109–116.
10. Kirkland, J.J. (1992). Superficially porous silica microspheres for the fast high-performance liquid chromatography of macromolecules, *Anal. Chem.*, **64**, 1239–1245.
11. Liekens, A., Denayer, J. and Desmet, G. (2011). Experimental investigation of the difference in B-term dominated band broadening between fully porous and porous-shell particles for liquid chromatography using the Effective Medium Theory, *J. Chromatogr. A*, **1218**, 4406–4416.
12. Salisbury, J.J. (2008). Fused-core particles: A practical alternative to sub-2 micron particles, *J. Chromatogr. Sci.*, **46**, 883–886.
13. Fekete, S., Fekete, J. and Ganzler, K. (2009). Shell and small particles; Evaluation of new column technology, *J. Pharm. Biomed. Anal.*, **49**, 64–71.
14. Guiochon, G. and Gritti, F. (2011). Shell particles, trials, tribulations and triumphs, *J. Chromatogr. A*, **1218**, 1915–1938.
15. Fekete, S. and Guillarme, D. (2013). Possibilities of new generation columns packed with 1.3 μm core-shell particles in gradient elution mode, *J. Chromatogr. A*, **1320**, 86–95.
16. Sanchez, A.C., Friedlander, G., Fekete, S. *et al.* (2013). Pushing the performance limits of reversed-phase ultrahigh- performance liquid chromatography with 1.3 μm core-shell particles, *J. Chromatogr. A*, **1311**, 90–97.
17. Marchetti, N., Cavazzini, A., Gritti, F. *et al.* (2007). Gradient elution separation and peak capacity of columns packed with porous shell particles. *J. Chromatogr. A*, **1163**, 203–211.
18. Kirkland, J.J., Truszkowski, F.A., Dilks, C.H. *et al.* (2000). Superficially porous silica microspheres for fast high-performance liquid chromatography of macromolecules, *J. Chromatogr. A*, **890**, 3–13.
19. Fekete, S., Ganzler, K. and Fekete, J. (2011). Efficiency of the new sub-2 μm core-shell (Kinetex) column in practice, applied for small and large molecule separation, *J. Pharm. Biomed. Anal.*, **54**, 482–490.

20. Fekete, S., Kohler, I., Rudaz, S. *et al.* (2014). Importance of instrumentation for fast liquid chromatography in pharmaceutical analysis, *J. Pharm. Biomed. Anal.*, **87**, 105–119.
21. Gallart-Ayala, H., Moyano, E. and Galceran, M.T. (2010). On-line solid phase extraction fast liquid chromatography–tandem mass spectrometry for the analysis of bisphenol A and its chlorinated derivatives in water samples, *J. Chromatogr. A*, **1217**, 3511–3518.
22. Gallart-Ayala, H., Moyano, E. and Galceran, M.T. (2011). Analysis of bisphenols in soft drinks by on-line solid phase extraction fast liquid chromatography–tandem mass spectrometry, *Anal. Chim. Acta*, **683**, 227–233.
23. Fekete, S. and Guillarme, D. (2013). Kinetic evaluation of new generation of column packed with 1.3 µm core-shell particles, *J. Chromatogr. A*, **1308**, 104–113.
24. Paramashivappa, R., Kumar, P.P., Vithayathil, P.J. *et al.* (2001). Novel method for isolation of major phenolic constituents from cashew (*Anacardium occidentale* L.) nut shell liquid, *J. Agric. Food Chem.*, **49**, 2548–2551.
25. Fekete, S., Berky, R., Fekete, J. *et al.* (2012). Evaluation of a new wide pore core-shell material (Aeris WIDEPORE) and comparison with other existing stationary phases for the analysis of intact proteins, *J. Chromatogr. A*, **1236**, 177–188.
26. Fekete, S., Berky, R., Fekete, J. *et al.* (2012). Evaluation of recent very efficient wide-pore stationary phases for the reversed-phase separation of proteins, *J. Chromatogr. A*, **1252**, 90–103.
27. Broeckhoven, K., Cabooter, D. and Desmet, G. (2013). Kinetic performance comparison of fully and superficially porous particles with sizes ranging between 2.7 µm and 5 µm: Intrinsic evaluation and application to a pharmaceutical test compound, *J. Pharm. Anal.*, **3**, 313–323.
28. Gritti, F. and Guiochon, G. (2013). Speed-resolution properties of columns packed with new 4.6 µm Kinetex-C18 core-shell particles, *J. Chromatogr. A*, **1280**, 35–50.
29. Lu, Y., Shen, Q., Dai, Z. *et al.* (2011). Development of an on-line matrix solid-phase dispersion–fast liquid chromatography–tandem mass spectrometry system for the rapid and simultaneous determination of 13 sulfonamides in grass carp tissues, *J. Chromatogr. A*, **1218**, 929–937.

30. Manchon, N., D'Arrigo, M., Garcia-Lafuente, A. *et al.* (2011). Comparison of different types of stationary phases for the analysis of soy isoflavones by HPLC, *Anal. Bioanal. Chem.*, **400**, 1251–1261.
31. Gallart-Ayala, H., Moyano, E. and Galceran, M.T. (2011). Fast liquid chromatography–tandem mass spectrometry for the analysis of bisphenol A-diglycidyl ether, bisphenol F-diglycidyl ether and their derivatives in canned food and beverages, *J. Chromatogr. A*, **1218**, 1603–1610.
32. Wang, J., Chow, W. and Cheung, W. (2011). Application of a tandem mass spectrometer and core-shell particle column for the determination of 151 pesticides in grains, *J. Agric. Food Chem.*, **59**, 8589–8608.
33. Rostagno, M.A., Manchon, N., D'Arrigo, M. *et al.* (2011). Fast and simultaneous determination of phenolic compounds and caffeine in teas, mate, instant coffee, soft drink and energetic drink by high-performance liquid chromatography using a fused-core column, *Anal. Chim. Acta*, **685**, 204–211.
34. Tölgyesi, A., Sharma, V.K. and Fekete, J. (2011). Development and validation of a method for determination of corticosteroids in pig fat using liquid chromatography–tandem mass spectrometry, *J. Chromatogr. B: Anal. Technol. Biomed. Life Sci.*, **879**, 403–410.
35. Wu, J., Qian, Y., Mao, P. *et al.* (2013). Separation and identification of phenolic compounds in canned artichoke by LC/DAD/ESI-MS using core-shell C18 column: a comparative study, *J. Chromatogr. B: Anal. Technol. Biomed. Life Sci.*, **927**, 173–180.
36. Pellati, F., Prencipe, F.P., Bertelli, D. *et al.* (2013). An efficient chemical analysis of phenolic acids and flavonoids in raw propolis by microwave-assisted extraction combined with high-performance liquid chromatography using the fused-core technology, *J. Pharm. Biomed. Anal.*, **81-82**, 126–132.
37. Tölgyesi, A. and Kunsagi, Z. (2013). Quantification of T-2 and HT-2 mycotoxins in cereals by liquid chromatography–multimode ionization–tandem mass spectrometry, *Microchem. J.*, **106**, 300–306.
38. Puchalska, P., Luisa Marina, M. and Concepcion Garcia, M. (2013). Development of a high-performance liquid chromatography–electrospray ionization–quadrupole–time-of-flight mass spectrometry methodology for the determination of three highly antihypertensive peptides in maize crops, *J. Chromatogr. A*, **1285**, 69–77.

39. Yáñez, K.P., Bernal, J.L., Nozal, M.J. *et al.* (2013). Determination of seven neonicotinoid insecticides in beeswax by liquid chromatography coupled to electrospray-mass spectrometry using a fused-core column, *J. Chromatogr. A*, **1285**, 110–117.
40. Gomez-Caravaca, A.M., Verardo, V., Toselli, M. *et al.* (2013). Determination of the major phenolic compounds in pomegranate juices by HPLC-DAD-ESI-MS, *J. Agric. Food Chem.*, **61**, 5328–5337.
41. Mao, J., Lei, S., Yang, X. *et al.* (2013). Quantification of ochratoxin A in red wines by conventional HPLC-FLD using a column packed with core-shell particles, *Food Control*, **32**, 505–511.
42. Esparza, X., Moyano, E., Cosialls, J.R. *et al.* (2013). Determination of naphthalene-derived compounds in apples by ultrahigh-performance liquid chromatography-tandem mass spectrometry, *Anal. Chim. Acta*, **782**, 28–36.
43. Pedrouzo, M., Borrull, F., Pocurull, E. *et al.* (2011). Drugs of abuse and their metabolites in waste and surface waters by liquid chromatography-tandem mass spectrometry, *J. Sep. Sci.*, **34**, 1091–1101.
44. Jessing, K.K., Juhler, R.K. and Strobel, B.W. (2011). Monitoring of artemisinin, dihydroartemisinin, and artemether in environmental matrices using high-performance liquid chromatography–tandem mass spectrometry (LC-MS/MS), *J. Agric. Food Chem.*, **59**, 11735–11743.
45. Shaaban, H. and Gorecki, T. (2012). Fast ultrahigh performance liquid chromatographic method for the simultaneous determination of 25 emerging contaminants in surface water and wastewater samples using superficially porous sub-3 µm particles as an alternative to fully porous sub-2 µm particles, *Talanta*, **100**, 80–89.
46. Echeverria, S., Borrull, F., Fontanals, N. *et al.* (2013). Determination of iodinated X-ray contrast media in sewage by solid-phase extraction and liquid chromatography tandem mass spectrometry, *Talanta*, **116**, 931–936.
47. Jerkovich, A.D., Mellors, J.S. and Jorgenson, J.W. (2003). The use of micrometer-sized particles in ultrahigh pressure liquid chromatography, *LCGC North Am.*, **21**, 600, 604, 606, 608, 610.

Chapter 3

Monolithic Columns in Fast Liquid Chromatography

Takeshi Hara,[a] Oscar Núñez,[b] Tohru Ikegami[c] and Nobuo Tanaka[d,e]

[a] Department of Chemical Engineering, Vrije Universiteit Brussel, Belgium
[b] Department of Analytical Chemistry, University of Barcelona, Spain
[c] Department of Biomolecular Engineering, Kyoto Institute of Technology, Japan
[d] GL Sciences Inc., Japan
[e] University of California, Davis, USA

3.1. Features of Monolithic Silica Columns: Rapid Separation Using Monolithic Silica Columns in Rod and Capillary Formats

3.1.1. *Fabrication of columns*

High-performance liquid chromatography (HPLC) has been applied for separation of a variety of compounds, and become an essential method for conducting qualitative and quantitative analysis today. Columns played a major role in the advances in HPLC.[1] Three kinds of separation media shown in Fig. 3.1, monolithic silica materials, sub-2 μm fully-porous silica particles, and fused-core (core-shell) silica particles,[2] have been developed recently to achieve faster and/or higher-efficiency separations compared to conventional silica particles. In this section the characteristic chromatographic properties of monolithic silica and the utility of the columns will be introduced.

The preparation procedures of monolithic silica materials have been established by Nakanishi and co-workers utilizing an organic alkoxysilane and a water-soluble polymer as starting materials.[3–6]

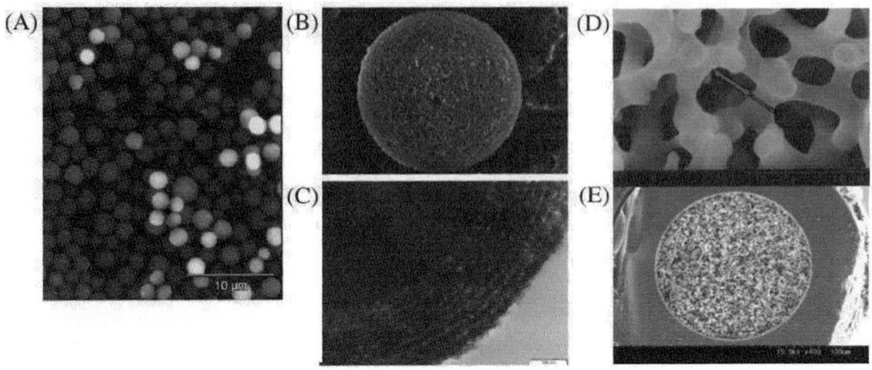

Figure 3.1. Structure of silica separation media for HPLC. (A) Sub-2 μm fully porous silica particles. Reproduced with permission from Cabooter, D., Billen, J., Terryn, H. *et al.* (2008). Detailed characterisation of the flow resistance of commercial sub-2 μm reversed-phase columns, *J. Chromatogr. A*, **1178**, 108–117). Copyright (2008) Elsevier. (B, C) Core-shell particles. Reproduced with permission from Gritti, F., Leonardis, I., Shock, D. *et al.* (2010). Performance of columns packed with the new shell particles, Kinetex-C18, *J. Chromatogr. A*, **1217**, 1589–1603). Copyright (2010) Elsevier. (D) Monolithic silica columns. The arrows indicate the size of the through-pore and the skeletons. Reproduced with permission from Tanaka, N., Kobayashi, H., Nakanishi, K. *et al.* (2001). A new type of chromatographic support could lead to higher separation efficiencies, *Anal. Chem.*, **73**, 420A–429A). Copyright (2001) American Chemical Society. (E) Monolithic silica capillary column. Reproduced with permission from Motokawa, M., Kobayashi, H., Ishizuka, N. *et al.* (2002). Monolithic silica columns with various skeleton sizes and through-pore sizes for capillary liquid chromatography, *J. Chromatogr. A*, **961**, 53–63. Copyright (2002) Elsevier.

They were successful in forming the co-continuous structure based on spinodal decomposition using sol-gel transition induced by the hydrolysis and polymerization of the silica precursors such as tetramethoxysilane (TMOS) and methyltrimethoxysilane (MTMS). Figure 3.2 shows the schematic for the typical preparation process of conventional monolithic silica columns in the polyethylene glycol (PEG) system.[7] After the gelation and solvent exchange process, the monolithic material encased in a polyether ether ketone (PEEK) tube can be employed as an HPLC column. A conventional monolithic silica column with an inner diameter (I.D.) of 4.6 mm, Chromolith™, was commercialized by Merck KGaA in 2000.[8]

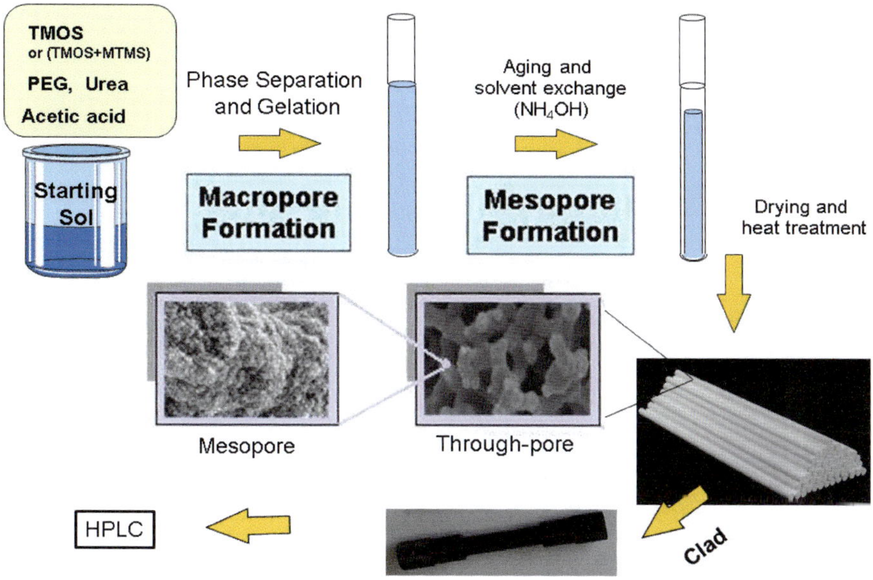

Figure 3.2. Preparation scheme of monolithic silica rod columns. Reproduced with permission from Tanaka, N., Kobayashi, H., Nakanishi, K. *et al.* (2001). A new type of chromatographic support could lead to higher separation efficiencies, *Anal. Chem.*, 73, 420A–429A. Copyright (2001) American Chemical Society.

Monolithic silica columns can also be prepared in fused-silica capillaries with an I.D. of 50–530 μm.[7,9–12] Monolithic silica capillary columns, including long ones, are accessible by a facile procedure, in comparison with particulate columns, which require frits to retain particles and high pressure to pack small particles into a capillary. Moreover, the preparation of a capillary column is feasible without cladding the monolithic silica structure with a PEEK resin to form a conventional-sized column, because the monolithic silica structure can be covalently bonded to the inner wall of a fused-silica capillary by formation of siloxane bonds. Compared to the conventional-sized columns, the capillary columns are expected to reduce the consumption of mobile phase and sample in an HPLC measurement, and conveniently incorporated in an on-line LC–MS system.[13–15] The simple fabrication and characterization of monolithic silica capillary columns accelerated the improvement in a preparation method resulting in the development of so-called second-generation monolithic silica columns in a capillary.[16] The improved preparation method and the properties of the product will be described in more detail in Section 3.2.

More recently developed was a new generation of Chromolith™ columns, possessing smaller domain size (a combined size of through-pore and skeleton (see Fig. 3.1(d)) than the first generation monolithic silica columns while achieving an increased structural homogeneity, particularly radial homogeneity, in the column.[17–19] It has been reported by Guiochon and co-workers that the new generation column gives a three-times-higher column efficiency and more symmetrical peaks than the first generation column,[19] while further improvement might still be possible.[20]

Surface modification of monolithic silica, including alkylsilylation to form a stationary phase, can be carried out by charging a solution of reagents into a column filled with the solid co-continuous structure of a monolith (in situ reaction). A radical polymerization method has also been applied for the surface modification using a variety of monomers which can react with the anchors immobilized on the silica surfaces, affording reversed phase, hydrophilic interaction liquid chromatography (HILIC), ion-exchange, and multi-mode stationary phase.[21–23] This section is intended to provide the fundamental properties of monolithic silica materials. For details, see Section 3.1.6.

3.1.2. *The support structure regarding through-pores, skeletons, and amount of silica in a column*

Comparison of a monolithic silica column with a particulate column reveals that the monolithic silica structure possesses a much larger ratio of through-pore size to silica skeleton size than the ratio of the size of interstitial space to particle size in a particulate column, resulting in unique chromatographic properties. The through-pore size/skeleton size ratios of monolithic silica columns are commonly in a range 1–2 as shown in Fig. 3.3, in comparison with 0.25–0.4 for particulate columns. In addition, for both the conventional-sized (MSR-1 and -2) and capillary columns (MSC-1, -2, and -3), Fig. 3.3

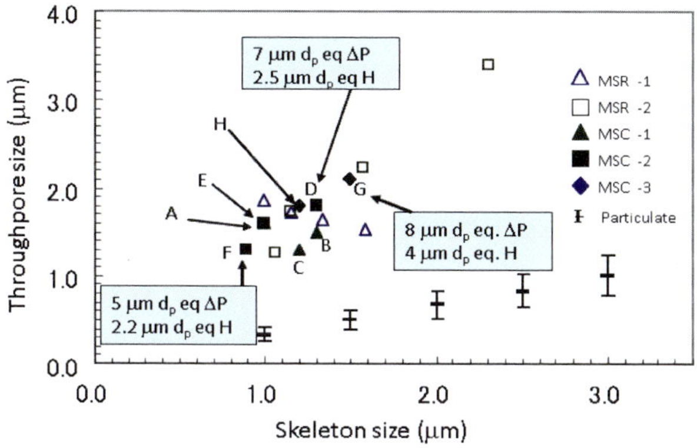

Figure 3.3. Plots of through-pore size against skeleton size for monolithic silica columns. Reproduced with permission from Nakanishi, K. and Tanaka, N. (2007). Sol-gel with phase separation. Hierarchically porous materials optimized for high-performance liquid chromatography separations, *Acc. Chem. Res.*, **40**, 863–873. Copyright (2007) American Chemical Society.

confirms that the through-pore size/skeleton size ratio of monolithic silica structure can be controlled by changing the preparation conditions, including the feed compositions.[4,5,16,24]

Table 3.1 compares the characteristic structures between monolithic silica columns and particulate columns. As described above, monolithic silica columns possess thin skeletons and large through-pores, associated with the large through-pore size/skeleton size ratio, which accordingly results in much higher column porosity than that of particulate columns. With respect to mesoporosity, it is seen that similar ranges of mesopore size and specific surface area are commonly found. Indeed, for both conventional-sized and capillary monolithic silica columns, it was reported that the column efficiency for small molecules is not critically affected by the size and volume of mesopores, which is related to steric hindrance for a solute inside mesopores (intraskeleton effect). Chromatographic properties of a column strongly depend on macroporous network structure (macropores corresponding to the interstitial space in a particulate column), domain size of a monolith (corresponding to particle size), and radial heterogeneity of the network structure (corresponding to

Table 3.1. Comparison of the structure of HPLC separation media

	Monolithic silica rod	Monolithic silica capillary	Particulate
Skeleton (μm)	0.7–1.5	0.9–2.0	1.3, 1.7, 2, 3, 5, -
Mesopore (nm)	10–25	10–25	8, 10, 30
Macropore (μm)	0.8–2.0	1.3–2.5	—
Through-pore size/ skeleton size ratio	1.0–1.5	1.5–2.2	(0.25–0.4)
Domain size (μm)	1.5–3.5	2.5–4.5	—
Surface area (m^2/g)	200–300	(300)	300–400
Total porosity	0.75–0.85	0.85–0.95	0.5–0.8
External porosity	0.55–0.65	0.65–0.75	0.4

packed bed). While particle size is used as a unit length for characterization of chromatographic properties of a particulate column, domain size has been used for a monolithic silica column.[5,25]

A clear difference exists between the two types of columns with respect to solute retention, because the absolute mesopore volume in a monolithic column is much smaller than that of a particulate column due to the smaller amount of silica in the monolithic columns.[26,27] The mesopore volume is related to the skeleton volume, and can be controlled in turn by controlling macropore and skeleton structures.[16] The results support the aforementioned: for monolithic silica, it is crucial to control the structures and the homogeneity regarding the macropores and the silica skeletons by investigating the preparation conditions to achieve desired performance.

3.1.3. *Column permeability, column efficiency, and improvement of preparation method*

3.1.3.1. *First-generation monolithic silica columns*

In terms of the HPLC performance of a monolithic silica column, Cabrera reported the results of assessment of column back-pressure (column permeability) and column efficiency using particulate columns and Chromolith™, a first-generation monolithic silica column of conventional size.[28] It was demonstrated that the monolithic silica column requires ~2.7 times lower pressure to maintain a certain flow rate of mobile phase than a column packed with 5 µm particles. Indeed, the values of column permeability (B_0) of 3.5 µm particles, 5 µm particles and the monolithic column, calculated by Eq. (3.1),[1,25] (where u is the linear velocity, η the viscosity of a mobile phase, L the column length, ε_t the total porosity) are 9.0×10^{-15} m^2, 2.0×10^{-14} m^2, and 5.4×10^{-14} m^2, respectively.

$$B_0 = \frac{\varepsilon_t u \eta L}{\Delta P} \tag{3.1}$$

Meanwhile, the monolithic silica column can yield an equivalent efficiency to that of a column packed with 3.5 µm particles (cf. Ref. 28). The domain size (~3 µm: sum of 1 µm silica skeleton and 2 µm through-pore) is supposed to dictate the column efficiency as a particle size in the case of a particulate column, as reported previously.[4] The above-mentioned comparison leads to the conclusion that the thin skeleton provides high column efficiency because of the fast mass transfer of a solute due to the short diffusion path, while the large through-pore contributes to the low column pressure drop. Therefore, for the chromatographic characteristics of silica monolith, this feature should be emphasized; a monolithic silica material is an attractive separation medium from the standpoint of achieving high column efficiency under the limitation of pressure available in a conventional HPLC system.

For the first-generation monolithic silica capillary columns, the column permeability from 10×10^{-14} m^2 to 100×10^{-14} m^2 was obtained, which was much higher than that of a particulate column with 5 µm particles (cf. Ref. 11). In addition, the results indicate that these columns showed higher column efficiency in a wide range of linear velocities compared to a particulate column packed with 5 µm particles. Typically the columns provided column efficiency equivalent to a column packed with 3–4 µm particles at optimum linear velocities, while showing column back-pressure comparable with that of a particulate column packed with 10 μm particles. Consequently, it was recognized that conventional-sized and capillary monolithic silica columns could provide higher column efficiency at lower pressure drop compared to a particulate column. However, reduction in column efficiency was noted at high liner velocities due to the contribution of large through-pores, or the A-term in Van Deemter plots, like a column packed with large-sized particles.

3.1.3.2. *Second-generation monolithic silica columns*

Ultrahigh pressure LC (UHPLC) utilizing a column packed with sub-2 μm particles was introduced early this century, and revolutionized chromatographic separations. Compared to a particulate column packed with sub-2 μm particles, the column efficiency of monolithic silica columns was still lower and hence it was desirable to reduce the domain size and increase the structural homogeneity of monolithic silica possessing a smaller domain size. The performance of monolithic silica capillary columns was improved by optimizing preparation conditions of monolithic silica capillary columns. Hara and co-workers increased the phase ratio by increasing the total silane concentration in the feed solution in comparison with that of the first-generation column, as shown in Table 3.2.[16]

As shown in Figs. 3.4A–3.4D using scanning electron microscopy (SEM), increasing the total silane (silica precursor) concentration in the preparation feed leads to a change of the morphology and porosity of a monolith. The second-generation capillary column (MS(100)-T1.4-A), prepared to have increased silica content, showed increased structural homogeneity (decreased irregularity of silica skeleton) compared to the first-generation monolith (MS(100)-T1.0-A),[16] as predicted by Desmet and co-workers.[29] Moreover, from Figs. 3.4E–3.4H it is evident that domain size can be decreased with an increase in PEG concentration in the feed solution, as reported previously.[4,5] Table 3.2 summarizes the preparation conditions for all the monolithic silica capillary columns shown in Figs. 3.4A–3.4H.

Pressure drop and plate height obtained for these columns are plotted against linear velocity in Figs. 3.5A and 3.5B, respectively,

Table 3.2. Feed compositions of monolithic silica capillary columns for high-speed separation.

Column	TMOS (mL)	PEG (g)	Urea (g)	CH_3COOH (mL)[a]
MS(100)-T1.0-A[b]	40	12.4	9.0	100
MS(100)-T1.4-A[c]	56	11.8	9.0	100
MS(100)-T1.6-A[c]	64	10.4	9.0	100
MS(100)-T1.8-A[c]	72	8.4	9.0	100
MS(100)-T1.0-B[b]	40	12.8	9.0	100
MS(100)-T1.4-BI[c]	56	11.7	9.0	100
MS(100)-T1.4-BII[c]	56	11.8	9.0	100
MS(100)-T1.4-BIII[c]	56	11.9	9.0	100

[a] 0.01M acetic acid aqueous solution. [b] Gelation temperature: 30 °C. [c] Gelation temperature: 25 °C.

for the capillary columns with different domain sizes (see Figs. 3.4E–3.4H). Compared to the capillary column (MS(100)-T1.0-B) prepared by an earlier method, the second-generation column (MS(100)-T1.4-BI), provided higher column permeability and higher efficiency at the same time, demonstrating that monolithic structures having higher homogeneity can be achieved under the improved preparation conditions. Furthermore, decreasing the domain size leads to further improvement in column efficiency. Accordingly MS(100)-T1.4-BIII, possessing the smallest domain size in the column series illustrated in Figs. 3.5A and 3.5B, showed plate height values expected for a column packed with ~2.5 μm silica particles with comparable column permeability to that of a column packed with 5 μm silica particles.

The feed composition for preparation of monolithic silica affects not only the column efficiency, but also chromatographic retentivity of monolithic silica. It was reported that increasing the total silane concentration in the feed resulted in an increase in the amount of stationary phase (e.g. C18 chains introduced by octadecylsilylation), leading

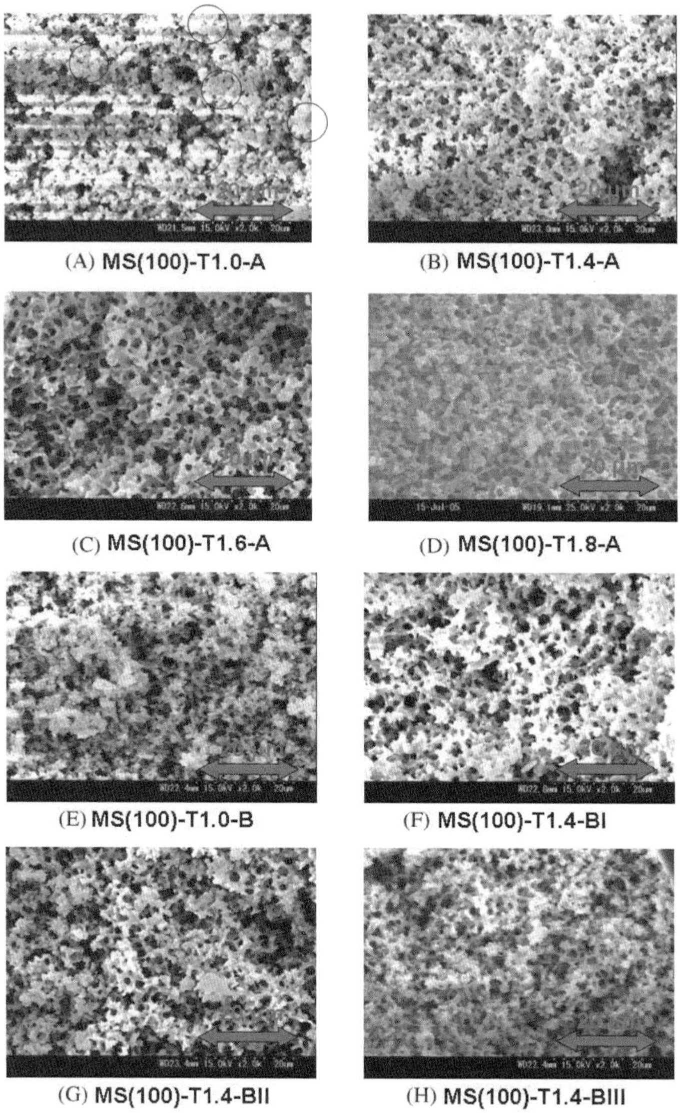

Figure 3.4. Scanning electron micrographs of monolithic silica columns with increased phase ratios and/or with smaller domain sizes. Scale bars correspond to 20 μm. Circles show examples of agglomerated skeletons in (A). Reproduced with permission from Hara, T., Kobayashi, H., Ikegami, T. *et al.* (2006). Performance of monolithic silica capillary columns with increased phase ratios and small-sized domains, *Anal. Chem.*, 78, 7632–7642. Copyright (2006) American Chemical Society.

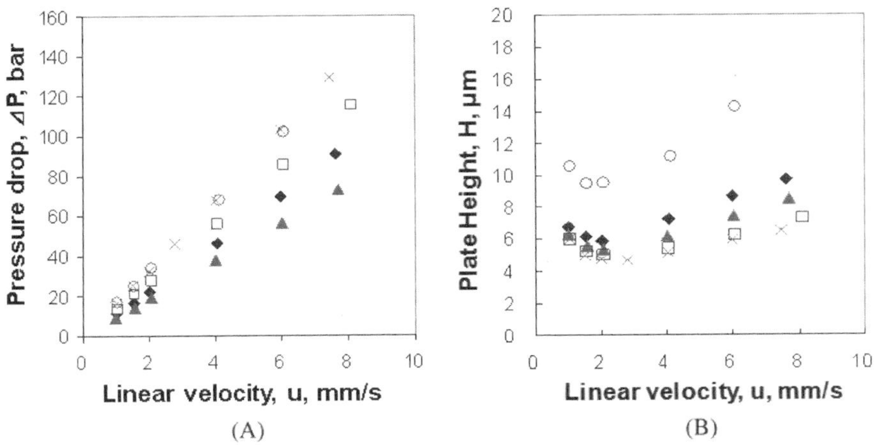

Figure 3.5. Plots of a column back-pressure (A) and plate heights (B) observed for ODS-modified monolithic silica columns against the linear velocity of the mobile phase. Mobile phase: Acetonitrile/water = 80/20. Temperature: 30°C. The pressures were normalized to a column length of 15 cm. Columns: Mightysil RP18 (o), MS(100)-T1.0-B (♦), MS(100)-T1.4-BI (▲), MS(100)-T1.4-BII (□), and MS(100)-T1.4-BIII (X). Reproduced with permission from Hara, T., Kobayashi, H., Ikegami, T. *et al.* (2006). Performance of monolithic silica capillary columns with increased phase ratios and small-sized domains, *Anal. Chem.*, 78, 7632–7642. Copyright (2006) American Chemical Society.

to enhanced retention ability in reversed-phase LC, since the total surface area of the silica in a column can be increased.[16] Additionally, it is known that the feed composition concerning the hybrid monolithic silica materials, prepared from a mixture of silica precursors, is directly related to the chromatographic behaviors for solutes. As an example, changing the concentration of MTMS in the MTMS/TMOS mixture influences the specific separation selectivity α(T/O) regarding triphenylene (T: planar compound) and o-terphenyl (O: nonplanar, bulky compound) because of the presence of methyl groups in MTMS, as shown in Fig. 3.6 (cf. Ref. 24). This proves that retention ability and selectivity for solutes in HPLC are not only derived from bonded stationary phase introduced by surface modification, but also strongly correlated with the nature of a hybrid silica material. It is interesting to introduce various silica precursors for preparing hybrid monolithic silica materials in order to effect a wide range of characteristic retention behaviors and separation selectivity.[30–32]

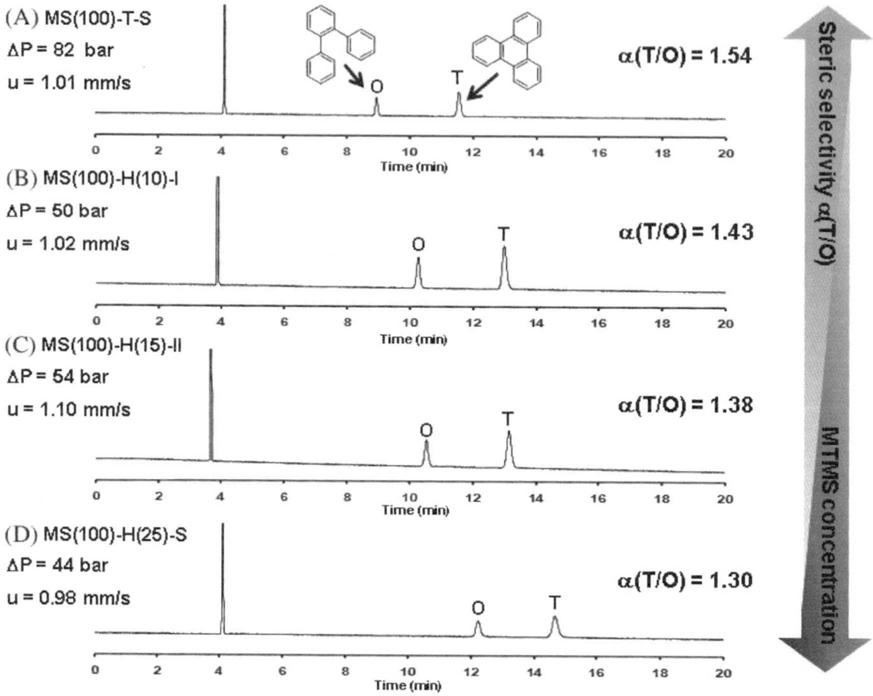

Figure 3.6. Chromatograms obtained for o-terphenyl (O) and triphenylene (T). Column: (A) MS(100)-T-IV 30.0 cm (effective length 25.0 cm) (B) MS(100)-Hy(10)-I 28.9 cm (effective length 23.9cm), (C) MS(100)-Hy(15)-II 29.5 cm (effective length 24.5 cm), and (D) MS(100)-Hy(25)-IV 29.4 cm (effective length 24.4 cm). Column diameter: 100 mm. Mobile phase: methanol/water = 80/20. Temperature: 30°C. Detection: 254 nm. The pressure drop, linear velocity, and steric selectivity α(T/O) are indicated. The MTMS content (%) in a feed solution is shown in parentheses following the monolithic columns. Reproduced with permission from Hara, T., Makino, S., Watanabe, Y. *et al.* (2010). The performance of hybrid monolithic silica capillary columns prepared by changing feed ratios of tetramethoxysilane and methyltrimethoxysilane, *J. Chromatogr. A*, **1217**, 89–98. Copyright (2010) Elsevier.

3.1.4. *Current performance of monolithic silica columns*

Improvement of column preparation procedure has been continued for conventional-sized monolithic silica columns also. Second generation Chromolith™ column has been introduced recently to provide much higher column efficiency (greater than 16,000 theoretical plates with a 10 cm column) compared to the first generation columns with improved peak symmetry. The macropore-skeleton network structure is reported to be more homogeneous than previous products, as mentioned earlier.[17–20]

MonoClad column, recently introduced by GL sciences, shows column efficiency close to that of a column packed with sub-3 µm core-shell particles.[33] Figure 3.7 shows the comparison of the chromatograms obtained for a prototype MonoClad C18 column (1.9 mm I.D., 5 cm) and Kinetex C18 column (2.6 µm particles, 2.1 mm I.D., 5 cm,) at flow rate of 0.4 mL/min in acetonitrile/water = 60/40. Figure 3.7A shows that a 5 cm monolithic silica column can produce 10,000 theoretical plates, much more than first generation monolithic silica columns can, and close to the column efficiency observed with second generation monolithic silica capillary columns. While first generation Chromolith™ column possesses through-pores of ~2.0 µm and mesopores of around 13 nm (surface area: 300 m^2/g), the through-pore size of the present MonoClad column is around 1.2 µm, with the mesopore size of 18 nm and surface area of 200 m^2/g. Furthermore, total porosity

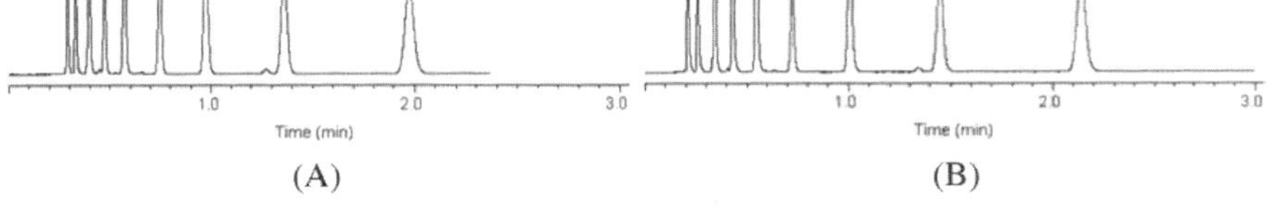

Figure 3.7. Chromatograms obtained for (A) monolithic silica column (MonoClad C18 prototype, 1.9 mm I.D., 5 cm) and (B) a column packed with 2.6 µm core-shell particles (Kinetex C18, 2.1 mm I.D., 5 cm). Mobile phase: acetonitrile/water = 60/40. Flow rate: 0.4 mL/min. Temperature: 40°C. Detection: 254 nm. Retention time and the number of theoretical plates are attached for the peaks of thiourea, acetanilide, CnH2n+1COPh (n = 1–3), naphthalene, and C_nH_{2n}+1COPh (5–7) in the order of elution.

and external porosity of the MonoClad column are ~75% and ~55% respectively, smaller than those of a first-generation Chromolith^{TM} column by about 10%. Thus, it is recognized that MonoClad columns possess a different structure and greater phase ratio between the stationary phase and the mobile phase in comparison with a common Chromolith^{TM} column.

It is noticeable that the monolithic column provided larger numbers of theoretical plates than the particulate column of core-shell particles in the earlier part of the chromatogram in Fig. 3.7, while slightly higher column efficiency was observed with the column of core-shell particles for the peaks eluted later. The difference may be attributed to the difference in the pore volume. The tendency is clearly seen in Van Deemter plots, shown in Fig. 3.8. In this figure, plate height values for unretained thiourea, naphthalene, and octanophenone are plotted against linear velocity of the mobile phase, acetonitrile/water (v/v) = 60/40 using a MonoClad column and Kinetex core-shell particles. Slightly larger plate height observed with Kinetex in the earlier part of the chromatogram may partly be

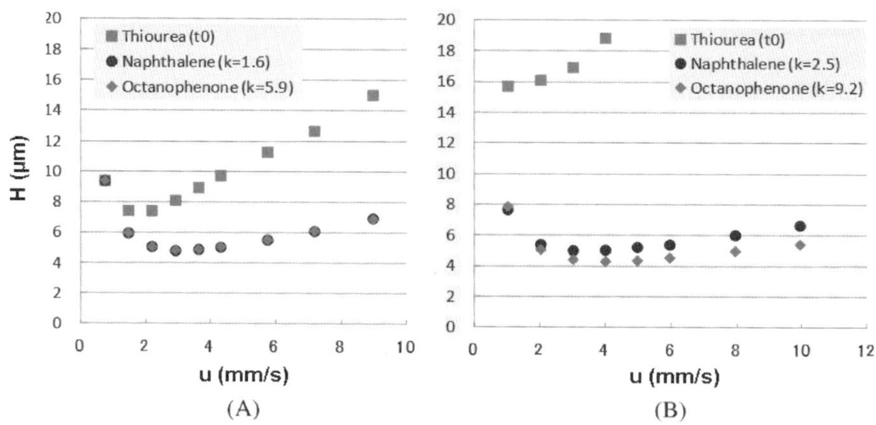

Figure 3.8. The Van Deemter plots obtained for (A) monolithic silica column (Monoclad C18 prototype, 1.9 mm I.D., 5 cm) and (B) a column packed with 2.6 µm core-shell particles (Kinetex C18, 2.1 mm I.D., 5 cm). Solutes: ■ thiourea, □ naphthalene, ● octanophenone. Other conditions are described in the caption of Fig. 3.7.

explained by the greater contribution of extra-column effect associated with the peaks having small elution volume.

It is interesting to note that the two columns, a prototype MonoClad-C18 (1.9 mm I.D., 5 cm) and a column of core-shell particles (2.6 μm particles, 2.1 mm I.D., 5 cm), provided similar retention range in terms of retention time, considering the difference in column diameter. The differences in retention factors were provided by the difference in the column void volume. The elution time of thiourea (the first peak, used as a void-volume marker) was ~35% larger on MonoClad column than on Kinetex, in spite of the smaller column size of the monolithic column. The pressure limit for use of the monolithic columns is 20 MPa in the case of Chromolith™ column.[34] 'Glass clad' monolithic columns are under development to increase pressure stability of monolithic silica columns.[6]

3.1.5. *Kinetic performance*

For examining the obtainable theoretical plate number N (column efficiency) per unit time (kinetic performance), the kinetic plots of using t_0 (the column dead time) and N, or $\log(t_0/N^2)$, under a given pressure limit has been suggested by Desmet and co-workers.[29,35] Figure 3.9 represents the comparison of the kinetic performance of monolithic silica columns with that of particulate columns packed with silica particles of different size (10 μm, 5 μm, 3 μm, 2 μm, and 1.4 μm) at 40 MPa.[16] In these kinetic plots, if a curve obtained for one column shows a smaller value of $\log(t_0/N^2)$ at the same value of $\log(N)$ compared to the other curves, it means that the former column is superior to the other columns in achieving faster and higher efficiency separations. Figure 3.9 includes the comparison of kinetic performance of monolithic silica columns and a column of core-shell particles. (The plots for MonoClad and Kinetex columns are based on results obtained with a 5 cm long column, 1.9 mm I.D and 2.1 mm I.D respectively, while the plots for fully porous particles are calculated based on the results obtained with a larger-sized column packed with 5 μm particles.)

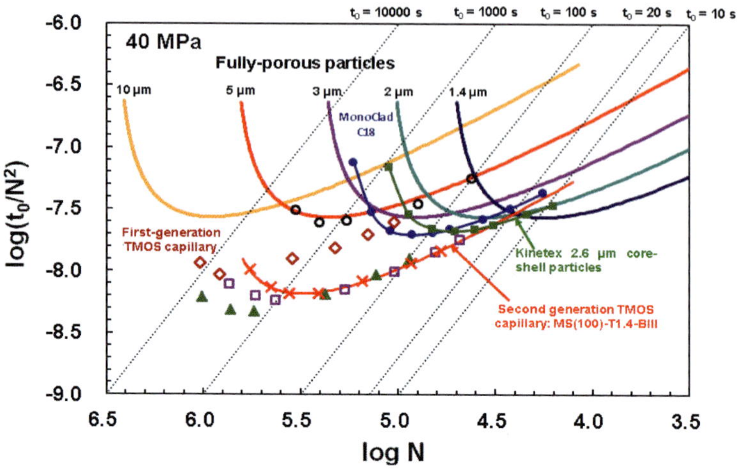

Figure 3.9. Plot of $\log(t_0/N^2)$ against $\log(N)$ for monolithic silica and particulate columns. The curves for particle-packed columns were obtained by assuming the following parameters: $\eta = 0.00046$ Pa·s, $f\varphi = 700$, $D_m = 2.22 \times 10^{-9}$ m^2/s, and Knox equation, $h = 0.65\nu^{1/3} + 2/\nu + 0.08\nu$. Maximum pressure: 40 MPa. The particle diameters for the particle-packed columns were 1.4 μm, 2 μm, 3 μm, 5 μm, and 10 μm. Experimental results (o) obtained with a Mightysil RP18 column (4.6 mm I.D., 15 cm long) packed with 5 μm particles are included (see Fig. 3.5).

From the comparison of the silica monolith and the particulate columns, the properties of monolithic silica columns can be recognized as follows:

- Under the limited pressure of 40 MPa, the optimal kinetic performance region (concave region) of the monolithic silica capillary columns gradually shifted from $N \approx 300{,}000$ to $N \approx 500{,}000$ with the increase in domain size, suggesting that a long monolithic capillary column would be very useful if high resolution separation (or a large number of theoretical plates) is needed.
- The second-generation monolithic silica columns of capillary can provide much higher kinetic performance than the first-generation products, achieving fast and high efficiency separation, and it is seen that the curves of the second-generation column (e.g. MS(100)-T1.4-BIII possessing the smallest domain size (2.2 μm)

in the series) merge with that for a column packed with 2–2.5 µm particles (fully-porous) in a range of N = 25,000–30,000.

- The conventional-sized monolithic silica column, Monoclad C18 (1.9 mm I.D., 5 cm), which shows comparable performance with a column packed with 2–2.5 µm in a range of $N \approx 50{,}000$, can result in higher overall performance than a first-generation column. However, in comparison with the second-generation capillary or core-shell particles (Kinetex: 2.6 µm particles, 2.1 mm I.D., 5 cm), it is evident that further improvement of the kinetic column performance of conventional-sized monolithic silica is still required for fast separation.

In addition, compared to the kinetic performance of core-shell silica particles with sub-3 µm or fully-porous particles below 2 µm, the kinetic performance of monolithic silica capillary columns is still inferior in the range of $N < 30{,}000$ (fast separation region) under 40 MPa (see Fig. 3.9). It is suggested that the preparation method of the monolithic silica columns still needs improvement and further study is required for the development of monolithic silica columns with a domain size smaller than 2.2 µm.

3.1.6. *Functionalization of monolithic silica columns*

Bare silica monolith columns can be used for HILIC separation,[36] but in many cases they are used after functionalization to form stationary phases on the silica surface. In previous sections of this chapter, features of monolithic silica columns modified with octadecyl group (C18 or ODS) were considered. In this section, several methods to functionalize monolithic silica columns with reversed-phase, HILIC, anion-exchange (AX), cation-exchange (CX), chiral separation, molecular imprint polymer (MIP), mixed mode and others, are introduced.

Functionalization of silica can be roughly categorized into two parts: one employs physical adsorption or coating using macromolecules on the surface of silica, while the other involves chemical modification of silanol groups to form Si–O–Si–C bonds using chlorosilanes or alkoxysilanes. Further change is applicable by functional group transformations, or by polymerization. Recently, one-pot procedures that involve formation of silica skeletons and radical polymerization of stationary phases have also been explored.[31,37–39] By physical adsorption of proteins or enzymes, monolithic materials can be used as chiral stationary phases[40] or on-line digesting devices of proteins.[41] Through a sol-gel process of titanium alkoxide on the silica surface, titania-coated monolithic silica was obtained, and used for the concentration of phosphate derivatives.[42] Table 3.3 summarizes the method of functionalization of monolithic silica, the column efficiency, and analytes.

3.1.7. *Advantages and disadvantages: roles of monolithic silica columns*

Following on from the introduction and previous publications, the advantages and disadvantages of monolithic silica columns are summarized below.

3.1.7.1. *Advantages associated with monolithic silica columns*

- High permeability and/or high column efficiency compared to a particulate column.
- Stable bed structure in high flow rate operations.
- Facile preparation of a long capillary column without using frits or a column packing process.
- Long capillary columns resulting in a very large number of theoretical plates, which contributes to high peak capacity and high-resolution separations.
- A wide range of surface modifications of the packed bed without the need for packing of products into a column (application of *in situ* modification of a preformed bed structure with proven performance for a new stationary phase).
- Possible operation of very-high-efficiency columns with conventional HPLC equipment without using UHPLC instruments.

Table 3.3. Preparation, column efficiency, and analytes of functionalized silica monolith columns.

Stationary phase	Column size	Modification process	Efficiency H (μm)	Analytes	Ref.
C_{30}	30 cm × 100 μm I.D.	Silane	25 at 0.4 mm/sec	Alkylbenzenes, carotenes	43
Poly(octadecyl methacrylate)	25 cm × 200 μm I.D.	Polymerization	8 at 1.0 mm/sec	Alkylbenzenes, substituted arenes	21
Poly(acrylamide)	38 cm × 100 μm I.D.	Polymerization	16 at 1.0 mm/sec	Nucleosides, saccharides	44
Poly(acrylic acid)	20 cm × 200 μm I.D.	Polymerization	9 at 0.8 mm/sec 20 at 6 mm/sec	Nucleosides, saccharides, peptides, protein digest	22
Sulfoalkylbetaine zwitterion	15 cm × 100 μm I.D.	Polymerization	6 at 1.0 mm/sec	Nucleosides	45
4° Ammonium functionalized latex	33.5 cm × 50 and 75 μm I.D.	Coating	20 at 1 mm/sec	Inorganic anions	46
3° and 4° Amine/ammonium possessing polymer	30 cm × 200 μm I.D.	Polymerization	8 at 0.82 mm/sec	Inorganic anions, nucleotides, proteins	23
Sulfonic acid	30 cm × 50 μm I.D.	Silane	CEC, 4 at 1 mm/sec	β-blockers	47
Iminodiacetic acid	10 cm × 4.6 μm I.D. based on Chromolith Performance Si	Silane	20 to 40 at 1 mm/sec	Alkali, alkaline earth, transition metal cations	48
Arylboronic acid	35 cm × 50 μm I.D.	One pot	Not available	Nucleosides, urine	37

(Continued)

Table 3.3. (*Continued*)

Stationary phase	Column size	Modification process	Efficiency *H* (μm)	Analytes	Ref.
L-Phenylalanin -amide	26.5 cm × 100 μm I.D.	Silane	CEC, 30 at 0.25 mm/sec	Dansyl amino acids	49
Avidin	20, 50 cm × 50 μm I.D.	Adsorption	CEC, 8, CLC, 15	Dansyl amino acids	40
β-Cyclodextrin	10 cm × 4.6 μm I.D. based on Chromolith Performance Si	Silane	23 at 1 mm/sec	Pharmaceutical drugs	50
3,5-Dimethylphenyl -carbamate derivative of cellulose	10 cm × 4.6 μm I.D. based on Chromolith Performance Si	Silane	Not available	Chiral ketones, alcohols, amine	51
Molecular imprint polymer	24.5 cm × 75 μm I.D.	Polymerization	24 at 0.2 mm/sec	Tröger's base, tetrahydropalmatine	52
RP/WAX mixed mode	10 cm × 4.6 μm I.D. based on Chromolith Performance Si	Silane	13 at 0.2 mm/sec	Alkylbenzenes, peptides, caffeine	53
Trypsin	2.5 cm × 4.6 μm I.D. based on Chromolith Flash	Silane, then trypsin immobilization	Not available	Digestion of myoglobin	41
TiO_2	8.3 cm × 4.6 μm I.D.	Coating	Not available	Nucleotides	42

3.1.7.2. *Disadvantages associated with monolithic silica columns*

- Limited commercial availability of silica support and chemically modified materials and columns.
- Limitation of column size and structures available (length, diameter, pore size).
- Restrictions in mobile phase solvent, temperature, and pressure for monolithic silica rod-type columns clad with PEEK resin.
- Lower column efficiency in very fast separation, smaller theoretical plate number N per unit time (i.e. N/t_0), in comparison with small-sized core-shell particles.
- Smaller retention factors under similar mobile phase conditions compared to particulate columns.
- One-by-one preparation of silica support, surface modification, and columns.
- Difficulty in the cladding process regarding a rod-type column (pressure limit, limited column length).

3.1.7.3. *Role of monolithic silica columns*

High performance columns are desired for two extremes or their combination: high-speed separations seeking large N per unit time, and very-high-efficiency separations seeking large absolute N. As described above, the development of monolithic silica columns for high-speed separations still presents some issues to be solved in terms of small and homogeneous domain structures, column fabrication (including cladding) and method transfer from a reference method obtained by using a particulate column. However, a large number of high-speed separations using short monolithic silica columns may be a suitable area of application for monolithic columns utilizing the stable bed structure; an increase in the pressure stability of conventional-sized monolithic columns is desirable for this purpose. If the current tendency of reduction of column size continues further, high-efficiency monolithic silica capillary columns of 0.5 mm I.D. and 0.3 mm I.D. are going to be developed to fill the gap between present capillary LC and UHPLC using 1–2 mm I.D. columns.

On the other hand, particularly given that monolithic silica columns can provide very high efficiency under relatively low pressure, ultrahigh-efficiency separation employing a very long capillary column may play an important role for specific applications such as 'omics' studies. For example, for a proteome analysis by a micro liquid chromatography–tandem mass spectrometry (μLC–MS/MS) system, Ishihama and co-workers recently demonstrated the identification of more than 2,200 proteins from *Escherichia coli* cells using a C18-modified long capillary column of 350 cm under pressure lower than 20 MPa, as shown in Fig. 3.10.[13]

Another area where monolithic columns may play a role is the preparation and use of a new stationary phase. A stable bed structure for which certain column efficiency is proved will be an attractive option for the preparation of various stationary phases. Although the preparation process may have some limitations associated with an in situ reaction, the column efficiency is guaranteed unless the stationary phase prepared hinders fast equilibration of solute molecules in the stationary phase. Development of a packing procedure for newly developed stationary phases based on particulate materials is sometimes a difficult task; however, monolithic

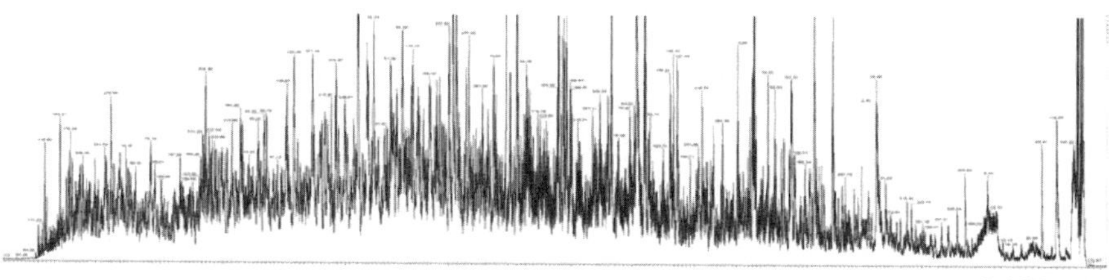

Figure 3.10. Base peak chromatograms for the analysis of *E. coli* cell lysate using a 350 cm long monolithic silica C18 column. Tryptic peptides in 4 μg of *E. coli* cell lysate were loaded onto the column. The mobile phases consisted of (A) 0.5% acetic acid and (B) 0.5% acetic acid in 80% acetonitrile. A linear gradient of 5–40% B delivered over 2,470 min was employed. Reproduced with permission from Iwasaki, M., Miwa, S., Ikegami, T. *et al.* (2010). One-dimensional capillary liquid chromatographic separation coupled with tandem mass spectrometry unveils the *Escherichia coli* proteome on a microarray scale, *Anal. Chem.*, 82, 2616–2620. Copyright (2010) American Chemical Society.

columns should be free from such problems. One-pot preparation of monoliths is expected to accelerate the development of new stationary phases, including those for HILIC applications.

3.2. Features of Organic Polymer Monolithic Columns

Monolithic materials possessing a continuous net-like skeleton can be prepared by using organic polymers, in addition to inorganic compounds like silica, titania and zirconia. Organic polymer monolith was first introduced by Hjertén and co-workers,[54] and many research groups (e.g. Svec and Fréchet, Schoenmakers, and Jandera) are significantly contributing to this field. Many good reviews are available, and several of the newest ones are cited here.[55–57] Of special interest is a review by Jandera focused polymer monolithic columns and their applications in food analysis.[55] Organic polymer monolithic materials are prepared by polymerization of monomers, a cross-linking agent, porogen, and a suitable solvent. In many cases, radical polymerization using initiators is involved, but ultraviolet and γ-ray irradiation are also employed to develop polymerization.[58] Monomers that are suitable for co-polymerization are useful for polymer monolithic materials; the variety of polymer skeleton composition is much wider than that of inorganic monoliths.[59] Typically, they are classified as polystyrene–co-divinylbenzene monoliths (PS–DVB), polyacrylate- and polymethacrylate-based monoliths (such as poly(butyl methacrylate-*co*-ethylene dimethacrylate) (BMA-EDMA)), polyacrylamide monoliths, and polyamine monoliths. Though there are several exceptions, many polymer monoliths have micro-globular structure, different from those of silica monoliths. As separation media, they can be prepared as thin disks, cylindrical rod-like columns, and capillary columns. If the monomers possess reactive functionalities, post-polymerization chemical modification is also possible to change separation characteristics (e.g. to change into cation- and anion-exchange columns, in addition to reversed-phase columns). Polymer stationary phases are stable in strong acidic and strong basic conditions, which compensates for the disadvantage of silica based stationary phases. Simultaneously, their tight network leads to slow mass transfer, which limits their use to the separation of proteins or synthetic polymers.[1]

Many researchers hold a stereotypical view of the separation efficiency of polymer monolith columns: that they are worse than those of silica monolith columns, and that their lower mechanical strengths and higher back-pressure make it impossible to use them for rapid separation. This view seems to be true, judging from a review of polymer monolith for reversed-phase HPLC, which summarized their separation efficiencies.[57] PS-DVB-based monolithic columns provide 13,500–83,200 plates/m, while methacrylate-based columns generate 20,000–67,000 plates/m for alkylbenzenes as analytes. It should be noted that researchers in this field often use the unit ‘plates/m’, but columns of 1 m or longer are difficult to prepare; thus, practical theoretical plates provided by a polymer monolith column are much smaller in many cases. Exceptionally high separation efficiency for polymer monolith columns (210,000[60] and 120,000[61] plates/m) were reported. A comparison of silica and polymer monolith columns in terms of column efficiency is summarized in Table 3.4. Readers should be careful to note that the column efficiency of these narrow-bore columns easily deteriorates due to the extra-column effect, retention of analytes, and characteristics of employed mobile phases; they were collected from different chromatographic conditions, and the direct comparison of their true performances might be difficult.

Long polymer monolithic columns are also reported. Tris (2,3-epoxypropyl) isocyanurate–4-[(4-aminocyclohexyl)methyl] cyclohexylamine polymer monolith (TEPIC–BACM) could be prepared in 150.5 cm × 100 μm I.D. dimension, and it generated 140,000–210,000 theoretical plates for the separation of alkylbenzenes in reversed-phase mode with 4 MPa back-pressure. Interestingly, the TEPIC–BACM column could be used in HILIC mode without further surface modification. Permeability of the columns changed in the range of 2.1×10^{-14} m^2 to 26.8×10^{-14} m^2, and its Brunauer–Emmett–Teller (BET) specific surface area was estimated to be 2.7 m^2/g. Morphology of the monolith was significantly different from common micro-globular structures.[60]

PS-DVB monolithic columns with column dimensions of 50 mm, 250 mm, and 1 m × 200 μm I.D. were prepared, and peak capacities in complex protein digests analysis by LC–MS/MS systems using

Table 3.4. Comparison between silica monolith and polymer monolith columns in terms of column efficiency.

Column	Column size	Efficiency (plates/m)	Analytes	Ref.
PDVB	250 mm × 100 μm I.D.	34,000 27,000	Benzene Hexylbenzene	62
BVPE	80 mm × 200 μm I.D.	72,000	Butyrophenone	63
BADMA	150 mm × 75 μm I.D.	35,000 to 61,000	Alkylbenzenes	64
LMA–EDMA	100 mm × 1.02 μm I.D.	50,000 to 70,000	Alkylbenzenes	65
GMA–EDMA–PCB–HEM	70 mm × 100 μm I.D.	120,000	Benzene	61
TEPIC–BACM	215 mm × 100 μm I.D.	218,000 167,000	Toluene Hexylbenzene	60
MonoBis, C18 monolithic silica column	100 mm × 1.0 mm I.D.	60,000 to 72,000	Alkylbenzenes	65
Chromolith® HighResolution RP-18e	100 mm × 4.6 mm I.D.	160,000	Butylbenzene	66
MonoClad C18-HS	250 mm × 3 mm I.D.	180,000 to 191,000	PAH	33
Kinetex C18	130 mm × 3 mm I.D.	218,000 to 219,000	PAH	33
L-column, C18, 3 μm	100 mm × 1.0 mm I.D.	65,000 to 75,000	Alkylbenzenes	65

PDVB: poly(divinylbenzene);
BVPE: 1,2-bis(p-vinylphenyl)ethane;
BADMA: bisphenol A dimethacrylate;
LMA–EDMA: poly(lauryl methacrylate-*co*-ethylene dimethacrylate);
GMA–EDMA–PCB–HEM: glycidyl methacrylate–ethylene dimethacrylate–[6,6]-phenyl-C61-butyric acid–2-hydroxyethyl methacrylate ester;
TEPIC–BACM: tris(2,3-epoxypropyl) isocyanurate–4-[(4-aminocyclohexyl)methyl]cyclohexylamine.

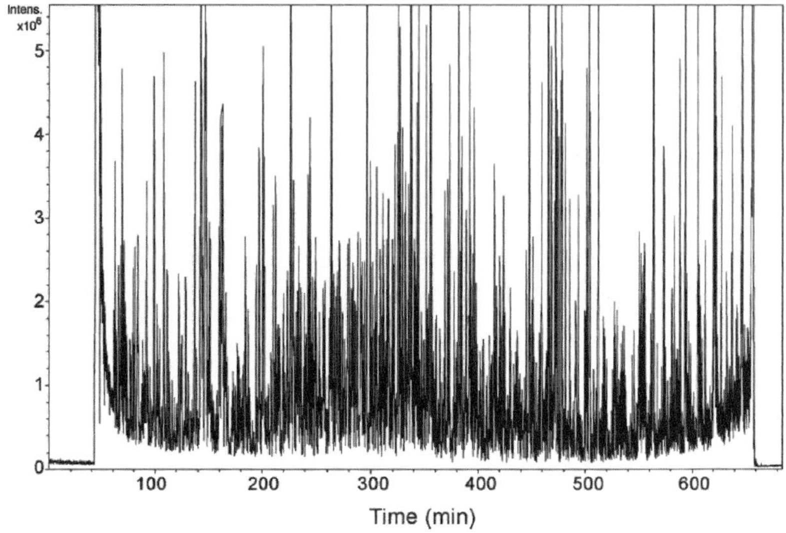

Figure 3.11. High-efficiency separation of an *E. coli* digest on a 1 m monolithic column operating at a flow rate 0.5 μL/min applying a 600 min gradient. 1 μL injection (concentration 2 μg/L). Reproduced with permission from Eeltink, S., Dolman, S., Detobel, F. *et al.* (2010). High efficiency liquid chromatography-mass spectrometry separations with 50 mm, 250 mm, and 1 m long polymer-based monolithic capillary columns for the characterization of complex proteolytic digests, *J. Chromatogr. A*, **1217**, 6610–6615. Copyright (2010) Elsevier.

them were discussed. The authors concluded that the 50 mm column gave larger peak capacity (485 by 200 min total analysis time) than the 250 mm column (370 by 180 min total analysis time). Use of the 1 m column in shallow gradient (600 min) resulted in the identification of 2,053 peptides and 283 proteins. As shown in Fig. 3.11, fine separation of peptides from *E. coli* digest was carried out on a 1 m PS-DVB monolithic column.[67]

One of the attractive applications of silica monolith columns is rapid separation under high flow rate, which has been difficult to carry out with polymer monolith columns. Recently, polymer monolith columns tolerant to high back-pressure were reported. A poly(octadecyl methacrylate-*co*-ethylene glycol dimethacrylate) (ODMA–EDMA) column prepared in a capillary format (200 mm × 250 μm I.D.) could be driven at a linear velocity of 100 mm/s with back-pressure of 33 MPa, though the resolution was not good enough to separate alkylbenzenes completely.[68] For rapid separation of proteins, an LMA–EDMA column (100 mm × 1.02 mm I.D.) was employed, and 10 proteins were separated within 2 min, providing more than 5,000 plates/10 cm.[65]

Polymer monolith columns could be employed at ultra-high pressure until 80 MPa. A commercially available PepSwift (PS-DVB) column (50 mm × 200 μm I.D.) connected to a UHPLC system separated protein digests at a flow rate 8.7 μL/min (80 MPa). As shown in Fig. 3.12, peptides were separated at conventional (A) and

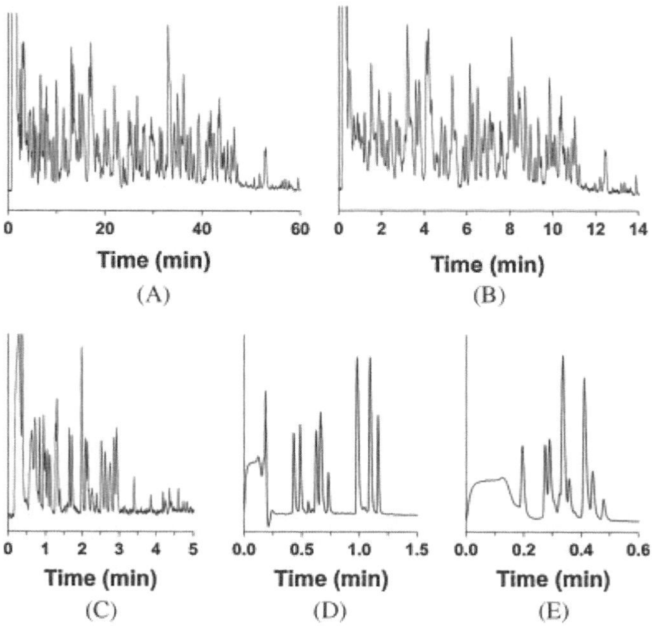

Figure 3.12. High-resolution peptide separations on a 50 mm × 200 μm I.D. polymer monolithic column (60 °C). (A) Six-protein mix digest separated using a conventional flow rate F = 2 μL/min and a 60 min gradient and (B) at a flow rate F = 8.7 μL/min (corresponding to the maximum system pressure of P_{max} = 80 MPa) while scaling the gradient volume (t_G = 14 min), (C) bovine serum albumin digest separated at P_{max} at F = 8.7 μL/min and t_G = 5 min, (D) and (E) cytochrome c digest separated at F = 8.7 μL/min in a 1 min and 14 s gradient, respectively. Reproduced with permission from Vaast, A., Nováková, L., Desmet, G. *et al.* (2013). High-speed gradient separations of peptides and proteins using polymer-monolithic poly(styrene-*co*-divinylbenzene) capillary columns at ultra-high pressure. J. Chromatogr. A, **1304**, 177–182. Copyright (2013) Elsevier.

UHPLC (B) conditions, with a peak capacity of 260 for conventional separation. In the case of cytochrome c digest, the experimentally determined peak capacity was around 100, but by a theoretical calculation fixing maximum pressure at 80 MPa, peak capacity can be estimated as 500, for a 2 to 4 coupled columns (i.e. longer column).[69] This kind of result shows high performance separation at ultrahigh pressure can be carried out using a commercially available polymer monolith column, though the degree of retention of small organic molecules must be considered when we apply the column to food or environmental analysis. Still, although the examples of the application of polymer monolith columns for this purpose are limited, there are several results, and they will be introduced in the next section.

3.3. Food and Environmental Applications

The unique structural properties of monolithic columns make them an excellent tool in the hands of analytical chemists, not only for fast separations but also for sample preparation. As has been previously commented in this chapter, their much higher external porosity compared to conventional particle-packed columns results in higher permeability and low-pressure drop with higher separation efficiency. Until now, monolithic columns have been applied to the analysis of different analytical matrices, such as pharmaceuticals, biofluids, food matrices, environmental samples, biochemical species, proteomics, etc.[51,70,71] Since they offer a great potential for the separation of complex mixtures as well as sample treatment (including sample extraction and clean-up steps), it can be expected that the interest in applying these columns will increase every year. For example, SiO_2/TiO_2 composite monolithic material was employed to pre-concentrate phospholipids,[72] and *N*-(2-aminoethyl)-3-aminopropyltrimethoxysilane modified silica monolith was used to concentrate aluminum ion in rainwater and fruit juice.[73]

In this section, a number of the food and environmental applications of monolithic columns used to achieve fast and high-efficiency chromatographic separations will be presented. Several reviews can be found in the literature dealing with the sample preparation using monolithic columns, nowadays including many type of surface modification, such as reversed-phase, anion- and cation-exchange, carboxylic acids and amides.[74–76]

Some relevant applications of monolithic columns in food analysis (including reversed-phase separation, HILIC separation, and 2D HPLC separation) and environmental analysis are summarized in Tables 3.5 and 3.6, respectively.

Chromolith columns manufactured by Merck are the monolithic silica columns most frequently used in food and environmental applications, as in many other application fields; they typically employ 100 mm × 4.6 mm I.D. columns, and in some cases shorter columns (50 mm)[78,113,119] or smaller I.D. (2.0 mm)[93,96,113,119] are proposed. Monolithic silica in capillary is commercially available, and can be applied for the analysis of amino acids.[77] Coupling two to ten monolithic columns to increase separation capacity and resolution has also been described, although in some cases this can produce a considerable increase in analysis time.[88,91,93,97,118] For instance, Rostagno *et al.*[93] developed a method for the analysis of the 12 main isoflavones in soybeans and derived foods by coupling two 100 mm × 4.6 mm I.D. Chromolith columns and working at a flow rate of 5 mL/min. As an example, Fig. 3.13 shows the chromatograms obtained when analyzing three soy-based products, with an analysis time lower than 10 minutes. The method developed was successfully applied to several soy food samples and spiked samples, achieving high chromatographic resolution (> 1.06), high reproducibility (relative standard deviations lower than 0.9%) and limits of quantitation in the range 0.80–1.96 mg/L.

Molecular imprint polymer (MIP) monolithic columns have also been reported for the analysis of several compounds in food samples, although these methods also presented fairly long analysis times. For example, MIP monolithic column was used to concentrate tetracyclines in honey and milk.[121] In addition, Sun *et al.* prepared theophylline-imprinted monolithic columns by an in situ thermal-initiated copolymerization technique, for the rapid separation of caffeine and theophylline.[79] As an example, Fig. 3.14 shows the scanning electron microscope (SEM) picture of the MIP monolithic column

Table 3.5. Application examples of monolithic columns for food analysis.

Analytes/Sample	Column	Mobile phase	Detector	Ref.
RP, 1D				
NBD-amino acid / mouse adrenal gland	MonoClad C18-HS, 150 mm × 3 mm I.D.	ACN–CB–$NaClO_4$, gradient, 2.0 mL/min	FL	77
Caffeine, theophylline, theobromine / coffee, cola	Chromolith SpeedRod 18e, 50 mm × 4.6 mm I.D.	ACN–water, 3.0 mL/min	UV	78
Caffeine, theophylline / green tea	Molecularly imprinted polymer monolith, 150 mm × 4.0 mm I.D.	ACN, 0.2–2.0 mL/min	DAD	79
Caffeine / tea, coffee, cocoa	hexyl methacrylate monolithic column, 150 mm × 0.53 mm I.D.	ACN–water, 41 μL/min	UV	80
Tocopherols, tocotrienols / vegetable oil	Chromolith CapRod RP-18e, 150 mm × 0.1 mm I.D.	ACN–MA–water–AA, 0.5 μL/min	DAD	81
Capsaicin, dihydrocapsaicin / hot sauces	Chromolith CapRod RP-18e, 300 mm × 0.1 mm I.D., and 150 mm × 0.1 mm I.D.	ACN–water–FA, gradient,	UV, MS	82
Carotenoids / corn	MonoCap C18, 350 mm × 0.2 mm I.D.	ACN–water, 60 μL/min	UV	83
Phenolic compounds / apple	Chromolith RP-18e, 100 mm × 4.6 mm I.D.	ACN–PB–MA, gradient, 2.5 mL/min	DAD	84

(*Continued*)

Table 3.5. (*Continued*)

Analytes/Sample	Column	Mobile phase	Detector	Ref.
Phenolic acids / fruits	Chromolith RP-18e, 100 mm × 4.6 mm I.D.	ACN–PB, gradient, 1.0 mL/min	DAD, MS	85
Ascorbic acid, glycoalkaloids, phenolic compounds / potato	Onyx column 100 mm × 4.6 mm I.D.	MA–water–AF, gradient, 2.9 mL/min	DAD, FL, MS	86
L-ascorbic acid / berry fruits	Onyx C18 column, 100 mm × 4.6 mm I.D.	ACN–PB, 2.5 mL/min	DAD	87
Daidzin, genistin and derivatives / soy extract	Chromolith RP-18e, 200 mm × 4.6 mm I.D.	ACN–water–AA, gradient, 3.0~4.0 mL/min	DAD	88
Rutin, hyperoside, isoquercitrin, quercitrin / St. John's Wort	Chromolith RP-18e, 100 mm × 4.6 mm I.D.	ACN–water, 2.0 mL/min	UV	89
Flavonoid aglycones / green tea, red wine, orange, propolis, *Ginkgo biloba*	Chromolith RP-18e, 100 mm × 4.6 mm I.D.	ACN–water, MA–water, 2.0 mL/min	DAD	90
Ginkgolide, bilobalide / *Ginkgo biloba*	Chromolith RP-18e, 200 mm × 4.6 mm I.D.	ACN–water–TFA, MA–water–TFA, 2-PA–water–TFA, THF–water–TFA, 1.0 mL/min	UV, ELSD	91
Flavonoids / tomato	Chromolith RP-18e, 100 mm × 4.6 mm I.D.	ACN–PB, 1.0 mL/min	UV, MS	92
Isoflavones / Soybean products	Chromolith RP-18e, 100 mm × 2 mm I.D.	MA–water–AA, gradient, 5.0 mL/min	DAD	93

(*Continued*)

Table 3.5. (*Continued*)

Analytes/Sample	Column	Mobile phase	Detector	Ref.
Carotene, lycopene / tomato, date, grapefruit, guava, papaya, watermelon	Chromolith RP-18e, 100 mm × 4.6 mm I.D.	ACN–MTBE, 1.0 mL/min	DAD	94
Anthocyanins / red cabbage	Chromolith RP-18e, 100 mm × 4.6 mm I.D.	ACN–water–FA, gradient, 4.0 mL/min	DAD	95
Allergen targeted peptides / nut-containing foods	Chromolith RP-18e, 100 mm × 2 mm I.D.	ACN–water–FA, gradient, 350 μL/min	MS	96
Polyprenols / *Eucommia ulmoides* leaves	Chromolith RP-18e, 100, 200, 500, 1000 mm × 4.6 mm I.D.	MA–2-PA–water, n-hexane–2-PA, gradient, 1.0, 3.0, or 4.0 mL/min	UV	97
Lipid / wheat flour, bread dough	Chromolith RP-18e, 100 mm × 4.6 mm I.D.	IO, AC–EA–AA, 2-PA–water, 1.4–3.0 mL/min	ELSD	98
Neutral and polar lipids / zooplankton	Chromolith RP-18e, 100 mm × 4.6 mm I.D.	IO–EA, AC–EA, 2-PA–water, gradient, 1.4–3.0 mL/min	ELSD	99
Peptides / spinach, pea leaves	PS–DVB monolith, 60 mm × 0.20 mm I.D.	ACN–water–TFA, gradient, 2 μL/min	MS	100
Nitrofuran veterinary drugs / animal feeds, farm water	Chromolith RP-18e, 100 mm × 4.6 mm I.D.	ACN–PB, 1.0 mL/min	UV	101
Sulfonamides / shrimp	Chromolith RP-18e, 100 mm × 4.6 mm I.D.	ACN–EA–PB, 1.5 mL/min	EC	102

(*Continued*)

Table 3.5. (*Continued*)

Analytes/Sample	Column	Mobile phase	Detector	Ref.
Sweeteners, antioxidants, preservatives / food and cosmetics	Chromolith Guard Cartridge RP-18e 5 mm × 4.6 mm I.D.	ACN–water, MA–water, 1.5–4.5 mL/min	UV	103
Fumonisins / foods, animal feeds	Chromolith RP-18e, 100 mm × 4.6 mm I.D.	MA–PB, 1.0 mL/min	FL	104
Aflatoxins / chilis, peanuts, rice	Chromolith RP-18e, 100 mm × 4.6 mm I.D.	ACN–MA–water, 1.0 mL/min	FL	105
Corticoids / animal feed	Chromolith RP-18e, 100 mm × 4.6 mm I.D.	ACN–water, 3.0 mL/min	DAD	106
Amnesic shellfish poisoning / shellfish	Chromolith RP-18e, 100 mm × 3.0 mm I.D.	ACN–water–FA, 1.75 mL/min	DAD	107
RP, 2D				
Flavones / beer, red wine	1D, Discovery HS PEG 150 mm × 4.6mm I.D., 2D, Chromolith RP-18e, 50 mm and 100 mm × 4.6 mm I.D.	1D, ACN–water, gradient, 0.1 mL/min, 2D, ACN–water, gradient, 2.0 mL/min	DAD	108
Vanillin, ferulic acid, liqustilide, 3-butylidenephthalide, linoleic acid / traditional Chinese medicine	1D, Kromasil-CN or Hypesil-SCX, 150 mm × 4.6 mm I.D., 2D, Chromonolith Speed ROD, 50 mm × 4.6 mm I.D.	1D, ACN–water–AA, gradient, 0.133 mL/min, 2D, MA–water–AA, gradient, 3.0 mL/min	DAD, MS	109

(*Continued*)

Table 3.5. (*Continued*)

Analytes/Sample	Column	Mobile phase	Detector	Ref.
Carotenoids / sweet orange essential oil, red orange juice	1D, Supelcosil LCSI, 300 mm × 1.0 mm I.D., 2D, Chromolith RP-18e, 100 mm × 4.6 mm I.D.	1D, n-Hexane–EA, 10 μL/min, 2D, ACN–2-PA–water, 4.7 mL/min	DAD	110
HILIC, 1D				
Peptides / rat liver	Zwitterionic-HILIC monolith, 270 mm × 0.1mm I.D.	ACN–water–AmA, gradient	MS/MS	111
Disaccharides /*Arabidopsis thaliana* leaves, corn, soybean	Polyacrylamide monolithic silica, 267 mm × 0.2 mm I.D.	ACN–water–AmA, gradient	UV, MS	112

Abbreviations: AA: acetic acid; AC: acetone; ACN: acetonitorile; AF: ammonium formate; AmA: ammonium acetate; CB: citrate buffer; CTAC: cetyltrimethylammonium chloride; DAD: diode array detector; EA: ethyl acetate; EC: electrochemical detector; ELSD: evaporative light scattering detector; EtOH; ethanol; FA: formic acid; FL: fluorescence detector; IO: isooctane; MA: methanol; MS: mass spectrometric detector; MTBE: methyl *tert*-butyl ether; NBD: 7-nitro-2,1,3-benzoxadiazolyl; 2-PA: 2-propanol; PB: phosphate buffer; PS-DVB: poly(styrene-*co*-divinylbenzene); TBA: tetrabutylammonium; TFA: trifluoroacetic acid; THF: tetrahydrofuran: UV: ultraviolet detection.

Table 3.6. Application examples of monolithic columns for environmental analysis.

Analytes/Sample	Column	Mobile phase	Detector	Ref.
Microcystins, nodularins / water	Chromolith FastGradient C18, 50 mm × 2 mm I.D.	ACN–water–FA, gradient, 1.0 mL/min	MS	113
Zinc pyrithione / environmental water	Chromolith RP-18e, 100 mm × 4.6 mm I.D.	MA–water–AmA, gradient	MS	114
Fluoroquinolones / surface waters from Mondego River	Chromolith RP-18e, 100 mm × 4.6 mm I.D.	MA–PB–TBA, 2.5 mL/min	FL	115
Pollutant phenols	Chromolith RP-18e, 100 mm × 4.6 mm I.D.	ACN–PB, 1.0–4.0 mL/min	DAD	116
Parabens / river water, wastewater	Chromolith RP-18e, 100 mm × 4.6 mm I.D.	ACN–water, gradient, 0.7 mL/min	UV	117
Pharmaceutical residues / river and potable water	Chromolith RP-18e, 200 mm × 4.6 mm I.D.	MA–water, gradient, 1.00–2.30 mL/min gradient	UV, MS	118
Drugs / surface water and wastewater	Chromolith Fast Gradient C18e, 50 mm × 2 mm I.D.	ACN–water–FA, 2.0 mL/min	UV	119
Inorganic anions / seawater	Cetrimide modified monolithic silica, 200 mm × 0.1 mm I.D.	Water–NaCl–CTMC, 5.6 μL/min	UV	120

Abbreviations: AA: acetic acid; AC: acetone; ACN: acetonitorile; AF: ammonium formate; AmA: ammonium acetate; CB: citrate buffer; CTAC: cetyltrimethylammonium chloride; DAD: diode array detector; EA: ethyl acetate; EC: electrochemical detector; ELSD: evaporative light scattering detector; EtOH; ethanol; FA: formic acid; FL: fluorescence detector; IO: isooctane; MA: methanol; MS: mass spectrometric detector; MTBE: methyl *tert*-butyl ether; NBD: 7-nitro-2,1,3-benzoxadiazolyl; 2-PA: 2-propanol; PB: phosphate buffer; PS-DVB: poly(styrene-*co*-divinylbenzene); TBA: tetrabutylammonium; TFA: trifluoroacetic acid; THF: tetrahydrofuran; UV: ultraviolet detection.

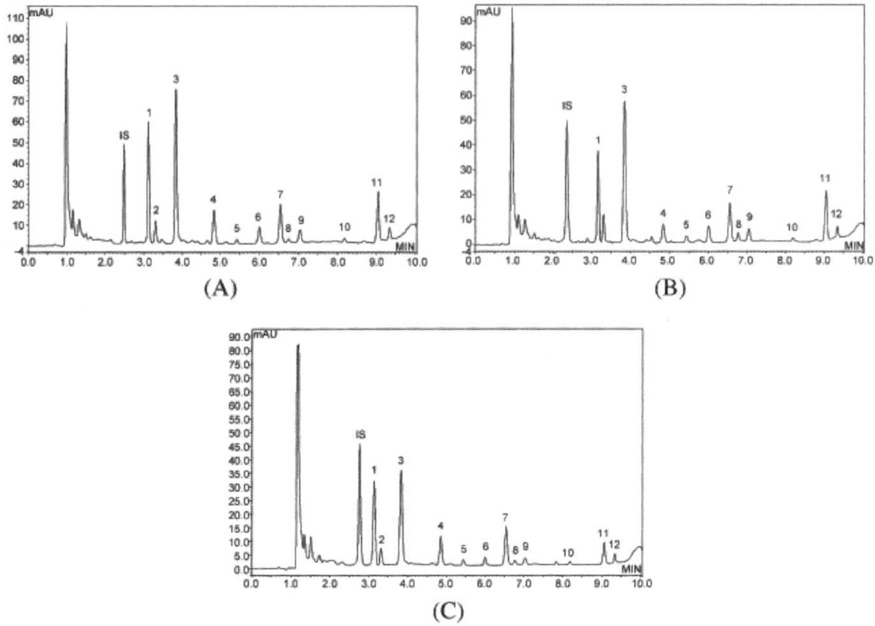

Figure 3.13. Chromatograms of soy food samples. (A) Soy flour, (B) texturised soy protein, and (C) soy fiber. Two monolithic columns, flow rate (5 mL/min), temperature, (35 °C), gradient of water (0.1% acetic acid) and methanol (0.1% acetic acid): 0 min (0% B), 2.0 min (31% B), 4.0 min (31% B), 5.0 min (35% B), 8.0 min (35% B), 9.5 min (100% B). Peak identification: 1, Daidzin; 2, Glycitin; 3, Genistin; 4, Malonyl-daidzin; 5, Malonyl-glycitin; 6, Acetyl-daidzin; 7, Malonyl-genistin; 8, Acetyl-glycitin; 9, Daidzein; 10, Glycitein; 11, Acetyl-genistin; and 12, Genistein. Reproduced with permission from Rostagno, M.A., Palma, M. and Barroso, C.G. (2007). Fast analysis of soy isoflavones by high-performance liquid chromatography with monolithic columns, *Anal. Chim. Acta*, **582**, 243–249. Copyright (2007) Elsevier.

and the chromatographic separation of caffeine and theophylline in green tea. Both compounds were fully separated under gradient elution conditions within an analysis time of 18 min. Regarding the use of monolithic columns coupled to mass spectrometry, several LC–MS methods using this kind of column are being proposed, although it seems that the extremely high flow-rates generally applied in monoliths makes the compatibility with mass spectrometry detection difficult. Some food[85,86,92,96,100,105] and environmental[113,114,118]

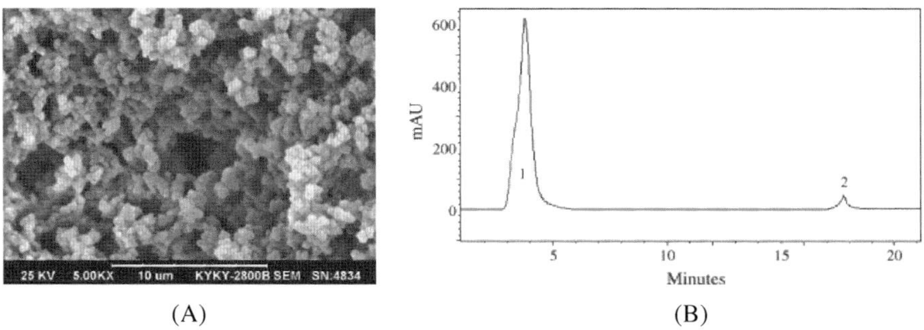

Figure 3.14. (A) Scanning electron microscope (SEM) picture of the monolithic MIP column. (B) Chromatographic separation of caffeine (1) and theophylline (2) in a green tea sample. Reproduced with permission from Sun, H.W., Qiao, F.X. and Liu, G.Y. (2006). Characteristic of theophylline imprinted monolithic column and its application for determination of xanthine derivatives caffeine and theophylline in green tea, *J. Chromatogr. A*, **1134**, 194–200. Copyright (2006) Elsevier.

applications of monolithic columns to LC–MS can be found in the literature. For instance, Bignardi *et al.*[96] compared the performance in terms of peak shape, resolution, analysis time and selectivity of both a C18 particle-packed column and a silica-based C18 monolithic silica column for the multi-allergen trace analysis in foods by liquid chromatography–electrospray ionization–linear ion trap–tandem mass spectrometry (LC-ESI-LIT-MS/MS). The chromatographic separation of ten targeted peptides from a matrix (cereals) tryptic digest on the C18 monolithic column is shown in Fig. 3.15. The chromatographic profile of the ten peptides on the monolithic column showed both excellent peak shape and retention time stability. Resolution values ranging from 0.8 to 12 were obtained.

The development of new LC–MS methods using monoliths will be a field to explore in depth in the future to achieve fast, sensitive and selective applications in both food safety and environmental analysis.

3.4. Summary and Conclusions

Monolithic columns possess attractive points as separation media, not only as chromatographic columns, but also as sample preparation

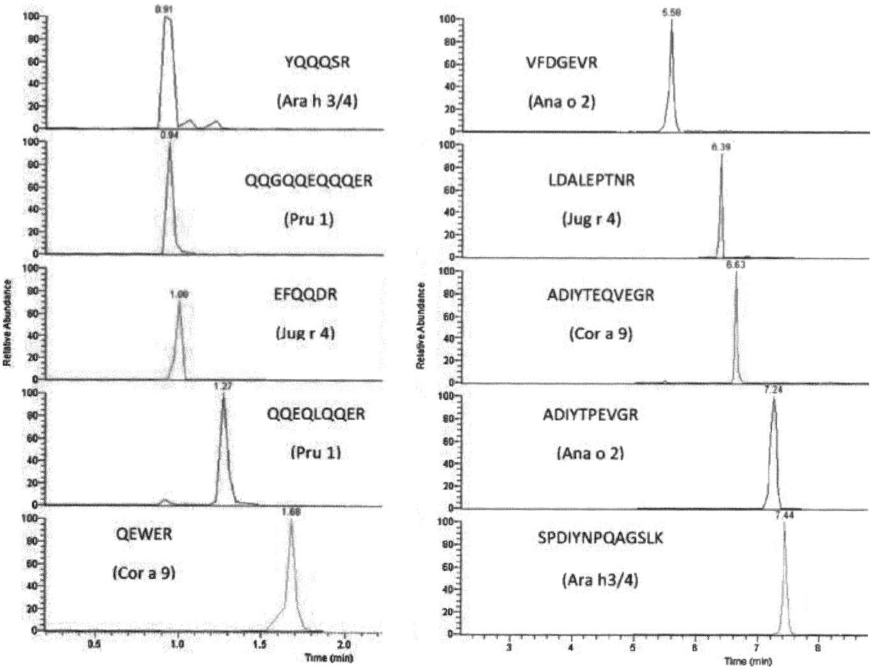

Figure 3.15. LC–ESI–LIT–MS/MS separation of ten targeted peptides from a matrix (cereals) tryptic digest on a C18 monolithic column (100.0 mm × 2.1 mm I.D.). Blank matrix was fortified with a mixture of five nuts 0.01% (w/w). Reproduced with permission from Bignardi, C., Elviri, L., Penna, A. *et al.* (2010). Particle-packed column versus silica-based monolithic column for liquid chromatography-electrospray-linear ion trap-tandem mass spectrometry multiallergen trace analysis in food, *J. Chromatogr. A*, **1217**, 7579–7585. Copyright (2010) Elsevier.

tools, as mentioned above. To use them effectively, researchers should know their advantages and disadvantages, as listed in Section 3.1.7. The variety of stationary phases on monolithic supports is significantly increasing in laboratories, and some examples should be commercially available in the future. The readers of this chapter should keep in mind the development of separation methods of the target analytes.

References

1. Neue, U.D. (1997). *HPLC Columns Theory, Technology, and Practice*, Wiley–VCH, New York.
2. Gritti, F., Leonardis, I., Shock, D. *et al.* (2010). Performance of columns packed with the new shell particles, Kinetex- C18, *J. Chromatogr. A*, **1217**, 1589–1603.
3. Nakanishi, K. (1997). Pore structure control of silica gels based on phase separation, *J. Porous Mater.*, **4**, 67–112.
4. Minakuchi, H., Nakanishi, K., Soga, N. *et al.* (1997). Effect of skeleton size on the performance of octadecylsilylated continuous porous silica columns in reversed-phase liquid chromatography, *J. Chromatogr. A*, **762**, 135–146.
5. Minakuchi, H., Nakanishi, K., Soga, N. *et al.* (1998). Effect of domain size on the performance of octadecylsilylated continuous porous silica columns in reversed-phase liquid chromatography, *J. Chromatogr. A*, **797**, 121–131.
6. Morisato, K., Miyazaki, S., Ohira, M. *et al.* (2009). Semi-micro-monolithic columns using silica rods with improved performance, *J. Chromatogr. A*, **1216**, 7384–7387.
7. Tanaka, N., Kobayashi, H., Nakanishi, K. *et al.* (2001). A new type of chromatographic support could lead to higher separation efficiencies, *Anal. Chem.*, **73**, 420A–429A.
8. Cabrera, K., Lubda, D., Eggenweiler, M.H. *et al.* (2000). A new monolithic-type HPLC column for fast separations, *J. High Resolut. Chromatogr.*, **23**, 93–99.
9. Ishizuka, N., Minakuchi, H., Nakanishi, K. *et al.* (1998). Chromatographic properties of miniaturized silica rod columns, *J. High Resolut. Chromatogr.*, **21**, 477–479.
10. Tanaka, N., Nagayama, H., Kobayahsi, H. *et al.* (2000). Monolithic silica columns for HPLC, micro-HPLC, and CEC, *J. High Resolut. Chromatogr.*, **23**, 111–116.
11. Motokawa, M., Kobayashi, H., Ishizuka, N. *et al.* (2002). Monolithic silica columns with various skeleton sizes and through-pore sizes for capillary liquid chromatography, *J. Chromatogr. A*, **961**, 53–63.

12. Motokawa, M., Ohira, M., Minakuchi, H. *et al.* (2006). Performance of octadecylsilylated monolithic silica capillary columns of 530 μm inner diameter in HPLC, *J. Sep. Sci.*, **29**, 2471–2477.
13. Iwasaki, M., Miwa, S., Ikegami, T. *et al.* (2010). One-dimensional capillary liquid chromatographic separation coupled with tandem mass spectrometry unveils *Escherichia coli* proteome on a microarray scale, *Anal. Chem.*, **82**, 2616–2620.
14. Eghbali, H., Sandra, K., Detobel, F. *et al.* (2011). Performance evaluation of long monolithic silica capillary columns in gradient liquid chromatography using peptide mixtures, *J. Chromatogr. A*, **1218**, 3360–3366.
15. Horie, K., Sato, Y., Kimura, T. *et al.* (2012). Estimation and optimization of the peak capacity of one-dimensional gradient high performance liquid chromatography using a long monolithic silica capillary column, *J. Chromatogr. A*, **1228**, 283–291.
16. Hara, T., Kobayashi, H., Ikegami, T. *et al.* (2006). Performance of monolithic silica capillary columns with increased phase ratios and small-sized domains, *Anal. Chem.*, **78**, 7632–7642.
17. Cabrera, K. (2012). *A new generation of silica-based monoliths HPLC columns with improved performance.* [Online] LCGC Chromatography Online. Available at: http://www.chromatographyonline.com/lcgc/Article/A-New-Generation-of-Silica-Based-Monoliths-HPLC-Co/Article Standard/Article/detail/772076 [Accesses 1 April 2012].
18. Hormann, K., Müller, T., Bruns, S. *et al.* (2012). Morphology and separation efficiency of a new generation of analytical silica monoliths, *J. Chromatogr. A*, **1222**, 46–58.
19. Gritti, F. and Guiochon, G. (2012). Measurement of the eddy dispersion term in chromatographic columns: III. Application to new prototypes of 4.6 mm I.D. monolithic columns, *J. Chromatogr. A*, **1225**, 79–90.
20. Hormann, K. and Tallarek, U. (2013). Analytical silica monoliths with submicron macropores: Current limitations to a direct morphology–column efficiency scaling, *J. Chromatogr. A*, **1312**, 26–36.
21. Núñez, O., Ikegami, T., Kajiwara, W. *et al.* (2007). Preparation of high efficiency and highly retentive monolithic silica capillary columns for reversed-phase chromatography by chemical modification

by polymerization of octadecyl methacrylate, *J. Chromatogr. A*, **1156**, 35–44.
22. Horie, K., Ikegami, T., Hosoya, K. *et al.* (2007). Highly efficient monolithic silica capillary columns modified with poly(acrylic acid) for hydrophilic interaction chromatography, *J. Chromatogr. A*, **1164**, 198–205.
23. Watanabe, Y., Ikegami, T., Horie, K. *et al.* (2009). Improvement of separation efficiencies of anion-exchange chromatography using monolithic silica capillary columns modified with polyacrylates and polymethacrylates containing tertiary amino or quaternary ammonium groups, *J. Chromatogr. A*, **1216**, 7394–7401.
24. Hara, T., Makino, S., Watanabe, Y. *et al.* (2010). The performance of hybrid monolithic silica capillary columns prepared by changing feed ratios of tetramethoxysilane and methyltrimethoxy-silane, *J. Chromatogr. A*, **1217**, 89–98.
25. Guiochon, G. (2007). Monolithic columns in high-performance liquid chromatography, *J. Chromatogr. A*, **1168**, 101–168.
26. Gritti, F. and Guiochon, G. (2009). Mass transfer kinetic mechanism in monolithic columns and application to the characterization of new research monolithic samples with different average pore sizes, *J. Chromatogr. A*, **1216**, 4752–4767.
27. Hara, T., Mascotto, S., Weidmann, C. *et al.* (2011). The effect of hydrothermal treatment on column performance for monolithic silica capillary columns, *J. Chromatogr. A*, **1218**, 3624–3635.
28. Cabrera, K. (2004). Applications of silica-based monolithic HPLC columns, *J. Sep. Sci.*, **27**, 843–852.
29. Gzil, P., Vervoort, N.V., Baron, G. *et al.* (2004). General rules for the optimal external of LC supports, *Anal. Chem.*, **76**, 6707–6718.
30. Roux, R., Abi Jaoudé, M. and Demesmay, C. (2009). Improvement of chromatographic performances of in-situ synthesized hybrid C8 silica monoliths by reduction of structural radial heterogeneities, *J. Chromatogr. A*, **1216**, 3857–3863.
31. Zhang, Z., Wu, M., Wu, R. *et al.* (2011). Preparation of perphenyl-carbamoylated β-cyclodextrin-silica hybrid monolithic column with "one-pot" approach for enantioseparation by capillary liquid chromatography, *Anal. Chem.*, **83**, 3616–3622.

32. Wu, M., Wu, R., Zhang, Z. *et al.* (2011). Preparation and application of organic-silica hybrid monolithic capillary columns, *Electrophoresis*, **32**, 105–115.
33. Miyazaki, S., Takahashi, M., Ohira, M. *et al.* (2011). Monolithic silica rod columns for high-efficiency reversed-phase liquid chromatography, *J. Chromatogr. A*, **1218**, 1988–1994.
34. Ikegami, T., Dicks, E., Kobayashi, H. *et al.* (2004). How to utilize the true performance of monolithic silica columns, *J. Sep. Sci.*, **27**, 1292–1302.
35. Desmet, G., Clicq, D. and Gzil, P. (2005). Geometry-independent plate height representation methods for the direct comparison of the kinetic performance of LC supports with a different size or morphology, *Anal. Chem.*, 77, 4058–4070.
36. Kawachi, Y., Ikegami, T., Takubo, H. *et al.* (2011). Chromatographic characterization of hydrophilic interaction liquid chromatography stationary phases: Hydrophilicity, charge effects, structural selectivity, and separation efficiency, *J. Chromatogr. A*, **1218**, 5903–5919.
37. Li, Q.-J., Lue, C.-C., Li, H.-Y. *et al.* (2012). Preparation of organic-silica hybrid boronate affinity monolithic column for the specific capture and separation of cis-diol containing compounds, *J. Chromatogr. A*, **1256**, 114–120.
38. Chen, M.L., Zhang, J., Zhang, Z. *et al.* (2013). Facile preparation of organic-silica hybrid monolith for capillary hydrophilic liquid chromatography based on "thiol-ene" click chemistry, *J. Chromatogr. A*, **1284**, 118–125.
39. Yang, H.H., Chen, Y.H., Liu, Y.X. *et al.* (2013). One-pot synthesis of (3-sulfopropyl methacrylate potassium)-silica hybrid monolith via thiol-ene click chemistry for CEC, *Electrophoresis*, **34**, 510–517.
40. Liu, Z., Otsuka, K., Terabe, S. *et al.* (2002). Physically adsorbed chiral stationary phase of avidin on monolithic silica column for capillary electrochromatography and capillary liquid chromatography, *Electrophoresis*, **23**, 2973–2981.
41. Calleri, E., Temporini, C., Perani, E. *et al.* (2004). Development of a bioreactor based on trypsin immobilized on monolithic support for the on-line digestion and identification of proteins, *J. Chromatogr. A*, **1045**, 99–109.

42. Miyazaki, S., Miah, M.Y., Morisato, K. *et al.* (2005). Titania-coated monolithic silica as separation medium for high performance liquid chromatography of phosphorus-containing compounds, *J. Sep. Sci.*, **28**, 39–44.
43. Chen, Y.-S., Chen, J. and Jia, L. (2009). Study of triacontyl-functionalized monolithic silica capillary column for reversed-phase capillary liquid chromatography, *J. Chromatogr. A*, **1216**, 2597–2600.
44. Ikegami, T., Fujita, H. and Horie, K. (2006). HILIC mode separation of polar compounds by monolithic silica capillary columns coated with polyacrylamide, *Anal. Bioanal. Chem.*, **386**, 578–585.
45. Moravcova, D., Planeta, J., Kahle, V. *et al.* (2012). Zwitterionic silica-based monolithic capillary columns for isocratic and gradient hydrophilic interaction liquid chromatography, *J. Chromatogr. A*, **1270**, 178–185.
46. Hutchinson, J.P., Hilder, E.F., Macka, M. *et al.* (2006). Preparation and characterisation of anion-exchange latex-coated silica monoliths for capillary electrochromatography, *J. Chromatogr. A*, **1109**, 10–18.
47. Xie, C.-H., Hu, J.-W., Xiao, H. *et al.* (2005). Preparation of monolithic silica column with strong cation-exchange stationary phase for capillary electrochromatography, *J. Sep. Sci.*, **28**, 751–756.
48. Sugrue, E., Nesterenko, P. and Paull, B. (2004). Ion exchange properties of monolithic and particle type iminodiacetic acid modified silica, *J. Sep. Sci.*, **27**, 921–930.
49. Chen, Z.-L. and Hobo, T. (2001). Chemically L-phenylalaninamide-modified monolithic silica column prepared by a sol-gel process for enantioseparation of dansyl amino acids by ligand exchange-capillary electrochromatography, *Anal. Chem.*, **73**, 3348–3357.
50. Lubda, D., Cabrera, K., Nakanishi, K. *et al.* (2003). Monolithic silica columns with chemically bonded β-cyclodextrin as a stationary phase for enantiomer separations of chiral pharmaceuticals, *Anal. Bioanal. Chem.*, **377**, 892–901.
51. Chankvetadze, B., Ikai, T., Yamamoto, C. *et al.* (2004). High-performance liquid chromatographic enantioseparations on monolithic silica columns containing a covalently attached 3,5-dimethylphenylcarbamate derivative of cellulose, *J. Chromatogr. A*, **1042**, 55–60.

52. Ou, J.-J., Li, X., Feng, S. *et al.* (2007). Preparation and evaluation of a molecularly imprinted polymer derivatized silica monolithic column for capillary electrochromatography and capillary liquid chromatography, *Anal. Chem.*, **79**, 639–646.
53. Nogueira, R., Lubda, D., Leitner, A. *et al.* (2006). Silica-based monolithic columns with mixed-mode reversed-phase/weak anion-exchange selectivity principle for high-performance liquid chromatography, *J. Sep. Sci.*, **28**, 39–44.
54. Hjertén, S., Liao, J.-L. and Zhang, R. (1989). High-performance liquid chromatography on continuous polymer beds, *J. Chromatogr.*, **473**, 273–275.
55. Jandera, P. (2013). Advances in the development of organic polymer monolithic columns and their applications in food analysis: A review, *J. Chromatogr. A*, **1313**, 37–53.
56. Podgornik, A., Yamamoto, S., Peterka, M. *et al.* (2013). Fast separation of large biomolecules using short monolithic columns, *J. Chromatogr. B*, **927**, 80–89.
57. Liu, K., Aggarwal, P. and Lawson, J. S. (2013). Organic monoliths for high-performance reversed-phase liquid chromatography, *J. Sep. Sci.*, **36**, 2767–2781.
58. Umemura, T., Kojima, N. and Ueki, Y. (2008). Organic monoliths for high-performance reversed-phase liquid chromatography, *Bunseki Kagaku*, **57**, 517–529.
59. Svec, F. (2010). Porous polymer monoliths: Amazingly wide variety of techniques enabling their preparation, *J. Chromatogr. A*, **1217**, 902–924.
60. Hosoya, K., Hira, N., Yamamoto, K. *et al.* (2006). High-performance polymer-based monolithic capillary column., *Anal. Chem.*, **78**, 5729–5735.
61. Chambers, S.D., Holcombe, T.W., Svec, F. *et al.* (2011). Porous polymer monoliths functionalized through copolymerization of a C60 fullerene-containing methacrylate monomer for highly efficient separations of small molecules, *Anal. Chem.*, **83**, 9478–9484.
62. Hasegawa, G., Kanamori, K., Ishizuka, N. *et al.* (2012). New monolithic capillary columns with well-defined macropores based on poly(styrene-*co*-divinylbenzene), *ACS Appl. Mater. Interfaces*, **4**, 2343–2347.

63. Greiderer, A., Trojer, L., Huck, C.W. *et al.* (2009). Influence of the polymerisation time on the porous and chromatographic properties of monolithic poly(1,2-bis(p-vinylphenyl))ethane capillary columns, *J. Chromatogr. A*, **1216**, 7747–7754.
64. Li, Y.-Y., Tolley, H.D. and Lee, M.L. (2011). Preparation of monoliths from single crosslinking monomers for reversed-phase capillary chromatography of small molecules, *J. Chromatogr. A*, **1218**, 1399–1408.
65. Shu, S., Kobayashi, H., Kojima, N. *et al.* (2011). Preparation and characterization of lauryl methacrylate-based monolithic microbore column for reversed-phase liquid chromatography, *J. Chromatogr. A*, **1218**, 5228–5234.
66. Hormann, K., Mullner, T., Bruns, S. *et al.* (2012). Morphology and separation efficiency of a new generation of analytical silica monoliths, *J. Chromatogr. A*, **1222**, 46–58.
67. Eeltink, S., Dolman, S., Detobel, F. *et al.* (2010). High-efficiency liquid chromatography–mass spectrometry separations with 50 mm, 250 mm, and 1 m long polymer-based monolithic capillary columns for the characterization of complex proteolytic digests, *J. Chromatogr. A*, **1217**, 6610–6615.
68. Ueki, Y., Umemura, T., Iwashita, Y. *et al.* (2006). Preparation of low flow-resistant methacrylate-based monolithic stationary phases of different hydrophobicity and the application to rapid reversed-phase liquid chromatographic separation of alkylbenzenes at high flow rate and elevated temperature, *J. Chromatogr. A*, **1106**, 106–111.
69. Vaast, A., Nováková, L., Desmet, G. *et al.* (2013). High-speed gradient separations of peptides and proteins using polymer-monolithic poly(styrene-co-divinylbenzene) capillary columns at ultra-high pressure, *J. Chromatogr. A*, **1304**, 177–182.
70. Núñez, O., Gallart-Ayala, H., Martins C.P.B. *et al.* (2012). New trends in fast liquid chromatography for food and environmental analysis, *J. Chromatogr. A*, **1228**, 298–323.
71. Rigobello-Masini, M., Penteado, J.C.P. and Masini, J.C. (2013). Monolithic columns in plant proteomics and metabolomics, *Anal. Bioanal. Chem.*, **405**, 2107–2122.
72. Wang, S.-T., Wang, M.-Y., Su, X. *et al.* (2012). Facile preparation of SiO2/TiO2 composite monolithic capillary column and its application in enrichment of phosphopeptides, *Anal. Chem.* **84**, 7763–7770.

73. Zheng, F. and Hu, B. (2008). Novel bimodal porous N-(2-aminoethyl)-3-aminopropyltrimethoxysilane-silica monolithic capillary microextraction and its application to the fractionation of aluminum in rainwater and fruit juice by electrothermal vaporization inductively coupled plasma mass spectrometry, *Spectrochim. Acta, B: Atom. Spectr.*, **63B**, 9–18.
74. Namera, A., Nakamoto, A., Saito T. *et al.* (2011). Monolith as a new sample preparation material: Recent devices and applications, *J. Sep. Sci.*, **34**, 901–924.
75. Namera, A. and Saito T. (2013). Advances in monolithic materials for sample preparation in drug and pharmaceutical analysis, *Trends Anal. Chem.*, **45**, 182–196.
76. Samanidou, V.F. and Karageorgou, E.G. (2011). An overview of the use of monoliths in sample preparation and analysis of milk, *J. Sep. Sci.*, **34**, 2013–2025.
77. Song, Y.T., Funatsu, T. and Tsunoda, M. (2012). Rapid determination of amino acids in biological samples using a monolithic silica column, *Amino Acids*, **42**, 1897–1902.
78. Tzanavaras, P.D. and Themelis, D.G. (2007). Development and validation of a high-throughput high-performance liquid chromatographic assay for the determination of caffeine in food samples using a monolithic column, *Anal. Chim. Acta*, **581**, 89–94.
79. Sun, H.W., Qiao, F.X. and Liu, G.Y. (2006). Characteristic of theophylline imprinted monolithic column and its application for determination of xanthine derivatives caffeine and theophylline in green tea, *J. Chromatogr. A*, **1134**, 194–200.
80. Al-Othman, Z.A., Aqel, A., Alharbi, M.K.E. *et al.* (2012). Fast chromatographic determination of caffeine in food using a capillary hexyl methacrylate monolithic column, *Food Chem.*, **132**, 2217–2223.
81. Cerretani, L., Lerma-García, M.J., Herrero-Martínez, J.M. *et al.* (2010). Determination of tocopherols and tocotrienols in vegetable oils by nanoliquid chromatography with ultraviolet-visible detection using a silica monolithic column, *J. Agri. Food Chem.*, **58**, 757–761.
82. Poster, S. and Altmaker, S. (2013). Simple and fast quantifiction of capsaicinoids in hot sauces: using monolithic silica capillaries and LC-MS, *LC-GC Europe*, **26**, 326–329.

83. Zhang, S., Jia, L. and Wang, S.-J. (2010). Determination of β-carotene in corn by in-tube SPME coupled to micro-LC, *Chromatographia*, **72**, 1231–1233.
84. Chinnici, F., Gaiani, A., Natali, N. *et al.* (2004). Improved HPLC determination of phenolic compounds in cv. golden delicious apples using a monolithic column, *J. Agric. Food Chem.*, **52**, 3–7.
85. Biesaga, M., Ochnik, U. and Pyrzynska, K. (2007). Analysis of phenolic acids in fruits by HPLC with monolithic columns, *J. Sep. Sci.*, **30**, 2929–2934.
86. Shakya, R. and Navarre, D.A. (2006). Rapid screening of ascorbic acid, glycoalkaloids, and phenolics in potato using high-performance liquid chromatography, *J. Agric. Food Chem.*, **54**, 5253–5260.
87. Walker, P.G., Gordon, S.L., Brennan, R.M. *et al.* (2006). A high-throughput monolithic HPLC method for rapid vitamin C phenotyping of berry fruit, *Phytochem. Anal.*, **17**, 284–290.
88. Apers, S., Naessens, T., Van Den Steen, K. *et al.* (2004). Fast high-performance liquid chromatography method for quality control of soy extracts, *J. Chromatogr. A*, **1038**, 107–112.
89. Brenier-Maurel, C., Delaurent, C., Pages, G. *et al.* (2002). Optimization of St. John's wort flavonoid separation by reversed phase liquid chromatography on a silica-based monolithic column, *Chromatographia*, **56**, 277–282.
90. Repollés, C., Herrero-Martínez, J.M. and Ràfols, C. (2006). Analysis of prominent flavonoid aglycones by high-performance liquid chromatography using a monolithic type column, *J. Chromatogr. A*, **1131**, 51–57.
91. van Nederkassel, A.M., Vijverman, V., Massart, D.L. *et al.* (2005). Development of a *Ginkgo biloba* fingerprint chromatogram with UV and evaporative light scattering detection and optimization of the evaporative light scattering detector operating condition, *J. Chromatogr. A*, **1085**, 230–239.
92. Biesaga, M., Ochnik, U. and Pyrzynska, K. (2009). Fast analysis of prominent flavonoids in tomato using a monolithic column and isocratic HPLC, *J. Sep. Sci.*, **32**, 2835–2840.

93. Rostagno, M.A., Palma, M. and Barroso, C.G. (2007). Fast analysis of soy isoflavones by high-performance liquid chromatography with monolithic columns, *Anal. Chim. Acta*, **582**, 243–249.
94. Pól, J., Hyötyläinen, T., Ranta-Aho, O. *et al.* (2004). Determination of lycopene in food by on-line SFE coupled to HPLC using a single monolithic column for trapping and separation, *J. Chromatogr. A*, **1052**, 25–31.
95. Arapitsas, P. and Turner, C. (2008). Pressurized solvent extraction and monolithic column-HPLC/DAD analysis of anthocyanins in red cabbage, *Talanta*, **74**, 1218–1223.
96. Bignardi, C., Elviri, L., Penna, A. *et al.* (2010). Particle-packed column versus silica-based monolithic column for liquid chromatography-electrospray-linear ion trap-tandem mass spectrometry multiallergen trace analysis in foods, *J. Chromatogr. A*, **1217**, 7579–7585.
97. Bamba, T., Fukusaki, E., Nakazawa, Y. *et al.* (2004). Rapid and high-resolution analysis of geometric polyprenol homologues by connected octadecylsilylated monolithic silica columns in high-performance liquid chromatography, *J. Sep. Sci.*, **27**, 293–296.
98. Gerits, L.R., Pareyt, B. and Delcour, J.A. (2013). Single run HPLC separation coupled to evaporative light scattering detection unravels wheat flour endogenous lipid redistribution during bread dough making, *LWT–Food Sci. Tech.*, **53**, 426–433.
99. Graeve, M. and Janssen, D. (2009). Improved separation and quantification of neutral and polar lipid classes by HPLC-ELSD using a monolithic silica phase: Application to exceptional marine lipids, *J. Chromatogr. B: Anal. Technol. Biomed. Life Sci.*, **877**, 1815–1819.
100. Walcher, W., Oberacher, H., Troiani, S. *et al.* (2002). Monolithic capillary columns for liquid chromatography: Electrospray ionization mass spectrometry in proteomic and genomic research, *J. Chromatogr. B: Anal. Technol. Biomed. Life Sci.*, **782**, 111–125.
101. Viñas, P., Campillo, N., Carrasco, L., *et al.* (2007). Analysis of nitrofuran residues in animal feed using liquid chromatography and photodiode-array detection, *Chromatographia*, **65**, 85–89.
102. Sangjarusvichai, H., Dungchai, W., Siangproh, W. *et al.* (2009). Rapid separation and highly sensitive detection methodology for

sulfonamides in shrimp using a monolithic column coupled with BDD amperometric detection, *Talanta*, **79**, 1036–1041.
103. Garcia-Jimenez, J.F., Valencia, M.C., Capitan-Vallvey, L.F. *et al.* (2007). Simultaneous determination of antioxidants, preservatives and sweetener additives in food and cosmetics by flow injection analysis coupled to a monolithic column, *Anal. Chim. Acta*, **594**, 226–233.
104. Khayoon, W.S., Saad, B., Salleh, B. *et al.* (2010). A reversed phase high performance liquid chromatography method for the determination of fumonisins B1 and B2 in food and feed using monolithic column and positive confirmation by liquid chromatography/tandem mass spectrometry, *Anal. Chim. Acta*, **679**, 91–97.
105. Khayoon, W.S., Saad, B., Lee, T.P. *et al.* (2012). High performance liquid chromatographic determination of aflatoxins in chilli, peanut and rice using silica based monolithic column, *Food Chem.*, **133**, 489–496.
106. Muñiz-Valencia, R., Ceballos-Magana, S.G., Gonzalo-Lumbreras, R. *et al.* (2008). A liquid chromatography method using a monolithic column for the determination of corticoids in animal feed and animal feeding water, *Anal. Bioanal. Chem.*, **391**, 2683–2691.
107. Regueiro, J., Álvarez, G., Mauriz, A. *et al.* (2011). High throughput analysis of amnesic shellfish poisoning toxins in bivalve molluscs by dispersive solid-phase extraction and high-performance liquid chromatography using a monolithic column, *Food Chem.*, **127**, 1884–1891.
108. Cacciola, F., Jandera, P., Hajdú, Z. *et al.* (2007). Comprehensive two-dimensional liquid chromatography with parallel gradients for separation of phenolic and flavone antioxidants, *J. Chromatogr. A*, **1149**, 73–87.
109. Hu, L.-H., Chen, X.-G., Kong, L. *et al.* (2005). Improved performance of comprehensive two-dimensional HPLC separation of traditional Chinese medicines by using a silica monolithic column and normalization of peak heights, *J. Chromatogr. A*, **1092**, 191–198.
110. Dugo, P., Škeříková, V., Kumm, T. *et al.* (2006). Elucidation of carotenoid patterns in citrus products by means of comprehensive normal-phase × reversed-phase liquid chromatography, *Anal. Chem.*, **78**, 7743–7750.

111. Malerod, H., Rogeberg, M., Tanaka, N. *et al.* (2013). Large volume injection of aqueous peptide samples on a monolithic silica based zwitterionic-hydrophilic interaction liquid chromatography system for characterization of posttranslational modifications, *J. Chromatogr. A*, **1317**, 129–137.
112. Ikegami, T., Horie, K., Saad, N. *et al.* (2008). Highly efficient analysis of underivatized carbohydrates using monolithic-silica-based capillary hydrophilic interaction (HILIC) HPLC, *Anal. Bioanal. Chem.*, **391**, 2533–2542.
113. Neffling, M.-R., Spoof, L. and Meriluoto, J. (2009). Rapid LC-MS detection of cyanobacterial hepatotoxins microcystins and nodularins: Comparison of columns, *Anal. Chim. Acta*, **653**, 234–241.
114. Bones, J., Thomas, K.V. and Paull, B. (2006). Improved method for the determination of zinc pyrithione in environmental water samples incorporating on-line extraction and preconcentration coupled with liquid chromatography atmospheric pressure chemical ionization mass spectrometry, *J. Chromatogr. A*, **1132**, 157–164.
115. Pena, A., Chmielova, D., Lino, C.M. *et al.* (2007). Determination of fluoroquinolone antibiotics in surface waters from Mondego River by high performance liquid chromatography using a monolithic column, *J. Sep. Sci.*, **30**, 2924–2928.
116. Cledera-Castro, M., Santos-Montes, A., and Izquierdo-Hornillos, R. (2005). Comparison of the performance of conventional microparticulates and monolithic reversed-phase columns for liquid chromatography separation of eleven pollutant phenols, *J Chromatogr A*, **1087**, 57–63.
117. Noorashikin, M.S., Mohamad, S. and Abas, M.R.B. (2013). Cloud point extraction (CPE) of parabens using nonionic surfactant phase separation, *Sep. Sci. Technol.*, **48**, 1675–1681.
118. Bones, J., Thomas, K., Nesterenko, P.N. *et al.* (2006). On-line preconcentration of pharmaceutical residues from large volume water samples using short reversed-phase monolithic cartridges coupled to LC-UV-ESI-MS, *Talanta*, **70**, 1117–1128.
119. Baranowska, I. and Kowalski, B. (2012). A rapid UHPLC method for the simultaneous determination of drugs from different therapeutic groups in surface water and wastewater, *Bull. Environ. Contam. Toxicol*, **89**, 8–14.

120. Suzuki, A., Lim, L.W., Hiroi, T. *et al.* (2006). Rapid determination of bromide in seawater samples by capillary ion chromatography using monolithic silica columns modified with cetyltrimethylammonium ion, *Talanta*, **70**, 190–193.
121. Sun, X., He, X., Zhang, Y. *et al.* (2009). Determination of tetracyclines in food samples by molecularly imprinted monolithic column coupling with high performance liquid chromatography, *Talanta*, **79**, 926–934.

Chapter 4

High-Temperature Liquid Chromatography

Thorsten Teutenberg

Institute of Energy and Environmental Technology e.V. (IUTA), Germany

4.1. A Brief Definition of High-Temperature Liquid Chromatography

In recent years, the use of elevated temperatures in liquid chromatography has become more familiar, although many people still fear to increase the temperature because of unwanted side effects. In most laboratories, separations are usually performed at room temperature. In this setting, elevated-temperature liquid chromatography would mean the temperature ranges from 40 °C to 60 °C, whereas high-temperature liquid chromatography extends from 60 °C to 200 °C. The lower temperature limit of 60 °C is defined by the fact that some solvents that are used in liquid chromatography start to boil above that temperature,[1] while the upper temperature limit is dictated by the column stability. Most stationary phases are not stable if the temperature is increased above 150 °C, although it would be very easy to heat a column to a much higher temperature. In fact, only a few stationary phases can be used with an acceptable lifetime between 150 °C and 200 °C.[2]

In the literature, many different terms have been used. These are:

- Subcritical water chromatography[3–16]
- Elevated-temperature liquid chromatography[17–39]
- Superheated water chromatography[40–59]
- Hot eluent liquid chromatography[60]

- (Ultra)High-temperature liquid chromatography (HT-HPLC)[1,61–104,139]
- Thermal aqueous liquid chromatography (TALC)[105]
- And others[106]

As can be seem from the high number of publications, high-temperature liquid chromatography is very well characterized; however, for the routine use of this technique the availability of commercial instrumentation is of the utmost importance, and therefore the focus of this chapter is mainly laid on the hardware and software tools that are indispensable for a practical implementation of high-temperature liquid chromatography. First, however, I would like to point out the potential of working at high eluent temperatures.

4.1.1. *Using high eluent temperatures for increasing the separation speed*

Fast analyses seem to be of ever-greater importance, and temperature is one key variable that is able to significantly speed up the analysis time. The reason behind this is that eluent viscosity is strongly dependent on temperature.[89] In liquid chromatography, viscosity is reduced as temperature is increased — as opposed to gas chromatography, where an increase in viscosity is observed for an increase in temperature. A lower viscosity simply means that the pressure drop across the column at a constant mobile phase velocity is also reduced. Thus, much higher flow rates can be applied at elevated temperature until the maximum pressure of either the system or the stationary phase is reached. Figures 4.1 and 4.2 depict the dependence of the mobile phase viscosity on temperature for the solvent systems acetonitrile-water and methanol-water.

As can be clearly seen, the huge viscosity maximum at room temperature is reduced when the temperature is increased to 100 °C. In order to speed up a separation, 100 °C is a reasonable temperature limit because above that temperature only a few ultra-stable stationary phases are available that can be used over a prolonged period of

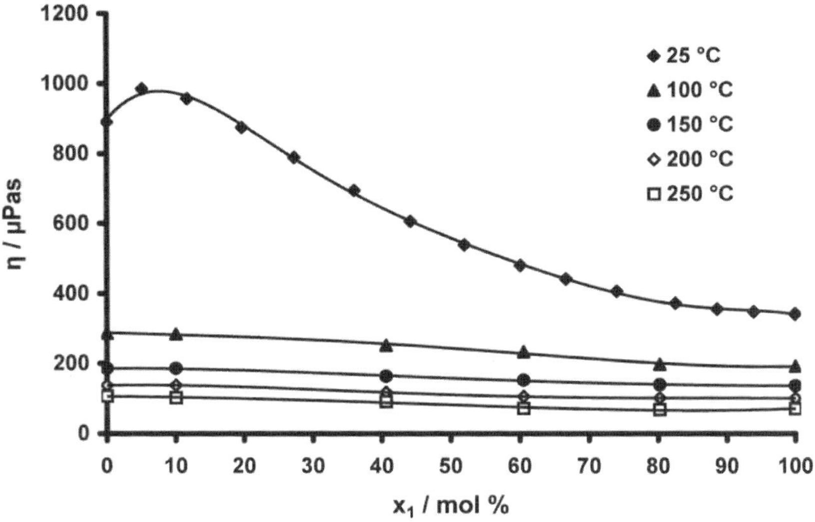

Figure 4.1. Experimentally determined viscosities of the binary mixture acetonitrile (1) water (2) at different temperatures and 100 bar. Redrawn with permission from Figure 7 from Teutenberg, T., Wiese, S., Wagner, P. *et al.* High-temperature liquid chromatography. Part II: Determination of the viscosities of binary solvent mixtures — Implications for liquid chromatographic separations. *J. Chromatogr. A*, **1216**, 8470. Copyright (2009) Elsevier.

time. A nice application is given in Fig. 4.3, which demonstrates the speed for an isocratic and isothermal separation at 90 °C on the Shimadzu Nexera LC-30 system.

In this application, a C-18 stationary phase consisting of 1.8 μm fully porous particles was used. The flow rate was increased until the maximum pressure of the column was reached. The UHPLC-system from Shimadzu is capable of generating a maximum pressure of about 1,300 bars, while the column can be used up to 1,200 bars. In order to avoid a flow rate far beyond 1 mL·min^{-1}, a column ID of 1 mm was chosen. The whole separation only took 12 seconds, which is even two seconds faster than the cycle time of the autosampler.

The group of Desmet and co-workers, as well as other authors, has intensively investigated the kinetic performance of columns at

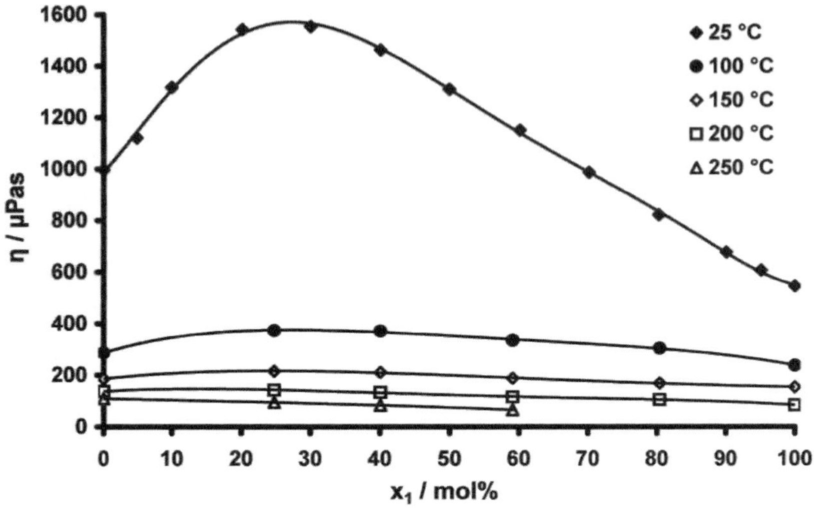

Figure 4.2. Experimentally determined viscosities of the binary mixture methanol (1)-water (2) at different temperatures and 100 bar. Redrawn with permission from Figure 5 from Teutenberg, T., Wiese, S., Wagner, P. *et al.* High-temperature liquid chromatography. Part II: Determination of the viscosities of binary solvent mixtures — Implications for liquid chromatographic separations. *J. Chromatogr. A*, **1216**, 8470. Copyright (2009) Elsevier.

elevated temperature and pressure. For a deeper understanding of the underlying theory, the reader is referred to these studies.[87,107,108]

4.1.2. *Using high eluent temperatures for modulation of solvent strength*

Besides the possibility of extremely speeding up a separation, increasing the temperature also affects the solvent properties of the mobile phase. In general, the mobile phase becomes more non-polar as the temperature is increased.[109] This has been used to replace organic co-solvents, which are usually necessary to increase the elution strength of the mobile phase by water. Figure 4.4 shows the dependence of the static permittivity of water on temperature.

As can be seen, the static permittivity of water drops from 80 at 20 °C to only 35 at 200 °C. Although methanol and acetonitrile have a static permittivity of 32 and 38 at 20 °C, respectively, it is difficult

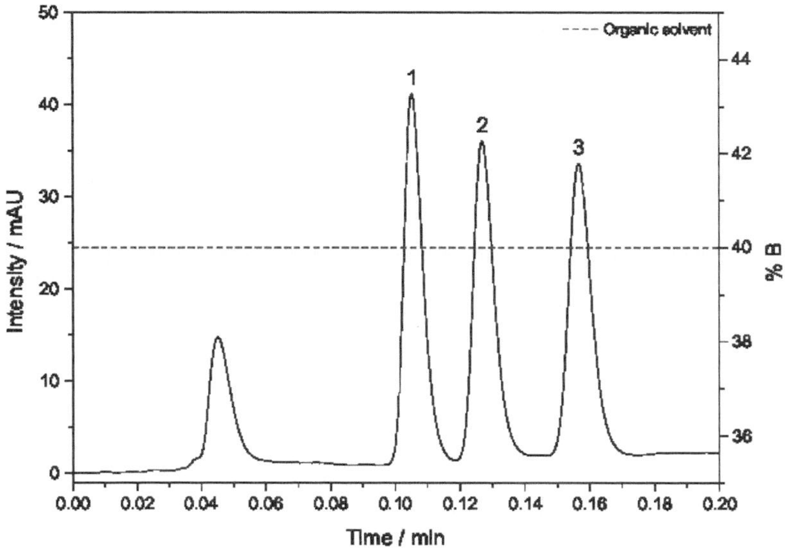

Figure 4.3. Chromatogram of the separation of three steroids and uracil. HPLC-system: Shimadzu Nexera LC-30A; chromatographic conditions — stationary phase: Agilent Zorbax StableBond C-18 (50 mm × 1.0 mm, 1.8 µm); chromatographic conditions — mobile phase: deionized water (A) and acetonitrile (B); flow rate: 1.1 mL·min^{-1}; detection: UV at 254 nm; analytes: (1) 19-nortestosterone, (2) testosterone, and epitestosterone; temperature: 90°C; pressure drop: 1,205 bars. Teutenberg, T. and Wiese, S. (2013), unpublished results.

to precisely correlate the polarity or elution strength of a water-only mobile phase to the elution strength of solvent mixtures comprised of water–methanol or water–acetonitrile. In general the elution strength of water at 200 °C is much lower than that of methanol or acetonitrile at ambient temperature. It is therefore not possible to elute very hydrophobic compounds from a reversed phase stationary phase with a water-only mobile phase. In Fig. 4.5 an example of a separation of selected steroids is given using a water-only mobile phase.

In order to obtain an elution of all compounds in the mixture, the temperature gradient starts already at a very high initial temperature. Moreover, a stationary phase with a C-4 modification has been used instead of the classical C-18 moieties, thus lowering the retentivity of the column. When a comparison is made between the

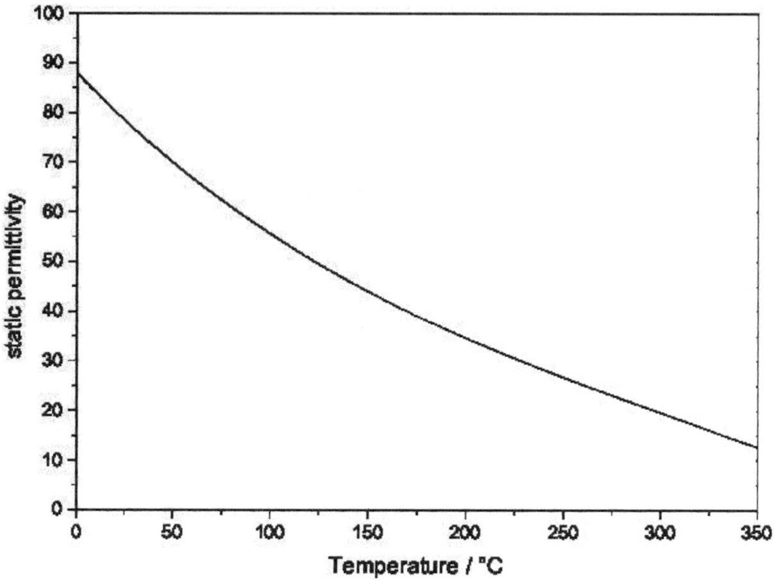

Figure 4.4. Dependence of the static permittivity of water on temperature.

separations depicted in Figs. 4.3 and 4.5 it is obvious that for a water-only mobile phase, much longer run-times result. Please note that a direct conversion of both methods is not possible, because different stationary phases and column geometries have been used. Nevertheless, the examples highlight that a complete substitution of the organic modifier in the mobile phase by water is possible if very high temperatures are applied.

To end this section it can be summarized that high-temperature liquid chromatography extends from 60 °C to 200 °C and that temperature can be used to either speed up a separation or to change the solvent properties of the mobile phase. This can be used to employ special hyphenation techniques which rely on a water-only mobile phase. Examples for this kind of hyphenation will be given in Section 4.5.

In the next section, the instrumental requirements to establish high-temperature liquid chromatography in a routine laboratory will be discussed.

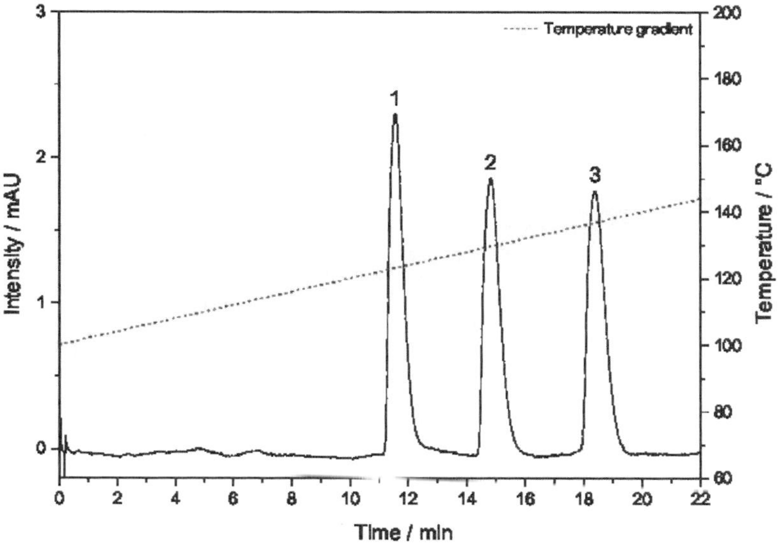

Figure 4.5. Chromatogram of the separation of three steroids. HPLC-system: Agilent 1200 (SL); chromatographic conditions — stationary phase: Waters XBridge BEH C-4 (50 mm × 2.1 mm, 3.5 μm); chromatographic conditions — mobile phase: deionized water; flow rate: 1.0 mL·min^{-1}; detection: UV at 254 nm; analytes: (1) 19-nortestosterone, (2) testosterone, and epitestosterone; temperature gradient: see figure; pressure drop: 132 bars. Teutenberg, T. and Wiese, S. (2012), unpublished results.

4.2. Instrumental Requirements

In principle, any HPLC system can be used for high-temperature liquid chromatography; the only prerequisite is a specially designed heating system which allows for operation at very high temperatures and also facilitates temperature gradients.[11] Figure 4.6 depicts a system on the basis of a contact heater, which is commercially available.[110]

There are three zones for which temperature can be independently adjusted. The first one is for eluent preheating to guarantee that the mobile phase entering the column will be thermally equilibrated with the stationary phase. It is very well documented that eluent preheating is a must at very high temperatures to reduce 'thermal mismatch band broadening', as this effect was termed by

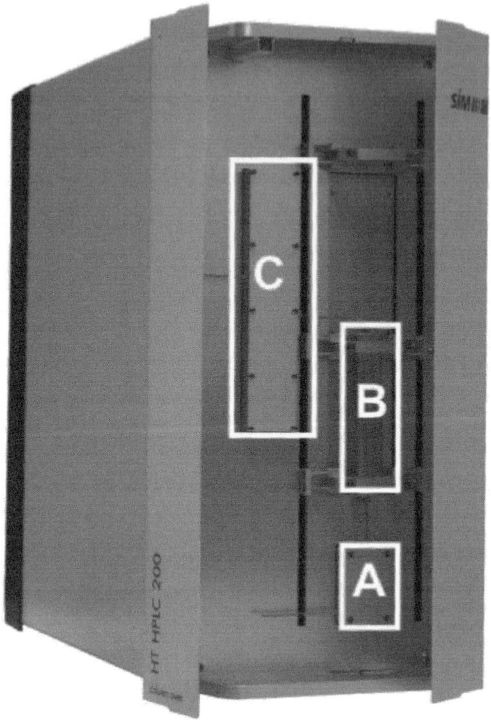

Figure 4.6. Modular heating oven for temperature programming based on contact heating. A: eluent preheating unit; B: column heating unit; C: eluent post column cooling unit. Copyright: SIM GmbH, Oberhausen.

Carr *et al.*[24] If eluent preheating is not taken into account, this can have a detrimental effect on the separation. Figure 4.7 depicts a comparison between two chromatograms at 60 and 90 °C, which were generated with and without eluent preheating on the Infinity 1290 system.

In many cases, the user is convinced that the stationary phase is degraded, although the peak broadening which is clearly visible in both chromatograms without eluent preheating occurs because of radial and axial temperature gradients across the column. Already at 60 °C, peaks appear to be broader when a preheating capillary is not used. This also leads to higher retention times for all analytes in the

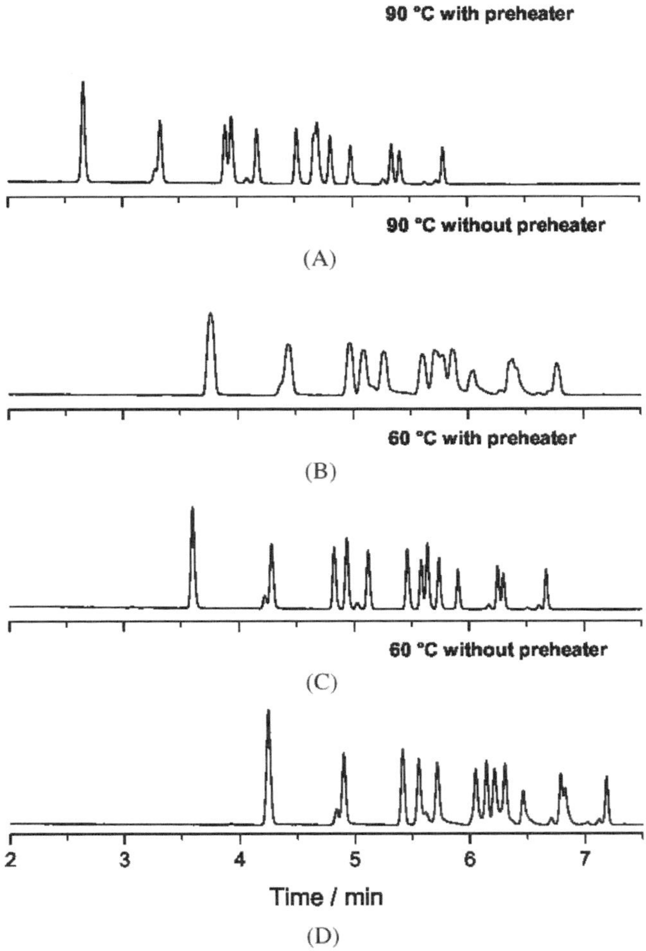

Figure 4.7. Effect of thermal mismatch band broadening on chromatographic efficiency. HPLC-system: Agilent 1290 Infinity; chromatographic conditions — stationary phase: Agilent Zorbax StableBond C-18 (50 mm × 2.1 mm, 1.8 µm); chromatographic conditions — mobile phase: deionized water (A) and acetone (B); flow rate: 0.5 mL·min^{-1}; detection: UV at 360 nm; analytes: commercially available mixture of thirteen aldehyde-2,4-dinitrophenylhydrazones (aldehyde-2,4-DNPH) and ketone-2,4-dinitrophenylhydrazones (ketone-2,4-DNPH); temperature: see figure; solvent gradient: 5 — 100% acetone in 10 minutes. Teutenberg, T. and Wiese, S. (2011), unpublished results.

mixture. At a temperature of 90 °C, the peak broadening becomes more pronounced. This means that there is a huge loss of the chromatographic efficiency. Therefore, the benefit of working at elevated temperatures to speed up the separation will be totally reversed if there is no eluent preheating.

The second module of the oven depicted in Fig. 4.6 is for column heating and can be adjusted to the column length. The use of short columns of 5 cm length is often recommended when sub 2 μm particle packed columns are used. However, some applications require longer columns up to 20 cm length which can also be incorporated in this system. The oven can be used either for isothermal operation up to 200 °C or in temperature gradient mode with a maximum heating rate of 30 °C·min^{-1}. During the system validation we were able to show that the temperature lag between the aluminum block and the stationary phase in the center of the column is only around one to two seconds.[111] A rapid cooling after a temperature gradient has been applied is possible because of the effective compressor cooling of this system.

The third module is for eluent cooling before the mobile phase enters the detector. Fluorescence detection is very sensitive to small temperature fluctuations and high eluent temperatures might even lead to a severe quenching of the signal. Therefore, the temperature of the eluent can be adjusted to the detector which is connected to the HPLC system.

In order to prevent the boiling of the mobile phase, a back-pressure regulator should be mounted behind the detector. When methanol is used at 200 °C, a back-pressure of about 40 bar will be sufficient to keep the eluent in the liquid state.[1] When a mass spectrometer is used, the transfer capillary from the column outlet to the ion source can be used as a linear restrictor to prevent the boiling of the mobile phase.

Many instrument vendors have now adopted the general concept of eluent preheating in their new systems. It has to be stressed that although some manufacturers specify that their ovens also support linear temperature gradients, the time for a re-equilibration after a temperature gradient is rather high, because an effective cooling of the stationary phase is not integrated. Most systems now also use contact heating, although the column is often not in close contact with the heating block. It is well known that heat transfer is optimal when there is a close contact between the heating block and the stainless steel encasement of the column.[112] If this is not the case, a rapid heating or cooling of the column is not possible.

Alternative concepts (e.g. water baths) have also been discussed in the scientific literature. However, these systems are never used in a routine environment because of safety concerns and contamination issues: in order to achieve high temperatures up to 200 °C, silicon oil has to be used as heating fluid, which might be dangerous to handle in a routine laboratory. In the event of leakage, the silicon oil bath or the stationary phase might get contaminated. Therefore, contact heaters are the first choice if applications that rely on temperature gradient elution have to be established.

4.3. Suitable Stationary Phases

4.3.1. *Silica-based stationary phases*

The column can still be considered the Achilles heel in high-temperature liquid chromatography. However, in recent years enormous progress has been made by the introduction of hybrid particle technology. Various approaches have been undertaken to shield the silica particle from hydrolytic dissolution. The mere encapsulation of the silica by an external polymer layer is often not appropriate for achieving high temperature stability.[113] In contrast, ethyl bridged hybrid silica particles have shown unparalleled stability at temperatures around 150 °C, as is depicted in Fig. 4.8.

As can be seen from the chromatograms before and after the stress test, there is virtually no change in the chromatographic conditions (Figs. 4.8C and (D)). However, a separation between dipropyl phthalate and naphthalene occurs after the stress test for the Gemini NX phase, which might be a result of a higher silanol activity because of some loss of C-18 moieties.

In the last ten years we have extensively investigated a lot of promising stationary phases for their application in high-temperature liquid chromatography and could only identify a handful of suitable

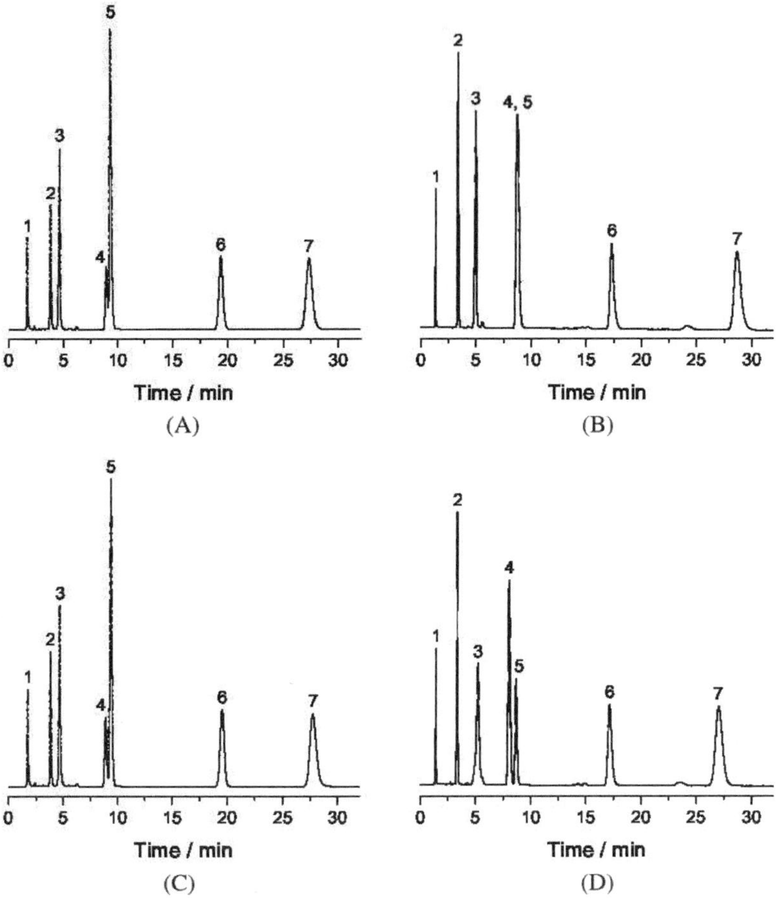

Figure 4.8. Comparison of column stability for two hybrid particles after exposure of the brand new columns to a mobile phase of water/methanol (90/10, v/v) at 150 °C for 25 hours. Test chromatograms were recorded at 25 °C using the Neue test mixture; (A) Waters XBridge, brand new; (B) Phenomenex Gemini NX, brand new; (C) Waters XBridge after 25 hours at 150 °C; (D) Phenomenex Gemini NX after 25 hours at 150 °C. Redrawn with permission from Figures 3a, 3b, 7a and 7b from Teutenberg, T., Hollebekkers, K., Wiese, S. *et al.* (2009). Temperature and pH-stability of commercial stationary phases, *J. Sep. Sci.*, 32, 1262–1274. Copyright (2009) Wiley-VCH Verlag GmbH & Co. KGaA, Weinhein.

materials. The majority of them are made up of silica-based reversed phase stationary phases, although some alternative materials should be mentioned. The interested reader is referred to the dedicated literature, where the test conditions and columns are described in detail.[2,113–115]

4.3.2. *Porous graphitic carbon*

Porous graphitic carbon (PGC) has been long known for its superior stability at extreme pH.[116] This phase, which consists completely of carbon bands, has no surface modifications that are prone to degradation or hydrolysis. It exhibits a very high retentivity, which is useful for the separation of very polar compounds that are often not retained on a classical RP stationary phase. However, the application of high eluent temperatures is often directed to hyphenation techniques which rely on a water-only mobile phase, as will be discussed in Section 4.5. In this respect, the high retentivity of a PGC stationary phase is quite contradictory to the aim of reducing the organic content in the mobile phase to enable special hyphenation techniques.

4.3.3. *Metal oxide stationary phases*

Other materials, like metal oxides that have been coated or cladded with polybutadiene or polystyrene, could not fulfil the initial euphoria felt when high-temperature liquid chromatography was young. Mainly, the group of Carr *et al.* has worked on the improvement of these phases, which consist of zirconium dioxide.[117,118] The titanium dioxide stationary phases developed by Sachtleben also never played a major role in routine analysis.[119] It is well known that the bare metal oxide phases will not dissolve in a water-only environment at high eluent temperatures; unfortunately, the cladding or coating with a polymer is detrimental to achieving a reversed phase stationary phase, which is often prone to degradation especially if binary mobile phases are used at high eluent temperatures. Also, although the manufacturer of the zirconia materials has significantly improved the polymer coating in recent years, resulting in stationary phases with a thinner polymer coating and leading to a better mass transfer. Unfortunately this also means that the phases are not as stable as before because the polymer is washed out a lot faster compared to the initial performance of the columns.

4.3.4. *Polymeric stationary phases*

Although polymeric stationary phases have long suffered from a swelling or shrinking when the content of the mobile phase is changed, polymeric phases made up of polystyrene divinylbenzene are nevertheless used for polymer analysis at temperatures up to 150 °C with tetrahydrofurane as mobile phase. However, a new material by Shodex seems to be a promising candidate for high-temperature reversed phase liquid chromatography. According to the manufacturer, the particle base material is a polyvinyl alcohol to which C-18 functional groups are "anchored" as described by Vanhoenacker *et al.*[95] This phase even produced symmetrical peak shapes after it was exposed for 25 hours with a mobile phase consisting of water-methanol (90/10, v/v) at a pH of 12.[2]

4.4. Retention Time Modeling

4.4.1. *Van't Hoff equation*

Method development is an important part in liquid chromatography. In many cases, method development is carried out simply by trial and error. This means that the user changes the parameters which influence the selectivity of the separation on the basis of his or her experience. However, this process is time-consuming and the best separation conditions will never be found.

Therefore, systematic method development should be applied. Fortunately, method development software that allows a systematic approach to method development is commercially available; DryLab and ChromSword are the software packages most frequently used. An overview of the general capabilities of these software packages is well beyond the scope of this chapter; the interested reader should consult the respective literature.[120–122] Instead, this section is focused on the possibility of using temperature as an active variable for method development in high-temperature liquid chromatography.

Basically, retention in liquid chromatography is governed by the van't Hoff equation (Eq. (4.1)), where T is the column temperature in K, k_i refers to the retention factor of the solute i, ΔH and ΔS are the enthalpy and entropy of transfer of the solute i from the mobile into the stationary phase, R is the ideal gas constant and β is the volume phase ratio of the stationary and mobile phase.

$$\ln(k_i) = -\frac{\Delta H}{R}\cdot\frac{1}{T} + \frac{\Delta S}{R} + \ln(\beta) \tag{4.1}$$

Moreover, the van't Hoff equation assumes that the enthalpy and entropy of transfer and the volume phase ratio are independent from temperature. If all analytes strictly obey the van't Hoff equation, a linear relationship is obtained between the natural logarithm of the retention factor ($\ln(k)$) and the inverse absolute temperature ($1/T$). Although there are some examples which highlight nonlinear van't Hoff behavior,[123,124] a linear functional relationship will lead to acceptable results in most cases. The interested reader is again referred to the respective literature dealing with different models for retention time predictions when temperature is used as the active variable in high-temperature liquid chromatography.[111,125,128] Also, some academic tools, which have been developed by the groups of Nikitas and Pappa-Louisi[127,128] as well as Cela (PREGA), are very helpful when simultaneous gradients of the organic solvent and temperature are employed. In the following sections, some examples that illustrate the effect of temperature either in isothermal or temperature gradient mode are given.

4.4.2. *Isothermal separations*

The first example depicted in Fig. 4.9 is the isothermal separation of a sulfonamide mixture using only water as the mobile phase.

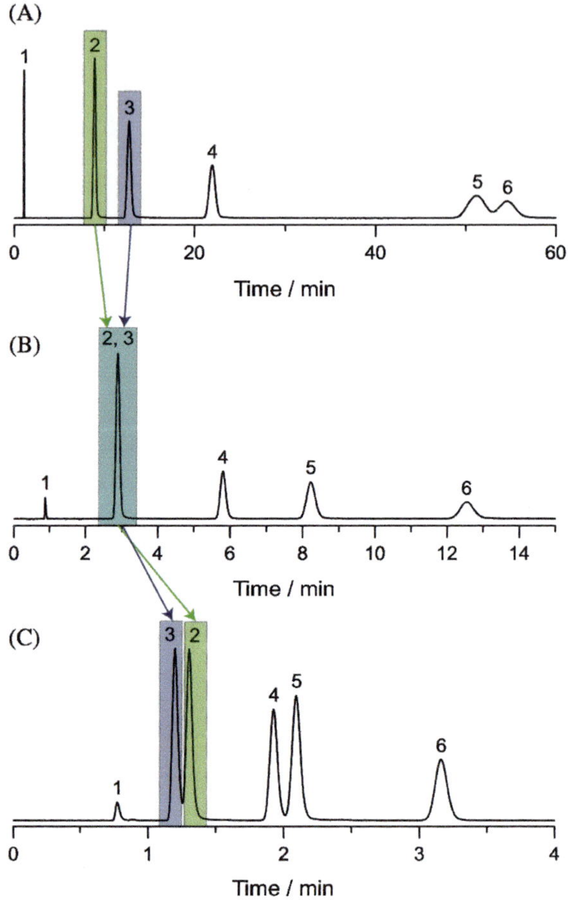

Figure 4.9. Isothermal separations of five sulfonamides and uracil. HPLC-system: Shimadzu LC-10A; chromatographic conditions — stationary phase: Waters XBridge C-18 (75 mm × 4.6 mm, 2.5 µm); chromatographic conditions — mobile phase: deionized water with 0.1% formic acid; flow rate: 1.0 mL·min^{-1}; detection: UV at 270 nm; analytes: (1) uracil, (2) sulfadiazine, (3) sulfathiazole, (4) sulfamerazine, (5) sulfamethoxazole, and (6) sulfamethazine; temperature: (A) 60 °C, (B) 120 °C, (C) 180 °C. Redrawn with permission from Figure 3 from Wiese, S., Teutenberg, T. and Schmidt, T.C. (2011). A general strategy for performing temperature-programming in high performance liquid chromatography — Prediction of segmented temperature gradients, *J. Chromatogr. A*, **1218**, 6898–6906. Copyright (2011) Elsevier.

At a temperature of 60 °C the separation of the first three sulfonamides is very good with a critical resolution (R_S) higher than 5.8, whereas R_S between sulfamethoxazole and sulfamethazine (peak pair 5/6) is inadequate (R_S = 0.9). Furthermore, at this temperature a long analysis time of approximately 60 minutes is observed, and the last peaks are eluted as broad bands. At a temperature of 120 °C the resolution between peak pair 5/6 is now very high (R_S = 7.2), whereas sulfadiazine and sulfathiazole completely co-elute. At this temperature, however, the analysis time was reduced to approximately 14 minutes. If the temperature is increased even further (e.g. to 180 °C), then an analysis time of about 3.5 minutes is obtained. Moreover, all sulfonamides were separated with a critical resolution of 1.1. The best separation is obtained at 80 °C with a critical resolution between sulfadiazine and sulfathiazole (peak pair 2/3) of 4.0 (see Fig. 4.10A).

This example clearly underlines that temperature has an influence not only on retention, but also on selectivity. Therefore, temperature should always be considered an important variable for tuning the selectivity of a separation.

4.4.3. *Temperature programmed separations*

The other option to achieve a good separation of the sulfonamide mixture is the use of temperature gradients as is shown in Fig. 4.10.

The temperature program for the separation depicted in Fig. 4.10B starts at 60 °C and consists of two linear temperature gradients with slopes of approximately 12 °C·min^{-1} and two isothermal holds at 115 °C and 180 °C. The separation of the sulfonamides was performed within 13 minutes with a critical resolution of 3.4 between sulfadiazine and sulfathiazole. When the temperature programmed elution is compared to the isothermal separation at 80 °C, the temperature programmed elution is superior in terms of analysis time. Moreover, a better signal-to-noise ratio is achieved for the last eluting compounds. In contrast, sulfamethoxazole and sulfamethazine elute as broad bands during the isothermal separation.

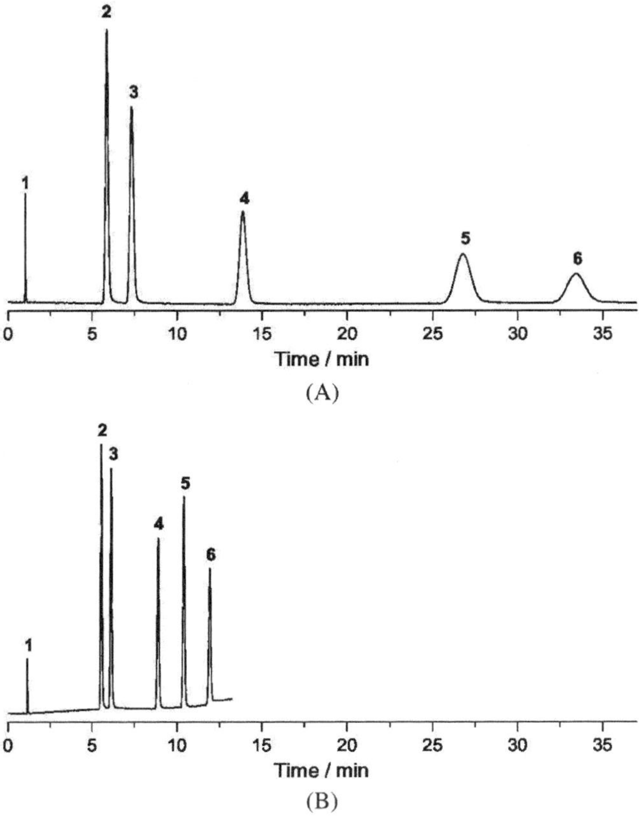

Figure 4.10. Comparison between the (A) isothermal and (B) temperature-gradient elution of selected sulfonamides. HPLC-system: Shimadzu LC-10A; chromatographic conditions — stationary phase: Waters XBridge C-18 (75 mm × 4.6 mm, 2.5 μm); chromatographic conditions — mobile phase: deionized water with 0.1% formic acid; flow rate: 1.0 mL·min^{-1}; detection: UV at 270 nm; analytes: (1) uracil, (2) sulfadiazine, (3) sulfathiazole, (4) sulfamerazine, (5) sulfamethoxazole, and (6) sulfamethazine; temperature: 80 °C for (A) and temperature gradient for (B) Reprinted with permission from Figure 5 from Wiese, S., Teutenberg, T. and Schmidt, T.C. (2011). A general strategy for performing temperature-programming in high performance liquid chromatography — Prediction of segmented temperature gradients, *J. Chromatogr. A*, **1218**, 6898–6906. Copyright (2011) Elsevier.

4.4.4. *Multi-gradient separations*

A more complex optimization procedure is the application of simultaneous temperature and solvent gradient programming. In the following example, a sulfonamide mixture was separated by either a linear solvent gradient or a combination of solvent and temperature gradients to improve the resolution of critical peak pairs as is shown in Fig. 4.11.

In Fig. 4.11A and B, the separation was carried out isothermally at a temperature of 70 °C and 90 °C, respectively. As can be seen, a co-elution is inevitable for the lower and higher start temperature. However, pronounced differences in terms of the selectivity can be observed. Therefore, a linear temperature gradient was supposed to yield a better resolution when applied simultaneously to the solvent gradient as is shown in Fig. 4.11C. However, the best overall resolution could be obtained when a more complex gradient profile for temperature and solvent composition was used, as shown in Fig. 4.11D. In addition, the flow rate was also increased to reduce the analysis time as far as possible.

4.5. Special Hyphenation Techniques

Although the gain in analysis speed is one of the main advantages of the application of high-temperature liquid chromatography, the huge potential of this method lies in the fact that special hyphenation techniques can be employed. Nearly all of these techniques rely on a mobile phase consisting of water or water with only a minimal portion of an organic co-solvent. The effect exploited here is that the static permittivity of water, or a binary mixture of water and an organic co-solvent, significantly decreases if the temperature is increased.[109]

4.5.1. *Isotope ratio mass spectrometry*

Isotope ratio mass spectrometry (IRMS) is used to distinguish between compounds which are either faked or are synthesized by different pathways, but which are chemically identical except for their abundance ratio of stable isotopes. More specifically, isotope

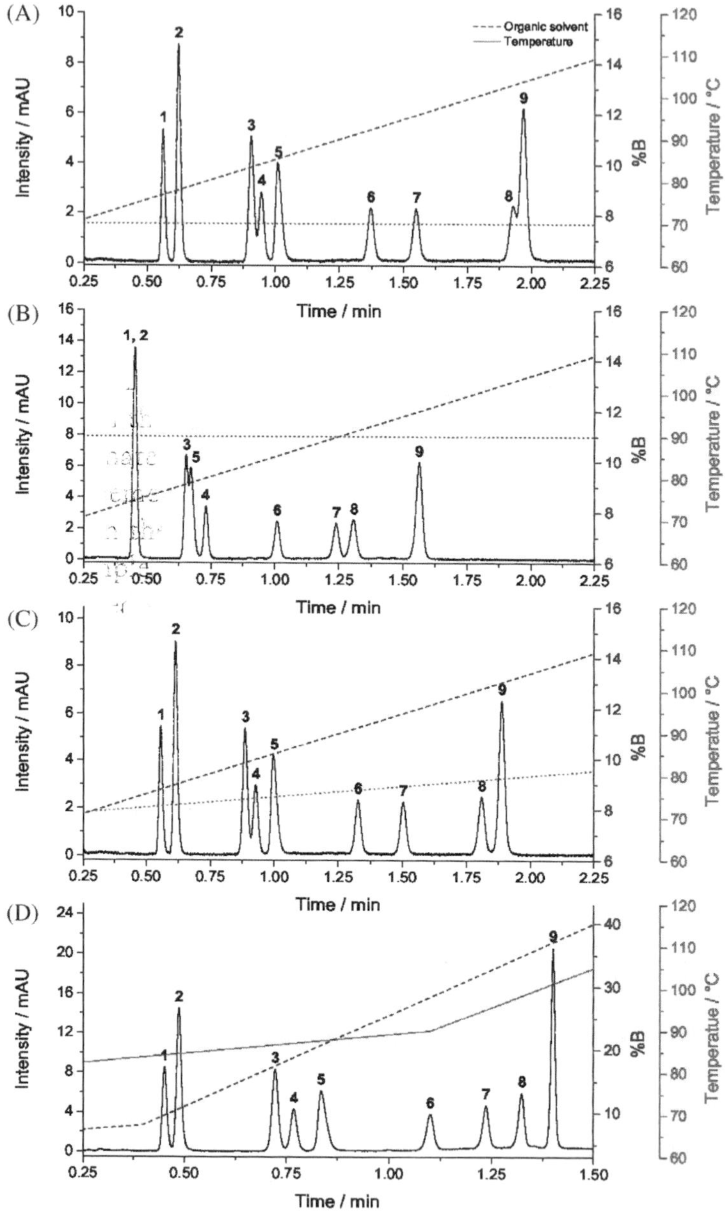

Figure 4.11. Chromatograms of the separation of eight sulfonamides and trimethoprim using temperature (red line) and solvent programming (blue line). Chromatographic conditions — stationary phase: Agilent Zorbax StableBond C-18

Figure 4.11. (Caption continued) (50 mm × 3 mm, 1.8 µm); chromatographic conditions — mobile phase: deionized water with 0.1% formic acid (A) and acetonitrile with 0.1% formic acid (B); flow rate: 1.4 mL·min^{-1} for A), B), and C); 1.7 mL·min^{-1} for D); detection: UV at 270 nm; analytes: (1) sulfadiazine, (2) sulfathiazole, (3) N^4-acetylsulfadiazine, (4) sulfamerazine, (5) trimethoprim, (6) N^4-acetylsul-famerazine, (7) sulfamethazine, (8) sulfamethoxazole, and (9) N^4-acetylsulfamet-hazine; temperature and solvent programming: see figure. Redrawn with permission from Figures 1 and 2 from Giegold, S., Teutenberg, T., Tuerk, J. *et al.* (2008). Determination of sulfonamides and trimethoprim using high temperature HPLC with simultaneous temperature and solvent gradient, *J. Sep. Sci.*, **31**, 3497–3502. Copyright (2008) Wiley-VCH Verlag GmbH & Co. KGaA, Weinhein.

ratio mass spectrometry is used to very precisely measure the relative abundances of the heavy and light isotopes of carbon. The rates at which heavier isotopes participate in chemical and physical processes are slightly different from those for lighter isotopes. This difference in rates leads to a subtle variation in the natural abundance of isotopes, owing to a variety of fractionation processes.

Isotope ratios are expressed relative to reference standards, rather than being reported as absolute isotope values. The isotope community has established the δ-notation, which is the difference in the $^{13}C/^{12}C$ isotope ratios of the sample and an internationally agreed standard normalized by the $^{13}C/^{12}C$ isotope ratio of the standard (Eq. (4.2)). The resulting $\delta\,^{13}C$ value is given in ‰.

$$\delta\,^{13}C = \left[\left(\frac{R_{\mathrm{Sample}}}{R_{\mathrm{Reference}}}\right) - 1\right] \times 1000 \tag{4.2}$$

Samples can be directly measured versus a reference gas, which is calibrated against the international reference.

In many studies, gas chromatography (GC) has been used for the hyphenation with an isotope ratio mass spectrometer. However, the drawback of GC–IRMS analysis is that many analytes of interest cannot be measured without derivatization. This procedure is not only extremely time-consuming, but also bears the risk of an isotope fractionation due to the derivatization process. If it is not possible to correct the measured $\delta^{13}C$ values for this isotope fractionation, the results are useless. In a recent instrumental development to overcome this limitation, a liquid chromatography (LC) interface for coupling high-performance liquid chromatography (HPLC) to IRMS has been introduced.[129] In this system, all compounds are quantitatively converted into CO_2 while the analyte is still dissolved in the aqueous liquid phase. The chemical oxidation is typically performed by peroxodisulfate under acidic conditions. The CO_2 is removed from the eluent and entrained into a flow of helium by a miniature separation unit. This helium stream passes a water trap system and is then directed to the ion source of the IRMS via an open split assembly.

4.5.2 LC Taste®

A very interesting process, which is known as LC Taste® and makes use of high-temperature HPLC, is the determination of gustatory active compounds in complex mixtures.[93,130]

The LC Taste® system uses the advantages of a separation based on high-temperature liquid chromatography and combines them with an in vivo detection of taste-active compounds by a sensory tester or sensory panel. Therefore, the analytical and sensory data can be correlated in a way similar to the hyphenation of gas chromatography and olfactometry. Water without any mobile phase additives is the preferred eluent because it is not toxic and will not interfere with the detection process of the sensory panel. However, the elution strength of water even at very high temperatures up to 200 °C is usually not sufficient for the elution of extremely hydrophobic components as was already outlined in Section 4.1.2. Therefore, the addition of organic modifiers that will increase the elution strength of the mobile phase is mandatory if these compounds have to be analyzed or eluted from the stationary phase. Toxic organic solvents (e.g. methanol or acetonitrile) are strictly forbidden if the eluate is directly tasted by a human being. However, ethanol is a very convenient co-solvent which — up to a certain concentration — has no negative impact on the sensory impression and also significantly enhances the elution strength of the mobile phase if solvent gradient elution is applied.

The concept of LC Taste® allows for the combined use of temperature and solvent programming. This is a decisive advantage because the elution strength can be increased by a simultaneous change of the temperature and the mobile phase composition as was described in Section 4.4.4. Moreover, there is a far greater chance to elute all compounds in the mixture within one chromatographic run. However, the ethanol concentration in the mobile phase should preferably be in the range from 5 to 30 weight percent. Therefore, a concomitant temperature gradient is necessary so that the concentration of ethanol will be as low as possible.

4.6. Applications Relevant to Environmental Analysis

In general, high-temperature liquid chromatography can be used for all applications dealing with environmental analysis, as outlined for the applications given in Section 4.4. Therefore, in this section I will not present additional chromatograms, but would rather like to comment on a problem that is related to residue analysis. As was already explained, by increasing the temperature of the mobile phase, the elution strength will also be increased. This is true for a water-only mobile phase as well as for binary mixtures comprised of water and an organic solvent, which are typically used in reversed phase liquid chromatography. Therefore, if a separation is carried out at elevated or high temperature in isothermal mode and the mixture contains very polar compounds, then there is no possibility for an on-column focusing or large volume injection of this mixture because the retention decreases as the temperature of the mobile phase increases. This can present a severe problem if very low limits of detection need to be reached, as is often the case in environmental trace analysis.[131] An isothermal separation at high temperatures might lead to a poor detection limit for compounds which are eluting with retention factors <2. Therefore, since the analysis of very polar compounds is of ever-greater importance, temperature gradient elution with a concomitant solvent gradient might be preferable; if the analysis starts at around 30 °C, the elution strength of the mobile phase is very low if only water is used as the initial solvent for sample injection as well as mobile phase for reversed phase liquid chromatography. The simultaneous increase of temperature, organic modifier and even flow rate during the elution step might then lead to very fast separations and peaks with narrow widths. Another advantage of this procedure might be that an excessively high pressure maximum can be avoided during the chromatographic run. This in turn will lead to a longer lifetime of the column.

For the application of high-temperature liquid chromatography in environmental applications, the temperature should not be increased above 100 °C. The reason is that the majority of all stationary phases on the basis of silica will rapidly degrade at higher temperatures. The main advantage of using high-temperature LC is to significantly reduce the analysis time and to reduce the viscosity and therefore pressure maximum which inevitably occurs at low eluent temperatures. The combination of high-temperature LC and UHPLC is therefore very favorable to increase the speed of the separation without a loss in efficiency, as demonstrated in Section 4.1.1 (see Fig. 4.3). Using simultaneous temperature and solvent gradient elution, a better focusing of polar analytes will be obtained at low temperatures so that lower limits of detection can be reached. Of course, the re-equilibration time after a temperature gradient needs to be considered.

4.7. Applications Relevant to Food Analysis

In principle, the same conclusions can be drawn for applications relevant to food analysis as for applications relevant to environmental analysis. However, in this section I would like to present some results which have been obtained by the application of high-temperature liquid chromatography hyphenated to isotope ratio mass spectrometry, as was already introduced in Section 4.5.1. This technique has drawn much attention to the possibility of authenticity control of food products. In recent years, there is a growing community of consumers who are willing to pay a higher price for food which is made up of natural ingredients instead of synthetic compounds. Unfortunately, some producers deliberately exploit the willingness of consumers to pay higher prices and mislabel their products, which may have been adulterated or not contain any natural ingredients at all. In this case, IRMS may help to identify fraud, thus increasing the pressure on food producers to correctly label their products. In the following part, I would like to give some examples which highlight the tremendous potential of this technique.

The analysis of sugars was one of the first applications of LC hyphenated to IRMS. The hydrophilic nature of sugars makes it possible to use water at ambient or elevated temperature to achieve an elution on an LC stationary phase.[132,133] Honey is a relatively expensive natural product, which contains carbohydrates, glucose and fructose in approximately equal amounts. Honey can then be easily corrupted with cheaper sugars such as high fructose corn syrup and sugar derived from corn, and therefore the detection of sugar adulteration is of major importance in order to protect consumers against fraud and to reveal fraudulent use of invert sugar syrups in the food industry.

In 2008, Elflein and Raezke published a paper where they described an LC–IRMS method for the detection of honey adulteration.[134] The authors of this study used a polymeric styrene divinylbenzene stationary phase loaded with cations (Phenomenex Rezek RCM, Ca^{2+}), which was operated at 55 °C using ultrapure water as eluent. Currently, this is the first and only worldwide accredited method for the detection of adulteration in honey.

Recently, Zhang *et al.* investigated sources of caffeine in tea, coffee and energy drinks by LC–IRMS applying high-temperature HPLC.[135] They used a C-18 XBridge column from Waters operated isothermally at 80 °C with water as the mobile phase and the heating system which is depicted in Fig. 4.6. It was shown that natural caffeine can be clearly distinguished from synthetically produced caffeine by its carbon isotope ratio. Several mislabeled products have been identified.

4.8. General Conclusions and Outlook

In order to close this chapter, I would like to make a few general conclusions and give a short outlook on the future prospects of HTLC. The huge potential of this technology can be seen in its ability to use detectors or hyphenation technologies that were not available for a direct coupling with liquid chromatography before. The examples which were chosen in this chapter to illustrate this statement have been mainly taken to show the potential benefit for environmental and food applications; however, there are other hyphenation techniques which have not been explicitly mentioned and are beyond the scope of this chapter. These include the hyphenation of HTLC with flame ionization detection[136,137] and the coupling of HTLC to a continuous-flow biochemical screening assay with electrospray ionization mass spectrometric detection.[138]

The software and hardware tools that are necessary to implement HTLC are commercially available. Therefore, the self-assembling of an HTLC system is no longer required, although many publications in the scientific literature suggest that this is still necessary. Unfortunately, the separation community is often very reluctant to implement new strategies and technologies, although examples from the computer and telecommunications market show that, for example, smart phones, tablets and other related electronic devices are used worldwide by an overwhelming number of people. Nonetheless, the implementation and integration of new separation devices and technologies in a routine laboratory requires sound theoretical knowledge about the processes involved. Furthermore, the laboratory personal that is actively working with these systems needs to be convinced of the benefits and advantages of new technologies. The reason why other separation techniques like micro- or capillary liquid chromatography are still not considered to be real options for increasing the sample throughput and minimizing costs and energy consumption can be attributed to misgivings similar to those for HTLC. I therefore hope that the examples and illustrations given in this chapter will encourage everyone to make active use of this emerging technology.

References

1. Teutenberg, T., Wagner, P. and Gmehling, J. (2009). High-temperature liquid chromatography. Part I: Determination of the vapour pressures of binary solvent mixtures — Implications for liquid chromatographic separations, *J Chromatogr. A*, **1216**, 6471–6480.

2. Haun, J., Oeste, K., Teutenberg, T., *et al.* (2012). Long-term high-temperature and pH stability assessment of modern commercially available stationary phases by using retention factor analysis, *J Chromatogr. A*, **1263**, 99–107.
3. Miller, D.J. and Hawthrone, S.B. (1997). Subcritical water chromatography with flame ionization detection, *Anal. Chem.*, **69**, 623–627.
4. Yang, Y., Jones, A.D. and Eaton, C.D. (1999). Retention behavior of phenols, anilines, and alkylbezenes in liquid chromatographic separations using subcritical water as the mobile phase, *Anal. Chem.*, **71**, 3808–3813.
5. Yang, Y., Jones, A.D., Mathis, J.A. *et al.* (2002). Flame ionization detection after splitting the water effluent in subcritical water chromatography, *J. Chromatogr. A*, **942**, 231–236.
6. Yang, Y., Lamm, L.J., He, P. *et al.* (2002). Temperature effect on peak width and column efficiency in subcritical water chromatography, *J. Chromatogr. Sci.*, **40**, 107–112.
7. He, P. and Yang, Y. (2003). Studies on the long-term thermal stability of stationary phases in subcritical water chromatography, *J Chromatogr. A*, **989**, 55–63.
8. Kondo, T. and Yang, Y. (2003). Comparison of elution strength, column efficiency, and peak symmetry in subcritical water chromatography and traditional reversed-phase liquid chromatography, *Anal. Chim. Acta*, **494**, 157–166.
9. Lamm, L.J. and Yang Y. (2003). Off-line coupling of subcritical water extraction with subcritical water chromatography via a sorbent trap and thermal desorption, *Anal. Chem.*, **75**, 2237–2242.
10. Yisong, S., Jen-Fon, J. and Weibing, Z. (2010). The method development of subcritical water chromatography, *Chin. J. Chromatogr.*, **23**, 238–242.
11. Teutenberg, T., Goetze, H.J., Tuerk, J. *et al.* (2006). Development and application of a specially designed heating system for temperature-programmed high-performance liquid chromatography using subcritical water as the mobile phase, *J. Chromatogr. A*, **1114**, 89–96.
12. Fogwill, M.O. and Thurbide, K.B. (2007). Rapid column heating method for subcritical water chromatography, *J. Chromatogr. A*, **1139**, 199–205.

13. Yang, Y. (2007). Subcritical water chromatography: A green approach to high-temperature liquid chromatography, *J. Sep. Sci.*, **30**, 1131–1140.
14. Fogwill, M.O. and Thurbide, K.B. (2008). Carbon dioxide modified subcritical water chromatography, *J. Chromatogr. A*, **1200**, 49–54.
15. Allmon, S.D. and Dorsey, J.G. (2009) Retention mechanisms in subcritical water reversed-phase chromatography, *J. Chromatogr. A*, **1216**, 5106–5111.
16. Allmon, S.D. and Dorsey, J.G. (2010). Properties of subcritical water as an eluent for reversed-phase liquid chromatography: Disruption of the hydrogen-bond network at elevated temperature and its consequences. *J. Chromatogr. A*, **1217**, 5769–5775.
17. Abbott, S., Achener, P., Simpson, R. *et al.* (1981). Effect of radial thermal gradients in elevated temperature high-performance liquid chromatography, *J. Chromatogr. A*, **218**, 123–135.
18. Chen, H. and Horvath, C. (1993). Rapid separation of proteins by reversed phase HPLC at elevated temperatures, *Anal. Method. Instrum.*, **1**, 213–222.
19. Le, X.C., Ma, M. and Wong, N.A (1996). Speciation of arsenic compounds using high-performance liquid chromatography at elevated temperature and selective hydride generation atomic fluorescence detection. *Anal. Chem.*, **68**, 4501–4506.
20. Li, J.W., Hu, Y. and Carr, P.W. (1997) Fast separations at elevated temperatures on polybutadiene-coated zirconia reversed-phase material, *Anal. Chem.*, **69**, 3884–3888.
21. Sheng, G.C., Shen, Y.F. and Lee, M.L. (1997). Elevated temperature liquid chromatography using reversed-phase packed capillary columns, *J. Microcolumn Sep.*, **9**, 63–72.
22. Le, X.C., Li, X.-F., Lai, V. *et al.* (1998). Simultaneous speciation of selenium and arsenic using elevated temperature liquid chromatography separation with inductively coupled plasma mass spectrometry detection, *Spectrochim. Acta B*, **53**, 899–909.
23. Neue, U.D. and Mazzeo, J.R. (2001). A theoretical study of the optimization of gradients at elevated temperature, *J. Sep. Sci.*, **24**, 921–929.

24. Thompson, J.D., Brown, J.S. and Carr, P.W. (2001). Dependence of thermal mismatch broadening on column diameter in high-speed liquid chromatography at elevated temperatures, *Anal. Chem.*, **73**, 3340–3347.
25. Jones, A. and Yang, Y. (2003). Separation of nonpolar analytes using methanol-water mixtures at elevated temperatures, *Anal. Chim. Acta*, **485**, 51–55.
26. Bidlingmeyer, B.A. and Henderson, J. (2004). Investigation of retention on bare silica using reversed-phase mobile phases at elevated temperatures, *J. Chromatogr. A*, **1060**, 187–193.
27. Coym, J.W. and Dorsey, J.G. (2004). Reversed-phase retention thermodynamics of pure-water mobile phases at ambient and elevated temperature, *J. Chromatogr. A*, **1035**, 23–29.
28. Zhu, C.R., Goodall, D.M. and Wren, S.A.C. (2004). Elevated temperature HPLC: Principles and applications to small molecules and biomolecules, *LCGC Eur.*, **17**, 530–540.
29. Ueki, Y., Umemura, T., Iwashita, Y. *et al.* (2006). Preparation of low flow-resistant methacrylate-based monolithic stationary phases of different hydrophobicity and the application to rapid reversed-phase liquid chromatographic separation of alkylbenzenes at high flow rate and elevated temperature, *J. Chromatogr. A*, **1106**, 106–111.
30. Vanhoenacker, G. and Sandra, P. (2006). Elevated temperature and temperature programming in conventional liquid chromatography — fundamentals and applications, *J. Sep. Sci.*, **29**, 1822–1835.
31. Xiang, Y., Liu, Y. and Lee, M.L. (2006). Ultrahigh-pressure liquid chromatography using elevated temperature, *J. Chromatogr. A*, **1104**, 198–202.
32. Hao, Z., Lu, C.Y., Xiao, B. *et al.* (2007). Separation of amino acids, peptides and corresponding Amadori compounds on a silica column at elevated temperature, *J. Chromatogr. A*, **1147**, 165–171.
33. Lippert, J.A., Johnson, T.M., Lloyd, J.B. *et al.* (2007). Effects of elevated temperature and mobile phase composition on a novel C18 silica column, *J. Sep. Sci.*, **30**, 1141–1149.
34. Sandra, P. and Vanhoenacker, G. (2007). Elevated temperature-extended column length conventional liquid chromatography to increase peak capacity for the analysis of tryptic digests, *J. Sep. Sci.*, **30**, 241–244.

35. Soliev, A., Quiming, N.S., Ohta, H. *et al.* (2007). Separation of ginsenosides at elevated temperature by ultrahigh-pressure liquid chromatography, *J. Liq. Chromatogr. R. T.*, **30**, 2835–2849.
36. Blackler, A.R., Speers, A.E. and Wu, C.C. (2008). Chromatographic benefits of elevated temperature for the proteomic analysis of membrane proteins, *Proteomics*, **8**, 3956–3964.
37. Hartonen, K. and Riekkola, M.L. (2008). Liquid chromatography at elevated temperatures with pure water as the mobile phase. *TrAC, Trends Anal. Chem.*, **27**, 1–14.
38. Grata, E., Guillarme, D., Glauser, G. *et al.* (2009). Metabolite profiling of plant extracts by ultrahigh-pressure liquid chromatography at elevated temperature coupled to time-of-flight mass spectrometry, *J. Chromatogr. A*, **1216**, 5660–5668.
39. Shaaban, H. and Gorecki, T. (2012). Optimization and validation of a fast ultrahigh-pressure liquid chromatographic method for simultaneous determination of selected sulphonamides in water samples using a fully porous sub-2 μm column at elevated temperature, *J. Sep. Sci.*, **35**, 216–224.
40. Smith, R.M. and Burgess, R.J. (1996). Superheated water: A clean eluent for reversed-phase high-performance liquid chromatography, *Anal. Commun.*, **33**, 327–329.
41. Smith, R.M. and Burgess, R.J. (1997) Superheated water as an eluent for reversed-phase high-performance liquid chromatography, *J. Chromatogr. A*, **785**, 49–55.
42. Ingelse, B.A., Janssen, H.-G. and Cramers, C.A. (1998). HPLC–FID with superheated water as the eluent: Improved methods and instrumentation, *J. High Res. Chromatogr.*, **21**, 613–616.
43. Chienthavorn, O. and Smith, R.M. (1999). Buffered superheated water as an eluent for reversed-phase high-performance liquid chromatography, *Chromatographia*, **50**, 485–489.
44. Smith, R.M., Burgess, R.J., Chienthavorn, O. *et al.* (1999). Superheated water: a new look at chromatographic eluents for reversed-phase HPLC, *LCGC*, **17**, 938, 940, 942, 944–945.
45. Wilson, I.D. (2000). Investigation of a range of stationary phases for the separation of model drugs by HPLC using superheated water as the mobile phase, *Chromatographia*, **52**, 28–34.

46. Fields, S.M., Ye, C.Q., Zhang, D.D. *et al.* Superheated water as eluent in high-temperature high-performance liquid chromatographic separations of steroids on a polymer-coated zirconia column, *J. Chromatogr. A*, **913**, 197–204.
47. Kephart, T.S. and Dasgupta, P.K. (2002). Superheated water eluent capillary liquid chromatography, *Talanta*, **56**, 977–987.
48. Tajuddin, R. and Smith, R.M. (2002). On-line coupled superheated water extraction (SWE) and superheated water chromatography (SWC), *Analyst*, **127**, 883–885.
49. Nakajima, R., Yarita, T. and Shibukawa, M. (2003). Analysis of alcohols by superheated water chromatography with flame ionization detection, *Bunseki Kagaku*, **52**, 305–309.
50. Yarita, T., Nakajima, R. and Shibukawa, M. (2003). Superheated water chromatography of phenols using poly(styrene-divinylbenzene) packings as a stationary phase, *Anal. Sci.*, **19**, 269–272.
51. Chienthavorn, O., Smith, R.M., Saha, S. *et al.* (2004). Superheated water chromatography–nuclear magnetic resonance spectroscopy and mass spectrometry of vitamins, *J. Pharm. Biomed. Anal.*, **36**, 477–482.
52. Coym, J.W. and Dorsey, J.G. (2004). Superheated water chromatography: a brief review of an emerging technique, *Anal. Lett.*, **37**, 1013–1023.
53. Chienthavorn, O., Smith, R.M., Wilson, I.D. *et al.* Superheated water chromatography–nuclear magnetic resonance spectroscopy of kava lactones, *Phytochem. Anal.*, **16**, 217–221.
54. Tajuddin, R. and Smith, R.M. (2005). On-line coupled extraction and separation using superheated water for the analysis of triazine herbicides in spiked compost samples, *J. Chromatogr. A*, **1084**, 194–200.
55. Yarita, T., Nakajima, R., Shimada, K. *et al.* (2005). Superheated water chromatography of low molecular weight polyethylene glycols with ultraviolet detection, *Anal. Sci.*, **21**, 1001–1003.
56. Smith, R.M. (2006). Superheated water: the ultimate green solvent for separation science, *Anal. Bioanal. Chem.*, **385**(3), 419–421.
57. Dugo, P., Buonasera, K., Crupi, M.L. *et al.* (2007). Superheated water as chromatographic eluent for parabens separation on octadecyl coated zirconia stationary phase, *J. Sep. Sci.*, **30**, 1125–1130.

58. Al-Khateeb, L. and Smith, R.M. (2008). Superheated water chromatography on phenyl bonded hybrid stationary phases, *J. Chromatogr. A*, **1201**, 61–64.
59. Smith, R.M. (2008). Superheated water chromatography — A green technology for the future, *J. Chromatogr. A*, **1184**, 441–455.
60. Kephart, T.S. and Dasgupta, P.K. (2000). Hot eluent capillary liquid chromatography using zirconia and titania based stationary phases, *Anal. Chim. Acta*, **414**, 71–78.
61. Teutenberg, T., Wiese, S., Wagner, P. *et al.* (2009). High-temperature liquid chromatography. Part III: Determination of the static permittivities of pure solvents and binary solvent mixtures — Implications for liquid chromatographic separations, *J. Chromatogr. A*, **1216,** 8480–8487.
62. Yamaki, S., Isobe, T., Okuyama, T. *et al.* Reversed-phase liquid chromatography on a microspherical carbon column at high temperature, *J. Chromatogr. A*, **728**, 189–194.
63. Djordjevic, N.M., Houdiere, F. and Fowler, P. (1998). High temperature and temperature programming in capillary HPLC, *Biomed. Chromatogr.*, **12**, 153–154.
64. Trones, R., Andersen, T., Hunnes, I. *et al.* Modified laser light-scattering detector for use in high temperature micro liquid chromatography, *J. Chromatogr. A*, **814**, 55–61.
65. Djordjevic, N.M., Fowler, P.W.J. and Houdiere, F. (1999). High temperature and temperature programming in high-performance liquid chromatography: instrumental considerations, *J. Microcolumn Sep.*, **11**, 403–413.
66. Molander, P., Trones, R., Haugland, K. *et al.* (1999). Aspects and applications of non-aqueous high temperature packed capillary liquid chromatography, *Analyst*, **124**, 1137–1141.
67. Trones, R., Andersen, T. and Greibrokk, T. (1999). Improved modification of a laser light-scattering detector for use in packed capillary high temperature liquid chromatography, *J. High Resolut. Chromatogr.*, **22**, 283–286.
68. Louden, D., Handley, A., Taylor, S. *et al.* (2001). High temperature reversed-phase HPLC using deuterium oxide as a mobile phase for the

separation of model pharmaceuticals with multiple on-line spectroscopic analysis (UV, IR, 1H-NMR and MS), *Analyst*, **126**, 1625–1629.

69. Sanagi, M.M., See, H.H., Ibrahim, W.A. *et al.* (2004). High temperature liquid chromatography of triazole fungicides on polybutadiene-coated zirconia stationary phase, *J. Chromatogr. A*, **1059**, 95–101.
70. Sanagi, M.M., Heng, S.H., Ibrahim, W.A.W. *et al.* (2004). High-temperature high-performance liquid chromatography of triazole fungicides on carbon-clad zirconia stationary phase, *Malays. J. Chromatogr.*, **6**, 55–66.
71. Albert, M., Cretier, G., Guillarme, D. *et al.* Some advantages of high temperature for the separation of pharmaceutical compounds with mass spectrometry detection, *J. Sep. Sci.* **28**, 1803–1811.
72. Guillarme, D. and Heinisch, S. (2005). Detection modes with high-temperature liquid chromatography — a review, *Sep. Purif. Technol.*, **34**, 181–216.
73. Vanhoenacker, G. and Sandra, P. (2005). High-temperature liquid chromatography and liquid chromatography–mass spectroscopy analysis of octylphenol ethoxylates on different stationary phases, *J. Chromatogr. A*, **1082**, 193–202.
74. Yang, X.Q., Ma, L.J. and Carr, P.W. (2005). High temperature fast chromatography of proteins using a silica-based stationary phase with greatly enhanced low pH stability, *J. Chromatogr. A*, **1079**, 213–220.
75. Pereira, L. (2006). Ultrahigh temperature in HPLC with a porous graphite carbon column, *LCGC N. Am. Appl. Noteb.*, **June 2006**, 75–77.
76. Riddle, L.A. and Guiochon, G. (2006). Separation of free sterols by high-temperature liquid chromatography, *J. Chromatogr. A*, **1137**, 173–179.
77. Shen, S., Lee, H., McCaffrey, J. *et al.* (2006). High-temperature high-performance liquid chromatography of substituted anilines using a C18 hybrid stationary phase, *J. Liq. Chromatogr. Related Technologies*, **29**, 2823–2834.
78. Stoll, D.R., Cohen, J.D. and Carr, P.W. (2006). Fast, comprehensive online two-dimensional high-performance liquid chromatography through the use of high temperature ultra-fast gradient elution reversed-phase liquid chromatography, *J. Chromatogr. A*, **1122**, 123–137.

79. Gritti, F. and Guiochon, G. (2007). Unusual behavior of the height equivalent to a theoretical plate of a new poroshell stationary phase at high temperatures, *J. Chromatogr. A*, **1169**, 125–138.
80. McNeff, C.V., Yan, B., Stoll, D.R. *et al.* (2007). Practice and theory of high temperature liquid chromatography, *J. Sep. Sci.*, **30**, 1672–1685.
81. Nguyen, D.T.T., Guillarme, D., Heinisch, S. *et al.* (2007). High throughput liquid chromatography with sub-2 μm particles at high pressure and high temperature, *J. Chromatogr. A*, **1167**, 76–84.
82. Pereira, L., Aspey, S. and Ritchie, H. (2007). High temperature to increase throughput in liquid chromatography and liquid chromatography–mass spectrometry with a porous graphitic carbon stationary phase, *J. Sep. Sci.*, **30**, 1115–1124.
83. Plumb, R., Mazzeo, J.R., Grumbach, E.S. *et al.* (2007). The application of small porous particles, high temperatures, and high pressures to generate very high resolution LC and LC/MS separations, *J. Sep. Sci.*, **30**, 1158–1166.
84. Giegold, S. (2008). Application of high-temperature–high-performance liquid chromatography (HT–HPLC): About the influence of temperature on the analysis time, selectivity and different detection systems [Dissertation], University of Siegen, Siegen.
85. Giegold, S., Teutenberg, T., Tuerk, J. *et al.* (2008). Determination of sulfonamides and trimethoprim using high temperature HPLC with simultaneous temperature and solvent gradient, *J. Sep. Sci.*, **31**, 3497–3502.
86. Gika, H.G., Theodoridis, G., Extance, J. *et al.* (2008). High-temperature-ultraperformance liquid chromatography–mass spectrometry for the metabonomic analysis of Zucker rat urine, *J. Chromatogr. B*, **871**, 279–287.
87. Heinisch, S., Desmet, G., Clicq, D. *et al.* (2008). Kinetic plot equations for evaluating the real performance of the combined use of high temperature and ultrahigh pressure in liquid chromatography — Application to commercial instruments and 2.1 and 1 mm ID columns, *J. Chromatogr. A*, **1203**, 124–136.
88. Teutenberg, T. (2009). Potential of high-temperature liquid chromatography for the improvement of separation efficiency — A review, *Anal. Chim. Acta*, **643**, 1–12.

89. Teutenberg T., Wiese, S., Wagner, P. *et al.* (2009). High-temperature liquid chromatography. Part II: Determination of the viscosities of binary solvent mixtures — Implications for liquid chromatographic separations, *J. Chromatogr. A*, **1216**, 8470–8479.
90. Causon, T.J., Shellie, R.A. and Hilder, E.F. (2009). High-temperature liquid chromatography with monolithic capillary columns and pure water eluent, *Analyst*, **134**, 440–442.
91. Causon, T.J., Nordborg, A., Shellie, R.A. *et al.* (2010). High-temperature liquid chromatography of intact proteins using organic polymer monoliths and alternative solvent systems, *J. Chromatogr. A*, **1217**, 3519–3524.
92. Tian, H., Xu, J., Ding, K. *et al.* (2009). Study of two-dimensional liquid chromatography with high temperature NPLC and room temperature RPLC, *Front. Chem. Chin.*, **4**, 196–201.
93. Reichelt, K.V., Peter, R., Paetz, S. *et al.* (2010). Characterization of flavor modulating effects in complex mixtures via high-temperature liquid chromatography, *J. Agric. Food Chem.*, **58**, 458–464.
94. Teutenberg, T. (2010). High-temperature liquid chromatography — a brief review about an emerging technique, *Chromatogr. Today*, **3**, 3–6.
95. Vanhoenacker, G., Pereira, A.D., Kotsuka, T. *et al.* (2010). Evaluation of a new polymeric stationary phase with reversed-phase properties for high-temperature liquid chromatography, *J. Chromatogr. A*, **1217**, 3217–3222.
96. Berta, R., Babjak, M. and Gazdag, M. (2011). A study of some practical aspects of high-temperature liquid chromatography in pharmaceutical applications, *J. Pharm. Biomed. Anal.*, **54**, 458–462.
97. de Vlieger, J.S.B., Giezen, M.J.N., Falck, D. (2011). High-temperature liquid chromatography hyphenated with ESI–MS and ICP–MS detection for the structural characterization and quantification of halogen-containing drug metabolites, *Anal. Chim. Acta*, **698**, 69–76.
98. Lee, D., Miller, M.D., Meunier, D.M. *et al.* Development of high-temperature comprehensive two-dimensional liquid chromatography hyphenated with infrared and light scattering detectors for characterization of chemical composition and molecular weight heterogeneities in polyolefin copolymers, *J. Chromatogr. A*, **1218**, 7173–7179.

99. Paproski, R.E., Liang, C. and Lucy, C.A. (2011). High-temperature normal-phase liquid chromatography of aromatic hydrocarbons on bare zirconia, *J. Chromatogr. A*, **1218**, 7948–7955.
100. Rogeberg, M., Wilson, S.R., Malerod, H. *et al.* High-efficiency, high-temperature separations on silica-based monolithic columns, *J. Chromatogr. A*, **1218**, 7281–7288.
101. Terol, A., Paredes, E., Maestre, S.E. *et al.* (2012). Rapid and sensitive determination of carbohydrates in foods using high-temperature liquid chromatography with evaporative light-scattering detection, *J. Sep. Sci.*, **35**, 929–936.
102. Terol, A., Ardini, F., Grotti, M. *et al.* (2012). High–temperature liquid chromatography — inductively-coupled plasma mass spectrometry for the determination of arsenosugars in biological samples, *J Chromatogr. A*, **1262**, 70–76.
103. Teutenberg, T. (2010). 'High-temperature liquid chromatography' in Smith, R.M. (ed.), *A User's Guide for Method Development*, Royal Society of Chemistry, Cambridge.
104. Teutenberg, T. (2011). 'High-temperature liquid chromatography' in Byrdwell, W.C. and Holcapek, M. (eds), *Extreme Chromatography: Faster, Hotter, Smaller*, AOCS Press, Urbana, IL.
105. Guillemin, C.L., Millet, J.L. and Dubois, J. (1981). Thermal aqueous liquid chromatography — the TALC technique, *J. High Resolut. Chromatogr.*, **4**, 280–286.
106. Kondo, T., Yang, Y. and Lamm, L. (2002). Separation of polar and non-polar analytes using dimethyl sulfoxide-modified subcritical water, *Anal. Chim. Acta*, **460**, 185–191.
107. Cabooter, D., Heinisch, S., Rocca, J.L. *et al.* (2007). Use of the kinetic plot method to analyze commercial high-temperature liquid chromatography systems — I: Intrinsic performance comparison, *J Chromatogr. A*, **1143**, 121–133.
108. Clicq, D., Heinisch, S., Rocca, J.L. *et al.* (2007). Use of the kinetic plot method to analyze commercial high-temperature liquid chromatography systems — II. Practically constrained performance comparison, *J. Chromatogr. A*, **1146**, 193–201.
109. Teutenberg, T., Wiese, S., Wagner, P. *et al.* High-temperature liquid chromatography — III: Determination of the static permittivities of

pure solvents and binary solvent mixtures — Implications for liquid chromatographic separations, *J. Chromatogr. A*, **1216**, 8480–8487.

110. SIM — Scientific Instruments Manufacturer GmbH: SIM HT HT HPLC 200 Column Oven. (2014). Available at: http://sim-gmbh.de/index.php/en/lc/sim-ht-column-oven.html [Accessed 15 January 2014].
111. Wiese, S., Teutenberg, T. and Schmidt, T.C. (2011). General strategy for performing temperature programming in high-performance liquid chromatography: prediction of linear temperature gradients, *Anal. chem.*, **83**, 2227–2233.
112. Verstraeten, M., Pursch, M., Eckerle, P. *et al.* (2011). Modelling the thermal behaviour of the low-thermal mass liquid chromatography system, *J. Chromatogr. A*, **1218**, 2252–2263.
113. Teutenberg, T., Hollebekkers, K., Wiese, S. (2009). Temperature and pH-stability of commercial stationary phases, *J. Sep. Sci.*, **32**, 1262–1274.
114. Teutenberg, T., Tuerk, J., Holzhauser, M. *et al.* Evaluation of column bleed by using an ultraviolet and a charged aerosol detector coupled to a high-temperature liquid chromatographic system, *J. Chromatogr. A*, **1119**, 197–201.
115. Teutenberg, T., Tuerk, J., Holzhauser, M. *et al.* Temperature stability of reversed-phase and normal-phase stationary phases under aqueous conditions, *J. Sep. Sci.*, **30**, 1101–1114.
116. West, C., Elfakir, C. and Lafosse, M. (2010). Porous graphitic carbon: A versatile stationary phase for liquid chromatography, *J. Chromatogr. A*, **1217**, 3201–3216.
117. Nawrocki, J., Dunlap, C., McCormick, A. *et al.* Part I: Chromatography using ultra-stable metal-oxide-based stationary phases for HPLC, *J. Chromatogr. A*, **1028**, 1–30.
118. Nawrocki, J., Dunlap, C., Li, J. *et al.* (2004). Part II: Chromatography using ultra-stable metal-oxide-based stationary phases for HPLC, *J. Chromatogr. A*, **1028**, 31–62.
119. Winkler, J. and Marme, S. (2000). Titania as a sorbent in normal-phase liquid chromatography, *J. Chromatogr. A*, **888**, 51–62.
120. Molnar, I., Rieger, H.J. and Monks, K.E. (2010). Aspects of the "Design Space" in high-pressure liquid chromatography method development, *J. Chromatogr. A*, **1217**, 3193–3200.

121. Monks, K., Molnar, I., Rieger, H.J. *et al.* (2012). Quality by design: Multidimensional exploration of the design space in high-performance liquid chromatography method development for better robustness before validation, *J. Chromatogr. A*, **1232**, 218–230.

122. Hewitt, E.F., Lukulay, P. and Galushko, S. (2006). Implementation of a rapid and automated high-performance liquid chromatography method development strategy for pharmaceutical drug candidates, *J Chromatogr. A*, **1107**, 79–87.

123. Teutenberg, T. (2009). Potential of high-temperature liquid chromatography for the improvement of separation efficiency — A review, *Anal. Chim. Acta*, **643**, 1–12.

124. Buckenmaier, S.M.C., McCalley, D.V. and Euerby, M.R. (2004). Rationalisation of unusual changes in efficiency and retention with temperature shown for bases in reversed-phase high-performance liquid chromatography at intermediate pH, *J. Chromatogr. A*, **1060**, 117–126.

125. Wiese, S., Teutenberg, T. and Schmidt, T. (2011). A general strategy for performing temperature-programming in high-performance liquid chromatography — Prediction of segmented temperature gradients, *J. Chromatogr. A*, **1218**, 6898–6906.

126. Wiese, S., Teutenberg, T. and Schmidt, T.C. (2012). A general strategy for performing temperature-programming in high-performance liquid chromatography — Further improvements in the accuracy of retention time predictions of segmented temperature gradients, *J. Chromatogr. A*, **1222**, 71–80.

127. Pappa-Louisi, A., Nikitas, P., Papachristos, K. *et al.* (2009). Multimode gradient elution in reversed-phase liquid chromatography: application to retention prediction and separation optimization of a set of amino acids in gradient runs involving simultaneous variations of mobile-phase composition, flow rate, and temperature, *Anal. Chem.*, **81**, 1217–1223.

128. Nikitas, P., Pappa-Louisi, A., Papachristos, K. *et al.* (2008). Theory and application of the two-mode gradient elution in liquid chromatography involving simultaneous changes in temperature and mobile-phase composition, *Anal. Chem.*, **80**, 5508–5514.

129. Krummen, M., Hilkert, A.W., Juchelka, D. *et al.* (2004). A new concept for isotope ratio monitoring liquid chromatography–mass spectrometry, *Rapid Commun. Mass Spectrom.*, **18**, 2260–2266.
130. Roloff, M., Erfurt, H., Kindel, G. *et al.*, inventors (2006). Process for the separation and sensory evaluation of flavours. Symrise GmbH & Co. KG, Holzminden, Germany, assignee.
131 Fischer, K., Fries, E., Koerner, W. *et al.* (2012). New developments in the trace analysis of organic water pollutants, *Appl. Microbiol. Biotechnol.*, **94**, 11–28.
132. Abramson, F.P., Black, G.E. and Lecchi, P. (2001). Application of high-performance liquid chromatography with isotope-ratio mass spectrometry for measuring low levels of enrichment of underivatized materials, *J. Chromatogr. A*, **913**, 269–273.
133. Cabañero, A.I., Recio, J.L. and Rupérez, M. (2006). Liquid chromatography coupled to isotope ratio mass spectrometry: A new perspective on honey adulteration detection, *J. Agric. Food Chem.*, **54**, 9719–9727.
134. Elflein, L. and Raezke, K.P. (2008). Improved detection of honey adulteration by measuring differences between C-13/C-12 stable carbon isotope ratios of protein and sugar compounds with a combination of elemental analyzer–isotope ratio mass spectrometry and liquid chromatography–isotope ratio mass spectrometry (delta C-13-EA/LC-IRMS), *Apidologie*, **39**, 574–587.
135. Zhang, L.J., Kujawinski, D.M., Federherr, E. *et al.* (2012). Caffeine in your drink: natural or synthetic?, *Anal. Chem.*, **84**, 2805–2810.
136. Young, E., Smith, R.M., Sharp, B.L. *et al.* (2012). Liquid chromatography–flame ionisation detection using a nebuliser/spray chamber interface. Part 1: Design and testing, *J. Chromatogr. A*, **1236**, 16–20.
137. Young, E., Smith, R.M., Sharp, B.L. *et al.* (2012). Liquid chromatography–flame ionisation detection using a nebuliser/spray chamber interface. Part 2: Comparison of functional group responses, *J. Chromatogr. A*, **1236**, 21–27.
138. de Boer, A.R., Alcaide-Hidalgo, J.M., Krabbe, J.G. *et al.* (2005). High-temperature liquid chromatography coupled on-line to a

continuous-flow biochemical screening assay with electrospray ionization mass spectrometric detection, *Anal. Chem.*, 77, 7894–7900.

139. Al-Khateeb, L.A. and Smith, R.M. (2009). High-temperature liquid chromatography of steroids on a bonded hybrid column, *Anal. Bioanal. Chem.*, **394**(5), 1255–1260.

Chapter 5

Hydrophilic Interaction Liquid Chromatography (HILIC) and Perfluorinated Stationary Phases

Cristina C. Jacob,[a] Héctor Gallart-Ayala[b] and Gonçalo Gamboa da Costa[c]

[a] BIOASTER Technological Research Institute, Metabolomics Core Facility, France

[b] LUNAM, Ecole Nationale Vétérinaire, Agroalimentaire et de l'Alimentation Nantes Atlantique (Oniris), USC 1329 INRA Laboratoire d'Etude des résidus et Contaminants dans les Aliments (LABERCA), France

[c] National Center for Toxicological Research, US Food and Drug Administration, Division of Biochemical Toxicology, USA

5.1. Introduction

Reverse-phase liquid chromatography, in particular the alkylsiloxane-bonded silica stationary phases, has been widely used in environmental and food analyses mainly due to its large applicability and ease of use. Despite the great versatility of these stationary phases, the separation of polar compounds is not always achieved and the variation of the stationary phase might be a useful option. In order to improve the chromatographic separation of polar compounds, HILIC and fluorinated reverse phases have been proposed for the analysis of certain polar analytes in environmental and food matrices.

This chapter will focus on the principles of HILIC and perfluorinated stationary phases. It includes a selection of representative work recently published. A discussion regarding the use of these stationary phases as chromatographic media, and their respective advantages and drawbacks, will be presented. Moreover, some applications of HILIC and perfluorinated stationary phases in food and environmental analysis will be discussed.

5.2. Hydrophilic Interaction Liquid Chromatography (HILIC)

High performance liquid chromatography is a common and well-suited separation technique, routinely used to solve analytical challenges due to its ability to fractionate complex mixtures of compounds possessing a wide range of polarities and acid-base properties.

Reverse-phase (RP) mode is the most frequently used in liquid chromatography. In RP mode, the stationary phase is non-polar, typically an alkyl-silica type bonded phase, whereas the mobile phase is a polar mixture of one or more organic solvents with water or buffer.[1] The retention increases with more lipophilic (hydrophobic) stationary phases and with decreasing concentration of organic solvents in the mobile phase. The non-polar solutes are therefore retained more strongly than polar ones. Given this retention mechanism, the methodology presents limited applicability to the analysis of small polar solutes.

Non-aqueous normal-phase (NP) liquid chromatography was predominantly used as the separation mode prior to the introduction of RP in the early 1970s, and it was mainly employed in thin-layer and low-pressure column liquid chromatographic techniques. In this case, the stationary phase is more polar than the mobile phase. Contrary to reverse-phase, the analyte retention increases as the polarity of the mobile phase decreases, and its mechanism relies on the competition between the solute and the mobile phase for localized polar adsorption centers on the adsorbent surface; an example of such adsorption centers is silanol groups on the surface of silica gel.[1] The use of normal-phase systems in modern laboratories is seen as environmentally unfriendly and expensive, in particular due to the need for disposal of potentially toxic eluents. Moreover, the use of alkanes and low-polarity solvents (e.g. chloroform and ethyl acetate) can result in poor analyte solubility with more hydrophilic compounds.

Aqueous normal-phase liquid chromatography had been used under different names prior to the introduction of the term 'hydrophilic interaction chromatography' (HILIC) by Andrew J. Alpert in

1990.[2] Alpert's work involved the use of a polar stationary phase, much like a normal-phase system, except the mobile phase was an aqueous-organic mixture, containing mainly higher proportions (>60%) of acetonitrile. Therefore, HILIC can be characterized as a chromatographic technique using a NP stationary phase in combination with a RP mobile phase.[1] Due to the more polar nature of the eluent, solubility issues of polar analytes associated with normal phase are minimized. Moreover, expensive ion pair reagents are not required in HILIC mode and, as discussed later in this chapter, HILIC may enhance analyte ionization when coupled to mass spectrometry, especially with the electrospray ionization (ESI) mode. As opposed to reverse-phase separation, in HILIC the gradient elution begins with a low-polarity organic solvent (e.g. acetonitrile) and polar analytes are eluted as the aqueous content increases.[3]

HILIC separation mode is becoming increasingly popular as a complementary method to reversed-phase chromatography for the analyses of polar and ionic compounds. This technique has in recent years been the subject of several reviews.[1,3–6]

Despite the great virtues of this technique, many analysts still feel uncomfortable when approaching it. Not only has the HILIC retention mechanism not been completely understood, but also a wide range of stationary phases is commercially available and it may be confusing when developing an analytical method. In order to enhance the range and the quality of possible applications of HILIC mode, the principles driving the separation and how they may be influenced by selection of both the stationary phase and chromatographic conditions should be understood.

5.2.1. *HILIC stationary phases*

The correct selection of a suitable adsorbent is a crucial step for the success of so-called 'solvent-generated' liquid–liquid chromatography, the category to which the basic HILIC mechanism belongs, as noted by Huber *et al.*[1,7] The family of HILIC stationary phases has since continually grown to suit specific separation problems. A common and useful classification of the HILIC stationary phases is based upon the presence of functional groups on the surface, and their charged state: they can be divided into unmodified bare silica gel and chemically bonded stationary phases, and these ones are classified as neutral, charged, or zwitterionic phases. Some HILIC stationary phases and respective properties are described in Table 5.1.

Over the last 15 years, HILIC columns have progressed through the implementation of second and third generations that involve mixed or multi-interaction solid phases. Nowadays traditional HILIC and its more sophisticated relatives are commercially available from many column manufacturers, resulting in a wider structural variation on HILIC stationary phases than on those found in RP systems. Perhaps due to this diversity, there is no single HILIC type that is considered as versatile a stationary phase as is C18 in RP mode.

The first HILIC applications were developed with unmodified bare silica gels (Type A and B) and used for carbohydrate separations.[8] These stationary phases are still among the most popular materials, in particular for pesticide analyses in food and environmental matrices.[9–11] Silica gels present silanol groups that are deprotonated at mobile-phase pH above 4–5 and can also work as cation-exchangers, leading to strong retention of charged basic solutes.

Bonded HILIC stationary phases are obtained by derivatization of polar functional groups with a support surface of silica or polymer. Due to the wide variety of polar functional groups that may be incorporated in HILIC stationary phases, they are conveniently classified as neutral, charged, or zwitterionic phases based on the charged state of the groups.

The polar functional groups in neutral stationary phases are in their neutral form in the pH range of 3–8 (a common mobile phase range used in HILIC), and therefore the retention is mainly due to polar interactions. Many HILIC stationary phases belonging to this class are commercially available and it includes a wide variety of functional groups, such as aspartamide, amide, diol, cross-linked diol, cyano, and cyclodextrin groups. Cyclodextrin phases are also used for HILIC chiral separations.[12]

Table 5.1. Polar stationary phases commonly used in HILIC mode.

Phase type	Phase name	Functional group structure	Column examples
Unmodified bare silica gels	Underivatized silica stationary phases that contain functional groups such as siloxanes and silanols	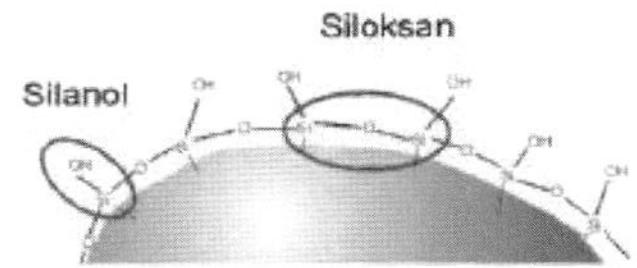	Atlantis Silica HILIC BEH HILIC
Neutral	Amide	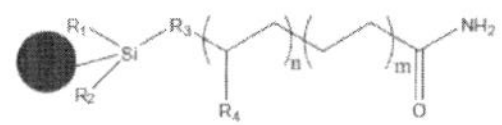	TSKgel Amide 80
		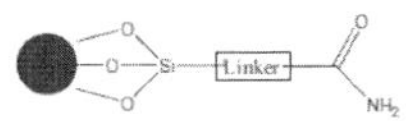	BEH Amide
	Aspartamide	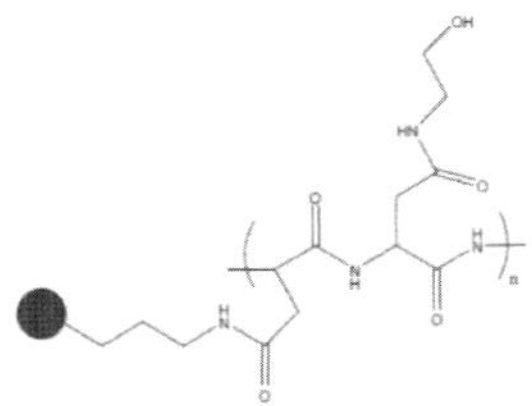	Polyhydroxyethyl A
	Cyano		YMC-Pack Cyano
	Diol	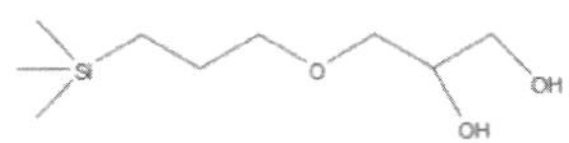	YMC-Pack Diol, Inertsil HILIC
	Mix-mode diol		Acclaim mix-mode HILIC 1

(*Continued*)

Table 5.1. (*Continued*)

Phase type	Phase name	Functional group structure	Column examples
	Cross-linked diol	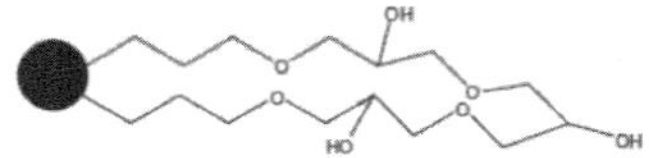	Luna HILIC
	Polyvinyl alcohol (PVA)	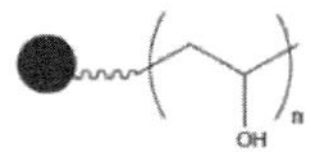	YMC PVA-Sil
Charged	Amino		YMC Pack amino
		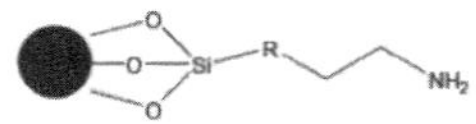	TSKgel NH2-100
	Polyamine	N/A	Luna NH2, YMC-Pack Polyamine II
	Imidazole	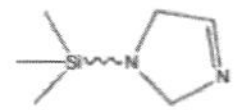	Sepax Polar-imidazole
	Triazole	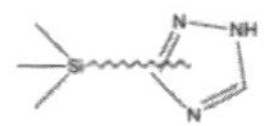	Cosmosil HILIC
	Poly(2-sulfoethyl)	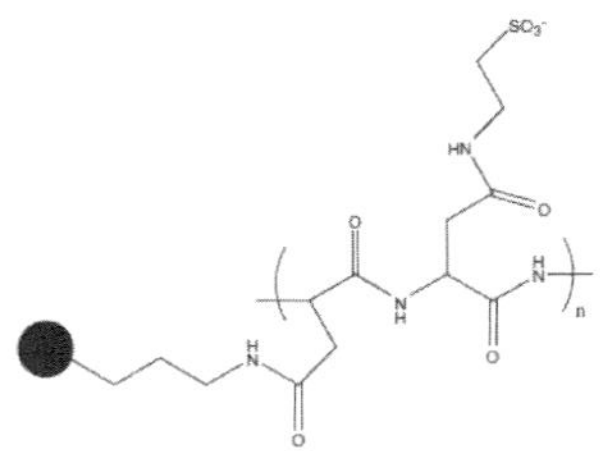	Polysulfoethyl A

(*Continued*)

Table 5.1. (*Continued*)

Phase type	Phase name	Functional group structure	Column examples
	Poly(aspartic Acid)	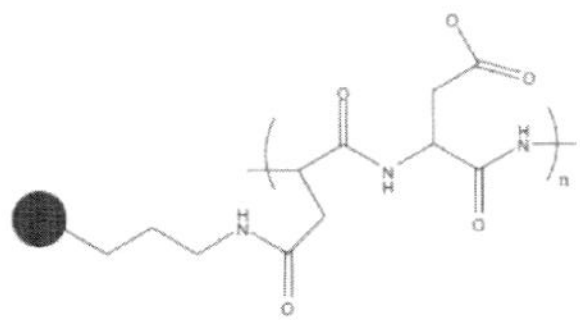	Poly CAT A
Zwitterionic	Sulfobetaine	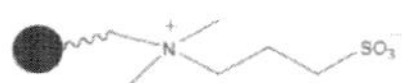	ZIC-HILIC
	Phosphocholine		Shiseido PC HILIC
	Obelisc N		Sielc Obelisc N

Adapted with permission from Guo, Y. and Gaiki, S. (2011). Retention and selectivity of stationary phases for hydrophilic interaction parameters in HILIC separations, *J. Chromatogr. A*, **1218**, 5920–5938. Copyright (2012) Elsevier.

The polar functional groups of charged stationary phases can be positively or negatively charged depending on the pH of the mobile phase. Therefore, charged analytes are separated based on ion exchange mechanisms in combination with hydrophilic partitioning. Amino phases are among the principal HILIC charged phases.[13] The functional group of these phases is usually an aminopropyl ligand with a primary amino group that is positively charged at common HILIC eluent pH values and shows high affinity for acid compounds largely based on an anion-exchange mechanism. In some instances this high affinity can lead to irreversible adsorption of the anion on the stationary phase. Due the high hydrophilicity of these phases, they can also be successfully used for the separation of neutral polar compounds.

Zwitterionic HILIC stationary phases are very versatile and to a certain degree they can be considered the HILIC all-purpose phases. In this case, zwitterionic ligands comprise groups with permanent positive and negative charges. These phases can be employed for the separation of neutral, acidic, and basic analytes, and also for the separation of inorganic ions, mainly due to their particular hydrophilicity and their modest ion-exchange properties.

5.2.2. *Retention mechanism*

The HILIC retention mechanism is likely to be complex and a comprehensive theoretical explanation for this phenomenon has not yet been presented. Although liquid–liquid partition mechanisms were originally proposed to explain HILIC retention, as pointed out by Alpert,[2] these mechanisms are thought to occur together with adsorption, ion exchange, hydrogen bonding, dipole–dipole, and even hydrophobic interactions, depending on the experimental conditions. The partition mechanism is based on the differential distribution of the analyte molecules between the organic-rich mobile phase and a water-enriched layer adsorbed onto the hydrophilic stationary phase. As a consequence, polar hydrophilic analytes are preferentially distributed into the water layer, and thus are strongly retained.[13]

A minimum of 2–3% water in the mobile phase is required for the creation of the water layer.[13] When the concentration of acetonitrile increases, water interacts more strongly with the surface of the polar stationary phase and, therefore, the retention of polar analytes is sustained. Increasing the water content has the effect of decreasing the difference in polarity between the bulk and the adsorbed layer, which results in the solubilization of the analyte in the mobile bulk phase, with ensuing elution.[13] However, the HILIC partitioning theory is based only on circumstantial evidence; indeed, the HILIC retention mechanism is much more complex, and beyond partitioning it also includes hydrogen donor interactions between neutral polar species, as well as electrostatic mechanisms with any of the charged HILIC stationary phases. The relative contributions of partitioning and surface adsorption mechanism are highly connected with the nature of the stationary phase (hydration and charge), the properties of the solutes (the kind and number of polar functional groups), and the mobile phase composition (pH, buffer concentration, column temperature, organic solvent).[3,5,14] Therefore, HILIC retention is more likely to be a multimodal mechanism.

5.2.3. *Practical aspects*

5.2.3.1. *Stationary phase selection*

A meticulous selection of the stationary phase is a crucial step in the development of a successful HILIC-based chromatographical separation. As outlined earlier in this chapter, in the last decades the range of HILIC stationary phases has expanded significantly, from the initial simple bare silica particles to a broad selection of new stationary phases. Commercially available phases now encompass: silica or polymer-based particles whose surfaces are modified with a number of functional groups (e.g. diol, amine, amide); combinations of multiple functional groups (e.g. ammonium and sulfonic groups in zwitterionic phases); or specialized substituents (e.g. chiral substituents). Furthermore, advances in material science have enabled the commercial availability of superficially porous and smaller sized particles, with sub-2 μm diameters now being common.

The combination of these innovations provides a remarkable range of selectivities, which is clear from the commercial offerings in today's

HILIC column market. These offerings have vastly expanded the range of analytes and matrices amenable to HILIC chromatography and have certainly contributed to the significantly increased use of HILIC in the peer-reviewed literature observed in the last decade (Fig. 5.1). Table 5.2 highlights recent examples of the application of HILIC chromatography in food and environmental analyses, demonstrating the highly varied nature of analytes and stationary phases used.

Although the current wealth of commercially available HILIC stationary phases opens new analytical avenues, the complexity of the separation mechanisms involved can make the selection of the optimal stationary phase for a particular analysis a challenging task. The chemical modification of the surface of the silica or polymeric particles adds potential new molecular interactions with the analytes, ranging from hydrogen bonding to ion pairing, anionic and cationic exchange, and chiral interactions, often resulting in mixed-mode retention mechanisms.[3] In a study addressing the selectivity, retention mechanism, and performance of several HILIC stationary phases for the separation of neutral, strongly acidic, and strongly basic analytes, McCalley concluded that the very different chromatographic characteristics evidenced by each column could not be explained by standard HILIC partition or adsorption mechanisms alone, and that ion exchange can play a significant role in the retention of ionized analytes.[15] While as a general rule it can be said that basic analytes tend to be better-retained in bare silica columns due

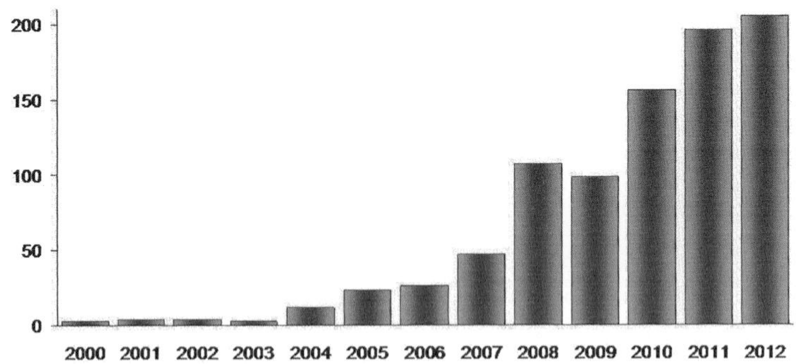

Figure 5.1. PubMed search results documenting the continuously growing research area of HILIC, based on the number of publications per year.

Table 5.2. Recent examples of application of HILIC chromatography in food and environmental analysis.

Analyte	Matrix	Column	Mobile phase/flow rate	Detection	Reference
Choline related compounds; phospholipids	Varied foods	Ascentis Express HILIC, 150 mm × 2.1 mm (2.7 μm), Sigma, St. Louis, MO, USA	A: acetonitrile; B: 10 mM ammonium formate in water pH 3.0; gradient, 200 μL/min	ESI–MS/MS	35
Artificial sweeteners	Wastewater; river water	Luna HILIC, 100 mm × 2.0 mm (3 μm, 200 Å), Phenomenex Inc., Torrance, CA, USA	A: acetonitrile/H_2O, 5 mM ammonium formate, pH 3.5; B: 5 mM ammonium formate in water; gradient, 200 μL/min	ESI–MS/MS	36
Aflatoxin M_1	Milk	Acquity BEH HILIC, 100 mm × 2.1 mm (1.7 μm), Waters Corp., Milford, MA, USA	A: acetonitrile; B: 25 mM acetic acid in water; gradient, 500 μL/min	ESI–MS/MS	26
Biotoxins	Shelfish	BEH Amide, 100 mm × 2.1 mm (1.7 μm), Waters Corp., Milford, MA, USA	A: acetonitrile, 0.1% formic acid; B: 2 mM ammonium formate in water, pH 3.5; gradient, 500 μL/min	ESI–HRMS	25
Aminoacids; peptides	Stewed beef juice	TSKgel Amide-80, 300 mm × 21.5 mm (10 μm), Tosoh Bioscience, Stuttgart, Germany	A: acetonitrile, 0.1% trifluoroacetic acid; B: Water, 0.1% trifluoroacetic acid; gradient, 6 mL/min	UV	37

(Continued)

Table 5.2. (*Continued*)

Analyte	Matrix	Column	Mobile phase/flow rate	Detection	Reference
Organic fungicides	Potatoes; cucumber	ZIC-pHILIC, 150 mm × 2.1 mm (5 μm), Merck, Darmstadt, Germany	5 mM ammonium acetate in 95:5 acetonitrile:water; isocratic, 200 μL/min	ESI–MS/MS	19
Pesticides	Varied foods	YMC-Pack Diol, 100 mm × 2.1 mm (5 μm), YMC Europe GmbH, Dinslaken, Germany	A: 5 mM ammonium formate and 0.1% acetic acid in 90:10 acetonitrile:water; B: 5 mM ammonium formate and 0.1% acetic acid in water; gradient, 200 μL/min	ESI–MS/MS	18
Gadolinium contrast agents	River water; plants	ZIC-HILIC, 250 mm × 2.1 mm (5 μm), Merck, Darmstadt, Germany	20 mM ammonium acetate in 65:35 acetonitrile:water (pH 7.3); isocratic, 150 μL/min	ICP–MS	38
Neurotoxin	River water; cyanobacteria films	ZIC-HILIC, 150 mm × 2.1 mm (3.5 μm), Merck Sequant AB, Umeå, Sweden	A: 0.1% formic acid in acetonitrile; B: 0.1% formic acid in water; gradient, 200 μL/min	ESI–MS/MS	27

(*Continued*)

Table 5.2. *(Continued)*

Analyte	Matrix	Column	Mobile phase/flow rate	Detection	Reference
Plant growth regulator	Meat	Xbridge HILIC, 150 mm × 2.1 mm (3.5 µm), Waters Corp., Wexford, Ireland	0.1% formic acid and 10 mM ammonium acetate in 40:60 water:acetonitrile; iscocratic, 100 µL/min	ESI–MS/MS	28
Carboxylic acids	Atmospheric aerosols	XBridge Amide, 100 mm × 3.0 mm (3.5 µm), Waters Corp., Milford, MA, USA	Ternary mobile phase consisting of acetonitrile, water and 100 mM ammonium acetate aqueous buffer at pH 5.0; gradient, 500 µL/min	ESI–MS/MS	29
Aromatic amines	Environmental water	Kromasil 100-5SIL 250 mm × 4.6 mm (5 µm), Eka Chemicals AB, Bohus, Sweden	85:15 [NaH_2PO_4–H_3PO_4 buffer (pH 1.5, containing 10 mM NaH_2PO_4)]:acetonitrile; isocratic, 1 mL/min	UV	30
Free estrogens and conjugates	River water	ZIC-pHILIC, 150 mm × 2.1 mm (5 µm), SeQuant AB, Umeå, Sweden	A: 95:5 acetonitrile: [aqueous ammonium acetate (5 mM, pH 6.80)]; B: 75:25 MeCN:[aqueous ammonium acetate (5 mM, pH 6.80)]; gradient, 150 µL/min	ESI–MS/MS	31

(Continued)

Table 5.2. (*Continued*)

Analyte	Matrix	Column	Mobile phase/flow rate	Detection	Reference
Biogenic amines	Tuna	Acquity BEH HILIC, 150 mm × 2.1 mm (1.7 μm), Waters Corp., Milford, MA, USA	A: ammonium formate buffer (50 mM, pH 3); B: acetonitrile; gradient, 750 μL/min	APCI–HRMS	39
Streptomycin	Apples	Atlantis Hilic, 150 mm × 2.1 mm (3 μm), Waters Corp., Milford, MA, USA	A: 0.05% formic acid in water; B: 0.05% formic acid in acetonitrile; gradient, 300 μL/min	ESI–MS/MS	32
Glycoside sweeteners	Soft drinks	Luna HILIC, 250 mm × 4.6 mm, Phenomenex, Aschaffenburg, Germany	80:20 acetonitrile:water; isocratic, 1.0 mL/min	UV	40
Gadolinium contrast agents	Wastewater effluent	Accucore HILIC column, 150 mm × 3 mm (2.1 μm), Thermo Fisher Scientific, Waltham, MA, USA	30:70 [50 mM ammonium formate in water pH 3.75]: acetonitrile; isocratic, 250 μL/min	ICP–MS	41
Oligosaccharides	Cow milk	Kinetex HILIC, 50 mm × 2.1 mm (1.7 μm), Phenomenex, Torrance, CA, USA	A: 50 mM ammonium acetate (pH 4.5); B: acetonitrile; gradient, 400 μL/min	ESI–MS/MS	24

(*Continued*)

Table 5.2. (*Continued*)

Analyte	Matrix	Column	Mobile phase/flow rate	Detection	Reference
Veterinary drugs	Chicken muscle	ZIC-HILIC, 150 mm × 2.1 mm (3.5 μm), Merck Sequant AB, Umeå, Sweden	A: 50 mM ammonium formate in water (pH 2.5); B: acetonitrile; gradient, 200 μL/min	ESI–MS/MS	20
Cyanotoxins	Cyanobacteria in aquatic environments	TSKgel Amide-80 HILIC, 150 mm × 2 mm, (3 μm), TOSOH Bioscience, San Francisco, CA, USA	A: 5 mM ammonium formate buffer (pH 3.5); B: 5 mM ammonium formate (pH 3.5) in 95:5 acetonitrile:water; gradient, 500 μL/min	ESI–MS/MS	33

to their higher stationary phase water and silanol contents, and that acidic compounds tend to be better retained in zwitterionic, amide, or diol phases, presumably due to the stationary phase lower unscreened silanol content and low acidity silica,[15] the work of other authors further corroborates the importance of specific interactions between the analytes and functional groups in the stationary phases.[16,17] Figure 5.2 illustrates the dramatic differences in resolution, retention time, and order of elution observed by McCalley for a range of

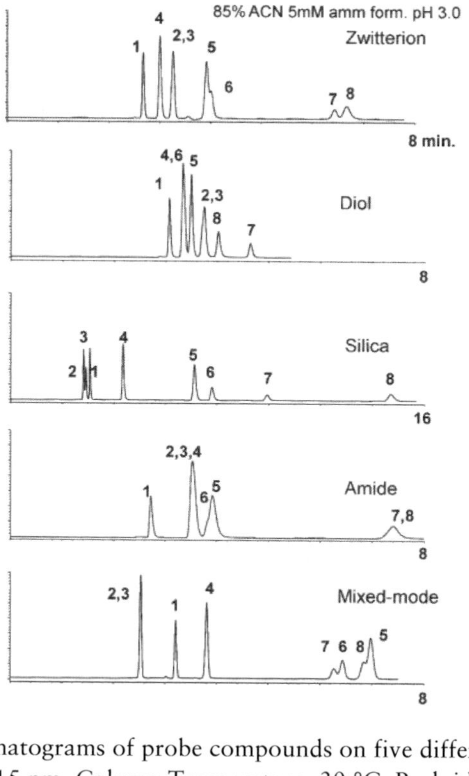

Figure 5.2. Chromatograms of probe compounds on five different HILIC phases. Detection: UV at 215 nm. Column Temperature: 30 °C. Peak identities: 1: phenol, 2: 2-naphthalenesulfonic acid; 3: p-xylenesulfonic acid; 4: caffeine; 5: nortriptyline; 6: diphenhydramine; 7: benzylamine; 8: procainamide. Flow rate 1 mL/min. Mobile phase: ACN-water (85:15 v/v) containing 5 mM ammonium formate pH 3.0. Adapted with permission from McCalley, D.V. (2010). Study of the selectivity, retention mechanisms and performance of alternative silica-based stationary phases for separation of ionized solutes in hydrophilic interaction chromatography, *J. Chromatogr. A*, **1217**, 3408–3417. Copyright (2010) Elsevier.

compounds eluted in different stationary phases with the same buffer and flow rate, highlighting the complexities involved in predicting the retention of a particular analyte in a particular stationary phase.

It can thus be concluded that in the absence of previous analytical reports in the literature for a particular analyte, the initial selection of the optimal HILIC stationary phase may necessarily involve an empirical approach by conducting a screening of several distinct stationary phases. This approach often seems unavoidable, as in the case of food and environmental analyses, where interfering compounds in complex matrices can substantially further complicate the choice of stationary phase.[18–20]

5.2.3.2. *Mobile phase selection*

While the origins of HILIC are thought to date back to work conducted by Martin and Synge who used water-saturated chloroform as the eluent to separate mixtures of amino acids on a purified silica column,[21] it later became clear that a water-immiscible solvent was not required to attain chromatographic separations in a silica stationary phase. In HILIC, and in contrast with that observed in reverse-phase chromatography, water is considered the 'strong' eluent, and organic solvents such as acetonitrile or acetone are considered 'weak' eluents. Alcohols such as methanol, ethanol, and isopropanol can also be used in HILIC, presenting intermediate eluotropic strengths, as follows:

Water > methanol > ethanol > isopropanol > acetonitrile > acetone.

Irrespective of the particular "weak" organic eluent used, the basic principle of the HILIC separation remains the same: the polar solutes partition between the water bound to the stationary phase and the mixture of water and organic solvent used in the mobile phase. The higher the percentage of water in the mobile phase, the larger the fraction of the polar solute in it, and thus the lower the retention in the stationary phase.

Of all 'weak' organic eluents used in HILIC, acetonitrile is undoubtedly the one that is most widely used, due to its excellent optical transparency in the ultraviolet range, good desolvation ability in mass spectrometer electrospray sources, low viscosity, and ability to afford superior peak shape with many analytes. Acetonitrile also presents the advantage of being an aprotic solvent, thus not competing with analytes for hydrogen bonding to polar sites in the HILIC particle surface, thereby enabling their retention in the stationary phase.[11] Given these notorious advantages over other solvents used in HILIC applications, it is not entirely surprising that all the recent examples of the application of HILIC in food and environmental sample analysis highlighted in Table 5.2 rely on the use of acetonitrile as an eluent.

Equally as important as the choice of solvents for HILIC is the choice of buffer salts in the mobile phase. While most HILIC stationary phases can be used with a simple binary system of water and organic solvent, better method robustness and peak shape are often achievable by the use of buffers in the eluent. The addition of buffers can enable significant changes in the selectivity of the columns by modulating the pH, and consequently the protonation of polar groups in the analytes and stationary phase. While on the analyte side small changes of pH above or below its pKa may determine its ionization, and thus its overall polarity/hydrophobicity, on the stationary phase side the selectivity of bare silica columns is strongly influenced by their ability to establish cationic interactions with the silica surface silanol groups. The choice of a mobile phase with a pH above 4–5 leads to increased silanol ionization, enhancing their interaction with charged basic analytes that become more strongly retained.[22] Conversely, negatively charged metabolites undergo electrostatic repulsion at the stationary phase level and their retention is reduced.

An important practical aspect to consider when selecting the pH of the mobile phase in HILIC is that bare silica columns cannot withstand high pH values due to the slow solubilization of silica, which ultimately leads to stationary phase loss. This problem is minimized in polymer-based stationary phases. For example the manufacturer-recommended pH range for Merck's zwitterionic silica-based

ZIC-HILIC column is pH 3–8 whereas its polymer-based counterpart ZIP-pHILIC has optimal pH range between 2–10 (Merck KGaA, Darmstadt, Germany).

Although the most dramatic effects of a buffer introduction in the HILIC mobile phase are directly related to its pH, the absolute concentration of the buffer salts can also play a role in analyte retention by further modulating the relative contributions of the adsorption and partition mechanisms in place. This is achieved by reducing or suppressing electrostatic interactions between the analyte and stationary phase through a competitive process with the buffer ions, and by the tendency of salts for partitioning to the water in the stationary phase surface, increasing its volume.[14]

One final aspect that needs to be considered when selecting a buffer for a HILIC application is the technology used to detect the analytes in the eluate. The majority of today's applications rely on the use of either ultraviolet-absorption-based detectors, or mass spectrometry. Thus, it is not unexpected that in the examples of recent food and environmental HILIC analyses highlighted in Table 5.2, most rely on the use of either ammonium acetate or ammonium formate buffers; both offer good transparency in the medium and long UV range, and also good volatility for electrospray ionization in mass spectral analyses.

5.2.3.3. *Sample preparation*

In contrast with reverse-phase chromatography, the HILIC retention mechanism relies on the use of relatively low concentrations of water in the mobile phase. It is not uncommon that for difficult-to-retain analytes the initial chromatographic conditions may require organic concentrations as high as 95%.[23] This requirement raises potential problems in the preparation of samples for HILIC. Band spreading and consequent loss of chromatographic resolution may occur as concentrations of water in the samples exceed those in the eluate. An example of such an effect in the food contaminant cyanuric acid is shown in Fig. 5.3.

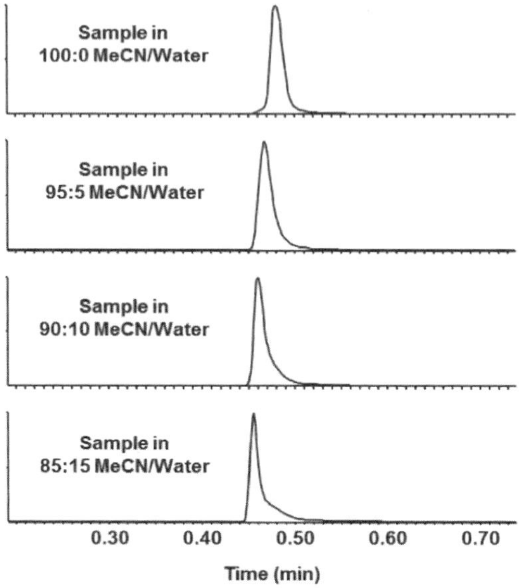

Figure 5.3. Comparison of the peak shape obtained in the HILIC analysis of a 200 ng/mL solution of cyanuric acid prepared with varying proportions of acetonitrile/water. In all cases 10 µL of sample were injected in a Waters Acquity I-class UPLC (Milford, MA, USA) coupled with a Waters Xevo TQ-S tandem mass spectrometer operated in negative electrospray ionization multiple reaction monitoring mode (UPLC–ESI–MS/MS). The samples were eluted in a 100 × 2.1 mm (1.7 µm particle size) Waters BEH HILIC Acquity column at 600 µL/min isocratically with 95:5 acetonitrile:water. The y-axis signal intensity scale is the same in all chromatograms. A notorious degradation of chromatographic resolution with peak tailing is observed with the increasing percentage of water in the sample.

Due to the often-complex matrices, the preparation of food and environmental samples routinely involves multiple extraction and purification steps (e.g. liquid–liquid partitions and solid-phase extraction procedures) using a wide range of solvents. In order to enable appropriate compatibility of the final sample with HILIC, the extraction procedures are typically designed to incorporate a high percentage of acetonitrile in the final extract[18–20,24,25] or involve a drying step and reconstitution in a percentage of acetonitrile that approaches or exceeds that of the initial chromatographic conditions.[26–33]

5.2.3.4. *HILIC–mass spectrometry*

Given its versatility, specificity, and sensitivity, in the last decade liquid chromatography coupled with mass spectrometry has become one of the analytical methods of choice in the area of food and environmental contaminants. As evidenced by recent representative applications in these areas (Table 5.2), the ionization methodology most commonly used to couple the liquid chromatographer to the mass spectrometer is by far electrospray ionization (ESI). In an ESI interface the eluate is flown through a capillary and subjected to high voltage, resulting in the production of a fine plume of charged droplets containing eluents and solutes. In order for the ionized solutes to be transferred to the mass spectrometer and undergo mass selective analysis, they need to be efficiently ionized and desolvated from the solvents used in the chromatography. Although most modern ESI sources feature the use of compressed nitrogen gas at high temperatures to assist the desolvation process, it still remains an instrumental challenge, in particular when high chromatographic flow rates are used, as is often the case in ultrahigh-pressure/high-performance liquid chromatography. Here, HILIC can offer substantial advantages, because the high percentages of organic solvent (most commonly acetonitrile) in the eluate enable a more efficient desolvation and ionization of the solutes in the eluate, in comparison with the highly aqueous eluate normally generated in reverse-phase chromatography. The ionization and desolvation process in the HILIC eluates, more efficient in relation to reverse-phase eluates, can result in sensitivity enhancements that are highly compound-dependent, with reductions in the limit of detection in the range of 10–1,700-fold being reported among the food and environmental analyses outlined in Table 5.2.[26,31,33] In a recent systematic comparison of electrospray responses of basic compounds in HILIC versus reversed-phase chromatography, Periat *et al.*[34] observed a four-fold median gain in sensitivity in HILIC, with some compounds undergoing signal enhancements as high as 100- to over 8,000-fold. The possibility of attaining such levels of signal improvement make HILIC an option well worth consideration, in particular in the area of food and environmental analysis, where the analyst is often required to attain very low levels of detection in order to comply with regulatory guidelines.

One practical aspect that needs to be considered in HILIC chromatography–electrospray mass spectrometry is the influence of the eluent buffer concentration on the signal intensity. As observed for the potential electrospray signal enhancement in HILIC, the influence of the buffer concentration on the signal intensity is also highly compound-dependent. Reports show that some compounds undergo signal enhancement[20,28] while others undergo signal suppression[19] in response to increasing concentrations of buffer in the eluent. Figure

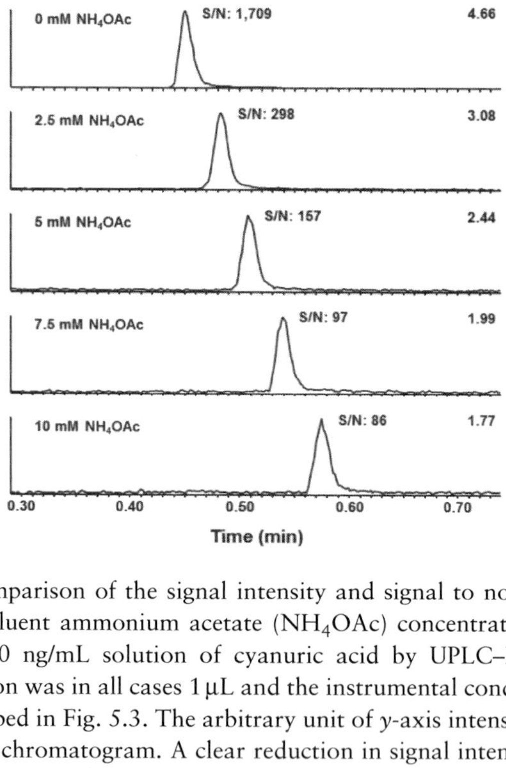

Figure 5.4. Comparison of the signal intensity and signal to noise ratio (S/N) as function of the eluent ammonium acetate (NH_4OAc) concentration in the HILIC analysis of a 200 ng/mL solution of cyanuric acid by UPLC–ESI–MS/MS. The volume of injection was in all cases 1 µL and the instrumental conditions were those previsouly described in Fig. 5.3. The arbitrary unit of *y*-axis intensity is noted in the top right of each chromatogram. A clear reduction in signal intensity (~2.6-fold) is observed with the ammonium acetate concentration increase from 0 mM to 10 mM. This signal intensity reduction is accompanied by a much more pronounced reduction (~20-fold) in the signal to noise ratio of the cyanuric acid peak, indicating that the buffer-related signal suppression was more pronounced in this analyte than in the background ions.

5.4 illustrates the signal intensity variation in the HILIC–electrospray tandem mass spectrometry analysis of cyanuric acid, where concentrations of ammonium acetate in the eluent range from 0 mM to 10 mM. While the cyanuric acid peak shape does not seem to be affected, and only minor changes are observed in its retention time, an overall signal intensity reduction of ~2.6-fold is observed. More importantly, a decrease of signal to noise ratio of almost 20-fold is registered when increasing the buffer concentration to 10 mM, suggesting that the signal suppression was more pronounced for cyanuric acid than for the background ions.

As an overall conclusion, while it is clear that the use of HILIC can be extremely advantageous in the analysis of certain compounds by mass spectrometry, it is also clear that a careful selection of the chromatographic conditions is required to maximize the potential of this technique.

5.3. Fluorinated Stationary Phases

In addition to HILIC, fluorinated bonded-silica stationary phases constitute another alternative for the retention and separation of polar compounds, as reviewed by Núñez *et al.*[6]

Organofluorines present a number of unique properties, such as high electronegativity, low polarizability, strong lipo- and hydrophobicity, and good thermal and chemical stability.[42] These properties led to the exploitation of organofluorines in liquid chromatography for the development of perfluoroalkyl- and perfluorophenyl-functionalized silica gels, and some polymer-supported fluorinated beads as stationary phases.[42] Some major differences are observed when compared with the traditional C8, C18, and phenyl reverse phases. Weaker retention for common hydrocarbon molecules; stronger retention for fluorinated compounds; resolution of fluorinated compound mixtures according to their fluorine content; increased retention for many polar and basic compounds with high percentages of organic solvent in the mobile phase; higher reproducibility; and longer lifetimes are some of the differences highlighted between the fluorinated and the non-fluorinated phases.[42] These observations emphasize the fact that the fluorinated phases are not simply chemically modified reverse phases, but rather a variety of different phases with great potential for new applications.

Perfluoroalkyl stationary phases have been used for the separation of halogenated compounds, and have shown good performance in terms of selectivity and separation of positional isomers and non-planar molecules.[6] However, this type of stationary phases is rarely used in food and environmental analyses. On the other hand, perfluorophenyl (PFP) stationary phases have been used for the analysis and separation of polar compounds in food and environmental matrices.[6] Perfluorophenyl stationary phases are commercially available from a number of vendors, as shown in Table 5.3.

As previously indicated for the HILIC stationary phases, the retention mechanism on pentafluorophenyl stationary phases is also not entirely elucidated. In addition to dispersive interactions available on traditional alkyl phases, the PFP stationary phases allow for dipole-dipole, π–π, charge transfer and ion-exchange interactions. Solutes with π-electrons will display a different retention behavior in these columns than on ordinary RP columns.[43] The π–π interactions between aromatic moieties of solutes and pentafluorophenyl ligands on the stationary phase are significant for retention of solutes on the PFP column, and are partially blocking the interactions between basic analytes and free silanol groups.[43]

Moreover, pentafluorophenyl propyl (PFPP) phases exhibit both reverse-phase and hydrophilic retention mechanisms for polar analytes, depending on the composition of the mobile phase. Table 5.4

Table 5.3. Some commercially available perfluorophenyl stationary phases.

Name	Stationary phase	Provider
Discovery F5 HS	Pentafluorophenyl propyl	Supelco
Hypersil Gold PFP Accucore PFP	Pentafluorophenyl	Thermo Fisher
Kinetex PFP	Pentafluorophenyl with TMS endcapping	Phenomenex
DB PFP Propyl	Pentafluorophenyl propyl	Restech
CSH Fluoro-phenyl	Propyl fluoro phenyl	Waters

Table 5.4. Examples of recent applications of PFPs stationary phases in food and environmental analyses.

	Compound	Matrix	Column	Conditions	Detection	Reference
Foodstuff and beverage analysis	Perfluorinated compounds	Diverse foodstuff	Fluorosep RP Octyl, 150 mm × 2.1 mm (5 µm), ES Industries, West Berlin, NJ, USA	A: 6.3 mM ammonium formate pH = 4 B: Methanol Flow rate: 0.3 mL/min Injection volume: 20 µL	MS/MS SRM ESI (−) ionization	50
	Seven medicinal ingredients	Health promoting food	Discovery HS F5, 50 mm × 2.1 mm, (3 µm), Supelco, Bellefonte, PA, USA	Formic acid ammonium formate in water: acetonitrile (60:40, v/v) Flow rate: 0.2 mL/min Injection volume: 2 µL	MS/MS SRM ESI (+,−) ionization	51
	Two UV ink photoinitiators isomers	Packaged foods	Discovery HS F5, 50 mm × 2.1 mm (3 µm), Supelco, Bellefonte, PA, USA	A: acetonitrile B: 25 mM formic acid–ammonium formate pH = 3.7 Flow rate: 0.3 mL/min Injection volume: 10 µL	MS/MS SRM ESI (+)	47
	Eleven UV ink photoinitiators	Packaged foods	Discovery HS F5, 50 mm × 2.1 mm (3 µm), Supelco, Bellefonte, PA, USA	A: acetonitrile B: 25 mM formic acid–ammonium formate pH = 2.7 Flow rate: 0.45 mL/min Injection volume: 10 µL	MS/MS SRM ESI (+)	48

(*Continued*)

Table 5.4. *(Continued)*

Compound	Matrix	Column	Conditions	Detection	Reference
Nine basic pharmaceuticals	Surface water	Varian PFP, 100 mm × 4.6 mm (5 μm), Varian, Sint-Katelijne-Waver, Belgium	A: 2 mM ammonium acetate and 2mM acetic acid in acetonitrile: water (5:95, v/v) B: 2 mM ammonium acetate and 2 mM acetic acid in acetonitrile: water (95:5, v/v) Flow rate: 1.0 mL/min Injection volume: 20 μL	MS/MS SRM ESI (+) ionization	52
Seven β-agonists	Animal feed and drinking water	Luna PFP, 100 mm × 2.0 mm, (3 μm), Phenomenex, Torrance, CA, USA	A: 10 mM acetic acid B: acetonitrile Flow rate: 0.3 mL/min Injection volume: 25 μL	MS/MS SRM ESI (+) ionization	46
Perfluoroalkyl carboxylate, sulfonate and sulfonamide isomers	Landfill leachate	Ascentis Express F5 PFP, 100 mm × 2.1 mm, (2.7 μm), Sigma-Aldrich, Oakville, ON, Canada	A: 20 mM ammonium formate/formic acid buffer B: Methanol Flow rate: 0.25 ml/min T: 30°C	MS/MS SRM ESI (−) ionization	49

summarizes examples of recent applications of food and environmental analyses, including the chromatographic and detection methods used.

The mobile phase typically used with PFP stationary phases is a methanol–water mixture, similar to that in reverse-phase mode, but other organic solvents such as acetonitrile or tetrahydrofuran can also be used for gradient elution.[42] At low content of organic modifier, analyte retention resembles that of classical reverse-phase mode. However, at high content of organic solvent the analyte behavior is more comparable to that of normal-phase mode. As illustrated in Fig. 5.5, the combination of reversed- and normal-phase behavior forms a 'U-shape' relationship between the retention time and the percentage of the organic solvent (methanol or acetonitrile).[44]

Although this retention mechanism is still under investigation, one possible explanation proposed by Zhang[42] is that high percentages of organic solvent in the mobile phase lead to wetting of the fluorinated phase. Therefore, more free silanols are accessible for interacting with polar solutes, showing a hydrophobic ion-exchange property. Some compounds may have a classical reversed-phase retention behavior at lower percentages of organic solvent while exhibiting hydrophilic interaction with high organic content. This behavior is caused by simultaneous hydrophobic or polar

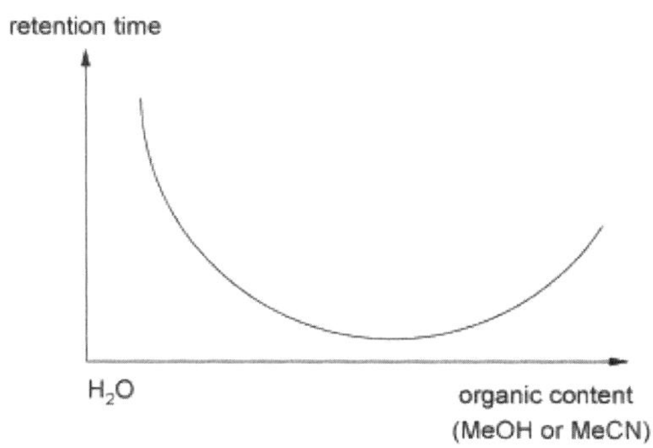

Figure 5.5. 'U-shape' retention of a basic compound the fluorinated stationary phase. Reproduced with permission from Bell, D.S. and Jones, A.D. (2005). Solute attributes and molecular interactions contributing to 'U-shape' retention on a fluorinated high-performance liquid chromatography stationary phase, *J. Chromatography A*, **1073**, 99–109. Copyright (2005) Elsevier.

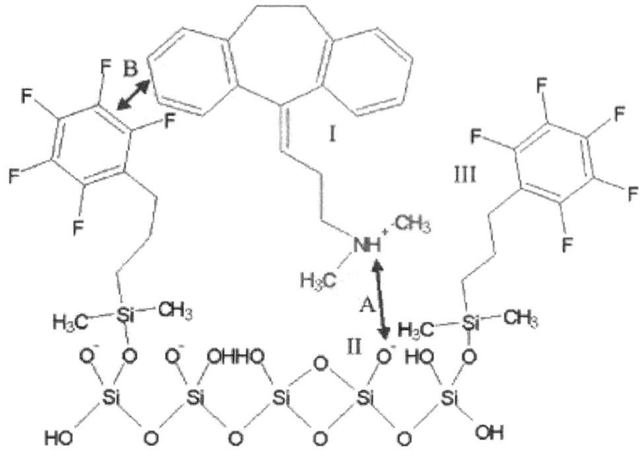

Figure 5.6. Potential interactions of protonated basic analytes with PFPP stationary phase. The protonated amitrptyline (I) molecule is shown to interact via ion-exchange (A) with ionized surface silanol (II) while an interaction via dispersive and polar mechanism (B) with the PFPP bonded phase (III) is simultaneously shown. Reproduced with permission from Bell, D.S. and Jones, A.D. (2005). Solute attributes and molecular interactions contributing to 'U-shape' retention on a fluorinated high-performance liquid chromatography stationary phase, *J. Chromatography A*, **1073**, 99–109. Copyright (2005) Elsevier.

interactions with the PFPP moiety and electrostatic interactions with the free silanol groups (Fig. 5.6).

Another important aspect to consider is that the 'U-shape' may depend on the mobile phase pH, which has been proven to be a valuable tool for the manipulation of basic analyte retention. In addition to controlling the analyte ionization state, the ionization state of the silanol surface is governed by mobile phase pH.[44]

One of the advantages of the PFP columns is the use of high percentages of organic solvents, such as acetonitrile or methanol, in the mobile phase. As previously discussed, a high percentage of organic solvent in the eluate improves the ESI ionization when coupled to mass spectrometry. These characteristics were used by Teixido *et al.*[45] for the analysis of 5-hydroxymethylfurfural in food samples in order to increase the retention of this compound and improve the ESI ionization efficiency at the same time. Juan *et al.*[46] also took

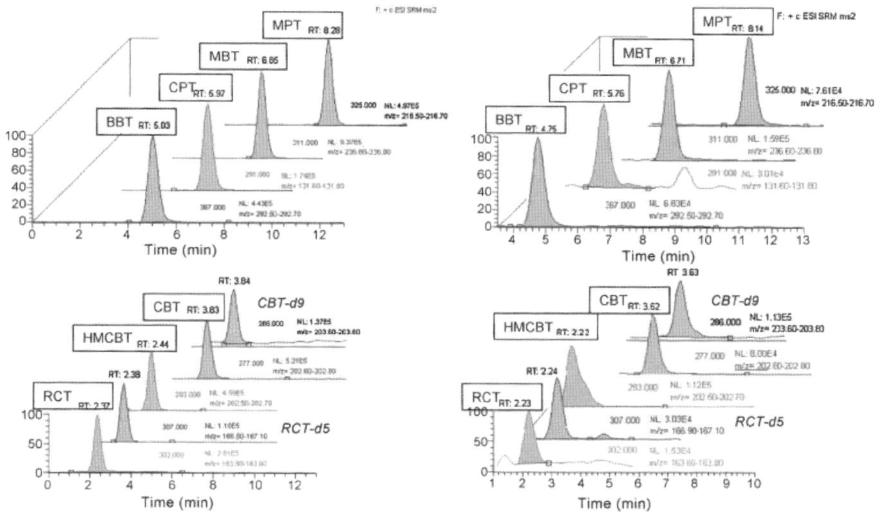

(A) Spiked Drinking water sample (B) Spiked Animal feed sample

Figure 5.7. Chromatograms of β-agonist-spiked samples analyzed with an optimized LC–MS/MS method using a PFP column: (A) drinking water spiked at 1.0 μg/L; (B) animal feed at spiked 0.8 μg/kg. Reproduced with permission from Juan, C., Igualada, C., Moragues, F. *et al.* (2008). Development and validation of a liquid chromatography tandem mass spectrometry method for the analysis of β-agonists in animal feed and drinking water, *J. Chromatogr. A*, **1217**, 6061–6068. Copyright (2008) Elsevier.

advantage of this feature in the analyses of β-agonists in animal feed and drinking water by using a PFP column and acetonitrile:water containing 10 mM acetic acid (80:20) as a mobile phase. Under these conditions, an excellent chromatographic resolution for the seven β-agonists and the two internal standards was obtained in less than 10 minutes. The respective chromatograms are shown in Fig. 5.7.

Gallart-Ayala *et al.*[47,48] successfully used a PFPP stationary phase for the analysis and separation of two UV ink photoinitiators in packaged food. In this case the multiple retention mechanisms presented by fluorinated stationary phases allowed the separation of two isomers of isopropylthioxanthone (2- and 4-ITX) in less than 10 minutes. Later on, the same stationary phase combined with porous-shell particles and operated at sub-ambient temperature was utilized for the simultaneous analysis of eleven UV photo initiators in packaged food. The chromatographic separation of ITX isomers was maintained.

Fluorinated phases are solvophobic towards organic and aqueous solvents.[42] Common lipophilic and hydrophilic analytes have low partition coefficients in the fluorinated stationary phases; therefore non-fluorinated compounds have short retention times. However, if the analyte is a fluorine-containing molecule, the solvophobicity has a tendency to repulse the fluorinated compound away from the mobile phase allowing it to participate in the fluorinated stationary phase, and therefore improving the chromatographic separations.[42] These capabilities were successfully employed by Benskin *et al.*[49] in the development of a LC–MS method where a fused-core PFP column was used for the analysis of perfluorooctane sulfonate (PFOS) and perfluorooctanoic acid (PFOA) together with the analysis of other perfluorinated compounds (PFCs) and unknown isomeric PFCs.

Although PFP stationary phases may offer an alternative for improvement of retention of polar compounds, these applications are still not well exploited in food and environmental analyses, and to date limited analytic methods are reported in the literature.

5.4. Summary

The analysis of low molecular weight polar compounds in complex matrices such as those stemming from food and environmental samples can pose substantial analytical challenges. Recent developments in the manufacture of chromatographic media have enabled the availability of a range of HILIC and fluorinated stationary phases that overcome a number of the shortcomings of more commonly used media, such as C18, enabling better retention and chromatographic resolution of polar compounds. As outlined in this chapter, these characteristics can provide dramatic sensitivity enhancements when coupled with electrospray ionization mass spectrometry. The growing popularity of HILIC and fluorinated phases is manifest in the increasing number of peer-reviewed publications based on these techniques. However, it is also clear that techniques such as HILIC can be much more sensitive to small variations in the experimental conditions than more commonly used analytical solutions based on the ubiquitous C18 phases. The objective of this chapter was to outline the main aspects that an analyst needs to consider during development of analytical methodologies based on HILIC and perfluorinated phases, in particular when facing challenges typically associated with the analysis of polar compounds in complex food and environmental matrices. The authors believe that the potential advantages of HILIC and perfluorinated phases clearly outweigh the requirement for a more careful experimental design, and hope to contribute to the increased use of these techniques in the years to come.

Disclaimer

The opinions expressed in this chapter do not necessarily reflect those of the U.S. Food and Drug Administration.

References

1. Jandera, P. (2011). Stationary and mobile phases in hydrophilic interaction chromatography: a review, *Anal. Chim. Acta*, **692**, 1–25.
2. Alpert, A.J. (1990). Hydrophilic-interaction chromatography for the separation of peptides, nucleic acids and other polar compounds, *J. Chromatogr. A*, **499**, 177–196.
3. Buszewski, B. and Noga, S. (2012). Hydrophilic interaction chromatography (HILIC) — a powerful separation technique, *Anal. Bioanal. Chem.*, **402**, 231–247.
4. Hemström, P. and Irgum, K. (2006). Hydrophilic interaction chromatography, *J. Sep. Sci.*, **29**, 1784–1821.
5. McCalley, D.V. (2010). Study of the selectivity, retention mechanisms and performance of alternative silica-based stationary phases for separation of ionised solutes in hydrophilic interaction chromatography, *J. Chromatogr. A*, **1217**, 3408–3417.
6. Núñez, O., Gallart-Ayala, H., Martins, C.P.B. *et al.* (2012). New trends in fast liquid chromatography for food and environmental analysis, *J. Chromatogr. A*, **1228**, 298–323.

7. Huber, J.F.K., Pawlowska, M. and Markl, P. (1990). Selection of the solid support — a crucial step for the realization of solvent generated liquid–liquid chromatography, *J. Chromatogr. A*, **500**, 257–280.
8. Linden, J.C. and Lawhead, C.L. (1975). Liquid chromatography of saccharides, *J. Chromatogr. A*, **105**, 125–133.
9. Esparza, X., Moyano, E. and Galceran, M.T. (2009). Analysis of chlormequat and mepiquat by hydrophilic interaction chromatography coupled to tandem mass spectrometry in food samples, *J. Chromatogr. A*, **1216**, 4402–4406.
10. van Nuijs, A.L.N., Tarcomnicu, I. and Covaci, A. (2011). Application of hydrophilic interaction chromatography for the analysis of polar contaminants in food and environmental samples, *J. Chromatogr. A*, **1218**, 5964–5974.
11. Bernal, J., Ares, A.M., Pól, J. *et al.* (2011). Hydrophilic interaction liquid chromatography in food analysis, *J. Chromatogr. A*, **1218**, 7438–7452.
12. Guo, Y. and Gaiki, S. (2011). Retention and selectivity of stationary phases for hydrophilic interaction chromatography, *J. Chromatogr. A*, **1218**, 5920–5938.
13. Greco, G. and Letzel, T. (2013). Main interactions and influences of the chromatographic parameters in HILIC separations, *J. Chromatogr. Sci.*, **51**, 684–693.
14. Karatapanis, A.E., Fiamegos, Y.C. and Stalikas, C.D. (2011). A revisit to the retention mechanism of hydrophilic interaction liquid chromatography using model organic compounds, *J. Chromatogr. A*, **1218**, 2871–2879.
15. McCalley, D.V. (2010). Study of the selectivity, retention mechanisms and performance of alternative silica-based stationary phases for separation of ionised solutes in hydrophilic interaction chromatography, *J. Chromatogr. A*, **1217**, 3408–3417.
16. Bicker, W., Wu, J., Lammerhofer, M. *et al.* (2008). Hydrophilic interaction chromatography in nonaqueous elution mode for separation of hydrophilic analytes on silica-based packings with noncharged polar bondings, *J. Sep. Sci.*, **31**, 2971–2987.
17. Lammerhofer, M., Richter, M., Wu, J. *et al.* (2008). Mixed-mode ion-exchangers and their comparative chromatographic characterization in reversed-phase and hydrophilic interaction chromatography elution modes, *J. Sep. Sci.*, **31**, 2572–2588.

18. Kittlaus, S., Schimanke, J., Kempe, G. *et al.* (2013). Development and validation of an efficient automated method for the analysis of 300 pesticides in foods using two-dimensional liquid chromatography–tandem mass spectrometry, *J. Chromatogr. A*, **1283**, 98–109.
19. Zhou, L., Liu, X., Kang, S. *et al.* (2013). A rapid determination method for ethylenethiourea in potato and cucumber by modified QuEChERS — high performance liquid chromatography–tandem mass spectrometry, *Food Chem.*, **138**, 1355–1359.
20. Chiaochan, C., Koesukwiwat, U., Yudthavorasit, S. *et al.* (2010). Efficient hydrophilic interaction liquid chromatography–tandem mass spectrometry for the multiclass analysis of veterinary drugs in chicken muscle, *Anal. Chim. Acta*, **682**, 117–129.
21. Martin, A.J. and Synge, R.L. (1941). A new form of chromatogram employing two liquid phases: A theory of chromatography. 2. Application to the micro-determination of the higher monoamino-acids in proteins, *Biochem. J*, **35**, 1358–1368.
22. Greco, G. and Letzel, T. (2013). Main interactions and influences of the chromatographic parameters in HILIC separations, *J. Chromatogr. Sci.*, **51**, 684–693.
23. Jacob, C.C. and Gamboa da Costa, G. (2011). Low-level quantification of melamine and cyanuric acid in limited samples of rat serum by UPLC–electrospray tandem mass spectrometry, *J. Chromatogr. B Analyt. Technol. Biomed. Life Sci.*, **879**, 652–656.
24. Fong, B., Ma, K. and McJarrow, P. (2011). Quantification of bovine milk oligosaccharides using liquid chromatography–selected reaction monitoring–mass spectrometry, *J. Agric. Food Chem.*, **59**, 9788–9795.
25. Blay, P., Hui, J.P., Chang, J. *et al.* (2011). Screening for multiple classes of marine biotoxins by liquid chromatography–high-resolution mass spectrometry, *Anal. Bioanal. Chem.*, **400**, 577–585.
26. Chen, W.L., Hsu, T.F. and Chen, C.Y. (2011). Measurement of aflatoxin M1 in milk by ultrahigh-performance liquid chromatography–tandem mass spectrometry, *J. AOAC Int.*, **94**, 872–877.
27. Combes, A., El Abdellaoui, S., Sarazin, C. *et al.* (2013). Validation of the analytical procedure for the determination of the neurotoxin beta-N-methylamino-L-alanine in complex environmental samples, *Anal. Chim. Acta*, **771**, 42–49.

28. Li, C., Jin, F., Yu, Z. *et al.* (2012). Rapid determination of chlormequat in meat by dispersive solid-phase extraction and hydrophilic interaction liquid chromatography (HILIC)–electrospray tandem mass spectrometry, *J. Agric. Food Chem.*, **60**, 6816–6822.
29. Kitanovski, Z., Grgic, I. and Veber, M. (2011). Characterization of carboxylic acids in atmospheric aerosols using hydrophilic interaction liquid chromatography tandem mass spectrometry, *J. Chromatogr. A*, **1218**, 4417–4425.
30. Li, R., Zhang, Y., Lee, C.C. *et al.* (2010). Development and validation of a hydrophilic interaction liquid chromatographic method for determination of aromatic amines in environmental water, *J. Chromatogr. A*, **1217**, 1799–1805.
31. Qin, F., Zhao, Y.Y., Sawyer, M.B. *et al.* (2008). Column-switching reversed phase–hydrophilic interaction liquid chromatography–tandem mass spectrometry method for determination of free estrogens and their conjugates in river water, *Anal. Chim. Acta*, **627**, 91–98.
32. Bohm, D.A., Stachel, C.S. and Gowik, P. (2010). Confirmatory method for the determination of streptomycin in apples by LC–MS/MS, *Anal. Chim. Acta*, **672**, 103–106.
33. Lajeunesse, A., Segura, P.A., Gelinas, M. *et al.* (2012). Detection and confirmation of saxitoxin analogues in freshwater benthic *Lyngbya wollei* algae collected in the St. Lawrence River (Canada) by liquid chromatography–tandem mass spectrometry, *J. Chromatogr. A*, **1219**, 93–103.
34. Periat, A., Boccard, J., Veuthey, J.L. *et al.* (2013). Systematic comparison of sensitivity between hydrophilic interaction liquid chromatography and reversed phase liquid chromatography coupled with mass spectrometry, *J. Chromatogr. A*, **1312**, 49–57.
35. Xiong, Y., Zhao, Y.Y., Goruk, S. *et al.* (2012). Validation of an LC–MS/MS method for the quantification of choline-related compounds and phospholipids in foods and tissues, *J. Chromatogr. B Analyt. Technol. Biomed. Life Sci.*, **911**, 170–179.
36. Ordonez, E.Y., Quintana, J.B., Rodil, R. *et al.* (2012). Determination of artificial sweeteners in water samples by solid-phase extraction and liquid chromatography–tandem mass spectrometry, *J. Chromatogr. A*, **1256**, 197–205.

37. Sonntag, T., Kunert, C., Dunkel, A. *et al.* (2010). Sensory-guided identification of N-(1-methyl-4-oxoimidazolidin-2-ylidene)-alpha-amino acids as contributors to the thick-sour and mouth-drying orosensation of stewed beef juice, *J. Agric. Food Chem.*, **58**, 6341–6350.
38. Birka, M., Wehe, C.A., Telgmann, L. *et al.* (2013). Sensitive quantification of gadolinium-based magnetic resonance imaging contrast agents in surface waters using hydrophilic interaction liquid chromatography and inductively coupled plasma sector field mass spectrometry, *J. Chromatogr. A*, **1308**, 125–131.
39. Self, R.L., Wu, W.H. and Marks, H.S. (2011). Simultaneous quantification of eight biogenic amine compounds in tuna by matrix solid-phase dispersion followed by HPLC–orbitrap mass spectrometry, *J. Agric. Food Chem.*, **59**, 5906–5913.
40. Wolwer-Rieck, U., Tomberg, W. and Wawrzun, A. (2010). Investigations on the stability of stevioside and rebaudioside A in soft drinks, *J. Agric. Food Chem.*, **58**, 12216–12220.
41. Lindner, U., Lingott, J., Richter, S. *et al.* (2013). Speciation of gadolinium in surface water samples and plants by hydrophilic interaction chromatography hyphenated with inductively coupled plasma mass spectrometry, *Anal. Bioanal. Chem.*, **405**, 1865–1873.
42. Zhang, W. (2008). Fluorocarbon stationary phases for liquid chromatography applications, *J. Fluorine Chem.*, **129**, 910–919.
43. Petruczynik, A. (2012). Effect of ionic liquid additives to mobile phase on separation and system efficiency for HPLC of selected alkaloids on different stationary phases, *J. Chromatogr. Sci.*, **50**, 287–293.
44. Bell, D.S. and Jones, A.D. (2005). Solute attributes and molecular interactions contributing to 'U-shape' retention on a fluorinated high-performance liquid chromatography stationary phase, *J. Chromatogr. A*, **1073**, 99–109.
45. Teixidó, E., Moyano, E., Santos, F.J. *et al.* (2008). Liquid chromatography–multi-stage mass spectrometry for the analysis of 5-hydroxymethylfurfural in foods, *J. Chromatogr. A*, **1185**, 102–108.
46. Juan, C., Igualada, C., Moragues, F. *et al.* (2010). Development and validation of a liquid chromatography–tandem mass spectrometry method for the analysis of β-agonists in animal feed and drinking water, *J. Chromatogr. A*, **1217**, 6061–6068.

47. Gallart-Ayala, H., Moyano, E. and Galceran, M.T. (2008). Liquid chromatography–tandem mass spectrometry (highly selective selected reaction monitoring) for the analysis of isopropylthioxanthone in packaged food, *J. Chromatogr. A*, **1208**, 182–188.
48. Gallart-Ayala, H., Núñez, O., Moyano, E. *et al.* (2011). Analysis of UV ink photoinitiators in packaged food by fast liquid chromatography at sub-ambient temperature coupled to tandem mass spectrometry, *J. Chromatogr. A*, **1218**, 459–466.
49. Benskin, J.P., Ikonomou, M.G., Woudneh, M.B. *et al.* (2012). Rapid characterization of perfluoralkyl carboxylate, sulfonate, and sulfonamide isomers by high-performance liquid chromatography–tandem mass spectrometry, *J. Chromatogr. A*, **1247**, 165–170.
50. Ballesteros-Gómez, A., Rubio, S. and van Leeuwen, S. (2010). Tetrahydrofuran–water extraction, in-line clean-up and selective liquid chromatography–tandem mass spectrometry for the quantitation of perfluorinated compounds in food at the low picogram per gram level, *J. Chromatogr. A*, **1217**, 5913–5921.
51. Inoue, S., Miyamoto, S., Ogasawara, M. *et al.* (2009). Simultaneous determination of medicinal ingredients in so-called health-promoting food using liquid chromatography–tandem mass spectrometry with a pentafluorophenyl stationary phase, *J. Health Sci.*, **55**, 183–191.
52. Van De Steene, J.C. and Lambert, W.E. (2008). Validation of a solid-phase extraction and liquid chromatography–electrospray tandem mass spectrometric method for the determination of nine basic pharmaceuticals in wastewater and surface water samples, *J. Chromatogr. A*, **1182**, 153–160.

Part 2

Advances in Fast Liquid Chromatography–Mass Spectrometry Methods

Chapter 6

On-Line Sample Pre-Treatment Procedures Applied to LC–MS

Tony Edge[a] and Joseph Herman[b]

[a] Thermo Fisher Scientific, Runcorn, United Kingdom
[b] Thermo Fisher Scientific, West Chester, Pennsylvania, United States of America

6.1. Introduction

Separations are often demonstrated using clean solutions, and as a result do not always reflect real-world samples, which come with many challenges associated with trying to remove matrix components, or which require pre-concentration. Generally, the matrix components are significantly different from the compounds that are being analysed so that relatively crude forms of sample preparation, such as filtration or centrifugation, can be utilised to clean up the primary sample. When analysing low concentration levels this approach is generally not applicable, in particular if mass spectrometry is being used as the detection method, since the matrix will have a significant effect on the detection of the analyte molecules.[1–5]

Typical matrices that analytical scientists need to analyse when dealing with food and environmental samples can include:

- Food — animal products, dairy products, vegetables, fruit, meat products;[6–10]
- Environmental — soil, plants, water, animals.[11–14]

The range of physiochemical properties[15,16] being analysed, from very polar molecules such as melamine, to very hydrophobic compounds such as fat-soluble vitamins, means that different approaches have to be considered. These compounds can also be endogenous or present as impurities, the latter due to either adulteration or contamination of the original sample, which presents further complications to the analysis.

6.2. Off-Line Approaches to Sample Preparation

The range of matrix types and the range of analytes mean that it is important to have the appropriate extraction technology to ensure that the sample is in a suitable format and that an appropriate amount of matrix is removed. There are a variety of different sample preparation techniques that are employed within the food, environmental and other industrial sectors:

- Centrifugation;[17]
- QuEChERS;[19–21]
- Protein precipitation;[18]
- Solid phase extraction;[22–25]
- Liquid–liquid extraction;[26,27]
- Support-assisted liquid–liquid extraction;[28,29]
- Filtration.[30–31]

There are other approaches that can be employed to process the raw sample prior to the final chromatographic quantitative determination of the analytes. The most common of these approaches will be discussed briefly.

6.2.1. *QuEChERS*

For environmental and food samples one of the most common approaches is that of QuEChERS, which is an acronym for Quick, Easy, Cheap, Effective, Rugged, and Safe. It was originally developed by Anastasiades *et al.* in 2003,[19] and is based on a liquid–liquid extraction, with an option for further clean-up using dispersive solid phase extraction (SPE). There are four variants:

- The original method;[19]
- Dispersive AOAC 2007.01;[32]
- Dual phase variation;[33]
- European method (EU CEN 15662 method).[34,35]

The approach is based around removing the matrix rather than isolating the compound of interest and so is used widely in markets where multiple compounds are being analysed. The different variants use slightly different reagents, each of which favours certain types of compounds or types of matrices. Although the technique is widely used, the recovery data produced tends to be quite variable, depending on the matrix and also the compound being analysed,[35] primarily due to the large applicability of the technique.

6.2.2. *Protein precipitation*

Protein precipitation is widely used within the pharmaceutical industry in the field of bioanalysis;[36–38] however, it has also been successfully applied to the analysis of a wide range of components within the field of environmental and food analysis.[39–41] Proteins' solubilities depend on several factors, one of which is the dielectric constant of the solution in which they are contained. There are different approaches that can be used to alter the dielectric constant, and thus precipitate the proteins in the solution. The most commonly employed approaches are:

- Addition of an organic solvent such as an alcohol or acetonitrile;
- Addition of acid, (TCA and TFA are common reagents);
- Addition of another chaotropic reagent, such as guanidine hydrochloride or urea;
- Addition of salt, causing salting out.

The procedure is quick, simple, and applicable to a wide range of compounds; however, it suffers because the extraction process focusses on the removal of proteins and thus is not applicable to samples that contain high levels of lipids or other lower-molecular-mass compounds. In addition, the protein precipitation process releases compounds that are highly protein bound to the matrix, which is important when the total amount of an analyte is desired. When only free or unbound concentrations of the analyte are required, protein precipitation would not be applicable.

6.2.3. *Liquid–liquid extraction*

Liquid–liquid extraction (LLE) is a method used to separate compounds based on their preferences for two different immiscible liquids, generally water and an organic solvent. To achieve a sufficient surface area to allow the compound of interest to diffuse into the organic solvent, the 'mix' must be shaken vigorously. Optimisation of the extraction procedure centres on altering the solubility of the compound in the extraction solvent, typically the organic layer, by changing the solvent or by adding salt. This extraction technique is notoriously difficult to automate. Typical solvents that are employed are hexane, methyl *tert*-butyl ether (MTBE), diethyl ether, and chloroform.[42–46]

A variant of this approach is support-assisted liquid–liquid extraction (SLE), which provides an easier-to-automate alternative. In this technique the initial aqueous phase is absorbed into the solid support particles. The particles used are diatomaceous earth, which consists of crushed fossilised skeletons of freshwater organisms and marine life (especially diatoms and other algae). It is mainly silica (SiO_2), which has a very large surface area relative to its size due to its highly porous structure. The water bonds to the particles via hydrogen bonding to the silica. The addition of the immiscible organic solvent results in a very large surface area between the two immiscible liquids, over which the diffusion of analyte between the aqueous and organic phase occurs. The organic solvent is not retained by the highly polar silica surface and so is readily eluted from the cartridge.[47–49]

6.3. How a Mass Spectrometer Works and Ion Suppression

The use of mass spectrometry (MS) coupled to liquid chromatography (LC) is now commonplace in many laboratories and in particular with on-line analysis of food and environmental samples.[49–58] Although it is now a routine technology, it took the Nobel laureate John Fenn to determine the best way of coupling these two analytical techniques together.[59] The challenge associated with this marriage is that the mass spectrometer works best under a vacuum with a pure gaseous compound, whereas the LC component works with very dilute solutions of compound mixed in with large amounts of mobile phase (not necessarily organic and usually a mixture of aqueous and organic). This presents many challenges, since the mobile phase has to be separated from the compound of interest, and also the compound of interest has to be ionized. An understanding of this process is important to better understand one of the major challenges associated with on-line detection using MS, namely ion suppression.[60]

So called 'matrix effects'[61–63] are well-recognised for their potential to distort the analytical data if not properly accounted for and either physically removed by the sample preparation or handled by the chromatography. These matrix effects arise because of the complexity of the matrix, which for a biological fluid, can contain several tens of thousands of different compounds with a very wide range ($>10^7$) of concentrations.[64] Each of the endogenous compounds can, and does, vary from sample to sample.[65] Many of these compounds will be ionisable under the conditions used for mass spectrometry, which results in:

- competition for the available charge in the ion source of the mass spectrometer;[66]
- enhancement of the ionisation capabilities of other compounds;[67]
- reduction in solvent evaporation.[68]

There also exist other processes, including space charge effects, micelle formation, and gas phase interactions,[69] that can cause variable responses from the mass spectrometer.

The variability in matrix composition potentially means that the degree of ionisation will vary from one sample to another, with possible adverse effects on the analysis of target analytes. Therefore, it is critical that the compound is resolved from any endogenous materials that produce matrix effects in order to reduce or eliminate ion suppression within the mass spectrometer source.

If ion suppression is obtained then there are several parts of the overall assay that can be altered to reduce the effect:

- the chromatography;
- the extraction;
- the MS conditions.

When performing on-line analysis, optimisation of these parameters can be achieved in a succinct manner, which is not the case when analysing samples in an off-line mode.

6.3.1. *Measuring matrix effects*

There are two common approaches employed in many laboratories. The first is to use post column infusion. The setup is as in Fig. 6.1. This approach is very qualitative but gives the separation scientist an idea of the effect that the ion suppression is having on the overall assay and indicates good and bad regions for the compound to elute in.

The second approach is to compare a standard made in solution (i.e. matrix free) with that of a blank matrix that has been extracted and then had the same nominal amount of compound added after the extraction (a so-called overspike).[63] The only difference between the two samples is that matrix components are present in the latter. From these data the extent of the matrix effect can be quantified. It is also necessary to understand that matrix effects are different

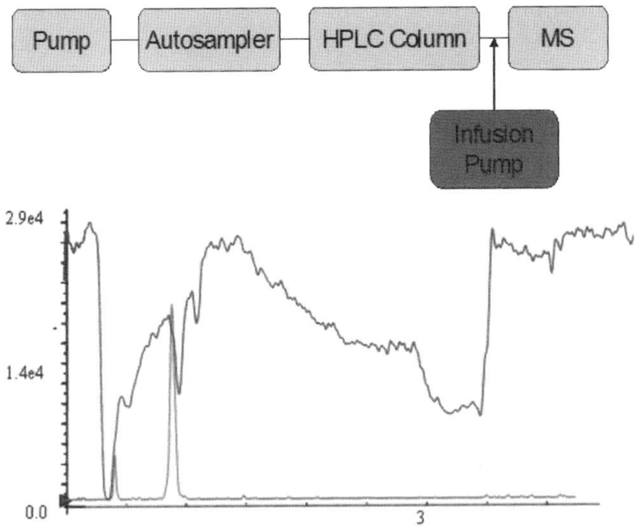

Figure 6.1. Schematic diagram of post column infusion, with the resultant chromatographic trace.

from recovery. Thus, recovery is calculated using data obtained using solutions that do not contain matrix components, typically aqueous solutions. This recovery of the assay will be affected by issues such as solubility, limited elution of the analyte, and breakthrough.

6.4. Other Matrix Effects

Within the on-line food and environmental sectors ion suppression is a major challenge associated with the matrix; however, this is not the only challenge that faces the analytical scientist when dealing with these types of samples.

6.4.1. *Method robustness*

One of the challenges in food and environmental analysis is the very large number of components that can be found in what would be deemed to be a relatively simple matrix; another is the effect that this can have on the analyte, either directly or through the extraction methodology. For example, within the food industry the quantification of fat-soluble vitamins (A, D, E and K) is commonplace.[70–72] The analysis of these compounds in fatty dairy products is made more difficult by the strong interaction with lipids, resulting in a lipid–vitamin bond that is not disrupted by more conventional extraction solvents. This is further complicated by the ion suppression effects that are commonly observed with lipids, which will be discussed in detail later. There are different approaches that can be employed but one of the most common is the saponification of the original sample,[73] which results in the hydrolysis of the fatty ester linkages, and another is the removal of the triacylglycerides by a similar process, the latter forming a large proportion of the matrix. This process is typically performed using a strong solution of potassium or sodium hydroxide, which releases the carotenoids, retinoids, tocopherols, and vitamin D compounds; however, this solution can cause the vitamins to break down, in particular vitamin E. If biological samples are now considered then there are several tens of thousands of compounds that are present, with concentration ranges from percentage levels down to low pg/mL. For most separation scientists it is of little interest what actually forms the bulk of the sample, but it is something that should be considered when using highly selective detection techniques such as mass spectrometry, where it is well known that the ionisation efficiency varies depending on the matrix concentration, which results in a variable detector response.

6.4.2. *System robustness*

The robustness of the analytical system will be compromised when using matrix-loaded samples. It is feasible to analyse an individual compound dissolved in the mobile phase for many thousands of injections; however, this is not the case when analysing matrix-loaded samples. Since most of the compounds that will be discussed will be analysed using a chromatographic technique, it is worthwhile stressing some of the disadvantages of not having an appropriate extraction process in place in terms of chromatographic performance, namely that:

- the matrix can affect the surface of any component in the HPLC system;
- the matrix can block up interstitial spaces on the stationary phase;
- the matrix can be detected by the detector (in the case of UV) and hide the response of the compound of interest;
- the matrix reduces the lifetime of the chromatographic system;
- in LC–MS, although the matrix may not be detected by the detector, it still can affect the signal, since the signal strength can decrease as the MS gets dirty.

6.4.3. *Carryover*

Within the field of quantification, one of the biggest limitations of an analytical procedure is the dynamic range of the assay. Memory effects or 'carryover' can have consequential effects in many areas where separation science is used.[74–76] The issue that arises is that unless the entire sample is removed from the analytical system, the subsequent analysis will have residual compound from the previous injection, which could potentially lead to inaccurate data being produced.

There are approaches that can be utilised to mitigate these effects, such as:

- running samples in concentration order — not very practical with unknown samples;
- addition of extra blank samples to reduce the amount of carryover seen by the real samples;
- dilution of samples into a limited calibration range — this, however, relies on knowledge of the sample concentration.

One approach that has been applied successfully to reducing the carryover is to 'isolate and eliminate.'[77] This is a process that has to be done systematically by first sequentially removing components from the chromatographic system to determine where the source of the carryover is. The second stage is then to determine what needs to happen to remove it from the system. This could be done by physically replacing a faulty or contaminated component, or by altering a wash solvent or mobile phase.

In terms of the nature of the carryover there are two main possible states:

- it can be removed with a very few blank injections;
- it is persistent and does not decay appreciably with repeat injections.

The former is preferable since there is a suggestion that the approach to removing the compound is working, whereas the latter findings suggests that there is an inherent problem with the chromatographic system, probably contamination.

If contamination is suspected then it is important that all potential sources of contamination are removed to ensure that the issue is addressed; often the cause can be an out-of-date buffer solution. Usually the contamination in expired buffers is from bacterial growth such that the addition of anti-bacterial compounds to the mobile phase can reduce the likelihood of bacterial growth in the mobile phase. For example, adding 2% acetonitrile to any aqueous solution will keep the mobile phase free of bacteria.

If the carryover is not from a contamination source, then further isolation of the chromatographic system has to be performed. The first component to isolate is the autosampler valve. The injection of a high standard with the chromatographic system set up in its original state, followed by a 'blank' injection with the autosampler removed from the chromatographic system will determine if the carryover is coming from the autosampler or from the rest of the system. There are currently many different autosamplers on the market; however, they all inject sample in a similar manner. The movement of the liquid into a valve may vary (some use syringes while others use flow through needles; some use air while others use liquid; some use pressure/vacuum while others do not), but they all inject in the same way. Central to the design is the use of a two-position port valve, which is used to transfer the sample from an injection device to the fluidic system of the chromatography system.

For simplicity, an autosampler that uses a syringe to draw up the sample and then inject this sample into a sample loop will be discussed; however, the principles of the approach can be readily applied to any type of autosampler. In this type of system the sample is drawn directly into the barrel of a syringe and then the sample is injected onto the chromatographic system using a two-position valve, connected with a sample loop. Other systems may use a piece of inert tubing between a syringe or metering device and the sample, but the basic concepts are very similar. Investigating the components that the sample comes into contact with (Fig. 6.2), it can be seen that there are several locations where the sample can be effectively trapped.

6.4.3.1. *The syringe*

There is a substantial opportunity for the sample to be trapped/adsorbed on the glass surface of the syringe or the metallic surface of the needle. If the seal between the plunger and the glass syringe barrel is not tight enough then there is a possibility of either:

- air being drawn in as the plunger is filling the syringe, reducing the amount of sample present in the syringe;
- the sample leaking through the plunger seal and the glass barrel when the plunger is being pushed down, resulting in sample being deposited in the barrel and/or plunger. In terms of carryover this can be very significant.

6.4.3.2. *The injector valve*

This relates to the swept volume as the sample goes from the syringe to the injector loop. Within the valve there are regions that are not

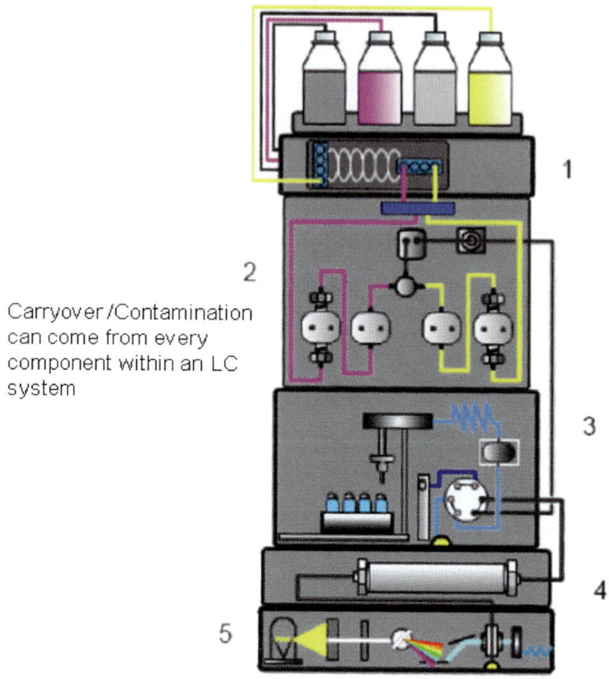

Figure 6.2. Schematic diagram of LC system. Every component within the system can be considered as a potential source of contamination or carryover. Reprinted with kind permission from Tony Taylor, Crawford Scientific & CHROMacademy.

properly swept due to the fluid dynamics of the arrangement, highlighted in Fig. 6.3, which shows the two possible positions of the injector valve. As the rotor rotates between the load and the inject position so the sample will come into contact with the stator and this will result in a possibility that the sample will be retained on the stator. Defects in the surface of the rotor or stator, such as scratches or small abrasions, can be a major cause for carryover in this scenario. If there are solubility issues in transferring the sample into the sample loop then it is possible that small deposits of the analyte will remain on the individual valve components. The same issues that are present with the autosampler valve will also be present with the switching valves that are employed.

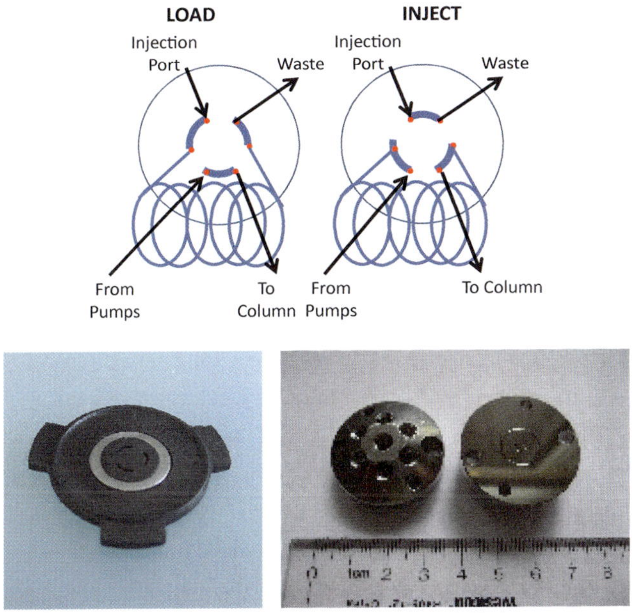

Figure 6.3. Design of most autosamplers rely on a valve (or stator), note that it is virtually impossible not to have a dead volume with this design.

6.4.3.3. *Carryover on the column*

There are other components where the sample can be very strongly retained and there is a range of reasons why this could occur. The column will typically have a very large surface area and so the possibility of some of the sample not eluting from the column in one elution cycle can be quite high. It is therefore possible that not all of the sample is removed from the column after it is injected. This can occur if the analyte has a limited solubility range in the mobile phase, prevalent with proteins and other macromolecules. If this is the case then that component will be available for elution on the next injection, which will result in an incorrect determination of the analyte concentration. Different columns will have different degrees of retentivity[78] which is further complicated by the inclusion of columns where matrix has been injected. Retained matrix can change the retention mechanism for the sample and potentially cause the creation of highly retentive sites which do not preferentially elute the compounds of interest.

Initially it is important to determine the nature of the carryover; is it at a persistent level or does it disappear after a few washes? This is a relatively easy test to perform; simply inject a high concentration followed by a series of blanks from the same vial. Obviously this raises the question that the blank could be contaminated. This can be tested using a variety of different sources of a blank and repeating the previous test. If the levels of analyte response do not drop then a source of possible contamination should be investigated.

It is one thing to isolate the carryover; however, it is also necessary to have an approach to remove the carryover. If the carryover is coming from the column, then changing in the mobile phase, increasing the amount of strong solvent, altering the pH, or altering the flow rate will all have an effect on the solubility of the analyte, and hence the levels of carryover. Also, it is worthwhile checking that the column has been installed correctly; improperly seated connections create a dead space that can retain analytes and cause carryover.

If the carryover is not coming from the column then further investigations can be made to reduce the chromatographic system further to isolate fewer and fewer components. The next easiest component to investigate is the syringe. Again the system is set up so that it is in a normal operating mode and then a high standard is injected. Following this, the syringe is removed or disabled so that it is not possible for sample to be introduced from the syringe. Thus carryover can only come from the valve, assuming that the column carryover has been checked. If the carryover is negligible then the carryover is coming from the one component that has been removed, which in this case is the syringe.

If the carryover is coming from the syringe then the syringe should be checked to make sure that there is a tight fit between the barrel and the plunger. There should also be a visual inspection of the syringe to ensure that there are no obvious aberrations in the walls of the barrel, on the plunger, or in the needle tip. Different washing solutions may be needed, if possible with your system; wash solvents in which the analyte has high solubility work best.

Similarly, if the carryover is coming from the valve then it is worth checking to see if there is any wear and tear on the stator or the rotor. In most cases these are easily accessible and can be viewed with ease. Occasionally it may be necessary to toggle the valve a couple times while running a wash solution as the mobile phase to remove carryover.

In both situations if there is no obvious physical manifestation then the choice of wash solvent and the volume of wash solvent become critical.[79] It is important that a wash solvent is chosen which ensures that there is maximum solubility of the compound obtained.[80] For organic compounds the work of Snyder[81] to categorise different solvents may be used, and then wash solvents may be employed that utilise the appropriate solvent to optimise the solvation of the compound thus reducing the carryover. If the physiochemical properties of the molecule are not understood then using a mixture of solvents that cover a range of solvent properties is a more generic approach that will give some success. The choice of pH can also be important, as for ionisable compounds this will result in the compound being in a charged or uncharged state, which will increase or reduce the solubility of the compound, as well as changing its adsorption properties.[82]

Figures 6.4–6.6 demonstrate three of the many different forms of carryover that can be found on any chromatographic system.

Figure 6.4 demonstrates the existence of carryover coming from the column. In this example a basic compound was being analysed using a low pH solvent mixture. The stationary phase was not well end-capped and as a result there was a considerable number of secondary interactions, which resulted in some tailing but also substantial amounts of carryover. Thus, the top standard with the subsequent blank overlaid is shown in Fig. 6.4a. The levels of carryover observed were nearly 10%, which made quantification virtually impossible. The original blank overlaid with another blank after the autosampler has been removed from the chromatographic system is shown in Fig. 6.4b. This clearly demonstrates that the carryover was not coming from the autosampler and other components need to be investigated instead. It was eventually determined that the carryover was in fact coming from the column. In this example altering the pH of the mobile phase above the pK_a of the analyte was investigated. This

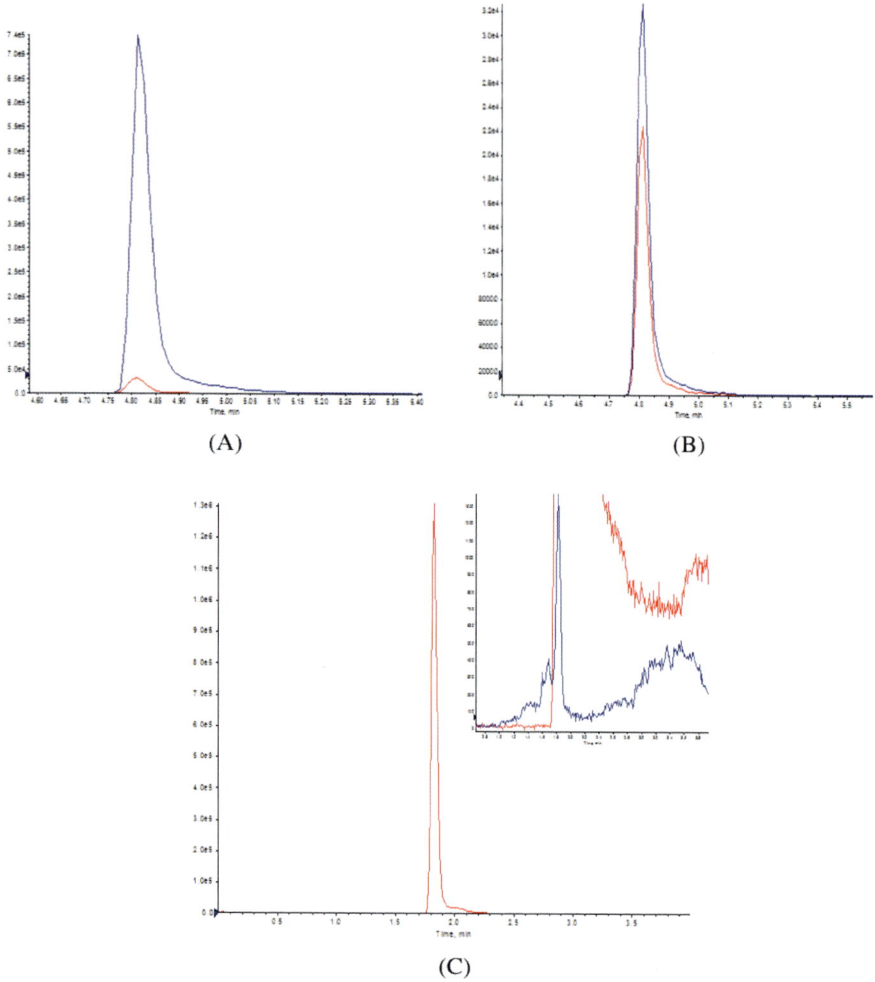

Figure 6.4. A: Top standard overlaid with a blank running with a low pH mobile phase. B: Previous blank overlaid with a blank injection with no autosampler connected with a low pH mobile phase. C: Top standard overlaid with a blank running with a high pH mobile phase. Reprinted with the kind permission of International Labmate Ltd.; first printed in *Chromatography Today*, **6**(3) (August/September 2013).

meant that there was no possibility of obtaining an ion exchange interaction and thus virtually eliminating the carryover (Fig. 6.4c).

Figure 6.5 was obtained during the development of a method for the analysis of a series of acetylcholinesterase inhibitors, in particular, edrophonium, neostigmine, and pyridostigmine. Employing the approach of isolate and then eliminate, it was determined that the carryover was actually coming from the autosampler valve. On inspection of the rotor and stator it became obvious that the issue was associated with a contaminated valve. Cleaning the stator in a solution of acidic methanol in a sonic bath and also replacing the rotor reduced the levels of carryover dramatically, as seen in Fig. 6.5c, which allowed the dynamic range of the assay to be extended. Care does have to be taken with this approach to ensure that more active adsorptive sites are not produced, as this will increase carryover levels.

Figure 6.6 very nicely demonstrates the importance of wash solvents on the chromatographic system. The mode of chromatography chosen was HILIC, in which the solvents' strengths are reversed when comparing to standard reversed phase chromatography. In this

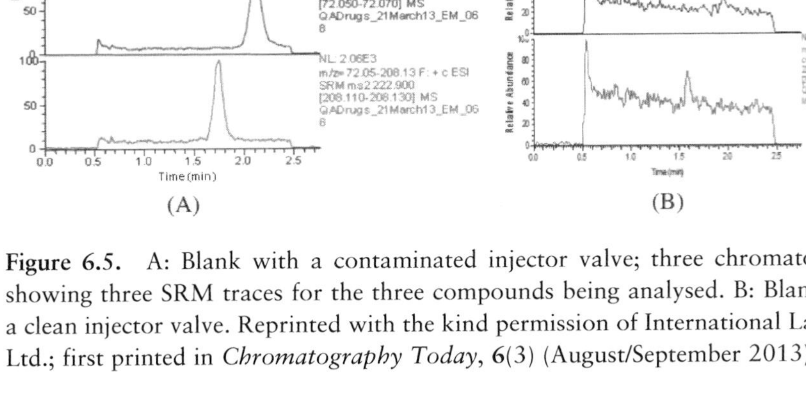

(A) (B)

Figure 6.5. A: Blank with a contaminated injector valve; three chromatograms showing three SRM traces for the three compounds being analysed. B: Blank with a clean injector valve. Reprinted with the kind permission of International Labmate Ltd.; first printed in *Chromatography Today*, **6**(3) (August/September 2013).

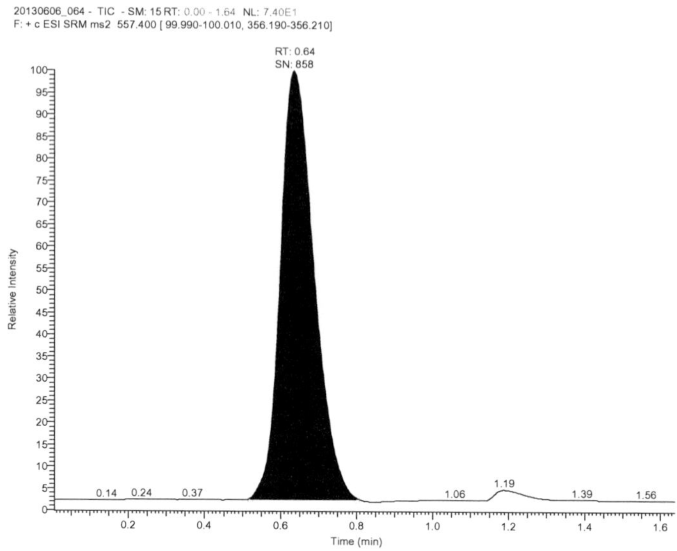

(A)

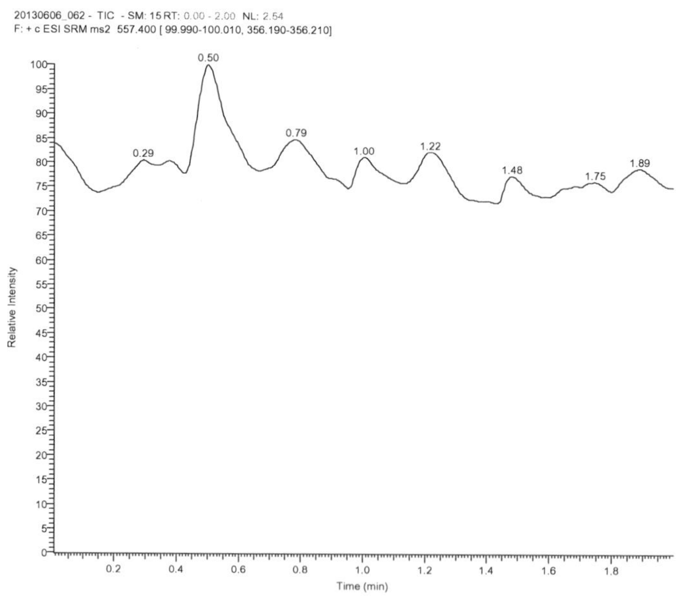

(B)

Figure 6.6. A: Blank following a top standard with inappropriate autosampler wash solvents. B: Blank following a top standard with appropriate autosampler wash solvents. Reprinted with the kind permission of International Labmate Ltd.; first printed in *Chromatography Today*, **6**(3) (August/September 2013).

scenario, after performing the tests that have been previously outlined it was evident that the carryover was coming from the autosampler and in particular was due to the wash solvents. In Fig. 6.6a, a blank is injected directly after the top standard with the wash solvent, which was left on the system from a reversed phased system (IPA:MeCN:Acetone 45:45:10). Figure 6.6b demonstrates the importance the wash solvents can have on the levels of carryover. In this situation the wash solvent was altered to water which is a much more appropriate solvent for very polar molecules, as there are fewer solubility issues.

However, this also raises another important fact, which is that there is no universal solvent to remove carryover; it really does depend on the molecule that is being investigated. Use of pH[83] and the Snyder triangle[81] can aid in the choice of a more applicable solvent, but in all cases where carryover is present the nature and the source of the carryover have to be considered in conjunction with the nature of the compound. The Synder triangle classifies solvents according to three parameters and this will allow the selection of the solvent that best matches the physiochemical properties of the analyte in order to ensure minimal carryover.

With traditional HPLC the sample is typically in a relatively clean state prior to being chromatographed; however, with on-line sample extraction the matrix can be permanently retained in the chromatographic system. The retained matrix changes the retentive properties of the column and the manner in which the analytes are retained and/or separated. Being aware of this means that appropriate steps can be taken to reduce the amount of matrix build up within the chromatographic system; typically this is done through a judicial wash regime.

6.5. On-Line Approaches

As with the off-line approaches discussed previously, the wide range of matrix types and analytes make the choice of on-line extraction technology critical to the success of removing issues arising from the matrix components. The primary on-line methodologies available for sample preparation that are employed within the food and environmental sectors are:

- turbulent flow chromatography;[86–97]
- solid phase extraction;[22–25,98–126]
- molecularly imprinted polymers;[127–169]
- restricted access media;[170–198]
- immunoaffinity columns.[199–201]

While other approaches exist, the most common of these approaches will be discussed next.

6.5.1. *Use of switching valves*

All of the on-line sample preparation techniques employ switching valves to allow the matrix components to be eluted to a waste stream. The switching valve can come in different configurations, with the 2-position 6-port being one of the most popular. The valve itself has several components, with different manufacturers having slightly different designs; however, the basics of the valve design are that it comprises of three major components:

- valve motor;
- rotor;
- stator.

The rotor and stator can be made of different materials and careful choice of the material can significantly reduce the levels of carryover.[83]

There are a variety of configurations that can be used to control the movement of the analyte and also of the matrix. In general, the more valves that are used the better the control and hence better removal of the matrix components and better chromatographic performance. However, the greater the number of valves that are employed the greater the probability of encountering carryover in one of these components.

6.5.2. *Discussion of different configurations available*

There are many different configurations that have been used;[122–125] the three most common configurations are described in detail below.

6.5.2.1. *Single-valve method*

There are four basic steps when using a single valve; load, wash, elute, and re-equilibrate the column. These steps are very comparable to the four basic steps involved in solid phase extraction (condition, load, wash, elute) and serve comparable purposes. In most cases the loading and washing steps take place simultaneously, but especially for SPE that is not always the case. Examples of single valve methods are represented in Fig. 6.7.

The sample is loaded onto the extraction column in a weak solvent to ensure that retention of the compound is achieved. A wash step uses the same valve position and in some cases this allows for a stronger wash solvent to be employed compared to the loading

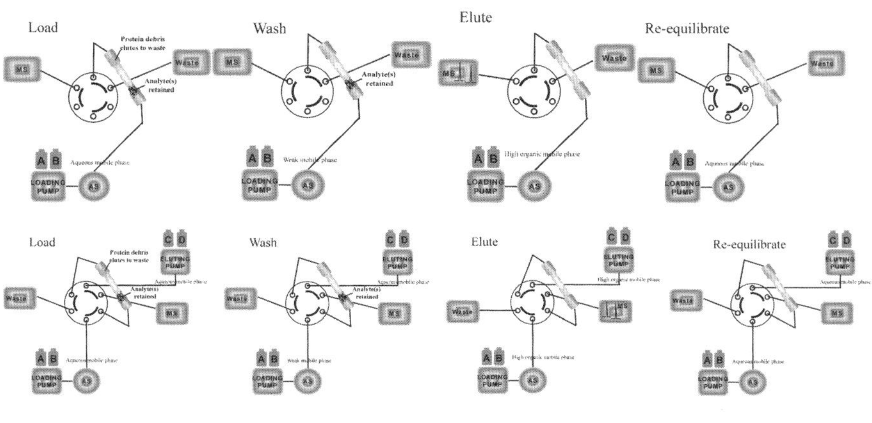

Figure 6.7. Diagram of single valve system using one and two pumps. Reprinted with permission from Edge, A. (2003). 'Turbulent flow chromatography in bioanalysis' in Wilson, I.D. (ed.), *Handbook of Bioanalytical Separations, Vol. 4, Bioanalytical Separations*, Elsevier, The Netherlands, pp. 91–127) Copyright (2003) Elsevier.

solvent. In both the load and the wash configurations, the valve is positioned such that all the eluent goes to the waste stream. Since the matrix has a broad spectrum of physiochemical properties, some components of the matrix will not be retained during this stage and will go directly to waste, effectively cleaning the sample or extracting it from the bulk matrix.

The next stage is the elution step, where the components retained after the initial loading step are eluted from the column with a high elutropic mobile phase. At this point the valve is switched so that the eluent stream now goes to the detector rather than to waste, either via an analytical column or directly. The use of a second column allows for greater separation of the analyte(s) and any remaining matrix components. The configuration of the valves is such that an optional second pump can be used in this step to flush out the autosampler and associated tubing, reducing the re-equilibration time and also reducing carryover. Any matrix components that are strongly retained will again be separated from the analyte(s). These components are generally washed off the column to waste at a later stage to improve the lifetime of the analytical system.

The final phase is to re-equilibrate the system to be ready for the next sample injection. The valve is re-positioned to its original position.

6.5.2.2. *Dual-valve method — no focussing*

As with the single valve method there are four basic steps; load, wash, elute and re-equilibrate of the columns. The tubing configuration for the dual valve method with no focussing is substantially more intricate than the single valve approach. Following the tubing diagrams given in Fig. 6.8, it can be seen that the function of the left-hand valve is to change the direction of the flow going through the extraction column, whereas the function of the right-hand valve is to change the eluent flow from the extraction column, either to waste or to the detector. The use of two valves allows greater flexibility and makes the method more applicable to a wider range of samples; however, it does also mean that there is a potential for greater carryover.

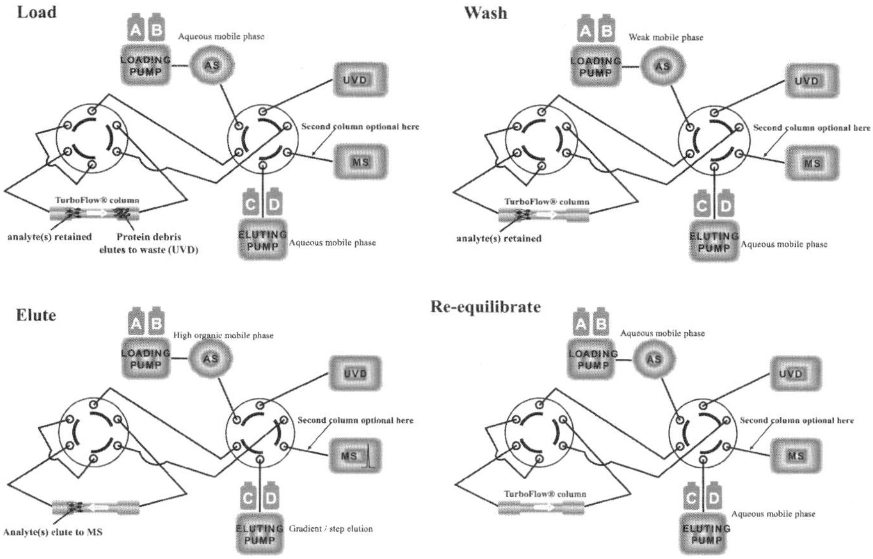

Figure 6.8. The three basic steps to using the dual valve with no focussing. Reprinted with permission from Edge, A. (2003). 'Turbulent flow chromatography in bioanalysis', in Wilson, I.D. (ed.), *Handbook of Bioanalytical Separations, Vol. 4, Bioanalytical Separations*, Elsevier, The Netherlands, pp. 91–127. Copyright (2003) Elsevier.

The sample is loaded onto the extraction column in exactly the same manner as for the one valve system. The advantage of the two valves is that the column can be back-flushed with the loading pump, using the same mobile phase as previously used. The washing of the extraction column can also be done in a forward or a reverse direction, by turning the left-hand valve.

The position of the right-hand valve determines where the eluent stream goes: either to waste or to the detector. Thus, in the elution step the right-hand valve is turned to allow the compounds of interest to be eluted off into the detector. In this arrangement the extracted sample moves directly onto the analytical column with no focussing step; instead the focussing of any components is achieved by careful selection of a chromatographic gradient and selection of the stationary phases. Typically, the analytical column is more retentive than the extraction column to allow for focussing of the analyte on the analytical column under the same conditions that elute the analyte from the clean-up column.

Finally, both valves are repositioned into their original starting positions to allow the system to be fully re-equilibrated, ready for the next sample.

6.5.2.3. *Dual-valve method — focussing*

To achieve chromatographic resolution the valve arrangement shown in Fig. 6.9 can be used. There are five steps to using this particular technique (load, wash, transfer, elute, re-equilibrate).

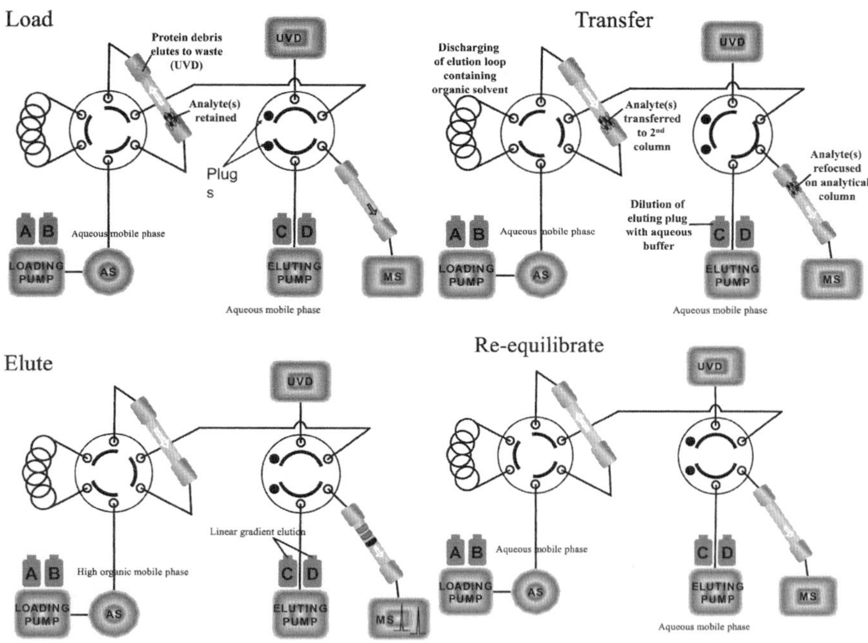

Figure 6.9. The four basic steps to using the Focus Mode with focussing on an analytical column. Reprinted with permission from Edge, A. (2003). 'Turbulent flow chromatography in bioanalysis' in Wilson, I.D. (ed.), *Handbook of Bioanalytical Separations, vol. 4, Bioanalytical Separations*, Elsevier, The Netherlands, pp. 91–127. Copyright (2003) Elsevier.

The first stage is to load the sample onto the extraction column, followed by a wash step; however, unlike the previous approach, it is not possible to reverse the flow through the column without discharging the contents of the sample loop, which are typically used to transfer the analyte from the extraction column to the analytical column.

Following the washing stage the sample is transferred onto an analytical column, using a strong solvent plug contained in the sample loop that was filled during the previous run. In this step both pumps are running a weak solvent which results in the analytes being retained on the analytical column as the solvent plug is diluted in the internal 'T'. Careful selection of the flow rates will ensure that the analytes are focussed at the top of the column, ready to elute the components from the analytical column. The use of a small plug of solvent also ensures that there is a significantly reduced possibility of the compounds starting to elute from the analytical column during the transfer step. The valves are rotated once more to allow for the second pump to be used to elute the compounds of interest from the analytical column. As opposed to the non-focussing configuration above, the same stationary phases can be used. There is a much larger range of column configurations that can be used and there is no need to critically assess the relative retentive properties of the two columns. This is the primary advantage of 'focussing'. The column is chosen only for the best chromatography and no consideration of the relative retentive properties of the two columns being used is required.

The fourth stage is to elute the compounds of interest from the top of the second column into the detector. During this step the clean-up column can be cleansed of any residual matrix components and the sample loop is filled with the appropriate mobile phase to perform the elution from the clean-up column on the next sample. A standard gradient separation can be performed with the added advantage that the compound of interest is focussed at the top of the analytical column, which is not the case with the other arrangements. Finally, the valve is returned to its original position, allowing the system to be re-equilibrated ready for the next injection.

6.6. On-Line SPE

6.6.1. *Theory of SPE*

Solid phase extraction (SPE) is a technique that can utilise a wide range of stationary phase sorbents: polar, non-polar, mixed-mode, polymeric and silica-based. The mixed-mode phases are the most interesting as these give the possibility of having two modes of interaction between a compound and the stationary phase, typically ion exchange and a hydrophobic mechanism.

6.6.2. *Some general considerations*

Although the technology is referred to as solid phase extraction, there are substantial similarities that can be readily applied to the field of chromatography.[126] A thorough understanding of the stationary phase mechanism derived from a chromatographic background will substantially help optimising the retention, washing and elution mechanisms. One obvious difference between the on-line SPE and a chromatography column is the type of samples that are injected. Due to the nature of the sample being injected it is necessary to consider the robustness of the stationary phase, and in particular to reduce the potential for blockages occurring on the column. The easiest approach to take for this is to make the particles bigger; not only does this make the interstitial particles larger, but it also means that larger frits can be employed within the column, which will substantially reduce the column blocking. The larger particles that are routinely employed in SPE will also affect the chromatographic properties, in particular the efficiency.[172–175]

6.6.3. *Mechanisms*

6.6.3.1. *Hydrophobic (reversed-phase)*

The mechanism of retention is the same as that in reversed-phase HPLC; therefore, water acts as a weak solvent and methanol or acetonitrile are strong solvents. The retention of the compounds of interest will depend on the log(P) or log(D) value. High positive values will be strongly retained by the stationary phase. The retention is usually easy to control; the difficulty is often the limit of the solvent strength that can be used to carry out any wash steps. If only very weak interference wash steps can be used, the final extract is likely to be quite dirty. There are a wide range of hydrophilic phases that can be chosen. The two major classes are the silica and polymeric based formats. The polymeric formats are in general more robust at extremes of pH, but some forms can suffer from swelling effects with the addition of organic solvents, which will reduce the chromatographic efficiency of the extraction column. In off-line SPE the polymeric forms benefit from the reduced adverse effects of drying that silica based formats suffer from. However, in the on-line scenario this is not an issue due to the continuous wetting of the phase.

6.6.3.2. *Ion exchange*

The retention of compounds in ion exchange requires that both the stationary phase and the analyte carry one or more charges, and they must be opposite in polarity. Elution can be carried out by using a counter-ion having a high affinity for the stationary phase (i.e. by displacement), but it is much more commonly done by removing the charge on either the stationary phase or the analyte (or both) by manipulation of the pH. It is for this latter reason that knowledge of the pK_a of the analyte (or the stationary phase for a weak exchanger) is important. To remove the charge from a basic compound the pH should be at least 2 pH units above the pK_a; for an acid compound the pH needs to be 2 pH units below the pK_a.

6.6.3.2.1. Strong cation exchange (SCX)

Strong cation exchangers have a sulfonic acid group as part of the stationary phase. This group is permanently negatively charged and will interact with cations, hence the term strong cation exchanger. The charge cannot be removed by altering the pH. They are used to extract basic analytes by manipulating the pH to add or remove charge from the compound of interest.

6.6.3.2.2. Strong anion exchange (SAX)

Strong anion exchangers have a quaternary ammonium group as part of the stationary phase. This group is permanently positively charged and will interact with anions, hence the term strong anion exchanger. The charge cannot be removed by altering the pH. They are used to extract acidic analytes by manipulating the pH to add or remove charge from the compound of interest.

6.6.3.2.3. Weak cation exchange (WCX)

Weak cation exchangers typically have a carboxylic acid group as part of the stationary phase. This group is ionised at pH values above the pK_a and is uncharged at pH values below its pK_a, hence the term weak cation exchanger. They are used to extract basic analytes by manipulating the pH to add or remove charge from either or both the compound of interest and/or stationary phase.

6.6.3.2.4. Weak anion exchange (WAX)

Weak anion exchangers typically have an amino group as part of the stationary phase. This group is ionised at pH values below the pK_a and is uncharged at pH values above its pK_a, hence the term weak anion exchanger. They are used to extract acidic analytes by manipulating the pH to add or remove charge from either or both the compound of interest and/or stationary phase.

6.6.3.2.5. Mixed mode

The term 'mixed mode' refers to phases that exhibit more than one primary interaction mechanism. Typically, reversed-phase interaction is coupled with one of the ion exchange interactions. All of the ion exchange mechanisms are available, SCX, SAX, WCX and

WAX. They have some advantages over the single mode phases, offering fast kinetics from the hydrophobic interaction combined with additional selectivity from the ion exchange mechanism. As an example, a typical way to use a mixed mode strong cation exchanger would be to have the sample containing basic drugs at a pH where the analytes are neutral, and load the sample in an aqueous environment so that the analytes are trapped by the hydrophobic interaction, which is fast. Then pass through some acid (e.g. formic acid 0.1 %) to ionise the basic compounds and invoke the ion exchange mechanism. A strong wash can then be carried out, typically 100 % methanol containing some formic acid. The analyte is still retained due to ionic interactions, but all the matrix components that only interact with the hydrophobic sites are completely removed, resulting in a very clean final extract. The elution step must disrupt both of the retention mechanisms, so in this example we would typically elute with methanol containing a low percentage of ammonia (2–5%). Utilising these phases can have some very positive effects in terms of removing matrix components, the significance of which will be discussed in subsequent sections.

6.6.4. *Practical considerations with on-line SPE*

6.6.4.1. *Optimisation of extraction process*

In this section a reversed-phase methodology is described. The basic principles outlined can be applied to all other phase types.

The optimum wash and elution conditions for an SPE method can be obtained in the following manner. Condition and equilibrate the stationary phase, which for a reversed-phase stationary phase will typically involve the addition of an organic solvent followed by re-equilibration using a weak eluting solvent such as water. Due to the nature of the on-line sample process the original sorbent conditioning step is often left out, since the stationary phase is rarely left in a non-wetted state; however, it is always good practice to allow at least one cycle to wet the stationary phase adequately to ensure good recovery. Where an ion exchange phase is being employed, the pH should be chosen to ensure that the compound of interest and the phase are both in a charged state. Apply the analyte in an aqueous medium and monitor the eluent with the detector, typically an MS. This will determine if the correct retention mechanism is being employed. The concentration of the analytes in the loading solution should be such that it can easily be detected with the detection system that is being employed. It should be noted that the sorbent will have a finite surface area, so it will be possible to overload the active surface sites, which will result in breakthrough of the analyte. For most applications within the environmental and food sectors the concentration levels are very low and as a consequence it is unusual to overload the sorbent material with the analyte, although overloading can also occur if the matrix is retained when using real samples; however, it is possible that if very large sample volumes are being used then the compounds can begin to elute from the sorbent, which may present an issue when analysing aqueous samples. Application of a gradient comprised of the weak solvent and the strong solvent will then allow the optimal wash and elution conditions to be determined. When using a mixed-mode phase for the trapping mechanism it is suggested that the ion exchange mechanism be deactivated by altering the pH before applying this action.

This process is reasonably accurate; however, there are a few points to be aware of for method development:

- If the total recovery in this experiment is low, the analyte must still be on the sorbent.
- Sample loading, both in terms of analyte concentration and also in terms of the sample volume, may cause some compounds to elute.
- Because it is done free of matrix, effects such as protein binding and matrix overloading are not taken into account. If your recovery is good, as previously discussed, which can be determined without matrix, then investigations into the following should be undertaken:
    - o Determine what matrix effects are occurring (binding, suppression, overloading etc.).

- o If protein binding is suspected then investigate approaches to disrupt the binding.
- o Investigate diluting the matrix to ensure that it is not being trapped on the extraction column and hence potentially overloading this column.
- o Alter the final form of chromatography before the detector as this could potentially highlight co-elution of a matrix component with the analyte.

6.6.4.2. *Examples of on-line SPE in the food and environmental areas*

There are many applications that utilise on-line SPE, and although it is not the predominant technology employed within food and environmental laboratories, it is certainly prevalent, aided by the introduction of dedicated instrumentation such as the Equan system (Thermo Scientific, San Jose, California) and also the Symbiosis/Prospekt (Spark Holland, Emmen, The Netherlands), which uses disposable cartridges placed in the fluidic pathway, as opposed to using the same extraction column for multiple samples.

6.6.4.2.1. Environmental applications

In recent years, the number of studies about the occurrence and fate of emerging contaminants such as pesticides, pharmaceuticals, personal care products, industrial chemicals, hormones, flame retardants, and disinfection by-products in the aquatic environment have increased steadily.[98–103] It should be noted that the increase in sensitivity associated with new generations of mass spectrometry have made this technology substantially more accessible to a wider audience since the amount of sample that is required is manageable with modern autosamplers. However, as the sensitivity of the detection system increases, the possibility of performing a direct analysis of the sample becomes more feasible, even though there may still be a requirement for some form of sample clean-up prior to detection when there is a large amount sample matrix present.

The ability to screen for multiple compounds at trace levels makes the use of on-line SPE LC–MS (either using a tandem quadrupole or an ion trap) very attractive. Huntscha *et al.* demonstrated that it was possible to analyse a large number of organic micropollutants,[98] with concentrations as low as 0.1 ng/L being detected. The authors used a novel extraction column that contained a combination of five commercially available sorbent materials. Oasis HLB 15 µm, (Waters, Milford, USA) was the first material in enrichment flow direction. The second material was a mixture of Strata X-AW (33 µm), Strata X-CW (25 µm, both from Phenomenex, Brechbühler AG, Schlieren, Switzerland) and Isolute ENV+ (70 µm, Biotage, Uppsala, Sweden) in a ratio of 1:1:1.5 (X-AW:X-CW:ENV+). This combination of sorbents was required due to the large physiochemical properties of the compounds being analysed. The compounds included pharmaceuticals, pesticides, biocides, and corrosion inhibitors, with acids, bases and neutral compounds being extracted. This study looked not just at the primary pollutant but also at some of the transformation products, which are invariably more-polar compounds. The analysis of transformation products is increasingly becoming of greater interest due to the toxicity of some of the transformation products[99] and their longevity within the environment.

Drugs of abuse (DoA) in water samples are also routinely analysed,[100–103] providing some interesting observations on the recreational habits of the inhabitants of large conurbations. Mendoza analysed up to nineteen DoAs and metabolites belonging to six different chemical classes (cocainics, amphetamine and amphetamine-type stimulants, opioids, lysergic compounds, cannabinoids, and benzodiazepines) from seven different river sources around the Madrid area in Spain. The data showed that the detected drug concentrations did not vary substantially according to the time collected but that the location where the sample was taken did have an effect on the detected drug concentration levels, with the largest variations being observed for the transformation products of cocaine and heroin (Benzoylecgonine and EDDP respectively). The levels detected, although not a threat to humans, could provide a threat to the aquatic life,[104] and there is a worrying trend that the levels are increasing.

6.6.4.2.2. Food applications

The greater complexity associated with food samples compared to water samples means that on-line SPE is not as widely used within the food analysis industry because there is a higher risk of issues associated with carryover and matrix effects from these types of samples.

One area where success has been seen is in the analysis of honey. Several applications have been developed for this particular type of matrix,[105–107] including the analysis of pesticides, sulphonamides and tetracylines. The analysis of tetracyclines was performed by Li,[105] using a 10-port valve and a 6-port valve arrangement. The 10-port valve was used to load the sample into a 1 mL sample loop, whereas the 6-port valve is used to initially load the sample onto the SPE and then to elute the analytes from the SPE to the analytical column. Using this approach and with a 5 g sample initially diluted with 25 mL 0.1 mol/L Na_2EDTA–McIlvaine buffer (pH 4.0), a limit of detection (LOD) for 5 different compounds in the range 5–40 ng/g was achieved with excellent mean recoveries (greater than 90% in all cases). This compared well against other on-line techniques referenced.[108–111]

Sulfonamides are used widely as an antibiotic for farm animals, but at high concentrations they can be harmful, causing diarrhoea, nausea, skin rash, headaches, and dizziness. As a consequence sulfonamides are routinely monitored within the food industry to ensure that food is safe for human consumption. Several methods have been developed for the analysis of this class of antibiotics in a variety of food categories including honey,[112] shrimp,[113] eggs[114] and pork.[114]

One such example is the analysis of sulphonamides in Chantarateepra,[113] which used electrochemical detection to great success for the analysis of seven sulfonamides (sulfaguanidine, sulfadiazine, sulfamethazine, sulfamonomethoxine, sulfamethoxazole, sulfadimethoxine and sulfaquinoxaline), although the lack of chromatographic resolution meant that sulfaguanidine and sulfadiazine could not be analysed uniquely. The quoted LODs ranged from 1–11 ng/mL. The shrimp had to be processed prior to analysis, and this was achieved by taking one gram of a homogeneous shrimp sample, adding 5 mL of Na_2EDTA–McIlvaine buffer solution, mixing well, and filtering. This potentially offers a more financially favourable approach to on-line SPE but still allows for the required detection limits to be obtained, which in the EU is specified at 100 ng/g.[115]

The over-use of herbicides and pesticides has meant that these pollutants have been detected in animal products. Presumably the pollutants came from digestion of the cereal products that were treated with these herbicides and pesticides. Gutiérrez Valencia[116,117] developed an assay for the analysis of a series of organophosphorous pesticides (parathionmethyl, fenitrothion, parathion, chlorfenvinphos, chlorpyrifos, diazinon, ethion, fenchlorphos and carbophenothion) in bovine tissue, following on from work by Kristenson[118] and Moliner-Martínez.[119] The interesting aspect of this work was the combination of on-line SPE with matrix solid-phase dispersion. A 50 mg sample was homogenised and blended with 200 mg of sorbent material and packed into a 30 mm × 8.0 mm I.D. stainless steel cartridge with a polytertafluoroethylene frit and 0.050 g of sorbent silica gel compressed in the bottom to act as a co-column. Fenitrothion was efficiently detected and quantified in an unhealthy liver sample at a concentration of 1.4 (±0.3) µg/g. However, the use of matrix solid phase dispersion with SPE leads to a more complicated valving arrangement (Fig. 6.10).

Other matrices that have been successfully analysed using on-line SPE are pear, tomato, wheat flour, coffee beans, smoked salmon, frankfurter, steak, and pork chop.[120,121] In all cases there was a need to pre-treat the sample prior to analysis, which in all cases involved the homogenisation of the raw sample followed by a crude extraction process, typically a liquid extraction. The analysis of liquids has also been successfully applied to the detection of a range of organic pollutants including milk, soft drinks and tea.[122–126] The pollutants include xenobiotic compounds that are added directly to the food (such as veterinary drugs, pesticides and herbicides), compounds that leach into the foods from packaging materials (such as bisphenol A[125]), as well as endogenous compounds, either at a naturally occurring level or where the food has been modified either additively, as in the case of some vitamins, or where it has been removed, as in the case of decaffeinated coffee.[126]

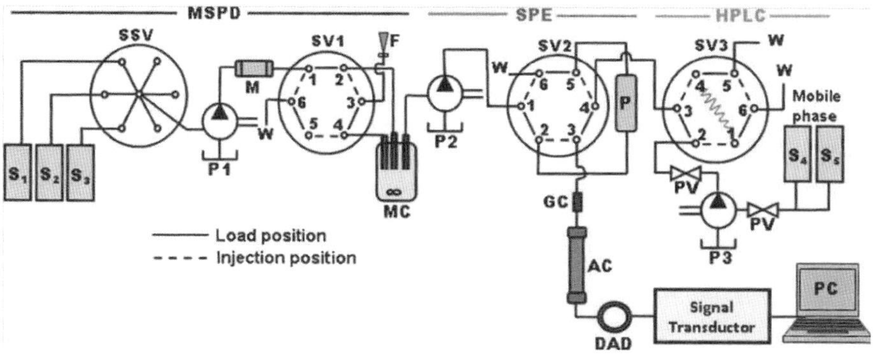

Figure 6.10. Valve configuration for the use of matrix solid-phase dispersion with SPE. Reprinted with permission from Gutierrez Valencia, T.M. and García de Llasera, M.P (2011). Determination of organophosphorus pesticides in bovine tissue by an on-line coupled matrix solid-phase dispersion-solid phase extraction-high performance liquid chromatography with diode array method, *J. Chromatogr. A*, **1218**, 6869–6877. Copyright (2011) Elsevier.

6.7. Turbulent Flow Chromatography

Turbulent flow chromatography (TFC) has courted a substantial amount of controversy since the original patent was submitted in 1988.[88] It has been the focus of a large court case and has also divided the academic community, with many chromatographers believing that the technique is simply a marketing scam. However, there have also been a substantial number of users who have benefitted from the technology, particularly in the field of bioanalysis, although there are also a growing number of publications coming from the food and environmental sectors.

6.7.1. *Theory of TFC*

The use of TFC allows direct injection of biological fluids onto a standard LC system[89] without a significant loss in chromatographic performance over many injections, typically approaching one thousand. TFC has been used widely in the field of bioanalysis, where it is seen as an effective combination of a size discrimination process, since it readily removes proteins from a sample,[89–97] and a trapping sorbent comparable to a solid-phase extraction. The removal of proteins is at least an order of magnitude better than conventional HPLC columns used in a standard chromatographic environment.[97] It has been suggested that the difference in performance is associated with the fluid dynamics of the mobile phase, and the subsequent interaction with the stationary phase.[97]

The development of the various band dispersion models within chromatography has resulted in the realisation that there is an optimal flow rate to perform an analysis, which, using a simple form of the Van Deemter equation,[202] is given by:

$$v = \sqrt{\frac{B}{C}}, \tag{6.1}$$

where B and C are constants, and ν is the reduced velocity, given by;

$$v = \frac{\mu_{mp} . d_p}{D_m}$$

where $\mu_{mp} = \frac{L}{t_0}$ is the average mobile phase velocity, t_0 is time for unretained totally permeating peak and $D_{\rm m}$ is the diffusion coefficient of analyte in the mobile phase.

Normally the B term dominates at low flow, but at high flow rates the C term is dominant. The C term relates to the mass transfer of the analyte through the mobile phase, to the stationary phase, and to any subsequent diffusion on the surface. Much theoretical work has been performed on the exact determination of this phenomenon.[203–206] Under laminar flow conditions, diffusional processes dominate the radial mass transfer through the mobile phase, since there are no other transport mechanisms available. Thus, as the flow increases, solute molecules cannot disperse evenly in a radial manner, resulting in band dispersion due to the different flow rate regime radially across the column. The original concept of using much higher velocities was that the flow enters an inertial driven flow regime where there is much better radial transport of solute molecules, which effectively eliminates the limiting *C* term.[207–215] Under these conditions, the large matrix molecules would not have enough time to diffuse into the pores, since it was thought that there was still a laminar layer around the stationary phase particles that the sample molecules would have to traverse. There was much work performed to demonstrate that the flow regime was not the same as was seen in a standard HPLC column and that indeed the flow was actually turbulent.[88,97,216] This has been questioned by some authors,[217,218] who have looked more closely at the original claims to understand better the fluid dynamics that are occurring within the column at the elevated flow rates used in this technique. Tallerek[217] has demonstrated that the flow regime is different from that used in traditional HPLC; however, this flow is classified as being non-Darcian (i.e. the pressure drop across the column is no longer linear to the flow rate, but the flow regime is still dominated by the viscous forces). Unfortunately, although it is agreed that the flow regime is different to that obtained using standard HPLC, there is still no real understanding of how this technique actually works, with very weak speculation being supplied by some authors,[218] although there is a substantial amount of experimental evidence demonstrating that the technique does work very effectively,[88–97] and that simply increasing the flow rate when under HPLC type fluid conditions would not be applicable.[89]

6.7.2. *Practical considerations with TFC*

6.7.2.1. *Examples of TFC in the environmental and food areas*

The first publications based on turbulent flow chromatography applied it in the field of sample preparation for the analysis of biological fluids and were by Ayrton.[89,90] The technique was successfully used in the analysis of a novel pyridine-isoquinolone compound and its deuterated analogue, under development by Glaxo–Wellcome at the time of the research. Subsequent to the original application of the technology there have been a substantial number of publications in this area, predominantly in the field of bioanalysis applied to clinical or pharmacokinetic studies.[91–97] There have been a limited number of applications in the field of food or environmental analysis although clearly the technique does lend itself to these fields.

6.7.2.1.1. Environmental samples

The first reported use of this technology in the field of environmental analysis, performed at the University of Leipzig,[219–221] has shown that substantial benefits can be achieved with the use of turbulent flow chromatography in the analysis of a range of water samples, including river and lake samples. Traditional methods of water analysis for organic contaminants, such as pesticides and herbicides, involve a lengthy analysis procedure with many different steps. Turbulent flow chromatography has allowed the analysis of environmental samples without the need of complex sample preparation. In the method developed it is possible to determine concentrations of contaminants in river water for all of the compounds listed between 1–125 ng/L in an analysis time of about 15 minutes.

Using the turboflow column as an in-line solid phase extraction cartridge allows large sample volumes to be loaded in a short period of time, which is critical for analysis of pollutants within an aqueous media. In the original work, the authors used the loading pump as the sampling device. This is clearly a limitation if multiple samples are being analysed, but highlighted the issue with the sensitivity of mass spectrometers when this technique was first being applied to these scientific disciplines. Typically the detection limits required for organic pollutants in water are in the range of pg/mL, a factor of 1000 lower than typically required for xenobiotic bioanalysis, although where endogenous compounds are being analysed then the concentration may be comparable for many compounds being monitored in the environmental sector. In order to obtain such sensitivity, solid phase extraction or liquid–liquid extraction has been used to pre-concentrate the sample volume from several hundred millilitres to a few hundred microlitres. The use of turboflow chromatography allows the sample to be loaded onto the extraction column substantially faster than with solid phase extraction. Results from Asperger[221]

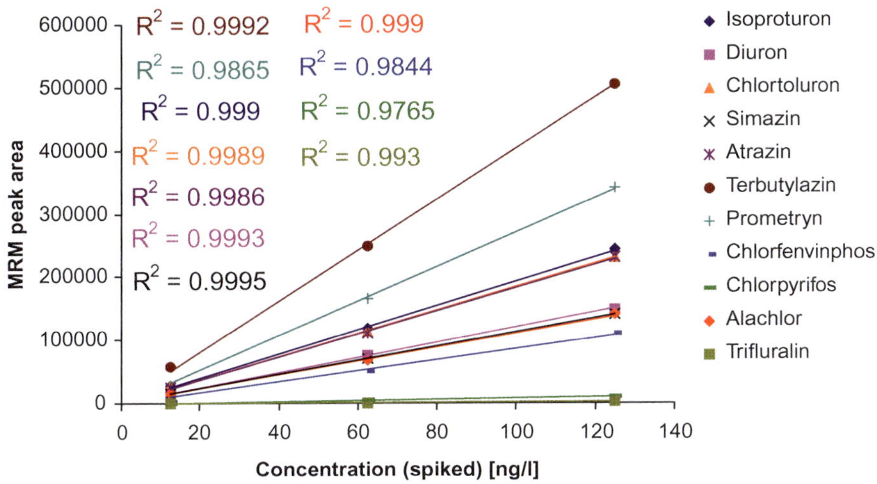

Figure 6.11. Calibration data for twelve compounds extracted from clean water; comparable data was also obtained for river water samples. Edge, A. (2003). 'Turbulent flow chromatography in bioanalysis,' in Wilson, I.D. (ed.), *Handbook of Bioanalytical Separations. Vol. 4: Bioanalytical Separations*, Elsevier, Amsterdam, pp. 91–127. Copyright (2003) Elsevier.

have shown that this approach can yield high recoveries, and very low limits of detection.

Using a single valve method these authors were able to detect 1 pg/mL of a pesticide mixture spiked into clean water, using an initial sampling volume of 10 mL. However, this technique is not applicable to non-polar compounds such as polycyclic aromatic hydrocarbons (PAH), chlorobenzenes, and chloronitrobenzenes, as these compounds do not ionise well using the LC–MS interface.

Other authors have also used this approach, notably López-Serna,[222] who demonstrated that using a combination of trapping column chemistries Cyclone P, C18-P XL and Cyclone MAX (Thermo Scientific, Franklin, USA), and using 2.5 mL (5 mL in negative mode) injection volume, 58 pharmaceutical compounds and their metabolites could be detected in a range of water types including ground water, river water, influent waste and also effluent waste. The work presented discussed the use of a single column chemistry and also the development of the coupled three column chemistries, along with the optimisation of the column I.D. to optimise the analytical performance. The development of greater detectors with improved sensitivity has allowed this approach to be more applicable. In the application developed by López-Serna,[222] a dual-valve, no-focussing approach was employed. Given the wide range of compounds being analysed, a focussed method approach would have been difficult to develop using an isocratic transfer from the trap column to the analytical column, however it would be feasible if a gradient was used to elute components from the trap column.

6.7.2.1.2. Food samples

Substantially more applications have been developed using TFC for the analysis of organic compounds in food products.[223–225] There are many notable pieces of work including the analysis of 36 compounds in chicken tissue by Bousova,[223] which followed on from the development of a method for the analysis of the same compounds in milk[224] and honey.[225] It is noteworthy that in all the publications the effect of the matrix has been investigated, either by varying the milk[224] or by a comparison of solvent and matrix matched samples. The compounds can be classified in accordance with the following compound groups: aminoglycosides, lincosamides, trimethoprim, macrolides, sulfonamides, tetracyclines, and quinolones.

The large number of compounds does cause problems optimising the chromatography, as invariably some compromises have to be made. Also the analysis of aminoglycosides that were present in the sample set meant that the use of an ion-pairing reagent was required, with heptafluorobutyric acid (HFBA) finally being chosen as the additive of choice.

Obviously with meat samples some initial pre-treatment of the sample has to be performed to allow it to be introduced into the chromatographic system in a liquid state. This was achieved using homogenisation of the original sample and then extraction of the homogenised chicken meat (0.5 g) with working internal standard solution (50 μL) and a solvent mixture of ACN:2% TCA (45:55, v/v) (450 μL) was added to the sample. The choice of the extraction solvent was based on the extraction efficiency and also the peak shape. The sample was shaken for 5 min on a vortex mixer equipped with a foam tube holder and then centrifuged at 12,000 rpm for 5 min. The supernatant was filtered through a nylon micro filter (0.45 µm pore size) directly into vials prior to analysis.

The method was validated for determination of thirty-six residues from seven different chemical classes of antibiotics according to EU Commission Decision 2002/675/EC. This entailed a quantifier ion and also a qualifier ion being monitored, as well as ensuring guidelines for linearity, accuracy and precision were adhered to.

Other compounds that have been monitored have been melamine in milk formula,[226] where a dual-valve, no-focussing method was used using a cyclone MCX TFC column (Thermo Scientific, Franklin, USA), and due to the polarity of the analyte, a zwitterionic HILIC approach was employed. The original sample was initially treated with acetonitrile causing removal of the proteins, and since an ion exchange mechanism was being employed for the trapping this did not cause breakthrough of the analyte. The data obtained compared very favourably with an off-line SPE approach and proficiency samples demonstrated that this approach was able to successfully distinguish between melamine and cyanuric acid. A parallel TFC approach was shown to be 15 times faster than the off-line approach.

The detection of enrofloxacin and ciprofloxacin in a variety of meat products has also been reported.[227] A single-valve approach was employed using a 1 mm × 50 mm Cyclone column (Thermo Scientific, Franklin, USA), with a monolithic column providing further enhanced chromatographic resolution. As with the other approaches presented, the meat sample sourced from pig, cattle, rabbit and turkey, with the tissues being monitored including liver, kidney, muscle and skin fat, was initially homogenised in the presence of an extraction solvent 6 mL (acetonitrile:water 1:1 (v/v)) + 0.1 mL formic acid for 1 g of sample). The LOQ was quoted as 25 µg/kg obtained from the validated methods.

Ates[228] demonstrated recently that a dual-valve-with-focussing TFC approach could be used as a screening technique for the analysis of 15 plant and fungal metabolites in wheat, maize and animal feed. A mixed-mode strong cation exchange TFC column, Cyclone MCX-2 (Thermo Scientific, Franklin, USA) was used to trap the compounds of interest, with chromatographic resolution of the compounds being achieved using a Hypersil GOLD column (Thermo Scientific, Runcorn, UK). The use of the Exactive™ Orbitrap™ high-resolution benchtop mass spectrometer (Thermo Fisher Scientific, Bremen, Germany) highlights how the developments in the field of mass spectrometry can be coupled to the development of on-line extraction technologies.

Finally, Mottier[229] and Aguilera-Luiz[230] both reported the development of an assay for a series of veterinary drugs in honey. Mottier carried out quantitative analysis of 16 quinolones in honey using TFC coupled on-line to LC–MS/MS, whereas Aguilera-Luiz used a dual-valve-with-focussing approach to analyse a broader range of antibiotics. The method developed by Mottier involved a pre-treatment of the sample by dilution with H_2O followed by filtration and transfer of an aliquot into a vial. Sample extraction time was 4.5 min, while the overall analysis took 18.5 min. Recovery of the method ranged from 85% to 127%, while the LOD of the method was 5 μg/kg. Mean recoveries presented by Aguilera-Luiz obtained at three concentration levels (5 μg/kg, 10 μg/kg and 50 μg/kg), ranging from 68% to 121% for most compounds. Repeatability (intraday precision) and interday precision (expressed as relative standard deviation, RSD) were <25% for most compounds. Limits of quantification (LOQs) ranged from 5 μg/kg to 50 μg/kg and limits of identification (LOIs) from 0.1 μg/kg to 50 μg/kg.

6.8. MIPs

With the awarding of the Nobel Prize to Cram, Lehn, and Pederson in 1987, the term 'molecular recognition' has been accepted all over the world.[130] The concept of molecular recognition and chemistry[131] is a powerful tool for the understanding of physiological and pharmacological phenomena. Indeed, molecular recognition is the basis of many biological functions, and the synthesis of molecules capable of molecular recognition is attracting a great deal of attention in the fields of biotechnology, medicine and bioanalytical science.

Most of the techniques discussed here relate to extraction of the analytes using the physiochemical properties of the molecules and using this to differentiate between the matrix components and the analytes. Although this will remove some of the matrix components, because of the complexity of the samples being analysed, there will still be a considerable opportunity for molecules that have similar physiochemical properties within the matrix to be present within the final extracted sample. Since these compounds will have similar physical and chemical properties to the analyte molecule, there is a good possibility that they will present issues in the detection system resulting primarily in some form of suppression, which as has been demonstrated previously has a substantial effect on the validity of the quantification of the assay. One technique which is unique in its application to the pre-treatment of a complex sample matrix is the use of molecularly imprinted polymers (MIPs), which relies on an affinity interaction which is more selective than the approach employed by SPE, RAM and TFC.

6.8.1. *Theory and manufacture of MIPs*

MIPs are prepared by the polymerisation of functional and cross-linking monomers in the presence of a template. The monomer is chosen so that it will initially complex with the template molecule, either through a single-monomer template molecule complex or through a multiple-monomers template complex. During the subsequent polymerisation process these complexes become spatially fixed in the highly cross-linked polymer network. Once the polymerisation process has been completed the template molecules are removed. This part of the process has shown weaknesses, since in the original templating part of the process the concentrations of the template molecules are relatively high and as a consequence they are quite difficult to completely remove, which causes severe challenges when the technology is applied to quantitative analysis. Template bleeding has been observed in many scenarios and is often still associated as one of the challenges of this technology.

A variety of approaches have been employed to eliminate template bleeding, by the application of aggressive wash solvents to the final polymer, and with the use of different extraction techniques including:

- methanol or ethanol, each with either acid or base;
- alternating acidic and basic wash conditions;
- microwave-assisted extraction of the template molecule;
- thermal annealing;
- Soxhlet extraction;
- supercritical fluid extraction.

Certainly these approaches do remove a large proportion of the template molecules; however, if the final MIP is to be used for the determination of low-level concentrations of a compound then the bleed may still be an issue. There have been several studies that have done extensive investigations of post polymerisation washes, the most notable of which is by Ellwanger,[132] but the final conclusion from these studies is that it is not feasible to completely remove the template molecule and that the best approach for the manufacture of the MIP is to use an analogue template molecule. Also with a very aggressive wash technique it was noted that there was some polymer degradation, with a resulting loss of selectivity.

Once the MIP has been washed, the resulting polymer then comprises an imprint of the original template molecule. The imprint possesses both a topological and a chemical memory of the template molecule. This allows the polymer to selectively bond to the imprint molecule, or a molecule that contains the same relevant spatial and chemical functionality, from a complex mixture.

There are two principally different approaches that have been utilised for the manufacture: the non-covalent and the covalent. The former is based on metal coordination interactions, and the latter uses reversible covalent bonds to link the monomer to the template molecule. The former approach is the most popular within the analytical chemistry field, although both approaches do have strengths and weaknesses associated with them. It is generally understood that the covalent bonding approach lends itself to more homogeneous and more targeted sites; the metal coordination approach gives more flexibility in the range of functional groups that can be targeted. The latter approach is also easier to use, since the initial template molecule will undergo complex formation with the monomer prior to the polymerisation process. There have been developments in terms of the covalent bonding approach which take advantage of a covalent sacrificial spacer imprinting with non-covalent recognition protocol, as well as the possibility of using a stoichiometric non-covalent imprinting protocol. However, as with the covalent MIPs they do suffer from a general lack of applicability, and as a consequence also some template bleeding.

For the non-covalent bonding of the MIP, the polymer, once it has been synthesised, is generally ground down, with the resulting particulate matter then sized. However, the grinding process does produce irregular-shaped particles and also particles with a large particle-size distribution, which can cause some issues with the final extraction process, either due to low recovery or due to back-pressure issues when using an on-line approach. There are a variety of ways in which the particle size distribution can be improved, either by utilising the technologies that are prevalent within the chromatographic media industry or by using other approaches such as elutriation, where the sample is allowed to separate in a solvent under gravity (heavier particles, and hence larger, will preferentially migrate to the bottom of the vessel before the lighter particles). The approaches used by the chromatographic media organisations tend to be a little more technologically advanced than these approaches and employ separation devices which allow for much tighter control of the particle size distribution. However, the irregular-shaped particles that are produced during the grinding process do result in the inefficient packing of the on-line extraction columns. This was an issue that was addressed by the column manufacturers with silica-based columns several decades previously, but relied on the formation of spherical particles through a sol gel process, which is not directly applicable in the manufacture of the MIPs.

There have been investigations to determine if a better approach can be utilised for the manufacture of a more regular-shaped particle, and these include the approach of using dispersion polymerisation, where the initial reagents are solutions but as the polymerisation proceeds the resulting polymer is insoluble and precipitates out of the solution. One approach that has had some success is to use a silica particle as a substrate and then to graft on a thin film of the imprinted polymer.[133] This uses spherical porous silica particles, which can be synthesized in a controlled manner to obtain specific particle diameters, as demonstrated by the myriad of particles offered by the chromatographic media manufacturers.

Monodisperse imprinted beads can also be manufactured using a two-step swelling process. In this scenario, latex seed particles are suspended in an aqueous solution. The particles are initially swelled with a suitable solvent and then the monomer and the template molecules are added.[134]

For the on-line extraction of compounds, the MIP is generally manufactured using a surrogate of the analyte molecule, due to the issues associated with template bleeding, and the resultant lack of a quantifiable method. Using a surrogate analyte molecule as the template molecule can still result in template bleeding; however, it is feasible to resolve the template molecule from the analyte using a chromatographic separation.[143] This was the first assay highlighted as a possible solution to the template bleed issue with the extraction of sameridine where a close structural analogue was employed as the template molecule. Table 6.1 gives some examples of the type of template molecules that have been employed and also the analytical molecules that were under investigation.

6.8.2. *Practical considerations with MIPs*

6.8.2.1. *Optimisation of extraction process*

6.8.2.1.1. Load

As with the other technologies discussed, an awareness of the method-development process and the interactions that are occurring in the various steps is important to aid optimisation of the extraction process. A MIP can be best characterised as mixed-mode phase having polar and lipophilic surface functionality as well as the imprinted affinity sites. Thus there are several sites where the analyte molecule

Table 6.1. Different structural analogues being used as template molecules in MIP designs.

Analyte	Template
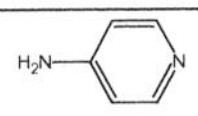	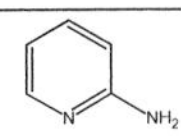
4 aminopyridine	2-aminopyridine
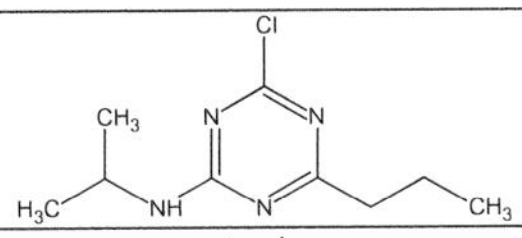	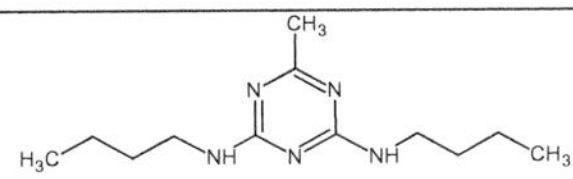
atrazine	dibutylmelamine
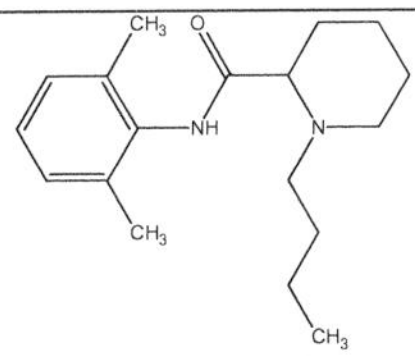	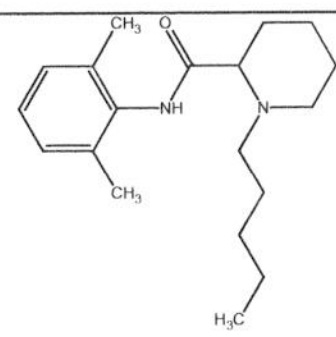
bupivacaine	pentivacaine
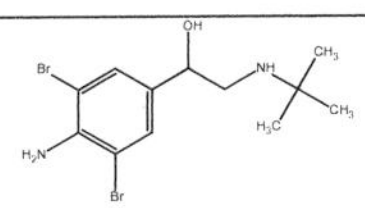	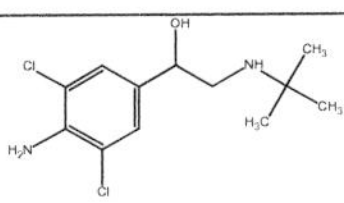
bromobuterol	Clenbuterol
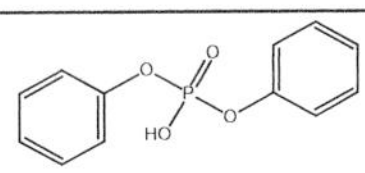	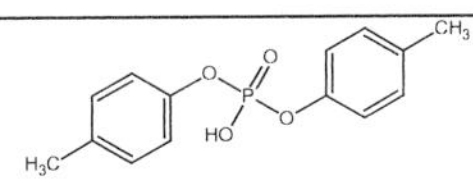
diphenylphosphate	diolylphoshate
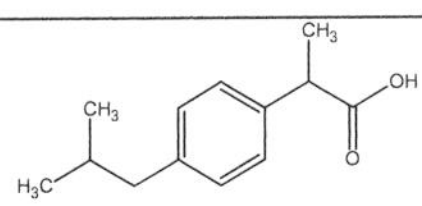	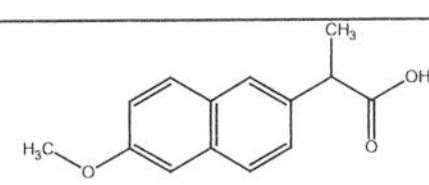
S-ibuprofen	S-naproxen
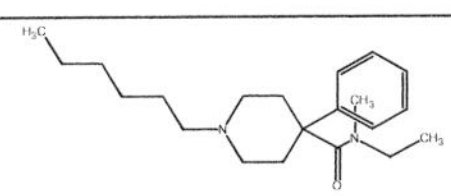	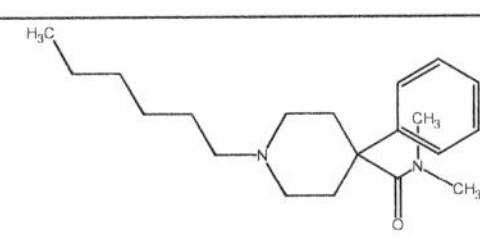
Sameridine	N-methyl-1-hexyl-N-methyl-4-phenylpiperidine-4-carboxamide

(*Continued*)

Table 6.1. (*Continued*)

Analyte	Template
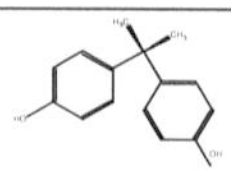	
Bisphenol A	Phenol
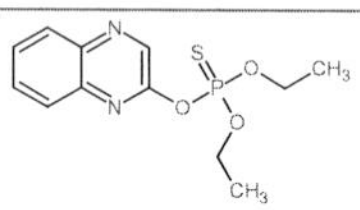	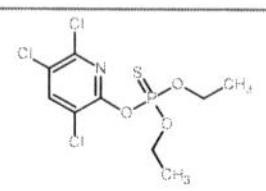
Quinalphos	Chlorpyrifos

can be effectively retained. In order to have the greatest specificity, it is necessary to ensure that the binding of the affinity site is substantially greater than that obtained with the more generic binding sites. If this is not the scenario then the extraction phase will have little selectivity. The advantage of this technology compared to the more generic solid phase extraction approach is that there is much greater selectivity; however, the phase has to be used correctly to ensure that this selectivity is optimised. To ensure that the selectivity associated with the affinity mechanism is dominant and that it is not masked by stronger polar and lipophilic interactions, the choice of the appropriate solvents is critical because it can alter the degree of interaction with each of the three mechanisms discussed.

To eliminate or reduce the degree of binding occurring through a lipophilic interaction with the polymer, organic solvent or a surfactant can be added.[135,136] There have been a range of surfactants that have been successfully employed to reduce the non-selective component of the retention of the analyte, including Tween 20, Triton X-100, and Brij 35.[137] Altering the pH can also affect the degree of non-specific binding.[137–139] However, the use of a detergent molecule can result in significant ion suppression and so some care has to be taken to ensure that this does not occur, either by ensuring that co-elution of the analyte and the detergent does not exist or that the detergent does not cause ion suppression.

6.8.2.1.2. Washing

Once the sample has been loaded onto the extraction cartridge it is necessary to wash any remaining non-specifically adsorbed matrix components from the phase. As with the other techniques mentioned, this requires a solvent that will disrupt the non-specific binding, in this case the hydrophobic interaction between the matrix components and the polymeric surface. Typical solvents that can be used as a suitable wash solvent are acetonitrile and dichloromethane. Unlike solid-phase extraction though, where the retention mechanism of the analyte molecule and the matrix components can be the same and so the composition of the solvent used becomes the discriminating factor, when using a MIP the change of the solvent will lead to a redistribution of the analyte molecules into the imprinted sites, where the mode of retention is a normal-phase one, based on strong hydrogen bonding and electrostatic interactions. The weaker retention mechanism of the bulk polymer for the non-targeted components is based on a hydrophobic interaction and as a consequence in the presence of an organic solvent this weaker bond is easily disrupted. For water analysis, the bulk polymer can be used to initially retain the analyte through a hydrophobic interaction; however, on switching the solvents over to the wash solvents the analyte molecules will become adsorbed to the MIP through a more selective binding mechanism.[140–142]

6.8.2.1.3. Elution

This is possibly the most difficult stage when using a MIP, due to the strength of the interaction between the binding sites and the analyte molecule, and thus to effectively disrupt this binding can require harsh conditions. This is something that has to be considered when deciding on which extraction technique to employ, since with some molecules these harsh conditions can cause a breakdown of the analyte of interest.[143,144] Compounds with amino functionalities on a methacrylic acid (MAA) MIP can be the most prone to degradation, since the elution solvents will typically consist of acetonitrile with high percentages of TFA, TEA, or acetic acid. Stronger reagents have also been used such as 5M sodium hydroxide, ethanol and heptane mixtures.[143] For neutral compounds and weak acids or bases, complete elution can occur with the use of a mixture of water and polar solvents.[144–146] This general technology can obviously be applied in an off-line as well as an on-line format.

6.8.2.2. *Examples of MIPs in the environmental and food areas*

6.8.2.2.1. Environmental

A variety of MIPs have been prepared for many of the standard pollutants found in the environment. Indeed the analysis of persistent organic pollutants is an ideal opportunity for MIPs, as many of the inherent costs associated with the development of the new imprinted phase can be offset by the continuing samples that are present, whereas the pharmaceutical industry can be more prone to changes due to the nature of that market. In particular, the analysis of phenolic compounds such as bisphenol A has generated some interest.[147] Ou[148] demonstrated that it was possible to add the MIP to a monolithic substrate by in situ polymerization using 4-vinylpyridine (4-VP) and ethylenedimethacrylate (EDMA) as functional monomer and cross-linker, respectively. Bisphenol A was used as the template molecules but the authors looked at six phenolic compounds in total. Comparisons were made between the imprinted monolith and the non-imprinted version and it was demonstrated that a substantial improvement in selectivity was obtained. The authors deduced that hydrophobic and hydrogen-bonding interaction played important roles in the recognition process, with the elution solvent being acidified acetonitrile.

Sambe[149] demonstrated that MIPs could be used for the analysis of methylthiotriazines in river water. In this example a RAM-MIP was employed for the synthesis using irgarol as the template molecule. Using a single-valve configuration and UV detection the authors were able to obtain good linearity and high levels of sensitivity for a range of compounds, including atrazine, propazine, simazine, ametryn, prometryn, simetryn, terbuthylazine and irgarol. The LOD was in the range 50–500 pg/mL (r > 0.999) with a 100 mL loading of a river water sample. The recoveries were also very good, in excess of 90% for all of the compounds being analysed.

Zhao *et al.*[150] developed an on-line MIP–SPE procedure coupled to HPLC for selective extraction of the four Sudan dyes in samples from tomato sauce, sausage, and water obtained from the Yellow River. The proposed method showed that the new MIP obtained using an attapulgite clay as the matrix was feasible in the determination of these Sudan dyes in real samples. The LODs were in the range of 1.0–3.0 ng/g for tomato sauce, 0.8–3.0 ng/g for sausage and 0.01–0.05 ng/mL for Yellow River water.

There have been a range of other MIPs that have been developed for other applications, including the analysis of diclofenac,[151] PAHs,[152] sulpride,[153] methylthiotriazine herbicide,[154] antiepileptics,[155] pirimicarb[156] and non-steroidal anti-inflammatory drugs,[157] all in different forms of water samples.

6.8.2.2.2. Food

As with the other on-line approaches, some rudimentary form of sample pre-treatment has to be applied to the analysis of food to convert the original sample into a format that is applicable to a liquid-handling autosampler. Once in a liquid format, the valve configurations that are employed are the same as the other techniques that have been discussed. However, although the literature has many examples of MIPs being used off-line, there are substantially fewer applications of this technology being applied to on-line sample analysis.

One such example[158] was successfully applied to the simultaneous multi-residue analysis of six tetracyclines in spiked milk and honey samples. Using tetracycline (TC) as the template, MAA as the functional monomer, ethylene glycol dimethacrylate (EGDMA) as the cross-linker, methanol as the solvent, cyclohexanol and dodecanol as the mixed porogenic solvents, the authors were able to synthesise a monolithic MIP that was able to extract the analytes of interest. The amount of analyte detected was in the range 0.1–5 mg/L in matrix, with recoveries greater than 70% for the majority of analytes tested. In an extensive optimisation study the authors present data highlighting the effect of buffer concentration on retention, and also the effect of the gradient elution on the optimisation of matrix removal. Data was also presented which demonstrated the benefits of this approach when compared to extraction using a non-imprinted monolithic column.

Cacho[159] demonstrated that MIP technology could be successfully incorporated into capillary electrochromatography for the analysis of thiabendazole in citrus samples. Using a monolithic MIP the authors were able to investigate the effects of mobile-phase composition, temperature and also the voltage across the column to optimise the extraction. Several other authors have also looked at the combination of MIPs with capillary electrochromatography.[160–166] It was noted that the mobile phase has a substantial effect on the imprinting factor (k_{MIP}/k_{NIP}, the relative dimensionless retention times of the imprinted and non-imprinted polymer). For the example given the use of acetonitrile as the solvent resulted in a substantially higher IF value compared to methanol. It was also noted that tailing was present, which is a common problem associated to MIP stationary phases, mainly due to the slow adsorption/desorption equilibrium. The authors looked at both lemon and orange peel/pulp and demonstrated that this approach was able to determine LODs of 0.04 mg/kg with very good recovery and precision being obtained.

Bjarnason[167] used a coupled-column system, consisting of a combination of a MIP and a C18-silica column, for the selective detection of triazine urine and apple extracts. The MIP showed good performance for selectively discriminating triazines from humic acid. Enrichment was observed in all cases, and triazine-enrichment factors of up to 100 could be recorded, with good extraction efficiency (74–77%).

A molecularly imprinted polymer (MIP) tailored for the HPLC determination of the fungicide thiabendazole (TBZ) was synthesised using a single preparative step by precipitation polymerisation in an acetonitrile/toluene co-solvent, using TBZ as template molecule, methacrylic acid as functional monomer and divinylbenzene-80 as a crosslinker. TBZ was shown to retain on the MIP, with the effects of different chromatographic parameters (e.g. temperature, flow-rate and elution solvents) on TBZ retention/elution studied. Under optimised conditions, the TBZ-imprinted column was used for the HPLC-fluorescence (HPLC-F) determination of TBZ directly from orange (both whole fruit and juice), lemon, grape, and strawberry extracts at low concentration levels in less than 15 min.[168]

Hantash *et al.*[169] were able to extract and detect carbaryl and its metabolites, derived from carbamate insecticides from apple homogenates. The apples were initially homogenised with phosphate buffer, pH 7.0 (50:50, w/w). The homogenate was centrifuged (3000 rpm; 15 min) and the supernatant filtered through a 0.45 µm Teflon filter. A single-valve experimental arrangement was employed for the analysis, with chromatographic resolution being supplied by a Gemini C18 column (Phenomenex). The apple homogenate was spiked at various concentrations ranging from 1 ng/mL to 8 ng/mL, with the coefficient of variation being less than 5 for all concentrations studied (N = 6).

6.9. Restricted-Access Media (RAM)

6.9.1. *Theory of RAM*

The development of HPLC packing materials for the direct injection of biological samples resulted in packing materials that protect partitioning bonded phases from protein contamination by preventing access to the bonded phases through size-exclusion mechanisms. This allows proteins to interact only with hydrophilic, non-adsorptive layers on the outer packing surfaces. Alternatively, small molecules penetrate the porous packing and gain full access to partitioning phases. There are other terms that are readily associated with this type of packing, included 'internal phase', 'internal surface', 'shielded', 'dual-zone', 'semi-permeable, and 'mixed-functional'. The 'restricted access' term can also refer to packing materials that have been coated with a hydrophilic bonded phase on large-pore silica, where macromolecules are not completely size-excluded.

There has been extensive work performed on a variety of RAM phases for biological fluids, and the technique is ideally suited for the field of bioanalysis, in particular pharmacokinetic and toxicokinetic studies.[170–172] There are a variety of considerations to ensure the design of a successful RAM, namely:

- The internal bonded-phase should be protected from irreversible adsorption of proteins and other macromolecules.
- The external surface of the packing material should be rendered non-adsorptive to macromolecules, allowing for the unwanted macromolecules to be eluted in one peak.
- The packing should allow partitioning of small molecules with the protected bonded phase with good diffusional mass transport.
- The bonded-phase should have sufficiently high selectivity to resolve analytes from related compounds or interfering endogenous small molecules.
- The retentivity of the bonded-phase should not be too strong to avoid the use of mobile phases that will not induce the coalescence/precipitation of any macromolecules.

Clearly, this does leave a possibility of a wide range of phases that would give the characteristic mixed mode interactions required to separate the small molecules and macromolecules. The most common of these types of phases include the protein-coated ODS phases,[173,174] the internal surface reversed-phase (ISRP) supports,[175,176] the shielded hydrophobic phases,[177,178] the semi-permeable surface phases,[178,179] the dual-zone material[180,181] and the mixed-functional phases.[182,183] The greatest amount of data has been published on the use of the ISRP columns and protein-coated columns; therefore, the rest of the discussion will focus on these types of columns.

Reversed-phase HPLC has been commonly used in the analysis of biological extracts, and in environments where sample throughput is a limiting factor there is a tendency to do little or no pre-treatment of plasma samples. This can result in plasma proteins denaturing on the stationary phase, accumulating in the interstitial packing space, and clogging the columns, causing them to fail due to over-pressurisation or due to the formation of poor peak shape caused by channelling effects. To circumvent these difficulties Yoshida and co-workers[184] coated the outer surfaces of large particulate (20–30 μm) octadecylsilane (ODS) silica, with 120 Å pore diameters, by purging packed ODS columns with bovine serum albumin (BSA) or rabbit plasma at pH 3 in methanol. The columns were then washed with methanol to remove any proteins that were not absorbed to the stationary phase. It is speculated that the protein, denatured by the methanol, coated the external surfaces of the ODS silica, and provided a means of attenuating the adsorption of injected sample proteins, while allowing small drug analytes to penetrate the pores.[185]

It was Hagestam and Pinkerton[186] who introduced the ISRP concept in 1985. High-performance ISRP packings are produced by first bonding a high coverage (300 μmol/g) hydrophilic phase to small-pore (80 Å), 5 μm diameter silica. This can be done with either glycerylpropyl (diol) groups[187] or aminopropyl groups capped with glycidol.[188] This then will form the outer non-adsorptive layer. The partitioning phase is then attached to the hydrophilic layer. Finally the bulk packing is treated with enzymes to remove the partitioning moieties only from the outer surfaces of the packing particulates, since the enzymes cannot penetrate the porous supports and reach a partitioning phase inside the silica particulates. A variety of molecular entities have been investigated as potential partitioning phases for ISRP packings.[175,176]

6.9.2. *Practical considerations with RAM*

6.9.2.1. *Optimisation of clean-up process*

The optimisation of the extraction process is very similar to that used for on-line SPE. The initial loading of the sample should be performed in an environment where any macromolecules are not retained on the column, either directly or due to the mobile phase causing precipitation. In general, these conditions will also benefit the retention of the small analyte molecules, and thus it is generally not an issue. In terms of ensuring that the analyte is effectively adsorbed to the inner surface of the substrate, it is necessary to ensure that the surface will be retentive enough in the mobile phase environment that is present during the loading stage. Once the analyte molecule is adsorbed to the internal surface of the sorbent, and the macromolecules are removed, the next stage is to wash the column to ensure that a degree of selectivity is maintained. This is performed and optimised in exactly the same manner as that proposed for the SPE previously. Finally the elution step has to be optimised, and, as with the SPE optimisation, careful selection of the solvent so that the elution strength is not too strong will ensure as clean an extract as possible. The variety of stationary phases available will make sure that a degree of optimisation can also occur with the stationary phase as well as the eluents.

6.9.2.2. *Examples of RAM in the environmental and food areas*

6.9.2.2.1. Environmental

Surprisingly, given that the technique is based on the removal of macromolecules, which are not prevalent in a wide range of environmental samples, the RAM technique has been applied in this area. One example is given by Barreiro,[189] who demonstrated the successful extraction of pantoprazole and lansoprazole enantiomers. The researcher used a BSA-protein-coated octyl silica (Luna, 10 μm particle size and 100 Å pore size (Phenomenex, Torrance, USA)) for the removal of the humic acids present in the river water samples. Using a modified single-valve approach they were able to inject 1 mL of sample onto the RAM column, which trapped the analytes, and then eluted the compounds from the extraction column onto a chiral column to allow discrimination of the two enantiomers using a simple acetonitrile water mobile phase. The total analysis time was 40 min but there was no sample preparation required, and the researchers were able to demonstrate that the technique could be used for real sample analysis. Other analytes found in water have also been analysed using this approach, including the analysis of s-triazines[190] and also omeprazole enantiomers.[191] For both applications the removal of humic acid was key to the success of the assay.

One interesting environmental application that is often quoted is the analysis of ten acidic pesticides[192] (bentazone denzothiadiazole, bromoxynil hydroxybenzonitrile, metsulfuron-methyl sulfonyl urea, 2,4-D phenoxy acid, MCPA phenoxy acid, 2,4-DP phenoxy acid, MCPP phenoxy acid, 2,3,5-T phenoxy acid, 2,4-DB phenoxy acid and MCPB phenoxy acid) in a range of different soils. The initial pre-treatment of the various soils was performed using microwave-assisted extraction, with 10 g of initial sample being analysed with 20 mL of solvent. A quarter of this solvent was then dried and reconstituted in 2 mL of solvent of which 400 μL was injected onto the system. Three valves and three pumps are utilised to obtain the extraction with the detection being performed by UV, which although not as selective as mass spectrometry does not suffer from the indirect matrix effects such as ion-suppression observed with the latter technique. Somewhat confusingly, the authors refer to the method as a multiple reaction monitoring (MRM) and compare it to a previous method that was for a single compound (single reaction monitoring (SRM)). The authors used principal-component analysis to determine correlations between soil type, age of spiked soil and recovery, with some correlation being observed between the different sets. The calculated LODs for all the compounds was in the low μg/kg range, demonstrating that this would be an effective analytical technique for real samples.

6.9.2.2.2. Food

There are several applications in the literature where RAM is used successfully.[193,194] Of particular note are examples of sulphonamides in milk and in eggs.[195] Based on previous work analysing the same compounds in milk,[177] Kishida demonstrated that it was possible to analyse sulfamonomethoxine, sulfadimethoxine, and their N^4-acetyl metabolites from eggs. Using 300 mg of egg that was pre-treated with 600 μL of 4 mol/L of ammonium sulfate and homogenised, 20 μL of the resultant supernatant was injected onto the analytical system. Using UV detection it was feasible to obtain LOD of 0.01–0.03 μg/g for all four compounds, with the extraction efficiencies greater than 90% for all compounds. As early as 1996, Ueno[196] was able to demonstrate the successful extraction of sulfamonomethoxine, miloxacin and oxolinic acid in serum and muscle of cultured fish using UV detection with almost 100% recoveries from serum, and greater than 70% recoveries quoted for muscle tissue.

Other compounds that have been analysed in milk include cephalexin and neomycin[197] and carbamazepine.[198] In all cases a protein-modified column was used as the RAM. It was noted by Lopes[198] that it was necessary to centrifuge the milk prior to analysis and then take aliquots from the middle portion of resulting sample (a thin fatty layer and a thin aqueous layer).

6.10. Discussion on Future and Other Technologies

The list of on-line sample preparation techniques discussed here in detail is not exhaustive and the level of detail given to the four main categories can be improved. In particular, the use of immunoaffinity columns has not been discussed, and there are a variety of examples of how this very selective — if somewhat costly — technology can be readily applied to the extraction of compounds in complex matrices. They are manufactured by binding an antibody group onto the surface of a substrate material, such as agrose, and the antibody then acts as an affinity ligand to selectively trap the compound of interest. The antibodies can be generated from the original protein or a peptide group. For example if an organism is immunised against a glutathione S-transferase (GST)-fusion protein, then it will produce antibodies against the fusion-protein, and possibly antibodies against the GST tag as well. The protein can then be covalently coupled to the solid support. The process of binding becomes very selective and this approach will invariably give a very clean extract. Most monoclonal antibodies have been purified using affinity chromatography based on immunoglobulin-specific Protein A or Protein G, derived from bacteria. Compounds trapped on an immunoaffinity column are typically eluted by changing salt concentrations, pH, charge, and ionic strength directly or through a gradient to resolve the components of interest.

The other significant area with development potential that has not been discussed is miniaturisation of the technology. This has often been discussed in conjunction with standard chromatography systems, and the possibilities that are offered by using on-line miniaturised systems are significant. It potentially allows for throw-away technology, which would allow analysis of samples in seconds, and offers the possibility of taking the sampling device to the samples rather than the conventional approach of bringing the samples to a central analytical facility.

The prospect of miniaturisation of on-line analytical systems using lab-on-a-chip devices has been present for quite a few years,[231–233] but has struggled to reach its full commercial potential. The use of column diameters less than 100 μm is becoming standard in the field of proteomics, and with this there has been a greater acceptance of the use of nano chromatography. Reasons for the greater level of acceptance of this technology include the better design of instrumentation and better connections of the columns to both the autosamplers and the detectors. The more robust technology may lend itself to on-line separations, although the miniaturisation of the valving technology is critical to the success of this. Other approaches have been employed, which have resulted in the reduction of the chromatographic assembly, most notably the lab-on-a-CD.[234] In this situation the separation is driven by the spinning of a disk which has a capillary column attached. UV detection can be utilised, although other detection techniques have also been used. This approach offers the potential to place a sample onto the CD, and simultaneously record the details of the sample electronically onto the same CD, which presents a unique solution to sample traceability.

6.11. Conclusion

The use of on-line technology is prevalent within the environmental and food industries, with an increase in its use arising as a result of the ever-increasing sensitivities associated with mass spectrometry. There are still some inherent challenges associated with this approach, primarily carryover; however, as analytical scientists become more aware of approaches in which this can be reduced, so the techniques can become more applicable. The increased pressure to reduce costs and also improve the quality of analysis ensures that there will be a growing number of laboratories that will continue to investigate and use on-line technologies. Improved designs of columns and the incorporation of valve systems into the mainstream vendor systems will also encourage users to take up this approach.

It has been demonstrated that as well as its challenges, there are distinct advantages to the on-line approaches, what with the reduction in the sample manipulation, and the allowing for the handling of reduced sample volumes. Within the water analysis field this has tremendous advantages over the more traditional liquid–liquid extractions, which will typically require a litre of sample. Investigation of the wider field of environmental analysis also demonstrates that there are advantages to be had over conventional approaches; in particular, the sample pre-concentration steps or blow-down steps can be very time consuming, whereas the use of an on-line approaches eliminates this step.

The use of mass spectrometry did initially threaten the routine use of sample preparation techniques, since it was thought that this highly sensitive and highly selective technique was not prone to matrix effects. However, a substantial amount of research in this area has demonstrated that the ionisation process is very definitely prone to matrix effects and as a consequence when dealing with complex samples it is still necessary to remove the bulk of components prior to analysing on the mass spectrometer. The most efficient manner to do this is to use a series of switching valves allowing the full automation of the analytical process. In particular, most foods are very complex, typically containing many tens of thousands of compounds, and thus effective sample preparation is critical with this type of sample to ensure that the mass spectrometer is not affected by co-elution of matrix components.

Where mass spectrometry is not used, the use of sample preparation is critical, and it has been demonstrated the advantages that the correct sample preparation technique can have on the analysis of a wide range of organic pollutants in a variety of food matrices, either animal-derived or based on fruit and vegetables. The variety of on-line technologies that are available to the analytical chemist can be bewildering; however, it has been demonstrated that technologies such as MIPS, SPE, TFC and RAM do have different mechanisms and thus intelligent choices can be made based on the financial burden and more importantly on the quality of the assay.

References

1. Ismaiel, O.A., Halquist, M.S., Elmamly, M.Y. *et al.* (2008). Monitoring phospholipids for assessment of ion enhancement and ion suppression in ESI and APCI LC/MS/MS for chlorpheniramine in human plasma and the importance of multiple source matrix effect evaluations, *J. Chromatogr. B,* **875**, 333–343.
2. Furey, A., Moriarty, M., Bane, V. *et al.* (2013). Ion suppression: A critical review on causes, evaluation, prevention and applications, *Talanta,* **115**, 104–122.
3. Li, X.Q., Yang, Z., Zhang, Q.H. *et al.* (2014). Evaluation of matrix effect in isotope dilution mass spectrometry based on quantitative analysis of chloramphenicol residues in milk powder, *Anal. Chimica Acta,* **807**, 75–83.
4. Yadav, M., Trivedi, V., Upadhyay, V. *et al.* (2012). Comparison of extraction procedures for assessment of matrix effect for selective and reliable determination of atazanavir in human plasma by LC–ESI-MS/MS, *J. Chromatogr. B*, **885–886**, 138–149.
5. Marchi, I., Viette, V., Badoud, F. *et al.* (2010). Characterization and classification of matrix effects in biological samples analyses, *J. Chromatogr. A*, **1217**, 4071–4078.
6. Socas-Rodríguez, B., Asensio-Ramos, M., Hernández-Borges, J. *et al.* (2013). Chromatographic analysis of natural and synthetic estrogens in milk and dairy products, *TrAC, Trends Anal. Chem.*, **44**, 58–77.
7. Navarre, D.A., Pillai, S. S., Shakya, R. *et al.* (2011). HPLC profiling of phenolics in diverse potato genotypes, *Food Chem.*, **127**, 34–41.
8. Murillo, E., Giuffrida, D., Menchaca, D. *et al.* (2013). Native carotenoids composition of some tropical fruits, *Food Chem.*, **140**, 825–836.

9. Du, W., Lei, C., Zhang, S. *et al.* (2014). Determination of clenbuterol from pork samples using surface molecularly imprinted polymers as the selective sorbents for microextraction in packed syringe, *J. Pharm. Biomed. Anal.*, **91**, 160–168.
10. Núñez, O., Moyano, E. and Galceran, M.T. (2005). LC–MS/MS analysis of organic toxics in food, *TrAC, Trends Anal. Chem.*, **24**, 683–703.
11. Petrović, M., Hernando, M.D., Díaz-Cruz, M.S. *et al.* (2005). Liquid chromatography–tandem mass spectrometry for the analysis of pharmaceutical residues in environmental samples: a review, *J. Chromatogr. A*, **1067**, 1–14.
12. da Silva Júnior, F.M.R., Rocha, J.A.V. and Vargas, V.M.F. (2009). Extraction parameters in the mutagenicity assay of soil samples, *Sci. Total Environ.*, 407, 6017–6023.
13. Pan, J., Zhang, C., Zhang, Z. *et al.* (2014). Review of online coupling of sample reparation techniques with liquid chromatography, *Anal. Chimica Acta*, **815**, 1–15.
14. Andreu, V. and Picó, Y. (2004). Determination of pesticides and their degradation products in soil: critical review and comparison of methods, *TrAC, Trends Anal. Chem.*, **23**, 772–789.
15. dos Reis, R.R., Sampaio, S.C. and de Melo, E.B. (2013). The effect of different log P algorithms on the modeling of the soil sorption coefficient of nonionic pesticides, *Water Res.*, **47**, 5751–5759.
16. Wenlock, M.C. and Barton, P. (2013), In silico physicochemical parameter predictions, *Mol. Pharmaceutics*, **10**, 1224–1235.
17. Krayukhina, E., Uchiyama, S., Nojima, K. *et al.* (2013). Aggregation analysis of pharmaceutical human immunoglobulin preparations using size-exclusion chromatography and analytical ultracentrifugation sedimentation velocity, *J. Biosci. Bioeng.*, **115**, 104–110.
18. Ji, C., Walton, J., Su, Y. and Tella, M. (2010). Simultaneous determination of plasma epinephrine and norepinephrine using an integrated strategy of a fully automated protein precipitation technique, reductive ethylation labeling and UPLC–MS/MS, *Anal. Chim. Acta*, **670**, 84–91.
19. Anastassiades, M., Lehotay, S.J., Stajnbaher, D. *et al.* (2003). Fast and easy multiresidue method employing acetonitrile extraction/partitioning and "dispersive solid-phase extraction" for the determination of pesticide residues in produce , *J. AOAC Int.*, **86**, 412–431.

20. Payá, P., Anastassiades, M. and Mack, D. (2007). Analysis of pesticide residues using the Quick Easy Cheap Effective Rugged and Safe (QuEChERS) pesticide multiresidue method in combination with gas and liquid chromatography and tandem mass spectrometric detection, *Anal. Bioanal. Chem.*, **389**, 1697–1714.
21. Wilkowska, A. and Biziuk, M. (2011). Determination of pesticide residues in food matrices using the QuEChERS methodology, *Food Chem.*, **125**, 803–812.
22. Huck, C.W. and Bonn, G.K. (2000). Recent developments in polymer-based sorbents for solid-phase extraction, *J. Chromatogr. A*, **885**, 51–72.
23. Nilsson, U.J. (2000). Solid-phase extraction for combinatorial libraries, *J. Chromatogr. A*, **885**, 305–319.
24. Picó, Y., Fernández, M., Ruiz, M.J. *et al.* (2007). Current trends in solid-phase-based extraction techniques for the determination of pesticides in food and environment, *J. Biochem. and Biophys. Meth.*, **70**, 117–131.
25. Ferrer, I. and Barceló, D. (1999). Validation of new solid-phase extraction materials for the selective enrichment of organic contaminants from environmental samples, *TrAC, Trends Anal. Chem.*, **18**, 180–192.
26. Pinto, M.I., Sontag, G., Bernardino, R.J. *et al.* (2010). Pesticides in water and the performance of the liquid-phase microextraction based techniques: A review, *Microchem. J.*, **96**, 225–237.
27. Pirsaheb, M., Fattahi, N. and Shamsipur, M. (2013). Determination of organophosphorous pesticides in summer crops using ultrasound-assisted solvent extraction followed by dispersive liquid–liquid microextraction based on the solidification of floating organic drop, *Food Control*, **34**, 378–385.
28. Majors, R. (2012). Supported liquid extraction: the best-kept secret in sample preparation, *LC-GC Europe*, **25**, 430–435.
29. Dong, H., Tang, H., Chen, D. *et al.* (2014). Analysis of 7 synthetic musks in cream by supported liquid extraction and solid phase extraction followed by GC–MS/MS, *Talanta*, **120**, 248–254.
30. Loayza-Muro, R. and Elías-Letts, R. (2007). Responses of the mussel *Anodontites trapesialis* (*Unionidae*) to environmental stressors: effect of pH, temperature and metals on filtration rate, *Environ. Pollut.*, **149**, 209–215.

31. Bubola, K.B., Koprivnjak, O. and Sladonja, B. (2012). Influence of filtration on volatile compounds and sensory profile of virgin olive oils, *Food Chem.*, **132**, 98–103.
32. AOAC Official Method 2007.01, Pesticide Residues in Foods by Acetonitrile Extraction and Partitioning with Magnesium Sulfate.
33. Kinsella, B., Whelan, M., Cantwell, H. *et al.* (2010). A dual validation approach to detect anthelmintic residues in bovine liver over an extended concentration range, *Talanta*, **83**, 14–24.
34. EN 15662 Version 2.2, Date 2008-04, Foods of plant origin — Determination of pesticide residues using GC–MS and/or LC–MS/MS following acetonitrile extraction/partitioning and cleanup by dispersive SPE–QuEChERS-method.
35. Lehotay, S.J., Son, K.A., Kwon, H. *et al.* (2010). A dual validation approach to detect anthelmintic residues in bovine liver over an extended concentration range, *J. Chromatogr. A*, **1217**, 2548–2560.
36. Venn, R.F. (2000). *Principles and Practice of Bioanalysis*, Taylor and Francis Inc., London/New York.
37. Watt, A.P., Morrison, D., Locker, K.L. *et al.* (2000). Higher throughput bioanalysis by automation of a protein precipitation assay using a 96-well format with detection by LC–MS/MS, *Anal. Chem.*, **72**, 979–984.
38. Wells, D. (2003). *High Throughput Bioanalytical Sample Preparation: Methods and Automation Strategies*, Elsevier, The Netherlands, pp. 199–254.
39. Hagerman, A.E. and Butler, L.G. (1978). Protein precipitation method for the quantitative determination of tannins, *J. Agric. Food Chem.*, **26**, 809–812.
40. Yamada, R., Kozono, M., Ohmori, T. *et al.* (2006). Simultaneous determination of residual veterinary drugs in bovine, porcine, and chicken muscle using liquid chromatography coupled with electrospray ionization tandem mass spectrometry, *Biosci. Biotech. Biochem.*, **70**, 54–65.
41. Wu, X., Xiong, E., Wang, W. *et al.* (2014). Universal sample preparation method integrating trichloroacetic acid/acetone precipitation with phenol extraction for crop proteomic analysis, *Nat. Protoc.*, **9**, 362–374.
42. 'Environmental analytical chemistry of PCBs' in Albaigés, J. (ed.) (1993), *Current Topics in Environmental and Toxicological Chemistry (Volume 16)*, Gordon and Breach Science.

43. Chen, S., Hoene, M., Li, J. *et al.* (2013). Simultaneous extraction of metabolome and lipidome with methyl tert-butyl ether from a single small tissue sample for ultra-high performance liquid chromatography/mass spectrometry, *J. Chromatogr. A*, **1298**, 9–16.
44. Escrivá, A., Esteve, M.J., Farré, R. *et al.* (2002). Determination of liposoluble vitamins in cooked meals, milk and milk products by liquid chromatography, *J. Chromatogr. A*, **947**, 313–318.
45. Tavakoli, L., Yamini, Y., Ebrahimzadeh, H. *et al.* (2008). Homogeneous liquid–liquid extraction for preconcentration of polycyclic aromatic hydrocarbons using a water/methanol/chloroform ternary component system, *J. Chromatogr. A*, **1196–1197**, 133–138.
46. Sun, L., Forni, S., Schwartz, M.S. *et al.* (2012). Quantitative determination of odanacatib in human plasma using liquid–liquid extraction followed by liquid chromatography–tandem mass spectrometry analysis, *J. Chromatogr. B*, **885–886**, 15–23.
47. Kujawski, M.W., Bargańska, Ż., Marciniak, K. *et al.* (2014). Determining pesticide contamination in honey by LC–ESI–MS/MS — Comparison of pesticide recoveries of two liquid–liquid extraction based approaches, *LWT-Food Sci. Technol.*, **56**, 517–523.
48. Sereshti, H., Heidari, R. and Samadi, S. (2014). Determination of volatile components of saffron by optimised ultrasound-assisted extraction in tandem with dispersive liquid–liquid microextraction followed by gas chromatography–mass spectrometry, *Food Chem.*, **143**, 499–505.
49. Schneider, N., Werkmeister, K. and Pischetsrieder, M. (2011). Analysis of nisin A, nisin Z and their degradation products by LC–MS/MS, *Food Chem.*, **127**, 847–854.
50. Chen, X-B. Wu, Y-L. and Yang, T. (2011). Simultaneous determination of clenbuterol, chloramphenicol and diethylstilbestrol in bovine milk by isotope dilution ultraperformance liquid chromatography–tandem mass spectrometry, *J. Chromatogr. B*, **879**, 799–803.
51. Spongberg, A.L. and Witter, J.D. (2008). Pharmaceutical compounds in the wastewater process stream in Northwest Ohio, *Sci. Total Environ*, **397**, 148–157.
52. Mottier, P., Parisod, V., Gremaud, E. *et al.* (2003). Determination of the antibiotic chloramphenicol in meat and seafood products by liquid chromatography–electrospray ionization–tandem mass spectrometry, *J. Chromatogr. A*, **994**, 75–84.

53. Rivera, S.M. and Canela-Garayoa, R. (2012). Analytical tools for the analysis of carotenoids in diverse materials, *J. Chromatogr. A*, **1224**, 1–10.
54. Horvath, K., Fairchild, J. and Guiochon, G. (2009). Optimization strategies for off-line two-dimensional liquid chromatography, *J. Chromatogr. A*, **1216**, 2511–2518.
55. Mondello, L., Tranchida, P.Q., Stanek, V. *et al.* (2005). Silver-ion reversed-phase comprehensive two-dimensional liquid chromatography combined with mass spectrometric detection in lipidic food analysis, *J. Chromatogr. A*, **1086**, 91–98.
56. Batt, A.L. and Aga, D.S. (2005). Simultaneous analysis of multiple classes of antibiotics by ion trap LC–MS/MS for assessing surface water and groundwater contamination, *Anal. Chem.*, 77, 2940–2947.
57. Alder, L., Greulich, K., Kempe, G. *et al.* (2006). Residue analysis of 500 high priority pesticides: Better by GC–MS or LC–MS/MS?, *Mass Spectrom. Rev.*, **25**, 838–865.
58. Kellmann, M., Muenster, H., Zomer, P. *et al.* (2009). Full scan MS in comprehensive qualitative and quantitative residue analysis in food and feed matrices: how much resolving power is required?, *J. Am. Soc. Mass Spectrom.*, **20**, 1464–1476.
59. Yamashita M. and Fenn, J.B. (1984). Electrospray ion source: Another variation on the free-jet theme, *J. Phys. Chem.*, **88**, 4451–4459.
60. Mei, H., Nardo, C., Xu, X. *et al.* (2002). Investigation of matrix effects in bioanalytical high-performance liquid chromatography–tandem mass spectrometric assays: application to drug discovery, *Rapid Commun. Mass Spectrom.*, **17**, 97–103.
61. Backe, W.J. and Field, J.A. (2012). Is SPE necessary for environmental analysis? A quantitative comparison of matrix effects from large-volume injection and solid-phase extraction based methods, *Environ. Sci. Technol.*, **46**, 6750–6758.
62. Sangster, T., Spence, M., Sinclair, P. *et al.* (2004). Unexpected observation of ion suppression in a liquid chromatography/atmospheric pressure chemical ionization mass spectrometric bioanalytical method, *Rapid Commun. Mass Spectrom.*, **18**, 1361–1364.

63. Matuszewski, B.K., Constanzer, M.L. and Chavez-Eng, C.M. (2003). Strategies for the assessment of matrix effect in quantitative bio-analytical methods based on HPLC–MS/MS, *Anal. Chem.*, 75, 3019–3030.
64. Liumbruno, G., A'Alessandro, A., Grazzini, G. *et al.* (2010). Blood-related proteomics, *J. Proteomics*, **73**, 483–507.
65. Wilkinson, G.R. (1997). The effects of diet, aging and disease-states on presystemic elimination and oral drug bioavailability in humans, *Adv. Drug Deliv. Rev.*, **27**, 129–159.
66. Larger, P.J., Breda, M., Fraier, D. *et al.* (2005). Ion-suppression effects in liquid chromatography–tandem mass spectrometry due to a formulation agent, a case study in drug discovery bioanalysis, *J. Pharm. Biomed. Anal.*, **39**, 206–216.
67. Ismaiel, O.A., Halquist, M.S., Elmamly, M.Y. *et al.* (2008). Monitoring phospholipids for assessment of ion enhancement and ion suppression in ESI and APCI LC–MS/MS for chlorpheniramine in human plasma and the importance of multiple source matrix effect evaluations, *J. Chromatogr. B*, **875**, 333–343.
68. Cole, R.B. and Harrata, A.K. (1993). Solvent effect on analyte charge state, signal intensity, and stability in negative ion electrospray mass spectrometry; implications for the mechanism of negative ion formation, *J. Amer. Soc. Mass Spectrom.*, **4**, 546–556.
69. Jessome, L.L. and Volmer, D.A. (2006). Ion suppression: a major concern in mass spectrometry, *LCGC North Am.*, **24**, 498–511.
70. Wallace, A.M., Gibson, S., de la Hunty, A. *et al.* (2010). Measurement of 25-hydroxyvitamin D in the clinical laboratory: current procedures, performance characteristics and limitations, *Steroids*, 75, 477–488.
71. Lu, B., Ren, Y., Huang, B. *et al.* (2008). Simultaneous determination of four water-soluble vitamins in fortified infant foods by ultra-performance liquid chromatography coupled with triple quadrupole mass spectrometry, *J. Chromatogr. Sci.*, 225–232.
72. Ferreiro-Vera, C., Priego-Capote, F. and Luque de Castroab, M.D. (2013). An approach for quantitative analysis of vitamins D and B9 and their metabolites in human biofluids by on-line orthogonal

sample preparation and sequential mass spectrometry detection, *Analyst*, **138**, 2146–2155.

73. Blake, C.J. (2007). Status of methodology for the determination of fat-soluble vitamins in foods, dietary supplements, and vitamin premixes, *J. AOAC Int.*, **90**, 897–910.
74. Asakawa, Y., Ozawa, C., Osada, K. *et al.* (2007). Reduction of carry-over in column-switching HPLC/MS system with automated system washing procedure for highly sensitive direct analysis of donepezil in dog plasma, *J. Pharm. Biomed. Anal.*, **43**, 683–690.
75. Armbruster, D.A. and Alexander, D.B. (2006). Sample to sample carryover: a source of analytical laboratory error and its relevance to integrated clinical chemistry/immunoassay systems, *Clin. Chim. Acta*, **373**, 37–43.
76. Hurtado-Sánchez, M.C., Romero-González, R., Rodríguez-Cáceres, M.I. *et al.* (2013). Rapid and sensitive on-line solid phase extraction–ultrahigh-performance liquid chromatography–electrospray–tandem mass spectrometry analysis of pesticides in surface waters, *J. Chromatogr. A*, **1305**, 193–202.
77. Clouser-Roche, A., Johnson, K., Fast, D. *et al.* (2008). Beyond pass/fail: a procedure for evaluating the effect of carryover in bioanalytical LC/MS/MS methods, *J. Pharm. Biomed. Anal.*, **47**, 146–155.
78. Tanaka, N., Kimata, K., Hosoya, K. *et al.* (1993). Stationary phase effects in reversed-phase liquid chromatography, *J. Chromatogr. A*, **656**, 265–287.
79. Vallano, P.T., Shugarts, S.B., Woolf, E.J. *et al.* (2005). Elimination of autosampler carryover in a bioanalytical HPLC–MS/MS method: a case study, *J. Pharm. Biomed. Anal.*, **36**, 1073–1078.
80. Dolman, S., Eeltink, S., Vaast, A. *et al.* (2013). Investigation of carry-over of peptides in nano-liquid chromatography–mass spectrometry using packed and monolithic capillary columns, *J. Chromatogr. B*, **912**, 56–63.
81. Snyder, L.R. (1974). Classification of the solvent properties of common liquids, *J. Chromatogr. A*, **92**, 223–230.
82. Shoghi, E., Fuguet, E., Bosch, E. *et al.* (2013). Solubility–pH profiles of some acidic, basic and amphoteric drugs, *Eur. J. Pharm. Sci.* **48**, 291–300.

83. Grant, R., Cameron, C., Mackenzie, S. *et al.* (2002). Generic serial and parallel on-line direct-injection using turbulent flow liquid chromatography–tandem mass spectrometry, *Rapid Commun. Mass Spectrom.*, **16**, 1785–1792.
84. Guo, F., Liu, Q., Shi, J-B. *et al.* (2014). Direct analysis of eight chlorophenols in urine by large-volume injection on-line turbulent flow solid-phase extraction liquid chromatography with multiple wavelength ultraviolet detection, *Talanta*, **119**, 396–400.
85. Marzinke, M.A., Breaud, A.R. and Clarke, W. (2013). The development and clinical validation of a turbulent-flow liquid chromatography–tandem mass spectrometric method for the rapid quantitation of docetaxel in serum, *Clin. Chim. Acta*, **417**, 12–18.
86. Herman, J.L. (2004). Generic approach to high throughput ADME screening for lead candidate optimization, *Int. J. Mass Spectrom.*, **238**, 107–117.
87. Hennion, M.-C. (1999). Solid-phase extraction: method development, sorbents, and coupling with liquid chromatography, *J. Chromatogr. A*, **856**, 3–54.
88. Quinn, H.M. and Takarewski, J.J. International Patent Number WO 97/16724.
89. Ayrton, J., Dear, G.J., Leavens, W.J. *et al.* (1999). The use of turbulent wflow chromatography–mass spectrometry for the rapid, direct analysis of a novel pharmaceutical compound in plasma, *Rapid Commun. Mass Spectrom.*, **11**, 1953–1958.
90. J. Ayrton, J., Plumb, R., Leavens, W.J. *et al.* (1998). Application of a generic fast gradient liquid chromatography–tandem mass spectrometry method for the analysis of cytochrome P450 probe substrates, *Rapid Commun. Mass Spectrom.*, **12**, 217–224.
91. Lim, H.K., Chan, K.W., Sisenwine, S. *et al.* (2001). Simultaneous screen for microsomal stability and metabolite profile by direct-injection turbulent-laminar flow LC–LC and automated tandem mass spectrometry, *Anal. Chem.*, **73**, 2140–2146.
92. Zimmer, D., Packard, V., Czembor, W. *et al.* (1999). Comparison of turbulent-flow chromatography with automated solid-phase extraction in 96-well plates and liquid–liquid extraction used as plasma sample

preparation techniques for liquid chromatography–tandem mass spectrometry, *J. Chromatogr. A*, **854**, 22–35.

93. Jemal, M. and Yuan-Qing, W. (1998). The use of high-flow high-performance liquid chromatography coupled with positive and negative ion electrospray–tandem mass spectrometry for quantitative bioanalysis via direct injection of the plasma/serum samples, *Rapid Commun. Mass Spectrom.*, **12**, 1389–1399.
94. Wu, J-T., Zeng, H., Qian, M. *et al.* (2000). Direct plasma sample injection in multiple-component LC–MS/MS assays for high-throughput pharmacokinetic screening, *Anal. Chem.*, **72**, 61–67.
95. Brignol, N., Bakhtiar, R., Dou, L. *et al.* (2000). Quantitative analysis of terbinafine (Lamisil®) in human and minipig plasma by liquid chromatography tandem mass spectrometry, *Rapid Commun. Mass Spectrom.*, **14**, 141–149.
96. Herman, J.L. (2002). Generic method for on-line extraction of drug substances in the presence of biological matrices using turbulent flow chromatography, *Rapid Commun. Mass Spectrom.*, **16**, 421–426.
97. Edge, A. (2003). 'Turbulent flow chromatography in bioanalysis', in Wilson, I.D. (ed.), *Handbook of Bioanalytical Separations Vol. 4: Bioanalytical Separations*, Elsevier, The Netherlands, pp. 91–127.
98. Huntscha, S., Singer, H.P., McArdell, C.S. *et al.* (2012). Multiresidue analysis of 88 polar organic micropollutants in ground, surface and wastewater using online mixed-bed multilayer solid-phase extraction coupled to high performance liquid chromatography–tandem mass spectrometry, *J. Chromatogr. A*, **1268**, 74–83.
99. Cui, N., Zhang, X., Xie, Q. *et al.* (2011). Toxicity profile of labile preservative bronopol in water: the role of more persistent and toxic transformation products, *Environ. Pollut.*, **159**, 609–615.
100. Mendoza, A., López de Alda, M., González-Alonso, S. *et al.* (2014). Occurrence of drugs of abuse and benzodiazepines in river waters from the Madrid Region (Central Spain), *Chemosphere*, **95**, 247–255.
101. Baker, D.R. and Kasprzyk-Hordern, B. (2011). Multi-residue determination of the sorption of illicit drugs and pharmaceuticals to wastewater suspended particulate matter using pressurised liquid extraction, solid phase extraction and liquid chromatography coupled with tandem mass spectrometry, *J. Chromatogr. A*, **1218**, 1620–1631.

102. Jurado, A., Mastroianni, N., Vàzquez-Suñé, E. *et al.* (2012). Drugs of abuse in urban groundwater: a case study — Barcelona, *Sci. Total Environ.*, **424**, 280–288.
103. Berset, J.-D., Brenneisen, R. and Mathieu, C. (2010). Analysis of licit and illicit drugs in waste, surface and lake water samples using large volume direct injection high performance liquid chromatography–electrospray tandem mass spectrometry (HPLC–MS/MS), *Chemosphere*, **81**, 859–866.
104. Farré, M., Pérez, S., Kantiani, L. *et al.* (2008). Fate and toxicity of emerging pollutants, their metabolites and transformation products in the aquatic environment, *TrAC, Trends Anal. Chem.*, **27**, 991–1007.
105. Li, J., Chen, L., Wang, X. *et al.* (2008). Determination of tetracyclines residues in honey by on-line solid-phase extraction high-performance liquid chromatography, *Talanta*, **75**, 1245–1252.
106. Sajid, M., Na, N., Safdar, M. *et al.* (2013). Rapid trace level determination of sulfonamide residues in honey with online extraction using short C-18 column by high-performance liquid chromatography with fluorescence detection, *J. Chromatogr. A*, **1314**, 173–179.
107. Schirmer, C. and Meisel, H. (2006). Synthesis of a molecularly imprinted polymer for the selective solid-phase extraction of chloramphenicol from honey, *J. Chromatogr. A*, **1132**, 325–328.
108. Pereira, A.V. and Cass, Q.B. (2005). High-performance liquid chromatography method for the simultaneous determination of sulfamethoxazole and trimethoprim in bovine milk using an on-line clean-up column, *J. Chromatogr. B*, **826**, 139–146.
109. Oliveira, R.V., Pietro, A.C.D. and Cass, Q.B. (2007). Quantification of cephalexin as residue levels in bovine milk by high-performance liquid chromatography with on-line sample cleanup, *Talanta*, **71**, 1233–1238.
110. Oliferova, L., Statkus, M., Tsysin, G. *et al.* (2007). On-line solid-phase extraction and high-performance liquid chromatography determination of polycyclic aromatic hydrocarbons in water using polytetrafluoroethylene capillary, *Talanta*, **72**, 1386–1391.
111. Yamamoto, E., Takakuwa, S., Kato, T *et al.* (2007). Sensitive determination of aspirin and its metabolites in plasma by LC–UV using on-line solid-phase extraction with methylcellulose-immobilized anion-exchange restricted access media, *J. Chromatogr. B*, **846**, 132–138.

112. Sajid, M., Na, N., Safdar, M. *et al.* (2013). Rapid trace-level determination of sulfonamide residues in honey with online extraction using short C-18 column by high-performance liquid chromatography with fluorescence detection, *J. Chromatogr. A*, **1314**, 173–179.
113. Chantarateepra, P., Siangproh, W., Motomizu, S. *et al.* (2012). Automated online solid-phase extraction coupled with sequential injection–HPLC–EC system for the determination of sulfonamides in shrimp, *Int. J. Electrochem.*, Vol. 2012, Article ID 862823, available at http://dx.doi.org/10.1155/2012/862823.
114. Fang, G.Z., He, J.X. and Wang, S. (2006). Multiwalled carbon nanotubes as sorbent for on-line coupling of solid-phase extraction to high-performance liquid chromatography for simultaneous determination of 10 sulfonamides in eggs and pork, *J. Chromatogr. A*, **1127**, 12–17.
115. Commission of the European Community. 1991. *The rules governing medical products in the European Community IV*, Brussels, Belgium.
116. Gutiérrez Valencia, T.M. and García de Llasera, M.P. (2011). Determination of organophosphorus pesticides in bovine tissue by an on-line coupled matrix solid-phase dispersion–solid phase extraction–high-performance liquid chromatography with diode array detection method, *J. Chromatogr. A*, **1218**, 6869–6877.
117. García de Llasera, M.P. and Reyes-Reyes, M.L. (2009). *Food Chem.*, **114**, 1510–1516.
118. Kristenson, E.M., Haverkate, E.G.J., Slooten, C.J. *et al.* (2001). Miniaturized automated-matrix solid-phase dispersion extraction of pesticides in fruit followed by gas chromatographic–mass spectrometric analysis, *J. Chromatogr. A*, **917**, 277–286.
119. Moliner-Martinez, Y., Campíns-Falcó, P., Molins-Legua, C. *et al.* (2009). Miniaturized matrix solid-phase dispersion procedure and solid-phase microextraction for the analysis of organochlorinated pesticides and polybrominated diphenylethers in biota samples by gas chromatography electron capture detection, *J. Chromatogr. A*, **1216**, 6741–6745.
120. Riediker, S., Obrist, H., Varga, N. *et al.* (2002). Determination of chlormequat and mepiquat in pear, tomato, and wheat flour using on-line solid-phase extraction (Prospekt) coupled with liquid chromatography–electrospray ionization tandem mass spectrometry, *J. Chromatogr. A*, **966**, 15–23.

121. Gosetti, F., Chiuminatto, U., Mazzucco, E. *et al.* (2011). Simultaneous determination of thirteen polycyclic aromatic hydrocarbons and twelve aldehydes in cooked food by an automated on-line solid phase extraction ultra high performance liquid chromatography tandem mass spectrometry, *J. Chromatogr. A*, **1218**, 6308–6318.
122. Kantiani, L., Farré, M. and Barceló, D. (2011). Rapid residue analysis of fluoroquinolones in raw bovine milk by online solid-phase extraction followed by liquid chromatography coupled to tandem mass spectrometry, *J. Chromatogr. A*, **1218**, 9019–9027.
123. Gallart-Ayala, H., Moyano, E. and Galceran, M.T. (2011). Analysis of bisphenols in soft drinks by on-line solid-phase extraction fast liquid chromatography–tandem mass spectrometry, *Anal. Chim. Acta*, **683**, 227–233.
124. Lin, S.-L., Lo, C.-Y. and Fuh, M.-R. (2012). Quantitative determination of perchlorate in bottled water and tea with online solid-phase extraction high-performance liquid chromatography coupled to tandem mass spectrometry, *J. Chromatogr. A*, **1246**, 40–47.
125. Petrovic, M., Eljarrat, E., López de Alda, M.J. *et al.* (2002). Recent advances in the mass spectrometric analysis related to endocrine disrupting compounds in aquatic environmental samples, *J. Chromatogr. A*, **974**, 23–51.
126. Salinas-Vargas, E.M. and Cañizares-Macías, P.M. (2014). On-line solid–phase extraction using a C18 minicolumn coupled to a flow injection system for determination of caffeine in green and roasted coffee beans, *Food Chem.*, **147**, 182–188.
127. Hu, X., Pan, J., Hu, Y. *et al.* (2009). Preparation and evaluation of propranolol molecularly imprinted solid-phase microextraction fiber for trace analysis of β-blockers in urine and plasma samples, *J. Chromatogr. A*, **1216**, 190–197.
128. Prasad, B.B., Tiwari, K., Singh, M. *et al.* (2008). Molecularly imprinted polymer-based solid-phase microextraction fiber coupled with molecularly imprinted polymer-based sensor for ultratrace analysis of ascorbic acid, *J. Chromatogr. A*, **1198**, 59–66.
129. Jandera, P. (2013). Advances in the development of organic polymer monolithic columns and their applications in food analysis — a review, *J. Chromatogr. A*, **1313**, 37–53.

130. Pedersen C.J. (1967). Cyclic polyethers and their complexes with metal salts, *J. Am. Chem. Soc.*, **89**, 7017–7036.
131. Mayes, A.G. and Mosbach, K. (1996). Molecularly imprinted polymer beads: suspension polymerization using a liquid perfluorocarbon as the dispersing phase, *Anal. Chem.*, **68**, 3769–3774.
132. Ellwanger, A., Berggren, C., Bayoudh, S. *et al.* (2001). Evaluation of methods aimed at complete removal of template from molecularly imprinted polymers, *Analyst*, **126**, 784–792.
133. Halhalli, M.H., Schillinger, E., Aureliano, C.S.A. *et al.* (2012). Thin-walled imprinted polymer beads featuring both uniform and accessible binding sites, *Chem. Mat.*, **24**, 2909–2919.
134. Karlsson, J., Andersson, L.I. and Nicholls, I.A. (2001). Probing the molecular basis for ligand-selective recognition in molecularly imprinted polymers selective for the local anaesthetic bupivacaine, *Anal. Chim. Acta.*, **435**, 57–64.
135. Souverain, S., Rudaz, S. and Veuthey, J.-L., (2004). Restricted access materials and large particle supports for on-line sample preparation: an attractive approach for biological fluids analysis, *J. Chromatogr. B*, **801**, 141–156.
136. Andersson, L.I., Abdel-Rehim, M., Nicklasson, L. *et al.* (2002). Towards molecular-imprint based SPE of local anaesthetics, *Chromatographia*, **55**, S65–S69.
137. Lu, C.-H., Zhou, W.-H., Han, B. *et al.* (2007). Surface-imprinted core-shell nanoparticles for sorbent assays, *Anal. Chem.*, **79**, 5457–5461.
138. Haupt, K., Dzgoev, A. and Mosbach. K. (1998). Assay system for the herbicide 2,4-dichlorophenoxyacetic acid using a molecularly imprinted polymer as an artificial recognition element, *Anal. Chem.*, **70**, 628–631.
139. Baggini, C., Giovannoli, C., Anfossi, L. *et al.* (2001). Molecularly imprinted solid-phase extraction sorbent for the clean-up of chlorinated phenoxyacids from aqueous samples, *J. Chromatogr. A*, **938**, 35–44.
140. Carro, E., Masque, N., Marcé, R.M. *et al.* (2002). Non-covalent and semi-covalent molecularly imprinted polymers for selective on-line solid-phase extraction of 4-nitrophenol from water samples, *J. Chromatogr. A*, **963**, 169–178.

141. Martin, P., Wilson, I.D., Morgan, D.E. *et al.* (1997). Evaluation of a molecular-imprinted polymer for use in the solid-phase extraction of propranolol from biological fluids, *Anal. Commun.*, **34**, 45.
142. Olsen, J., Martin, P., Wilson, I.D. *et al.* (1999). Methodology for assessing the properties of molecular imprinted polymers for solid phase extraction, *Analyst*, **124**, 467–471.
143. Andersson, L.I., Paprica, A. and Arvidsson, T. (1997). A highly selective solid-phase extraction sorbent for pre-concentration of sameridine made by molecular imprinting, *Chromatographia*, **46**, 57–62.
144. Haginaka, J. and Sanbe, H. (2000). Uniform-sized molecularly imprinted polymers for 2-arylpropionic acid derivatives selectively modified with hydrophilic external layer and their applications to direct serum injection analysis, *Anal. Chem.*, **72**, 5206–5210.
145. Matsui, J., Fujiwara, K., Ugata, S. *et al.* (2000). Solid-phase extraction with a dibutylmelamine-imprinted polymer as triazine herbicide-selective sorbent, *J. Chromatogr. A*, **889**, 25–31.
146. Ferrer, I., Lanza, F., Tolokan, A. *et al.* (2000). Selective trace enrichment of chlorotriazine pesticides from natural waters and sediment samples using terbuthylazine molecularly imprinted polymers, *Anal. Chem.*, **72**, 3934–3941.
147. Ikegami, T., Lee, W.S., Nariai, H. *et al.* (2004). Covalent molecular imprinting of bisphenol A using its diesters followed by the reductive cleavage with $LiAlH_4$, *J. Chromatogr. B*, **804**, 197–201.
148. Ou, J., Hu, L., Hu, L. *et al.* (2006). Determination of phenolic compounds in river water with on-line coupling bisphenol A imprinted monolithic precolumn with high performance liquid chromatography, *Talanta*, **69**, 1001–1006.
149. Sambe, H., Hoshina, K. and Haginaka, J. (2007). Molecularly imprinted polymers for triazine herbicides prepared by multi-step swelling and polymerization method: their application to the determination of methylthiotriazine herbicides in river water, *J. Chromatogr. A*, **1152**, 130–137.
150. Zhao, C., Zhao, T., Liu, X. *et al.* (2010). A novel molecularly imprinted polymer for simultaneous extraction and determination of sudan dyes

by on-line solid phase extraction and high-performance liquid chromatography, *J. Chromatogr. A*, **1217**, 6995–7002.

151. Dai, C.-M., Geissen, S.-U., Zhang, Y.-L. *et al.* (2011). Selective removal of diclofenac from contaminated water using molecularly imprinted polymer microspheres, *Environ. Pollut.*, **159**, 1660–1666.
152. Song, X., Li, J., Xu, S. *et al.* (2012). Determination of 16 polycyclic aromatic hydrocarbons in seawater using molecularly imprinted solid-phase extraction coupled with gas chromatography–mass spectrometry, *Talanta*, **99**, 75–82.
153. Kubo, T., Kuroda, K., Tominaga, Y. *et al.* (2014). Effective determination of a pharmaceutical, sulpiride, in river water by online SPE-LC–MS using a molecularly imprinted polymer as a preconcentration medium, *J. Pharm. Biomed. Anal.*, **89**, 111–117.
154. Sambe, H., Hoshina, K. and Haginaka, J. (2007). Molecularly imprinted polymers for triazine herbicides prepared by multi-step swelling and polymerization method: their application to the determination of methylthiotriazine herbicides in river water, *J. Chromatogr. A*, **1152**, 130–137.
155. Hoshina, K., Horiyama, S., Matsunaga, H. *et al.* (2009). Molecularly imprinted polymers for simultaneous determination of antiepileptics in river water samples by liquid chromatography–tandem mass spectrometry, *J. Chromatogr. A*, **1216**, 4957–4962.
156. Mena, M.L., Martínez-Ruiz, P., Reviejo, A.J. *et al.* (2002). Molecularly imprinted polymers for on-line preconcentration by solid phase extraction of pirimicarb in water samples, *Anal. Chim. Acta*, **451**, 297–304.
157. Hoshina, K., Horiyama, S., Matsunaga, H. *et al.* (2011). Simultaneous determination of non-steroidal anti-inflammatory drugs in river water samples by liquid chromatography–tandem mass spectrometry using molecularly imprinted polymers as a pretreatment column, *J. Pharm. Biomed. Anal.*, **55**, 916–922.
158. Li, J., Chen, L., Wang, X. *et al.* (2008). Determination of tetracyclines residues in honey by on-line solid-phase extraction high-performance liquid chromatography, *Talanta*, **75**, 1245–1252.
159. Cacho, C., Schweitz, L., Turiel, E. *et al.* (2008). Molecularly imprinted capillary electrochromatography for selective determination of thiabendazole in citrus samples, *J. Chromatogr. A*, **1179**, 216–223.

160. Lin, J.M., Nakagama, T., Uchiyama, K. *et al.* (1997). Enantioseparation of D,L-phenylalanine by molecularly imprinted polymer particles filled capillary electrochromatography, *J. Liq. Chromatogr. Rel. Technol.*, 27, 1489–1506.
161. Chirica, G. and Remcho, V.T. (1999). Silicate entrapped columns — new columns designed for capillary electrochromatography, *Electrophoresis*, 20, 50–56.
162. Sulitzky, C., Rückert, B., Hall, A.J. *et al.* (2002). Grafting of molecularly imprinted polymer films on silica supports containing surface-bound free radical initiators, *Macromolecules*, 35, 79–91.
163. Schweitz, L., Andersson, L.I. and Nilsson, S. (2002). Molecularly imprinted CEC sorbents: investigations into polymer preparation and electrolyte composition, *Analyst*, **127**, 22–28.
164. Liu, Z.S., Xu, Y.L., Wang, H.F. *et al.* (2004). Chiral separation of binaphthol enantiomers on molecularly imprinted polymer monolith by capillary electrochromatography, *Anal. Sci.*, **20**, 673–678.
165. Huang, Y.C., Lin, C.C. and Liu, C.Y. (2004). Preparation and evaluation of molecularly imprinted polymers based on 9-ethyladenine for the recognition of nucleotide bases in capillary electrochromatography, *Electrophoresis*, **25**, 554–561.
166. Nilsson, C. and Nilsson, S. (2006). Nanoparticle-based pseudostationary phases in capillary electrochromatography, *Electrophoresis*, **27**, 76–83.
167. Bjarnason, B., Chimuka, L. and Ramström, O. (1999). On-line solid-phase extraction of triazine herbicides using a molecularly imprinted polymer for selective sample enrichment, *Anal. Chem.*, **71**, 2152–2156.
168. Turiel, E., Tadeo, J.L., Cormack, P.A.G. *et al.* (2005). HPLC imprinted-stationary phase prepared by precipitation polymerisation for the determination of thiabendazole in fruit, *Analyst*, **130**, 1601–1607.
169. Hantash, J., Bartlett, A., Oldfield, P. *et al.* (2006). Use of an on-line imprinted polymer pre-column for the liquid chromatographic–UV absorbance determination of carbaryl and its metabolite in complex matrices, *J. Chromatogr. A*, **1125**, 104–111.
170. Cassiano, N.M., Barreiro, J.C., Moraes, M.C. *et al.* (2009). Restricted-access media supports for direct high-throughput analysis of biological fluid samples: review of recent applications, *Bioanalysis*, **1**, 577–594.
171. Cassiano, N.M., Lima, V.V., Oliveira, R.V. *et al.* (2006). Development of restricted-access media supports and their application to the direct

analysis of biological fluid samples via high-performance liquid chromatography, *Anal. Bioanal. Chem.*, **385**, 1462–1469.

172. Rbeida, O., Christiaens, B., Chiap, P. *et al.* (2003). Fully automated LC method for the determination of sotalol in human plasma using restricted access material with cation exchange properties for sample clean-up, *J. Pharm. Biomed. Anal.*, **32**, 829–838.
173. Yoshida, H., Morita, I., Tamai, G. *et al.* (1984). Some characteristics of a protein-coated ODS column and its use for the determination of drugs by the direct injection analysis of plasma samples, *Chromatographia*, **19**, 466–472.
174. Emara, S., Morita, I., Tamura, K. *et al.* (1994). Simultaneous determination of free and bound adriamycin, adriamycinol, adriamycinone and duanorubicin in plasma using column-switching technique and protein-coated pre-columns, *Talanta*, **41**, 1973–1980.
175. Hogendoorn, E.A., Dijkman, E., Baumann, B. *et al.* (1999). Strategies in using analytical restricted-access media columns for the removal of humic acid interferences in the trace analysis of acidic herbicides in water samples by coupled column liquid chromatography with UV detection, *Anal. Chem.*, **71**, 1111–1118.
176. Gasparrini, F., Ciogli, A., D'Acquarica, I. *et al.* (2007). Synthesis and characterization of novel internal surface reversed-phase silica supports for high-performance liquid chromatography, *J. Chromatogr. A*, **1176**, 79–88.
177. Kishida, K. (2007). Restricted-access media liquid chromatography for determination of sulfamonomethoxine, sulfadimethoxine, and their N4-acetyl metabolites in eggs, *Food Chem.*, **101**, 281–285.
178. Carla P., Desilets, C.P., Rounds, M.A. *et al.* (1991). Semipermeable-surface reversed-phase media for high-performance liquid chromatography, *J. Chromatogr. A*, **544**, 25–39.
179. Hogendoorn, E.A., van Zoonen, P., Polettini, A. *et al.* (1998). The potential of restricted-access media columns as applied coupled-column LC/LC–TSP/MS/MS for the high speed determination of target compounds in serum. Application to the direct analysis of salbutamol and clenbuterol, *Anal. Chem.*, **70**, 1362–1368.
180. Gustavson, K.E., DeVita, W.M., Revis, A. *et al.* (2000). Novel use of a dual-zone restricted access sorbent: Normal-phase solid-phase extraction

separation of methyl oleate from polynuclear aromatic hydrocarbons stemming from semi-permeable membrane devices, *J. Chromatogr. A*, **883**, 143–149.

181. Williams, D.E. and Kabra, P.M., (1990). Extended life for blood serum analysis columns using dual zone chromatographic materials, *Anal. Chem.*, **62**, 807–810.
182. Chiap, P., Rbeida, O., Christiaens, B. *et al.* (2002). Use of a novel cation-exchange restricted-access material for automated sample clean-up prior to the determination of basic drugs in plasma by liquid chromatography, *J. Chromatogr. A.*, **975**, 145–155.
183. Ye, Y., Bishop, A.M., Needham, L.L. *et al.* (2008). Automated on-line column-switching HPLC–MS/MS method with peak focusing for measuring parabens, triclosan, and other environmental phenols in human milk, *Anal. Chim. Acta*, **622**, 150–156.
184. Yoshida, H., Morita, I., Tamai, G. *et al.* (1984). Some characteristics of a protein-coated ODS column and its use for the determination of drugs by the direct injection analysis of plasma samples, *Chromatographia*, **19**, 466–472.
185. Asakawa, N., Ohe, H., Tsuno, M. *et al.* (1991). Liquid chromatography–mass spectrometry system using column-switching techniques, *J. Chromatogr. A*, **541**, 231–241.
186. Hagestam, H. and Pinkerton, T.C. (1985). Internal surface reversed-phase silica supports for liquid chromatography, *Anal. Chem.*, **57**, 1757–1763.
187. Haginaka, J., Wakai, J. and Yasuda, H. (1990). Synthesis of mixed-functional-phase silica supports for liquid chromatography and their applications to assays of drugs in serum, *J. Chromatogr. A*, **535**, 163–172.
188. Zhang, H.X., Liu, M.C. and Zhu, P.L. (2000). Preparation of a polyglycol-C8 bonded phase and its characteristics, *Chromatographia*, **51**, 437–442.
189. Barreiro, J.C., Vanzolini, K.L. and Cass, Q.B. (2011). Direct injection of native aqueous matrices by achiral–chiral chromatography ion trap mass spectrometry for simultaneous quantification of pantoprazole and lansoprazole enantiomers fractions, *J. Chromatogr. A*, **1218**, 2865–2870.
190. Önnerfjord, P., Barceló, D., Emnéus, J. *et al.* (1996). On-line solid-phase extraction in liquid chromatography using restricted-access

pre-columns for the analysis of s-triazines in humic-containing waters, *J. Chromatogr. A*, **737**, 35–45.

191. Barreiro, J.C., Vanzolini, K.L., Madureira, T.V. *et al.* (2010). A column-switching method for quantification of the enantiomers of omeprazole in native matrices of waste and estuarine water samples, *Talanta*, **82**, 384–391.
192. Hogendoorn, E.A., Huls, R., Dijkman, E. *et al.* (2001). Microwave-assisted solvent extraction and coupled-column reversed-phase liquid chromatography with UV detection: use of an analytical restricted-access medium column for the efficient multi-residue analysis of acidic pesticides in soils, *J. Chromatogr. A*, **938**, 23–33.
193. Menezes, M.L. and Felix, G. (1996). Analysis of organochlorine pesticides in plain milk using direct injection on an ISRP column, with column switching, *J. Liq. Chromatogr. Relat. Technol.*, **19**, 3221–3228.
194. Hummert, C., Luckas, B. and Siebenlist, H. (1995). Determination of chloramphenicol in animal tissue using high-performance liquid chromatography with a column-switching system and ultraviolet detection, *J. Chromatogr. B*, **668**, 53–58.
195. Kishida, K. and Furusawa, N. (2004). Application of shielded-column liquid chromatography for determination of sulfamonomethoxine, sulfadimethoxine, and their N4-acetyl metabolites in milk, *J. Chromatogr. A*, **1028**, 175–177.
196. Ueno, R. and Aoki, T. (1996). High-performance liquid chromatographic method for the rapid and simultaneous determination of sulfamonomethoxine, miloxacin and oxolinic acid in serum and muscle of cultured fish, *J. Chromatogr. B*, **682**, 179–181.
197. Agarwal, V.K. (1990). Solid phase extraction of sulfonamides using cyclobond-I cartridges, *J. Liq. Chrom. Rel. Technol.* **13**, 2475–2487.
198. Lopes, B.R., Barreiro, J.C., Baraldi, P.T. *et al.* (2012). Quantification of carbamazepine and its active metabolite by direct injection of human milk serum using liquid chromatography tandem ion trap mass spectrometry, *J. Chromatogr. B*, **889–890**, 17–23.
199. Hage, D.S. (1998). Survey of recent advances in analytical applications of immunoaffinity chromatography, *J. Chromatogr. B*, **715**, 3–28.

200. Van Emon, J.M. and Chuang, J.C. (2013). Development and application of immunoaffinity chromatography for coplanar PCBs in soil and sediment, *Chemosphere* **90**, 1–6.
201. Şenyuva, H.Z. and Gilbert, J. (2010). Immunoaffinity column clean-up techniques in food analysis: a review, *J. Chromatogr. B*, **878**, 115–132.
202. Van Deemter, J.J., Zuiderweg, F.J. and Klinkenberg, A. (1956) Longitudinal diffusion and resistance to mass transfer as causes of nonideality in chromatography, *Chem. Eng. Sci.*, **5**, 271–289.
203. Bird, R.B., Stewart, W.E. and Lightfoot, E.N. (1960). *Transport Phenomena*, Wiley, New York, chapters 5–6.
204. Aris, R. (1956). On the dispersion of a solute in a fluid flowing through a tube, *Proc. R. Soc. London, Ser. A*, **235**, 67–77.
205. Tichacek, L.J., Barkelew, C.H. and Baron, T. (1957). Axial mixing in pipes, *AIChE J.*, **3**, 439–442.
206. Reynolds, O. (1883). An experimental investigation of the circumstances which determine whether the motion of water shall be direct or sinuous, and the law of resistance in parallel channels, *Roy. Soc. Phil. Trans.*, **35**, 84–99.
207. Giddings, J.C. (1962). Nature of gas phase mass transfer in gas chromatography, *Anal. Chem.*, **34**, 1186–1192.
208. Knox, J.H. (1999). Band dispersion in chromatography — a new view of A-term dispersion, *J. Chromatogr, A*, **831**, 3–15.
209. Knox, J.H. and Scott, H.P., (1983). B and C terms in the Van Deemter equation for liquid chromatography, *J. Chromatogr. A*, **282**, 297–313.
210. Taylor G.I. (1953). Dispersion of soluble matter in solvent flowing slowly through a tube, *Proc. R. Soc, Ser. A*, **219**, 186–203.
211. Giddings, J.C. (1963). Evidence on the nature of eddy diffusion in gas chromatography from inert (nonsorbing) column data, *Anal. Chem.*, **35**, 1338–1341.
212. Van Deemter, J.J. (1957). '2nd informal symposium of the gas chromatography discussion group, Cambridge', quoted in Purnell, J.H. (Ed.) (1962). *Gas Chromatography*, Wiley, London, p. 128.
213. Giddings, J.C. (1965). *Dynamics of Chromatography–Part 1*, Marcel Dekker, New York/London.

214. Gritti, F. and Guiochon, G. (2013). The Van Deemter equation: assumptions, limits, and adjustment to modern high-performance liquid chromatography, *J. Chromatogr. A*, **1302**, 1–13.
215. Gritti, F. and Guiochon, G. (2012). Mass transfer kinetics, band broadening and column efficiency, *J. Chromatogr. A*, **1221**, 2–40.
216. Edge, A. (2002). CAST, **March/April**, 10–16.
217. Hlushkou, D. and Tallarek, U. (2006). Transition from creeping via viscous-inertial to turbulent flow in fixed beds, *J. Chromatogr. A*, **1126** 70–85.
218. Khirevich, S., Höltzel, A., Seidel-Morgenstern, A. *et al.* (2009). Time and length scales of eddy dispersion in chromatographic beds, *Anal. Chem.* **81**, 7057–7066.
219. Asperger, A., Efer, J. and Engewald, W. (2001, April). Poster presented at 10th Symposium on Handling of Environmental and Biological Samples in Chromatography, Mainz, Germany.
220. Bahl, K., Efer, W., Asperger, A. *et al.* (2002). Poster presented at HTC, Bruges, Belgium.
221. Asperger, A., Efer, J., Koal, T. *et al.* (2002). Trace determination of priority pesticides in water by means of high-speed on-line solid-phase extraction–liquid chromatography–tandem mass spectrometry using turbulent-flow chromatography columns for enrichment and a short monolithic column for fast liquid chromatographic separation, *J. Chromatogr. A.*, **960**, 109–119.
222. López-Serna, R.,Petroví, M. and Barceló, D. (2012). Direct analysis of pharmaceuticals, their metabolites and transformation products in environmental waters using on-line TurboFlow™ chromatography–liquid chromatography–tandem mass spectrometry, *J. Chromatogr. A*, **1252**, 115–129.
223. Bousova, K., Senyuva, H. and Mittendorf K. (2013). Quantitative multi-residue method for determination antibiotics in chicken meat using turbulent flow chromatography coupled to liquid chromatography–tandem mass spectrometry, *J. Chromatogr. A*, **1274**, 19–27.
224. Stolker, A.A.M., Peters, R.J.B., Zuiderent, R. *et al.* (2010). Fully automated screening of veterinary drugs in milk by turbulent-flow chromatography and tandem mass spectrometry, *Anal. Bioanal. Chem.*, **397**, 2841–2849.

225. Gremaud, E. and Guy, P.A. (2008). Quantitative high-throughput analysis of 16 (fluoro)quinolones in honey using automated extraction by turbulent-flow chromatography coupled to liquid chromatography–tandem mass spectrometry, *J. Agric. Food Chem.*, **56**, 35.
226. Roach, J.A.G., DiBussolo, J.M., Krynitsky, A. *et al.* (2011). Evaluation and single laboratory validation of an on-line turbulent flow extraction–tandem mass spectrometry method for melamine in infant formula, *J. Chromatogr. A*, **1218**, 4284–4290.
227. Feitosa-Felizzola, J., Temime, B. and Chiron S, (2007). Evaluating on-line solid-phase extraction coupled to liquid chromatography–ion trap mass spectrometry for reliable quantification and confirmation of several classes of antibiotics in urban wastewaters, *J. Chromatogr. A*, **1164**, 95–104.
228. Ates, E., Godula, M., Stroka, J. *et al.* (2014). Screening of plant and fungal metabolites in wheat, maize and animal feed using automated on-line clean up coupled to high-resolution mass spectrometry, *Food Chem.*, **142**, 276–284.
229. Mottier, P., Parisod, V. and Turesky, R.J. (2000). Quantitative determination of polycyclic aromatic hydrocarbons in barbecued meat sausages by gas chromatography coupled to mass spectrometry, *J. Agric. Food Chem.*, **48**, 1160–1166.
230. Aguilera-Luiz, M.M., Romero-González, R., Plaza-Bolaños, P. *et al.* (2013). Rapid and semiautomated method for the analysis of veterinary drug residues in honey based on turbulent-flow liquid chromatography coupled to ultrahigh-performance liquid chromatography–Orbitrap mass spectrometry (TFC–UHPLC–Orbitrap-MS), *J. Agric. Food Chem.*, **61**, 829–839.
231. Geschke, O. Klank. H. and Telleman, P. (eds) (2004). *Microsystem Engineering of Lab-on-a-chip Devices*, 1st ed., John Wiley & Sons, London.
232. Herold, K.E. and Rasooly, A. (eds) (2009). *Lab-on-a-Chip Technology: Fabrication and Microfluidics*, Caister Academic Press, Norfolk.
233. Manz, A., Graber, N. and Widmer, H.M. (1990). Miniaturized total chemical analysis systems: a novel concept for chemical sensing, *Sensor Actuat. B-Chem.*, **1(1)**, 244–248.
234. Myers, P. (2011). Separations-on-a-chip? Scientists at Liverpool put a different spin on things!, *Chromatogr. Today*, **Aug/Sept.**, 22.

Chapter 7

Ambient Mass Spectrometry: Food and Environmental Applications

Tiina J. Kauppila and Anu Vaikkinen

Faculty of Pharmacy, University of Helsinki, Finland

7.1. Ambient Mass Spectrometry Techniques

Ambient mass spectrometry is a diverse family of techniques designed for the direct analysis of compounds from sample surfaces. The techniques allow the direct analysis of unconventional bulk samples, such as whole tablets, plant parts or tissue sections. The analyses take place outside the mass spectrometer, in ambient pressure, which speeds up the sampling. Often, the analyses can be performed in mere seconds, without any sample preparation, which is a significant advantage when compared to conventional analysis methods. This chapter aims to present the reader with an insight into the most popular ambient MS techniques, desorption electrospray ionization (DESI)[1] and direct analysis in real time (DART),[2] together with a third technique, desorption atmospheric pressure photoionization (DAPPI), and their food and environmental applications.[3] For readers interested in more details on the different ambient MS techniques and their applications, we refer to recent reviews on the subject.[4,5]

7.1.1. *Desorption electrospray ionization (DESI)*

In DESI, charged droplets from a capillary, assisted by a strong gas flow, are electrosprayed at the surface under study (Fig. 7.1). As the charged droplets hit the sample, they dissolve and pick up analyte molecules from the surface. The analyte molecules enter the

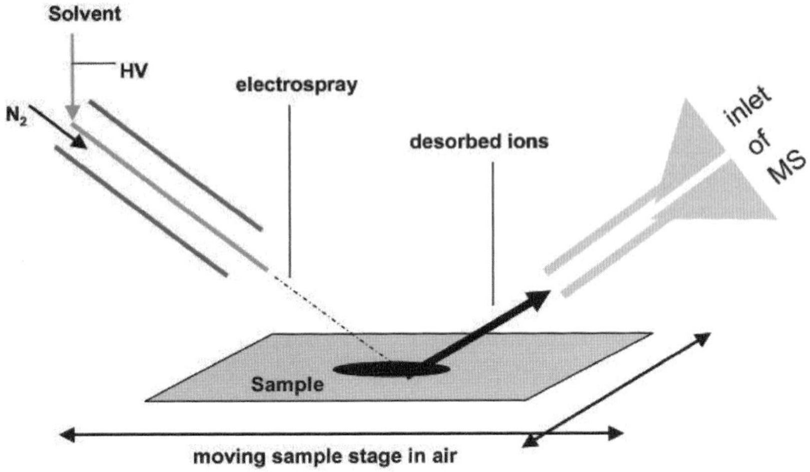

Figure 7.1. Schematic of a typical DESI experiment. Reprinted with permission from Nielen, M.W.F., Hooijerink, H., Zomer P. *et al.* (2011). Desorption electrospray ionization mass spectrometry in the analysis of chemical food contaminants in food, *Trends Anal. Chem.*, **30**, 165–180. Copyright (2011) Elsevier.

charged droplets, after which the ionization proceeds similarly to conventional electrospray. Similarly to electrospray ionization (ESI), polar compounds of all sizes are efficiently ionized in DESI. In positive ion DESI, the analytes typically form proton, metal or ammonium adducts, while deprotonated molecules are observed in negative-ion mode. Large molecules can form multiply-charged ions, similarly to ESI.

A number of parameters are important to optimize in DESI, such as the spray solvent composition and flow rate, the nature of the sampling surface, and the angles and distances between the sprayer, the sampling surface, and the MS inlet.[7] The solvents used in DESI are typically a mixture of water and a polar organic solvent, such as methanol or acetonitrile. Often, mild volatile acids, such as acetic or formic acid, are added to the spray in positive-ion mode to enhance the formation of protonated analyte molecules. Sometimes, adduct-forming agents can be added to the solvent to aid the ionization of compounds that are not easily ionized via protonation or deprotonation reactions (reactive DESI).[8] According to Takáts *et al.*, the DESI surface should be nonconductive to avoid neutralization of the charged droplets at the surface.[7] Polymer surfaces, such as polymethylmethacrylate (PMMA) and Teflon (PTFE) are examples of often-used DESI substrates: PTFE as a highly electronegative polymer gives excellent signal stability in negative-ion mode, while PMMA gives better performance in positive-ion mode. Other frequently used DESI substrates are glass and paper. The affinity of the analytes towards the surface should not be too high. The ion yield can sometimes be increased by the heating of the surface.

7.1.2. *Direct analysis in real time (DART)*

In DART (Fig. 7.2), a stream of gaseous metastables desorbs and ionizes molecules from the sample surface.[2] The reactive gas is supplied via a probe, where the neutral gas is first exposed to a glow discharge.[2,9] The glow discharge produces ions, electrons and excited-state species. The charged species are removed,[2] and the gas is heated if necessary. The gas stream is typically aimed directly at the MS inlet and the sample is exposed (dipped) to the stream. Different automated sample introduction devices are commercially available for this purpose, including holders for melting point capillaries, cotton swabs, tablets, and mesh screens.

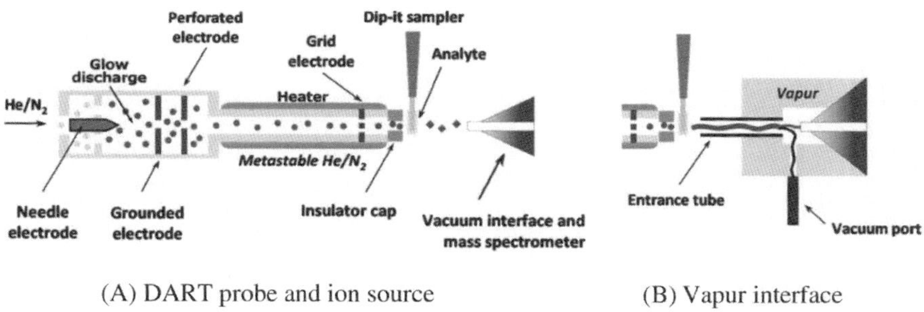

Figure 7.2. (A) DART probe and ion source, and (B) Vapur interface. Reprinted with permission from Hajslova, J., Cajka, T. and Vaclavik, L. (2011). Challenging applications offered by direct analysis in real time (DART) in food-quality and safety analysis. *TrAC Trends Anal. Chem.*, 30, 204–218. Copyright (2011) Elsevier.

The electronically or vibrationally excited gas species provide energy for the ionization reactions.[2,9,10] Analytes can be ionized directly from the sample surface by Penning ionization, as shown in Scheme 7.1, Reaction 1.[2] Because air and water are present in the ambient source and they typically react very rapidly with the metastables, the role of direct analyte ionization is most likely low. When helium, the most common DART reaction gas, is used, it is excited to the (2^3S) electronic state (19.8 eV), and can react with ambient water (clusters, Scheme 7.1, Reaction 2).[2,11,12] The water (cluster) ions protonate analyte molecules (M) if the proton affinity (PA) of the analyte is above the PA of water (clusters, Scheme 7.1, Reaction 3). When the humidity of the source is kept low, ambient oxygen can have a role in the ionization process.[9] Other reactant ions of interest in DART are ambient NO^+ and NH_4^+ ions.[2,9] NO^+ has been suspected to cause analyte oxidation,[9] while NH_4^+ produces analyte adduct ions (Scheme 7.1, Reaction 4),[2] and thus ammonia is often used as a dopant to enhance the ionization of, for example, triglycerides.[13–18]

While the ionization mechanism in DART is somewhat established, the desorption and ion-transport mechanisms are not as well studied. In many cases, the desorption process is thermal, because the higher heating temperature of the reaction gas increases the analyte signals.[19–21] The transmission efficiency of the ions is determined by interactions of fluid dynamics, heat transfer, and electrostatic phenomena within the sampling region.[22] The transmission efficiency can be improved by additional pumping near the MS inlet, as in the commercial Vapur interface, which is placed between the DART source and the mass spectrometer as depicted in Fig. 7.2B.

Scheme 7.1. The most common reactions occurring in the DART source (From Ref. 2, 9, 11). G = gas, G* = gas metastable, N = ambient, additive or matrix neutral, M = analyte molecule.

$G^* + N$	$\rightarrow$	$N^{+\cdot} + G + e-$	(1)
$He(2^3S) + nH_2O$	$\rightarrow$	$[(H_2O)_{n-1}H]^+ + OH^- + He(1^1S)$	(2)
$[N_n+H]^+ + M$	$\rightarrow$	$MH^+ + N_n$	(3)
$NH_4^+ + M$	$\rightarrow$	$[M+NH_4]^+$	(4)

7.1.3. *Desorption atmospheric pressure photoionization (DAPPI)*

DAPPI uses heated gas and solvent flow to thermally desorb the analytes from the sample surface, after which the ionization takes place in the gas phase, by photon-initiated reactions (Fig. 7.3).[3] DAPPI uses a heated microchip, which delivers hot, vaporized dopant and nebulizer gas to the sampling surface. The hot vapor flow causes the thermal desorption of the analytes from the sample surface, which are then ionized in the gas phase by photon-initiated reactions. A vacuum ultraviolet krypton discharge lamp is used to deliver 10 eV photons, which can ionize any compounds that have ionization energies (IEs) below the energy of the photons. The ionization reactions are similar to those in atmospheric pressure photoionization (APPI); the initial reaction is the photoionization of the dopant (Scheme 7.2, Reaction 1). Typical DAPPI (and APPI) dopants are toluene, acetone or anisole, which all have IEs below 10 eV. The dopants form

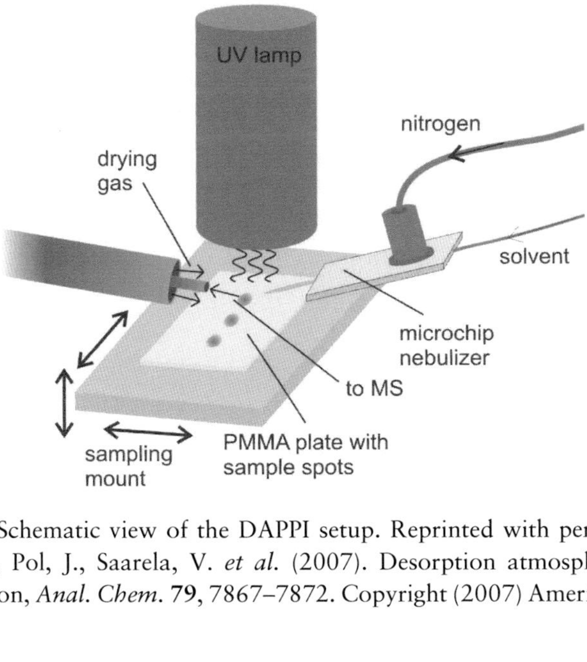

Figure 7.3. Schematic view of the DAPPI setup. Reprinted with permission from Haapala, M., Pol, J., Saarela, V. *et al.* (2007). Desorption atmospheric pressure photoionization, *Anal. Chem.* **79**, 7867–7872. Copyright (2007) American Chemical Society.

Scheme 7.2. Ionization reactions in positive- (Reactions 1–3) and negative-ion DAPPI (Reactions 1, 4–8).[23] S = solvent/dopant, M = analyte, IE = ionization energy, PA = proton affinity, EA = electron affinity, $\Delta_{acid}G$ = gas-phase acidity.

$S + h\nu$	$\rightarrow$	$S^{+\bullet} + e^-$		(1)
$S^{+\bullet} + M$	$\rightarrow$	$S + M^{+\bullet}$,	if IE(M) < IE(S)	(2)
$S^{+\bullet} + M$	$\rightarrow$	$[S - H]^\bullet + [M + H]^+$,	if PA(M) > PA(S)	(3)
$O_2 + e^-$	$\rightarrow$	$O_2^{-\bullet}$		(4)
$M + O_2^{-\bullet}$	$\rightarrow$	$M^{-\bullet} + O_2$,	if EA(M) > EA (O_2)	(5)
$M + e^-$	$\rightarrow$	$M^{-\bullet}$,	if EA(M) > 0 eV	(6)
$M + O_2^{-\bullet}$	$\rightarrow$	$[M - H]^- + HO_2^\bullet$,	if $\Delta_{acid}G(M) < \Delta_{acid}G(HO_2^\bullet)$	(7)
$M + O_2^{-\bullet}$	$\rightarrow$	$[M - H + O]^- + OH^\bullet$		(8)

molecular ions (or protonated molecules in the case of acetone), which can react further with the analytes through charge exchange or proton transfer (Scheme 7.2, Reactions 2 and 3, respectively). Typically, compounds that have higher PAs tend to form protonated molecules, while compounds with low PAs are ionized via the charge exchange route. In negative-ion DAPPI, the analytes can form negative molecular ions, deprotonated molecules or oxidation products (Scheme 7.2, Reactions 4–8).

Similarly to APPI, DAPPI can be used to ionize completely non-polar compounds and therefore the polarity range that can be achieved is wider than that with, for example, DESI. The thermal desorption is highly efficient for small molecules, but it prevents the analysis of large, low-volatility and/or thermolabile compounds.

7.2. Food Analysis

7.2.1. *Pesticides and fungicides*

Ambient MS screening of pesticides and fungicides from fruit and vegetable surfaces has been studied by several groups as summarized in Table 7.1. In the most simple cases, the fruit peels are analyzed directly, as in the studies by Garcia-Reyes *et al.* and Berchtold *et al.*, who used DESI for the direct analysis of agrochemicals from various

Table 7.1. Ambient MS in the study of pesticides and fungicides in food.

MS method	Analytes	Matrix	Sample preparation	Results	Ref.
DESI–portable CIT	DEET, alachlor, atrazine	Spiked cornstalk leaves, tomato	None	10 ng amounts detected, also MS/MS	27
DART–TOF	Azoxystrobin, picoxystrobin, dimoxystrobin, kresoxim-methyl, pyraclostrobin, trifloxystrobin	Milled wheat grains	Extracted with ethyl acetate for 2 min, filtered, evaporated and reconstituted in ethyl acetate	Quantitative method, R^2 0.9860-0.9978 at 6-1200 μg/kg, recoveries 78–92% and RSDs 8–15%	31
DESI–LIT	Azoxystrobin, picoxystrobin, dimoxystrobin, kresoxim-methyl, pyraclostrobin, trifloxystrobin	Milled wheat grains	Extracted to methanol in a microextraction pipet tip	Insufficient recovery with the extraction method, therefore not all the analytes could be detected	31
DESI–IT	16 agrochemicals and their metabolites	Fruit peel, fruit extract (orange, lemon, apple, green pepper, persimmon, grapefruit, tomato, pear, and grape)	Fruit extracts: modified QuEChERS, Fruit peel: none	LODs 1 pg, but 3–15 times higher in the presence of matrix; sensitivity sufficient for detecting the max. Residue levels allowed by EU pesticide regulation requirements; RSDs below 15%	24

(Continued)

Table 7.1. (*Continued*)

MS method	Analytes	Matrix	Sample preparation	Results	Ref.
DAPPI–IT	Imazalil	Orange peel	None	Imazalil detected from conventionally produced, but not from organically produced oranges	28
DART–Orbitrap	240, 140, 132 and 60 typical pesticides per fruit	Spiked apple, kiwi, peach, tomato	The fruit was swabbed using a foam disk, disk analyzed	Results depended highly on the sample surface, storage decreased pesticide detection only mildly	32
DART–Orbitrap	Thiram, ziram	Spiked pear, strawberry	Modified QuEChERS	LOD 0.1 and 1 mg/kg (below EU MRL), semi-quantitative	33
DESI–LIT	Thiram, ziram	Spiked pear, strawberry	Surface extraction	Only thiram detected, LOD 0.15 ng, semi-quantitative	33
DART–Orbitrap	132 pesticides	Spiked apple, grape and orange	Polyurethane foam was wetted with solvent and used to swab the samples, analyte release by 3 min temperature gradient of DART stream	86% of analytes were detected at 2 ng/g level (10 ng/g for grape), RSDs 30–59%	17

(*Continued*)

Table 7.1. (*Continued*)

MS method	Analytes	Matrix	Sample preparation	Results	Ref.
DART–Orbitrap	11 pesticides	Orange, blueberry, cabbage, tomato, bell pepper	Modified QuEChERS	Roughly > 100 ng/g levels detected	17
DART–Orbitrap	Dimethoate, methamidophos, malathion	Spiked cherry tomato, baby carrot, navel orange, peach	Swabs soaked in MeOH and entire spiked fruit/ vegetable surface was swapped	Signal depends on the sample	30
DART–Orbitrap	Imazalil	Apple, orange	None	Method linear at 1–2500 ng, $R^2 > 0.99$, LOD 300 μg/kg, 20% interday RSD, good agreement between DART and UHPLC	29
DESI–IT	Chlorpropham	Potato skin	None	LOD 6.5 mg/kg, signal variation 12%	25

fruits and fruit/vegetable extracts,[24] and chlorpropham herbicide from spiked potato skin, respectively.[25] In Ref. 24, imazalil was detected from lemon skin, and imazalil and thiabendazole were detected from grapefruit skin. Besides the direct analysis, the authors also showed analysis of fruit and vegetable samples after a modified QuEChERS protocol.[26] The limits of detection (LODs) were studied from spiked tomato or orange extracts, and they were at 1 pg at best. The use of DESI for the screening of banned substances was demonstrated by analyzing green pepper peel spiked with isofenphos-methyl, a non-authorized organophosphorus insecticide. Finally, the qualitative analysis of agrochemicals from 20 authentic fruit and vegetable extracts by DESI–MS/MS was performed. The experiments enabled the confirmation of agrochemicals in authentic samples at concentration levels that were 50–100 times lower than the maximum residue levels (MRL) allowed by the European Union pesticide regulation requirements. Quantitation of imazalil residues was also undertaken using an isotopically labeled standard. The DESI data was in agreement with data obtained by an LC–MS reference method, with relative standard deviation (RSD) values consistently below 15%. DESI has also been mounted on a custom-built, portable MS, which was used for direct analysis of cornstalk leaves spiked with DEET, alachlor and atrazine.[27] 10 ng amounts of the analytes were successfully analyzed individually and in mixture. The portable MS instrument could also be used in MS/MS mode, which is important for reliable identification of the compounds, and such an instrument could thus be highly useful in field analysis of agrochemicals. Pesticides have also been studied directly from fruit using DAPPI and DART, by Luosujärvi *et al.*[28] and Farré *et al.*,[29] respectively. Luosujärvi *et al.* analyzed the peels of organic and conventionally produced oranges directly with DAPPI and detected imazalil from the latter. Farré *et al.* detected and quantified imazalil from apple and orange peels, and compared the results obtained with direct DART analysis to those obtained with UHPLC–MS and DART–MS from methanol extracts of the peels. The authors found that both sampling methods in DART showed adequate linearities and had limits of quantification (LOQs) of 1 ng, which corresponds to ~300 µg/kg of apple. All the DART results agreed with those obtained with UHPLC–MS, although linearity and RSDs were better with UHPLC–MS.

Analyses almost as rapid can be achieved when the surface of the fruit is sampled by a swab that is subsequently analyzed by ambient MS. For example, Crawford and Musselman studied screening of dimethoate, methamidophos, and malathion from cherry tomatoes, baby carrots, navel oranges and peaches by wiping and DART–Orbitrap.[30] They were able to detect all the studied analytes at 10–100 times below the US Environmental Protection Agency (EPA) tolerance levels. In a more comprehensive study, Edison *et al.* spiked apples, grapes and oranges with 132 common pesticides at a 2 ng/g (or 10 ng/g for grape) level.[17] A 3 min temperature gradient of the DART metastable stream was applied to partially separate the analytes by their volatilities and 86% of the studied pesticides were detected. DART-Orbitrap was also compared with HPLC–MS/MS for the analysis of QuEChERS extracts of authentic field samples. Roughly put, DART was able to detect the pesticides present at > 100 ng/g level (determined by HPLC), while six pesticides present at lower levels were detected only by HPLC–MS/MS. In another study, Edison *et al.* found the surface texture of fruits to affect the results of DART–Orbitrap analysis, as the hairy surface texture of pears and kiwis lead to physical degradation of the swab material.[32] They also found that ambient MS analyses could provide reliable results irrespective of storage time of the produce, because when the fruits were stored in a refrigerator three days and eight days before the analysis, 80–93% of the studied pesticides could still be detected.

More extensive sample preparation methods have frequently been found to be necessary to enable the ambient MS analyses. For example, Cajka *et al.* compared DESI and DART for the analysis of dithiocarbamate fungicides thiram and ziram.[33] The samples were pear and strawberry, spiked with the dithiocarbamate standards. The samples were homogenized and extracted using liquid–liquid extraction (LLE), solid-phase extraction (SPE) or a modified QuEChERS protocol. The authors also tried to detect spiked thiram directly from pear leaves by DESI–MS, but a much lower signal was obtained than from the same standard solution on a glass slide, which was thought to be due to the different surface and interference by the matrix. For both DART and DESI, the extraction step was found necessary for the detection, and ziram could only be analyzed by DART. Comparison of DART–TOF (resolving power 5000 FWHM) and DART–Orbitrap (resolving power 25,000) showed that high-mass resolving power was necessary for the analysis of thiram, because of isobaric background peaks that increased the LOD in the TOF analysis.

Schurek *et al.* studied DART–TOF and DESI–LIT–MS/MS to control six strobilurin fungicides in wheat.[31] Direct DART analysis of milled wheat enclosed in homemade envelopes could be used to detect the analytes, but the protocol was improved by LLE. With prochloraz as the IS, quantitation was achieved with linear range at 6–1200 μg/kg, R^2 between 0.986 and 0.998, recoveries between 78% and 92%, and RSDs of 8–15%. The LOQs were below EU requirements for all studied compounds, except for kresoxim-methyl. For six wheat grain samples, the quantitative DART procedure took only 1.5 h compared with the 5 h procedure with LC–MS/MS. The RSDs were worse in DART (6–17% compared with 2–4%), but quantitative results agreed, making DART a feasible option for the analysis of large sample batches. DESI–LIT–MS/MS analysis was explored for analyte identification. The qualitative results obtained with DESI were in agreement with those obtained with DART, and a sufficient number of identification points for the EU requirements could be obtained from the MS/MS studies.

7.2.2. *Food chemistry*

Because the mass analysis in ambient MS techniques occurs immediately after the sample is probed, these methods can be useful for the study of labile compounds that do not survive traditional sample preparation protocols. For example, Cody and coworkers have studied the chemistry of volatile compounds from different *Allium* (onion) species by puncturing them with melting point capillaries and analyzing the capillaries immediately by DART.[34–36] This method was able to find several short-lived compounds that were proposed to appear when the *Allium* are wounded or crushed, but which have not been previously observed experimentally. Butanethial S-oxide and a series of butyl or 1-butenyl thiosulfonates were found in a study of the ornamental *A. siculum* and *A. tripedale*,[36] and 2-propenesulfenic acid, 2-propenesulfinic acid, and trisulfane S-oxide in a study of *A. sativum* (garlic).[35]

These contributions also report the DART–MS profiles of leek (*A. porrum*), elephant garlic (*A. ampeloprasum*), onion (*A. cepa*), and Chinese chive (*A. tuberosum*). Later, Li used a confined sampling interface to direct volatiles more efficiently to DART–MS analysis (Fig. 7.4).[39] He reported two orders of magnitude improved ionization

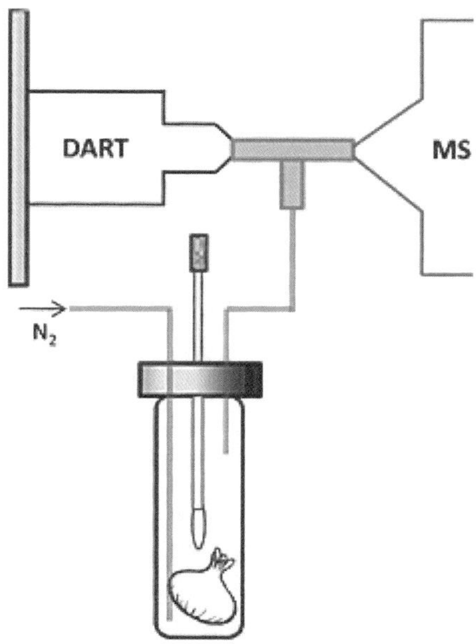

Figure 7.4. Schematic diagram of the confined DART ion source and the experimental setup for studies of volatile compounds. Reprinted with permission from Li, Y. (2012). Confined direct analysis in real time ion source and its applications in analysis of volatile organic compounds of *Citrus limon* (lemon) and *Allium cepa* (onion), *Rapid Commun. Mass Spectrom.*, **26**, 1194–1202. Copyright (2012) John Wiley and Sons.

efficiency compared to the open DART source, and studied the volatiles of lemon and onion (*A. cepa*). The sampling interface contained a mounted blade designed specifically for the release of volatile compounds from onion, and enabled the study of the release kinetics. *Allium* species garlic[38] and tearless and tearful onions[40] have also been studied using DESI. Phenotyping of tearless and tearful onion leaves and bulbs was achieved by DESI and proton transfer reaction–MS with headspace sampling.[40] The detected analytes were somewhat different with the two techniques, which was suggested to be due to more efficient detection of early release compounds with DESI, thanks to the speed of analysis. DESI–MS leaf compound profiles also allowed the rapid distinction of a variety of onion cultivars to aid plant-breeding selections.

Due to the rapid nature of ambient MS, it has been used to characterize the biochemical changes occurring in various foods due to cultivation conditions,[14,18] processing,[47,48] storage,[48] and heat treatment.[15] For example, Vaclavik *et al.* used DART to study chemical (oxidative) changes in vegetable oils due to heating.[15] The native and heat-processed oils were diluted in toluene and analyzed by transmission mode DART using mesh screens for sample introduction. In positive-ion mode, triacylglycerols (TAGs), fragments of the TAGs, and plant sterols were observed, while free fatty acids were observed in negative-ion mode. As expected, the heat-treatment led to the appearance of oxidation products in the spectra. A principal component analysis (PCA) analysis of 45 selected ions in the DART spectra was able to differentiate the studied oil types (olive oil, rapeseed oil, soybean oil, and sunflower oil), and within each oil type, separation due to different heat treatment times was seen. For simplicity, the authors proposed using the DART–MS signal of oxidized linoleoyl-dioleoylglycerol (LOO, normalized by the signal of LOO) as a marker of the heat treatment, as it corresponded to the amount of TAG polymers in the samples (analyzed by size exclusion chromatography with refractometric detection).

Cajka *et al.* studied common carp (*Cyprinus carpio L.*) to monitor the effect of feeding practice on fish meat quality, and to establish a rapid analytical procedure that could identify the production system of marketed fish based on metabolic profiles.[18] The muscle of the fish was homogenized, liquid–liquid extracted with water/hexane emulsion, and centrifuged. Both liquid phases were collected and analyzed by DART–TOF. Strong signals for small acids and other small polar molecules like histidine and creatine were observed in the water phase, while the organic phase showed signals for TAGs and free fatty acids. Data sets from both phases were used to create two PCA and orthogonal partial least squares discriminant analysis (OPLS-DA) models of naturally fed versus cereal-fed fish. The model for the nonpolar extracts gave better distinction between the two classes; 100% accurate classification was obtained by OPLS-DA. The naturally fed fish were characterized by lower amount of the most prominent TAGs, and a higher abundance of TAGs with polyunsaturated fatty acids and odd-chain fatty acids. Recently, almost identical metabolomics fingerprinting and chemometric analysis methods were used to study chicken legs to assess feed fraud with banned bone meal.[14] Also in this case, the authors observed that the DART–MS–OPLS-DA and PCA classification of the chicken is highly effective, but the TAG profiles of chicken legs depended on the season, and different feed preparation methods complicated the classification.

Some quantitative DART–MS analyses of food ingredients have been presented (e.g. analysis of caffeine in coffee products;[49] study of release of cyclohexanecarboxamide, N-ethyl-5-methyl-2-(1-methylethyl) (a taste-refreshing cooling agent in gum) into saliva;[52] study of acidic phytohormones in fruit juices;[50] and quantitation of isoflavones in soybeans.[51]) In the case of the coffee products and saliva, very rapid analysis protocols were achieved, as the only sample preparation method needed was sample dilution, but the quantitation of the isoflavones required hydrolysis and extraction, and the juice samples were extracted with single drop liquid–liquid–liquid microextraction. The above-mentioned and other food chemistry related applications of DESI, DART and DAPPI are listed in Table 7.2.

Table 7.2. Ambient MS applications in the study of food chemistry.

MS method	Analytes and samples	Notes	Ref.
Analytical method development			
DART–Orbitrap	Humulenes and acids in beer	DART used to detect which compounds were removed from sample by precipitation	37
DART–TOF	TAG in fish and shrimp	Screening of sample extracts for highly disturbing matrix compounds	16
Identification of biochemicals			
DESI–QqQ	Allicin from garlic	Cysteine added in the spray solvent; an allicin-cysteine complex observed	38
DART–TOF	Short-lived volatile compounds in *Allium*		35, 36
DART–TOF	Volatile compounds in onion and lemon	New confined sampling interface developed	39
DESI–LIT	Analysis of sulfur volatiles from onion leaves and bulbs	Phenotyping of tearless and tearing onions	40
DART–Orbitrap	Polyphenols in *Bergenia crassifolia* (herbal medicine) green leaf extracts	*Bergenia crassifolia* studied with HRMS for the first time	41
DART–TOF	Polyphenols in elderberry fruit	Preliminary identification of antiviral polyphenols	42

(*Continued*)

Table 7.2. (*Continued*)

MS method	Analytes and samples	Notes	Ref.
DESI–LIT	Studying the chain length and the degree of unsaturation of triglycerides from edible oils and margarines	Reactive DESI by addition of ammonia in the spray solvent: triglycerides detected as ammonium adducts	43
DAPPI–IT, DESI–IT	Analysis of butter, and fish oil and α-tocopherol vitamin capsules	DAPPI was more sensitive than DESI for cholesterol (butter) and α-tocopherol; both techniques showed fatty acids from the fish oil capsules	44
Nano-DESI–LIT	Fast profiling of anthocyanins and their aglycans in wine	Distinct profiles acquired for different vintages and cultivars	45
DESI–LIT imaging	Analysis of hydroxynitrile glucosides from barley leaf tissue	Imaging analysis	46
Monitoring of biochemical changes			
DART–TOF	Extracts of rice bran	The effect of a new processing method on the chemical profile of the rice bran was studied	47
DART–TOF	5-hydroxymethylfurfural in honey and caramel	Heat processing and storage of foods monitored	48
DART–Orbitrap	TAGs and sterols in olive oil, rapeseed oil, soybean oil, and sunflower oil	Characterization of changes in oils due to heat induced oxidation	15

(*Continued*)

Table 7.2. (*Continued*)

MS method	Analytes and samples	Notes	Ref.
DART–TOF	TAGs and free fatty acids in carp (*Cyprinus carpio L.*) muscle	Influence of diet on fish studied	18
Quantitation of food incredients			
DART–TOF	Caffeine in coffee products	Sample diluted with MeOH, method linear at 0.1–10 μg/mL, RSDs ≤ 5%, good correlation between HPLC and DART results	49
DART–QTOF	Phytohormones in freshly squeezed fruit juice	Sample preparation by single drop liquid–liquid–liquid extraction, method linear for two orders of magnitude with R^2 0.991–0.996, RSDs 6.9–14%	50
DART–Orbitrap	Isoflavones in soybeans	Quantitative results were obtained after hydrolysis of isoflavone glycosides	51
Product development			
DART–QqQ/IT	Cyclohexanecarboxamide and N-ethyl-5-methyl-2-(1-methylethyl) in chewing gum	Determination of release kinetics of chewing gum constituents to saliva. Results quantitative and agree with LC-MS	52

7.2.3. *Authenticity assessment*

Ambient MS is able to produce mass spectrum fingerprints very rapidly, which makes it a feasible tool for authenticity assessment purposes. Saka *et al.*, for example, used DART-MS to screen counterfeit dietary supplements for identification of toxic compounds, which were suspected to have caused the death of an unwary consumer.[53] Out of seven different tablets studied, the DART analysis was able to preliminarily identify the active ingredient in four, greatly helping the subsequent confirmatory LC–MS analysis. Hrbek *et al.* studied cheese adulteration with plant oils by DART–Orbitrap.[54] The samples were quickly extracted with toluene before the analysis, and positive-ion DART was used to detect TAGs from the toluene extract. Since plant oils are characterized by TAGs with longer-chain fatty acids than those in dairy, the adulteration was easily confirmed by the presence of TAG ions at *m/z* 840–910. The authors estimated that 1% adulteration could be detected by the method, and when home-prepared cheeses with known rapeseed, sunflower and soy bean oil contents were studied, the signals for characteristic plant oil TAGs correlated with the adulteration level ($R^2 = 0.976–0.994$). Hartmanova *et al.* used DESI for the fast profiling of anthocyanins and their aglycans in wine.[45] The authors used a nanospray capillary in their DESI source and thus called their setup 'nano-desorption electrospray (nano-DESI)' (not to be confused with another ambient MS technique called nanospray DESI[55]). Differently from conventional DESI, nano-DESI does not use nebulizer gas and is reported to produce very small primary droplets. Distinct profiles of main anthocyanins in wine samples, two vintages (2005 and 2007) of three cultivars (Alibernet, Neronet and Rubinet), were successfully acquired. Nano-DESI could also distinguish between wines of the different cultivars and wine mixtures. Nano-DESI was suggested to be useful for detection of adulteration of wine by illegal wine mixing or by coloring using anthocyanins extracted from other fruits. Anthocyanins could also be identified from wine grapes, chokeberries and elderberries, and from wine stains on a cotton fabric, which could be useful in forensic applications.

In many cases, chemometric tools like PCA, linear discriminant analysis (LDA), and PLS–DA have been utilized with fingerprinting, without the need to identify the observed ions. For example, DART and DESI profiles and chemometrics have been used to study the authenticity and/or quality of beer,[56] animal fats,[57] edible oils,[20,43] milk,[54] organically grown tomatoes and peppers,[58] and cubeb fruit.[59] In Ref. 21, the LDA models of the DART–MS profiles of edible oils were able to detect hazelnut oil, a typical adulterant of extra virgin olive oil, at levels of 15% (nonpolar extracts with TAGs) and 6% (polar extracts). In the study of animal fats, the TAG profiles proved very characteristic, and 10% (w/w) of beef tallow in lard, 5% lard in beef tallow, and 10% of beef in pork or vice versa could be detected.[57] In both these cases, DART–MS presents a feasible, very rapid method for the reliable screening of food, and we expect it to find use especially during adulteration epidemics that have been recurrent in markets around the world during the recent decades. In more complex cases, however, the identification may not be as straightforward. For example, Cajka *et al.* studied processing the DART–TOF-MS spectral data with PCA, LDA, PLS–DA, and artificial neural networks with multilayer perceptrons (ANN-MLP) to identify beer origin.[56] The only sample preparation method applied was degassing the beers for 5 min. Due to the complexity of the samples, the chemometric models allowed the correct classification of test samples in only 82.8–98% of the cases, with ANN-MLP model leading to slightly better results than LDA and PLS–DA. When comparing DART–MS with HS–SPME–GC–TOF-MS used in their previous study on beer recognition,[60] the authors found the DART analysis an order of magnitude quicker than the one employing GC, while the chemometric results were better for the GC data, and the prediction ability was 84.8–100% with the same set of models and methods used in the DART study. The studies of organically versus conventionally produced tomatoes,[58] peppers,[58] and milk[54] found the year or season of production to affect more the ambient MS fingerprints than the farming method, reminding of the importance of finding the right markers for the chemometric analyses.

7.2.4. *Quality control*

While ambient MS could provide a good tool for the real-time quality control of the food supply chain from production to table, only a few studies exploring its use in quality control have been reported.[48,61–63] Chernetsova and Morlock[62] and Rajchl *et al.*[48] have studied 5-hydroxymethyl furfural (HMF) in honey[48,62] and caramel.[48] HMF is formed during heating and prolonged storage in the decomposition of monosaccharides, and the maximum limit in honey set by EU is 40 mg/kg. Chernetsova and Morlock reported that the quantitative analysis of real samples was hindered by sucrose, which formed ions at the same nominal *m/z* with all observed HMF ions.[62] Rajchl *et al.*, on the other hand, used a high-resolution MS instrument and found that disturbing artifacts were formed only in negligible amounts.[48] Thus Rajchl *et al.* were able to analyze HMF in real samples starting from (LOQ) 3 mg/kg for honey and 4 mg/kg for caramel. With the aid of isotopically labeled IS, DART–MS quantitation correlated reasonably with that of HPLC ($R = 0.978$). However, PCA analysis of the DART–MS spectra showed separation of the native and heated samples, but not solely based on the signal of HMF, implying that HMF as the only heat treatment indicator may not be sufficient.

Fraser *et al.* presented a comprehensive case study on the direct DART–MS quality control of two batches of oolong tea production, which is typically monitored by organoleptic evaluation.[61] Eighteen samples were taken at different stages of the thirty-six-hour production process. Positive-ion DART–MS showed caffeine as the base peak with other minor ions, but a single positive ion (*m/z* 363) was found indicative of the manufacturing process by visual inspection of the spectra. In addition, 3 more marker ions were revealed by Pearson correlation matrix. The relative intensity of the said ions increased during fermenting, peaked at its final phase, and finally decreased during de-greening and subsequent production stages. The markers could be identified by accurate MS and MS^n of the samples and standards.

Van Berkel *et al.* combined thin-layer chromatography (TLC) with DESI–MS for the qualitative screening of goldenseal alkaloids in dietary supplements.[63] The analytes were separated on a TLC plate, and subsequently analyzed by DESI–MS, directly from the plate. White and UV light were used during the DESI measurements to aid in observation of the analytes on the TLC plate. The method was used for the qualitative screening of six commercially available goldenseal dietary supplements (goldenseal root or capsule, see Fig. 7.5.), which were extracted prior to the TLC separation. In one sample, alkaloids were detected that strongly suggested

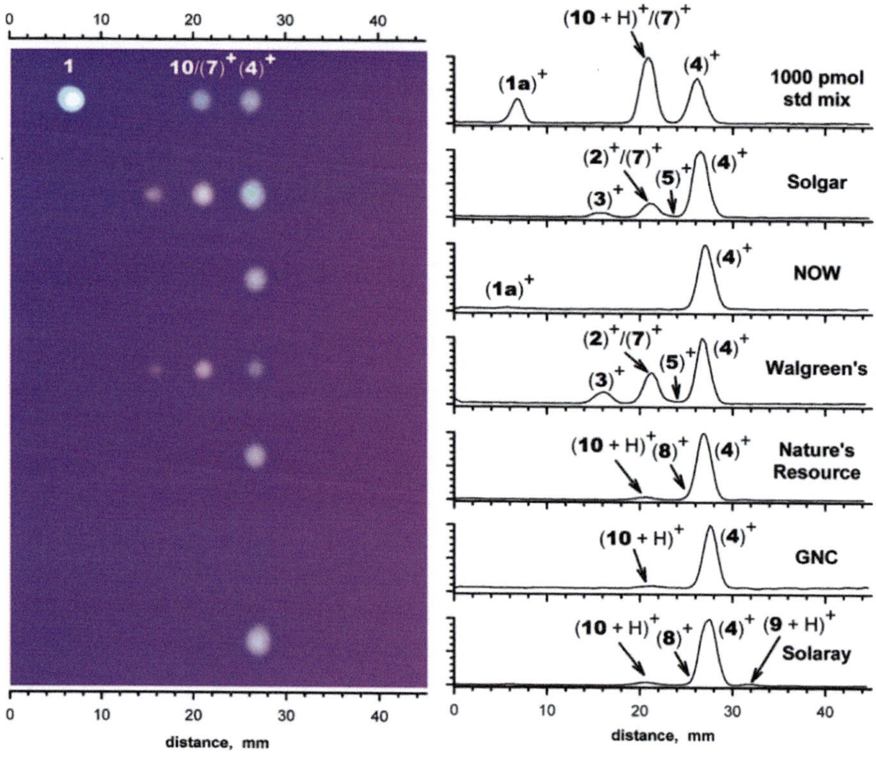

Figure 7.5. Left: Photograph of normal-phase TLC separations of a mixture of goldenseal alkaloid standards and six different goldenseal supplements on a single glass-based plate. Right: The corresponding summed extracted ion current profiles for all the alkaloid species observed during the different analyses. Numbers 1–10 represent different goldenseal alkaloids. Reprinted with permission from Van Berkel, G.J., Tomkins, B.A. and Kertesz, V. (2007). Thin-layer chromatography/desorption electrospray ionization mass spectrometry: investigation of goldenseal alkaloids, *Anal. Chem.* 79, 2778–2789. Copyright (2007) American Chemical Society.

the presence of at least one additional herb not declared on the product label. The results obtained with DESI were compared to those obtained with fluorescence detection and they were in reasonable agreement.

7.2.5. Food packaging/tableware

Due to recent frequent cases of additive contamination in food, rapid ambient MS-based screening techniques for plastic food-packaging materials have been developed.[64–67] Rothenbacher and Schwack used DART to study additives of poly(vinyl chloride) (PVC), which is typically used in the lids of glass jars.[65] Model plastisols with varying amounts of additives were custom-manufactured for the study. Many of the additives could be detected at 1% level, while strong matrix suppression was observed for epoxidized soybean and linseed oils. While in this study the analyte identification was based on characteristic ions in single quadrupole MS scan spectra, more accurate identification could be achieved by high resolution MS or MS^n. One such automated MS/MS method for phthalic acid esters was reported in a study of common PVC household items by DART.[68] DART–TOF has also been used to study the transfer of common print ink photoinitiators to the food contact surface of food-packaging materials,[66] and to detect additives in plastic food packaging.[64] Mattarozzi *et al.* used DESI to monitor the migration of melamine from plastic tableware.[67] Migration of melamine was tested by exposing the tableware 3 times for 2 h to pre-warmed food simulant, after which small amounts of the simulant were placed on a hydrophobic surface for DESI analysis, or analyzed by LC–ESI–MS. ^{13}C-melamine was used as the internal standard for quantitation. The LODs and LOQs for the DESI method were 5.0 µg/kg and 9.6 µg/kg, respectively, and the linearity ranged from the LOQ to 5 mg/kg. Finally, the authors analyzed 44 authentic samples of old and new plastic tableware using the DESI–MS and LC–ESI–MS methods. Similar results were obtained with both methods: highest melamine concentrations were detected in new tableware, while in old tableware melamine was absent, suggesting that it is gradually released from the tableware during its continuous usage. Two samples showed melamine concentrations above the legal limit (2.5 mg/kg).

7.2.6. *Food safety*

Because ambient MS is often much quicker than conventional MS techniques, it has been studied to help the screening and identification of contaminated food and drink products.[13,19,69–74] Lin *et al.* used DESI–MS for the speciation of ten different arsenic compounds from environmental and food samples.[75] The samples included arsenic-polluted animal feed (authentic sample) and vegetables (spiked cabbage, celery and tomato). The authors suggest that the method could be applied to in situ environmental monitoring of arsenic pollution, such as caused by arsenic pesticides, animal-feed additives, herbicides or wood treatment. Self and Wu used DART–Orbitrap to screen seven phthalates from beverages and food items due to a recent contamination epidemic in Taiwan.[73] In drinks, the analytes were detected starting at 0.5–1.0 µg/mL level, and in nutraceutical samples at 0.5–50 µg/g level. Vaclavik *et al.* used DART–Orbitrap for the quantitative analysis of 24 mycotoxins in wheat and maize.[72] Standards of 11 of the studied compounds were ionized efficiently, and these compounds were chosen for further examination. Next, spiked cereal samples were extracted with a modified QuEChERS protocol. Even after the extraction, the DART–Orbitrap signals of the mycotoxins suffered from matrix effects, and were 11.6–39.0% of those of standards in pure solvents. Nevertheless, the mycotoxins could be analyzed starting from 50–150 µg/kg, meaning that the regulated deoxynivalenol and zearalenone could be monitored at the EU set maximum levels for the cereal matrices. The comparison of DART–MS and UHPLC–MS analyses of certified reference materials showed that the results were accurate and in reasonable agreement. However, the analysis of the most dilute sample for zearalenone showed UHPLC to be more sensitive than DART, which did not detect the analyte, and the RSDs were lower in the UHPLC analysis of all reference samples.

Due to the recent crises in China, many methods to rapidly analyze melamine have been developed. For example, Vaclavik *et al.* used DART–TOF to quantitate melamine and cyanuric acid in solid milk containing products.[19] They first extracted the samples with methanol:5.0% formic acid (50:50) to break down melamine cyanurate complexes and reduce matrix effects. As in a few other ambient MS studies,[69,76,77] the formation of hydroxymethyl furfural was found to interfere with the analysis of melamine, because the *m/z* of their MH^+ ions are only 0.0337 units apart. Interestingly, Dane and Cody have shown that melamine can be selectively ionized in DART with argon as the metastable gas and acetylacetone and pyridine as reagent gases.[69] In this case, hydroxymethyl furfural is not ionized. Nevertheless, Vaclavik *et al.* found their DART–TOF method with He metastable gas to provide LODs below EU action limits of 2.5 mg/kg, namely 170 µg/kg and 450 µg/kg for melamine and cyanuric acid in milk powder, respectively.[19] The method was linear with $R^2 \geq 0.997$ from LOQ to 50 mg/kg; linearity was limited to these low concentrations due to the formation of the melamine cyanurate complex. Comparison of DART–TOF, LC–MS/MS and an enzyme-linked immunosorbent assay (ELISA) in the analysis of melamine in contaminated dried milk, powdered dried cheese, and condensed milk showed the ELISA test to overestimate the melamine concentrations, but the DART and LC–MS/MS results were in adequate agreement.

Moravcova *et al.* compared DART–Orbitrap and UHPLC–ESI–Orbitrap for the analysis of 3-chloropropane-1,2-diol esters, food-processing contaminants in vegetable oils.[13] They observed that direct analysis of diluted oils was not feasible with either of the techniques due to rapid contamination of the system (UHPLC) and severe signal suppression (DART), and thus the samples were first purified with column chromatography. The lowest analyte levels that could be analyzed by DART were roughly ten times higher than by UHPLC–ESI–MS. Linearity was better with the UHPLC method, which also did not require an isotopically labeled IS like DART. Certain isomeric analytes could not be resolved by either of the techniques, but with DART, some analytes with 2-unit nominal mass difference also suffered from overlapping of the M+2 isotope peak. However, the DART analysis took approximately 20 s per sample, while the UHPLC run took 8.5 min, and comparison of the quantitative results showed reasonable correlation of the two techniques. Thus, DART was concluded to be suitable for the screening of the most contaminated samples (40–118 µg/kg), while UHPLC allowed the accurate determination of analyte levels starting from 2–5 µg/kg.

Ambient MS methods have also been highly popular in forensic studies.[78] D'Aloise *et al.* showed an example of the suitability of DESI–MS for the analysis of flunitrazepam, a date-rape drug, from alcoholic beverages.[74] They spiked drinks, such as soda, juice, beer and liquor, with different amounts of flunitrazepam, and a constant amount of clonazepam as the internal standard. 20 µL of the liquid sample was applied on a chromatographic paper and analyzed by DESI–MS before drying. The linear range of the method was 3–20 µg/ml with R^2 of 0.997–0.998. The method was applied for high-throughput analysis of six samples within six minutes without any observable sample carry-over. The authors concluded that DESI–MS shows potential as a rapid, sensitive, and selective technique for forensic analysis of spiked beverages, which are typical evidence of drug-facilitated sexual assault and robbery cases.

7.3. Environmental Analysis

Ambient MS using DESI, DART and DAPPI has also found applications of environmental interest. Some examples are listed in Table 7.3 and discussed below.

7.3.1. *Analysis of toxic compounds from contaminated surfaces*

Luosujärvi *et al.* showed the suitability of DAPPI in the analysis of completely nonpolar PAHs.[28] They used direct DAPPI–MS for the analysis of a soil pellet spiked with phenanthrene, chrysene, and benzo[k]fluoranthene (10 µg/g of each in dry soil), and a blank soil pellet. All three compounds were clearly observed in the spectrum of

Table 7.3. Environmental ambient MS applications.

MS method	Sample	Analytes	Sample preparation	Notes	Ref.
Soil and other surfaces					
DAPPI–IT	Soil pellet	phenanthrene, chrysene, and benzo[k] fluoranthene	None, but the sample was spiked with standards	Analytes observed as M+. ions	28
DAPPI–IT	Circuit board	tetrabromobisphenol	None		28
DART–TOF	Drywall	sulfur-containing chemicals	None	Presence of sulfur-containing compounds was confirmed, but the compounds were not identified	79
DART–TOF	Concrete driveway, mirror	Aspirin, caffeine	Surface swabbed by cotton wipe, wipe analyzed	Air-dispersed chemicals could be mapped rapidly for remediation purposes	81
DART–TOF	Glass, paint, tile, wood door, rug and other household surfaces	methamphetamine, amphetamine, pseudoephedrine, ketamine, phencyclidine, heroin, cocaine and fentanyl, morphine, THC and nicotine	Surface swabbed by cotton wipe, wipe analyzed	LODs were below 12 US state minimum decontamination limits, but the type of the sample had a great effect on the DART signal	82

(*Continued*)

Table 7.3. *(Continued)*

MS method	Sample	Analytes	Sample preparation	Notes	Ref.
Water					
DESI–IT	Groundwater	Explosives (authentic); atrazine, alachlor and acetochlor (spiked)	LLE, SPE and analysis from wetted filter paper	LLE and SPE gave LODs between 346 ppb and 16.3 ppm for the explosives, 300 ppb for spiked acetochlor	83
DESI–Orbitrap	Wastewater treatment plant effluent	Pharmaceuticals and personal-care products (PPCP)	Thin-film microextraction, direct analysis from the blades		84
DESI–IT + thermal assistance	Spiked drinking water	PPCP	Continuous flow of liquid sample directed on a heated (~220 °C) platform	LOD at ppq level achieved for carbamazepine; dynamic range for citalopram was 20–7500 ppt, R2 = 0.9964 and RSD = 6–13%	85
DESI–IT + thermal assistance	Tap water	Cosmetic analytes: benzinide, benzyl salicylate, Orange II, resorcinol	Continuous flow of liquid sample directed on a heated (~220 °C) platform	The LODs ranged from low ppt to low ppb, quantitative data with modest linearity	86

(Continued)

Table 7.3. (*Continued*)

MS method	Sample	Analytes	Sample preparation	Notes	Ref.
DART–TOF	Natural waters	UV filters	Stir-bar sorptive extraction, DART analysis directly from the stirbars	LODs were 20–40 ng/L, R > 0.959, linear range 50 ng/L or 100 ng/L to 1000 ng/L, RSDs 5–30 % at 500 ng/L level	87
DAPPI–IT	Wastewater	Spiked benso[a] pyrene, testosterone and verapamil	Pieces of PDMS soaked in the aqueous sample, extraction for 24 h while shaking	LODs 10–60 nM, linear range 1.5 orders of magnitude	
Volatile organic compounds and atmospheric aerosols					
DESI–IT	Laboratory generated biomass burning aerosols and atmospheric aerosols collected from city air	oxalic and oleic acid		External calibration: different amounts of the analyte standards deposited on a quartz filter surface: linear range of 5 orders of magnitude with R2 > 0.99 and RSD 10–15%	88

(*Continued*)

Table 7.3. (*Continued*)

MS method	Sample	Analytes	Sample preparation	Notes	Ref.
DESI–Orbitrap	Laboratory generated organic aerosols	limonene secondary organic aerosols and their aging products	Aerosols collected on Teflon filter and analyzed directly by DESI	Highly conjugated nitrogen-containing species with different optical properties detected	89
DART–TOF	Eucalyptus leaves	VOC, e.g. terpenes, sesquiterpenes	None	Alcohols and terpenes detected at different desorption temperatures	90
DART–TOF	Hybrid poplar	VOCs, pyrolysis products	Sample was ground into saw dust and pyrolyzed, volatiles were analyzed	Release of the VOCs could be easily monitored at different pyrolysis temperatures	91
DART–QTOF	Laboratory generated organic aerosols	Single-component aerosols of alkanes, alkenes, acids, esters, alcohols, aldehydes and amino acids	Aerosols delivered via a (heated) transfer tube directly to the DART metastable stream	Reactions of aerosols could be monitored	92, 93

(*Continued*)

Table 7.3. (*Continued*)

MS method	Sample	Analytes	Sample preparation	Notes	Ref.
Species identification by chemotyping					
DART–TOF	Agarwood (*Aquilaria*) sawdust, wood chips, and perfumes	chromones	Scented oils removed by extraction to methanol; samples washed by water and extracted with methanol for 2 days	6–17 chromone ions found in agarwood samples by DART analysis; effective identification criteria for the agarwood products: in the analysis of 25 other scented woods, no false positives were found	94
DART–TOF	Wood slivers of *Dalbergia nigra, D. spruceana, Swartzia tomentosa, Phoebe porosa* and *Machaerium scleroxylon*		None	PCA and LDA analysis of the wood spectra compared, 100%, 91.24% and 96.15% classification accuracies achieved	95

(*Continued*)

Table 7.3. (*Continued*)

MS method	Sample	Analytes	Sample preparation	Notes	Ref.
DART–TOF	White oak (*Quercus alba*) and northern red oak (*Quercus rubra*)	Small organic acids	Small pieces (1mm × 3 mm) were cut from the samples for the analysis	PCA and LDA with 100% classification accuracy between while and northern red oak	96
Marine organisms					
DESI–IT	Tropical seaweed tissue	bromophycolides	The algal samples were preserved in 10% formalin/sea water and stored at room T, rinsed in ultrapure water and allowed to dry before the analysis	Sensitivity improved by Cl- adduct formation	97

(*Continued*)

Table 7.3. (*Continued*)

MS method	Sample	Analytes	Sample preparation	Notes	Ref.
DESI–IT imaging	Tropical seaweed tissue	bromophycolides and other antifungals	The algal samples were preserved with 10% formalin in seawater, were affixed to PTFE substrates and kept moist with seawater	Light microscopy revealed cell integrity before and after DESI measurements → a physically nondestructive method for analysis of compounds from intact biological surfaces	98
DESI–IT imaging	Red alga *Phacelocarpus* neurymenolides	neurymenolide A	A blade removed from the living alga, no pretreatment	Bleaching of coral colonies was found to be caused by allelopathy by neurymenolide A from seaweed	99

the spiked soil pellet, as shown in Fig. 7.6. The analysis of soil with high organic content shows the applicability of the method in the analysis of real samples, since PAHs (with three rings or more) tend to accumulate in nature into the humified organic part of soil. Curtis *et al.* presented a comprehensive case study, where DART–MS was used to confirm possible contamination in imported drywall.[79] Pieces of the samples and reference material were held with tweezers and moved in the DART metastable stream for 1 min each. Negative-ion mode mass spectra of the imported drywall showed peaks for inorganic sulfur ions, which could be identified based on accurate mass and isotope patterns. Although the DART methodology was unable to identify the substance responsible for releasing the sulfur-containing ions, the simple nature of the analysis and short analysis time makes DART a rapid tool for preliminary investigation of unknown contaminants.

While these two examples showed that ambient MS could be a helpful tool in qualitative screening, the works by Grange[80,81] and Grange and Sovocool[82] explore its quantitative applications. In these studies, contaminated surfaces were sampled with cotton

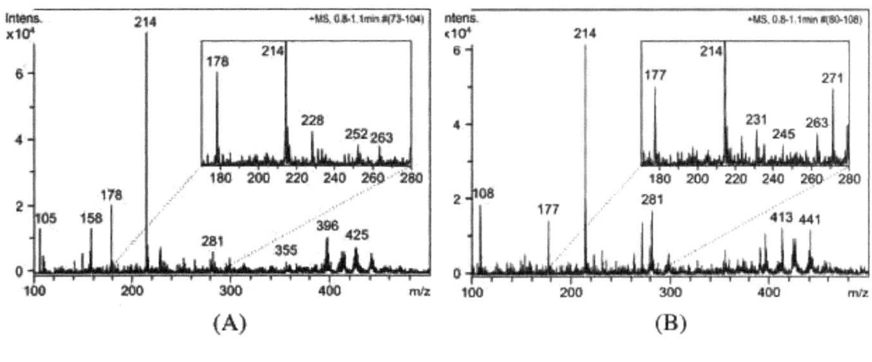

Figure 7.6. DAPPI mass spectra obtained from (A) spiked soil pellet and (B) blank soil pellet in positive-ion mode with toluene (10 mL/min) as the spray solvent. In (A), ions originating from benzo[k]fluoranthene, chrysene, and phenathrene are seen at *m/z* 252 ($M^{+\cdot}$), 228 ($M^{+\cdot}$), and 178 ($M^{+\cdot}$), respectively. The amount of each PAH compound was 10 mg/g of soil. Reproduced with permission from Luosujärvi, L., Kanerva, S., Saarela, V. *et al.* (2010). Environmental and food analysis by desorption atmospheric pressure photoionization-mass spectrometry, *Rapid Commun. Mass Spectrom.*, **24**, 1343–1350. Copyright (2010) John Wiley and Sons.

swabs and the swabs were analyzed by DART–MS. The earliest contribution[81] was devoted to the study of experimental procedures, such as different wiping techniques, suitable solvents for wetting the swabs, and data-analysis techniques. The more recent studies focus on evaluating possible application of DART in contamination sites, for example in 'meth labs' (i.e. places where methamphetamine is synthesized), and EPA National Priority List targets like Superfund toxic waste dump sites. Due to the high analyte amounts in the contamination scenarios, sufficiently low LODs could easily be achieved (e.g. one-fourth of the lowest U.S. state decontamination limit for methamphetamine (0.025 μg/100 cm^2)).[82] Unfortunately, significant carry-over was frequently observed, and sometimes the sampled material even clogged the MS.[80,81] Easy automated-cleaning strategy[81] and temperature and sampling speed optimization[80] were shown to reduce the carry-over, and the clogging could be eliminated by additional pumping of the sampling region by a commercial Vapur interface,[80] although this increased the carry-over effect. In these applications, DART–MS was found to be semi-quantitative, with two to three decades of linear dynamic range and RSDs of a little over 20%. While DART was not applied to study real contamination sites, the thorough simulations in the studies imply that DART could be used to categorize contamination levels in a spatially resolved manner, and the data could be used to plan, monitor, and document remediation activities. However, the sampling surface was found to greatly affect the analyte recovery to the swabs, which questions the quantitativity of the swab-sampling method in general. While not yet covered in the literature, continued development of the methodology could lead to on-site analyses with portable mass spectrometers facilitating real-time cleaning validation and action planning, and enable direct examination of the contaminated surfaces.

7.3.2. *Water*

Although the nature of ambient MS methods is more suited to the analysis of solid surfaces, there are several successful applications that have been developed for analyzing environmental water samples. The emphasis has been on how to concentrate the interesting analytes from high volumes of liquid to a small spot that could be analyzed by ambient MS. Mulligan *et al.* used DESI for the analysis of explosive residues from contaminated groundwater from an army ammunition area.[83] Efficient preconcentration of the analytes from authentic groundwater samples was achieved by LLE and SPE, with LODs at 346 ppb to 16.3 ppm. Best sensitivity was achieved using LLE (low ppb–ppm level), but an advantage of the SPE method was that the membrane could be removed from the cartridge and directly exposed to analysis by DESI without a separate elution step. It would also make possible the extraction of the samples on site, followed by DESI analysis in laboratory, without having to move large volumes of contaminated water from the field to the lab. In another contribution by Strittmatter *et al.*, thin-film microextraction (TFME) blades were used for extracting various pharmaceuticals and personal care products (PPCP) from wastewater.[84] LODs in the low ng/L range were achieved, although serious matrix effects were observed when treated wastewater was analyzed (signal reduction 90% and 50% for triclosan and carbamazepine, respectively). The method was tested for screening of 100 mL of wastewater treatment plant effluent, and several PPCP compounds were detected, such as beta-blockers, nonsteroidal anti-inflammatory drugs, and UV filters.

Campbell *et al.* introduced an additional capillary to the DESI source, through which a continuously flowing (1–90 μL/min) aqueous sample was delivered to a heated platform (~220 °C) for the DESI–MS analysis.[85] The design was tested for multicomponent analysis of spiked PPCP contaminants from drinking water, and the LODs in general were at ppt levels. However, for carbamazepine LOD of 900 ppq was achieved, which can be considered a notable achievement for an ambient MS method that does not use timely sample preparation and chromatographic separation. The enhancement was explained by formation of smaller progeny droplets, concentration of analytes in the droplets, and faster achievement of the Rayleigh limit due to more efficient evaporation of solvent from the droplets, and thus more efficient formation of gas-phase analyte ions. The method also showed quantitative ability, since the dynamic range for citalopram in tap water was reported to be 20–7500 ppt, with R^2 of 0.9964 and RSDs of 6–13%, even without internal standard.

Haunschmidt *et al.* developed a sample preparation method for DART to screen natural waters for seven organic UV filters common in sunscreen products.[87] Samples of 250 mL were preconcentrated with stir bar sorptive extraction for 4 hours, and the PDMS stir bars were exposed to the DART stream without a separate elution step. LODs were 20–40 ng/L depending on the analyte, and the method was semi-quantitative with $R > 0.959$ for the calibration curves ranging from 50 ng/L or 100 ng/L to 1000 ng/L and RSDs of 5–30% at 500 ng/L level. Analysis of an authentic lake water sample collected at an area used for leisure activities revealed contamination with benzophenone-3 and octocrylene. A reference analysis with thermal desorption GC–MS gave comparable results for the two analytes, but also detected three additional UV filters that were present at concentrations below the LODs of the DART method. While this study did not thoroughly evaluate the possible matrix effects or isobaric interferences, it shows that DART offers a rapid screening method to detect aquatic contamination. PDMS was also used as the SPE material in a DAPPI study of aqueous samples: pieces of PDMS were soaked in the sample for 24 hours, after which the DAPPI analysis took place directly from the surface of the PDMS pieces.[100] The feasibility of the method in the analysis of 100 nM concentrations of testosterone, anthracene and verapamil from 100 mL wastewater and MilliQ water samples was demonstrated. The signals for the analytes were only slightly reduced in the case of the wastewater samples and the wastewater matrix did not give disturbing background-ion signals at the studied *m/z* range. However, the long extraction time required restricts the applicability of the method, although several samples can in theory be extracted simultaneously.

7.3.3. *Volatile organic compounds and atmospheric aerosols*

Biogenic volatile organic compounds (VOCs) and atmospheric aerosols have become of great interest in recent decades, when their roles in climate change have been unraveled. VOCs can be reliably studied by GC–MS and PTR–MS, while the development of analytical methods for aerosol particle research is currently underway to find the best protocols for the task. Due to the ease of sampling, ambient MS could be a suitable way to study VOCs in situ as they are released. For example, Maleknia *et al.* used DART to study eucalyptus leaf volatiles.[90] They applied high DART gas temperatures to simulate the release of VOCs in wild fires. Alcohols were detected at lower temperatures, while high temperatures released terpenes from the leaves. In a study by Jones *et al.*, a micropyrolyzer was built between the DART probe and the MS inlet to study poplar biomass.[91] While the emphasis of the study was to understand the pyrolysis process of biomass for energy production purposes, the results suggest that the experimental set-up is a good tool for the study of temperature dependent release of VOCs.

In the case of aerosols, ambient MS is a simple way to study the particles, could omit the risk of chemical modifications during sample preparation, and is not limited by the transmission efficiency of aerodynamic lenses used with vacuum MS techniques. Li *et al.* used DESI for the quantification of oxalic and oleic acids in aerosols.[88] Quantitation was achieved externally, by depositing different amounts of analyte standards on a quartz filter surface. Calibration showed a linear range of 5 orders of magnitude with $R^2 > 0.99$ and RSD 10–15%. The method was used to analyze laboratory-generated biomass burning aerosols from combustion of rice straw, and atmospheric aerosols collected from city air (12 h). The quantitative results obtained with DESI were consistent with those obtained with ion chromatography (IC) and GC–MS. In a contribution by Laskin *et al.*, DESI–Orbitrap was used for the chemical characterization of laboratory generated organic aerosols.[89] Secondary organic aerosols (SOA) were produced from the ozonolysis of limonene vapors and collected on Teflon surface, which were then exposed for analysis by DESI (Fig. 7.7A). Chemical aging of the limonene SOA was studied by exposing the Teflon surface with the collected aerosol sample to ammonia vapor. As a result, highly conjugated nitrogen-containing species were formed, which were observed to change the optical properties of the sample and were thought to be responsible for the abstraction of visible light (Fig. 7.7B). The same samples were also studied by ESI–MS after extraction from the Teflon surface. Due to

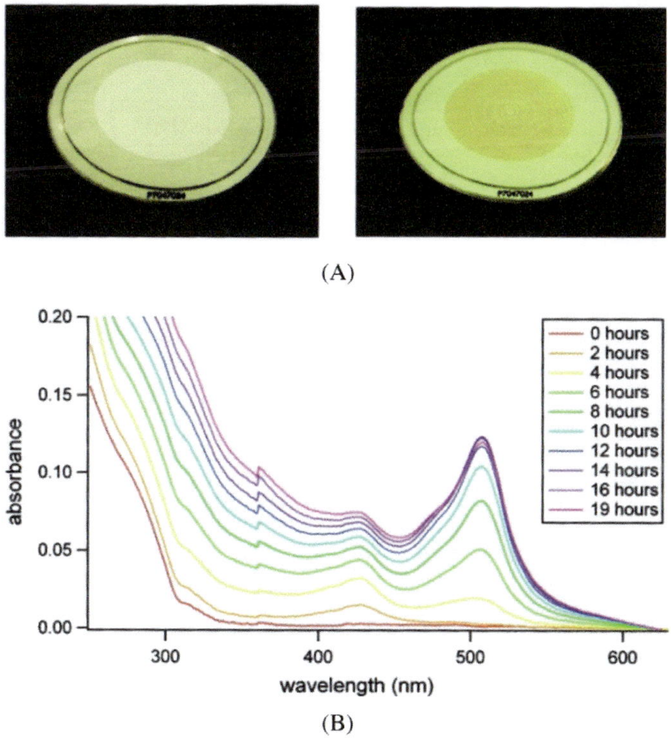

Figure 7.7. (A) Photographs of the fresh (left) and aged (right) limonene secondary organic aerosols (LSOA) samples on a Teflon filter; (B) UV-visible spectra of LSOA, aged on CaF_2 window in the presence of NH_3(g). Reprinted with permission from Laskin, J., Laskin, A., Roach, P.J. *et al.* (2010). High-resolution desorption electrospray ionization mass spectrometry for chemical characterization of organic aerosols, *Anal. Chem.*, 82, 2048–2058. Copyright (2010) American Chemical Society.

the shorter interaction time of the analytes and the solvent, DESI was shown to allow the analysis of chemically labile species that could not be detected with ESI–MS.

Nah *et al.* studied single-component aerosols of alkanes, alkenes, acids, esters, alcohols, aldehydes and amino acids by delivering them directly to the DART metastable stream (Fig. 7.8).[92] They showed that analytes were desorbed and ionized from the aerosol particle surface, as the observed signal depended on the total particle surface area. When the aerosol transfer tube was heated, complete aerosol particles could be studied. In a follow-up study, it was found that 1–10 nm of the particle surface is desorbed by the DART metastable gas, depending on its temperature and physicochemical properties of the aerosols (e.g. volatility and surface area.)[93] DART was successfully applied to determine the reaction rate of oleic acid aerosol particles with ozone.[92] However, DART could only detect the ~100–140 nm single component aerosol particles at mass-concentration levels similar to those of multicomponent particles in urban areas. Thus, in its current state, this DART methodology is best suited for studies of synthetic aerosols, and the study of environmental species may not be feasible.

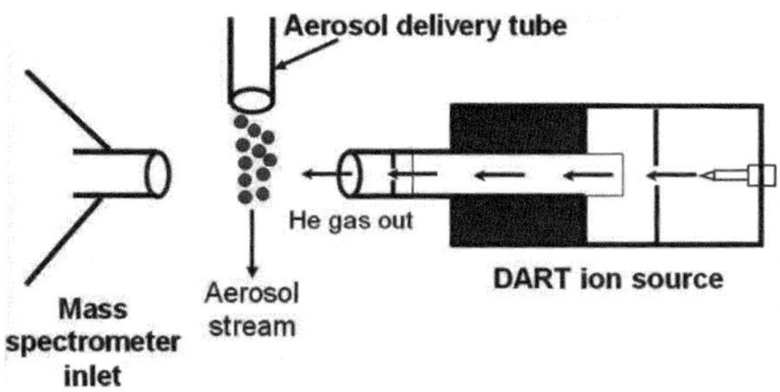

Figure 7.8. Sample introduction of aerosol particles for DART–MS studies. Reprinted with permission from Nah, T., Chan, M., Leone, S.R. *et al.* (2013). Real time in situ chemical characterization of submicrometer organic particles using direct analysis in real time–mass spectrometry, *Anal. Chem.*, 85, 2087–2095. Copyright (2013) American Chemical Society.

7.3.4. *Species identification by chemotyping*

Identification of wood is often based on morphological examinations, which are hindered by the similarity of closely related species and destruction of anatomical features during processing. It has been studied whether DART could complement or replace the morphological identification methods. In a study of endangered agarwood (*Aquilaria*),[94] the DART–MS identification was based on *Aquilaria* chromones. A variety of wood products were examined, including sawdust, wood chips, and perfumes. Since the solid products are often coated with scented oils, the samples were extracted with methanol to remove the potentially disturbing substances, washed, and the chromones were extracted for two days and analyzed by DART. Almost identical results were achieved by analyzing the samples directly without the extraction step, but this quicker method was not further exploited. The results enabled the writers to propose effective identification criteria for the agarwood products; in the analysis of 25 other scented woods, no false positives were found, and examination of 151 commercial products suspected of containing agarwood found 0–76% positives, depending on sample type. In a study of *Dalbergia*,[95] the aim was to identify *Dalbergia nigra* from other *Dalbergia* taxa, especially the anatomically similar *D. spruceana*, to help trade regulation enforcement, which only prohibits the trade of *D. nigra*. Slivers of wood were analyzed by placing them in the DART metastable stream for 6 s. When the spectra of *D. spruceana* and *D. nigra* were compared, they appeared different, but contained the same ions. To confirm the visual observations, PCA and LDA models were created based on 18 *D. nigra* and 20 *D. spruceana* sample spectra. In leave-one-out cross-validation, PCA and LDA gave 100% and 91.24% classification accuracy, respectively. A second model was created to distinguish common trade timbers *Swartzia tomentosa*, *Phoebe porosa* and *Machaerium scleroxylon* from the endangered *D. nigra*. Here, LDA gave 96.15% classification accuracy (Fig. 7.9). In a similar experiment, Cody *et al.* differentiated white oak (*Quercus alba*) from northern red oak (*Quercus rubra*), which responds differently to wood processing

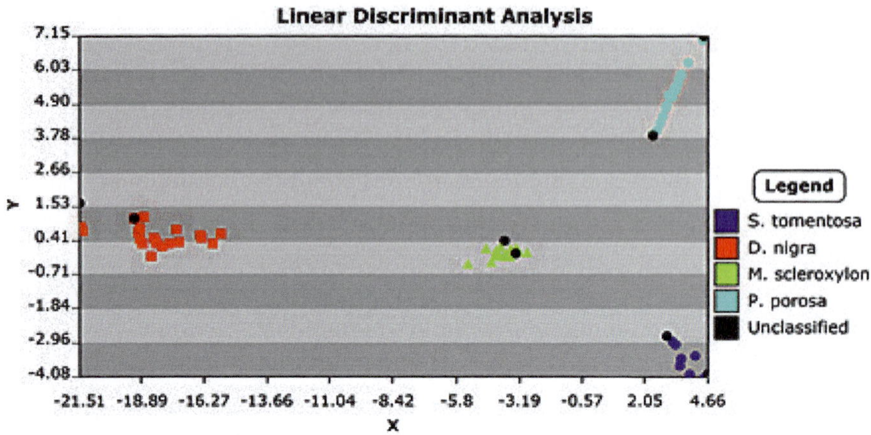

Figure 7.9. Linear discriminant analysis of commonly traded *Swartzia tomentosa*, *Phoebe porosa* (imbuia), *Machaerium scleroxylon* (pau ferro), and *Dalbergia nigra*. Reprinted with permission from Lancaster, C. and Espinoza, E. (2012). Analysis of select Dalbergia and trade timber using direct analysis in real time and time-of-flight mass spectrometry for CITES enforcement, *Rapid Commun. Mass Spectrom.*, **26**, 1147–1156). Copyright (2012) John Wiley and Sons.

protocols.[96] DART–MS with PCA and LDA gave 100% classification accuracy.

7.4. Conclusions

Based on this extensive body of work, it can be concluded that ambient MS is a rapid and reliable method for the screening of compounds from food and environmental matrices. DESI, DART, DAPPI and other ambient MS techniques could be highly useful in routine screening analyses of food products in control laboratories, for example in food scandal cases. Other examples where the techniques can give valuable information are the fingerprinting of food and drink products for quality control and authenticity assessment purposes, and recognition of endangered species in biodiversity protection, provided that suitable marker ions are known. Also, ambient MS may prove a valuable tool in field MS as the portable instruments develop. Of course, ambient MS sets certain requirements to the MS instrument that is used. Since there is no chromatographic separation involved in ambient MS, there is also no retention time information, which could be utilized as an additional tool for the identification of the compounds. Therefore, high mass-resolving power of the mass spectrometer is often necessary to mitigate isobaric interferences and MS/MS studies may be needed for reliable identification. There are various examples of successful quantitative or at least semi-quantitative analysis with ambient MS. In most cases this has been achieved using an isotopically labeled internal standard and sample preparation. The qualitative and quantitative results obtained with ambient MS have mainly been similar to those obtained with conventional reference methods. Although the LODs, linear range, and repeatability of ambient MS are typically poorer than with conventional GC–MS or LC-MS methods, this is compensated by the much faster analysis time, which can be as little as seconds/sample in the case that the direct analysis without sample preparation is feasible.

Requirements set by authorities still prevent the implementation of ambient MS in some application areas, such as control of pesticide residues in food. The US EPA, EU, and nationally set tolerance levels/MRLs for pesticides and fungicides are based on the analysis of the whole sample and given in µg/kg of produce, and therefore require the homogenization of the whole sample, since large variations in the surface concentration of the compounds in real-life samples are likely. Direct analysis from the fruit or vegetable surface by ambient MS without any sample preparation would significantly speed up the analysis process, thus establishment of MRLs that are applicable for ambient MS would be justified.

References

1. Takáts, Z., Wiseman, J.M., Gologan, B. *et al.* (2004). Mass spectrometry sampling under ambient conditions with desorption electrospray ionization, *Science*, **306**, 471–473.
2. Cody, R.B., Laramee, J.A. and Durst, H.D. (2005). Versatile new ion source for the analysis of materials in open air under ambient conditions, *Anal. Chem.*, 77, 2297–2302.

3. Haapala, M., Pol, J., Saarela, V. *et al.* (2007). Desorption atmospheric pressure photoionization, *Anal. Chem.*, **79**, 7867–7872.
4. Monge, M.E., Harris, G.A., Dwivedi, P. *et al.* (2013). Mass spectrometry: recent advances in direct open air surface sampling/ionization, *Chem. Rev.*, **113**, 2269–2308.
5. Harris, G.A., Galhena, A.S. and Fernández, F.M. (2011). Ambient sampling/ionization mass spectrometry: applications and current trends, *Anal. Chem.*, **83**, 4508–4538.
6. Ifa, D.R., Wu, C., Ouyang, Z. *et al.* (2010). Desorption electrospray ionization and other ambient ionization methods: current progress and preview, *Analyst*, **135**, 669–681.
7. Takáts, Z., Wiseman, J.M. and Cooks, R.G. (2005). Ambient mass spectrometry using desorption electrospray ionization (DESI): instrumentation, mechanisms and applications in forensics, chemistry and biology, *J. Mass Spectrom.*, **40**, 1261–1275.
8. Cotte-Rodriguez, I., Takats, Z., Talaty, N. *et al.* (2005). Desorption electrospray ionization of explosives on surfaces: sensitivity and selectivity enhancement by reactive desorption electrospray ionization, *Anal. Chem.*, **77**, 6755–6764.
9. Cody, R.B. (2009). Observation of molecular ions and analysis of nonpolar compounds with the direct analysis in real time ion source, *Anal. Chem.*, **81**, 1101–1107.
10. Song, L., Dykstra, A.B., Yao, H. *et al.* (2009). Ionization mechanism of negative ion–direct analysis in real time: a comparative study with negative ion–atmospheric pressure photoionization, *J. Am. Soc. Mass Spectrom.*, **20**, 42–50.
11. Song, L., Gibson, S.C., Bhandari, D. *et al.* (2009). Ionization mechanism of positive-ion direct analysis in real time: a transient microenvironment concept, *Anal. Chem.*, **81**, 10080–10088.
12. Domin, M.A., Steinberg, B.D., Quimby, J.M. *et al.* (2010). Routine analysis and characterization of highly insoluble polycyclic aromatic compounds by direct analysis in real time mass spectrometry (DART), *Analyst*, **135**, 700.
13. Moravcova, E., Vaclavik, L., Lacina, O. *et al.* (2012). Novel approaches to analysis of 3-chloropropane-1,2-diol esters in vegetable oils, *Anal. Bioanal. Chem.*, **402**, 2871–2883.

14. Cajka, T., Danhelova, H., Zachariasova, M. *et al.* (2013). Application of direct analysis in real time ionization–mass spectrometry (DART–MS) in chicken meat metabolomics aiming at the retrospective control of feed fraud, *Metabolomics*, **9**, 545–557.
15. Vaclavik, L., Belkova, B., Reblova, Z. *et al.* (2013). Rapid monitoring of heat-accelerated reactions in vegetable oils using direct analysis in real time ionization coupled with high resolution mass spectrometry, *Food Chem.*, **138**, 2312–2320.
16. Kalachova, K., Pulkrabova, J., Drabova, L. *et al.* (2011). Simplified and rapid determination of polychlorinated biphenyls, polybrominated diphenyl ethers, and polycyclic aromatic hydrocarbons in fish and shrimps integrated into a single method, *Anal. Chim. Acta*, **707**, 84–91.
17. Edison, S.E., Lin, L.A., Gamble, B.M. *et al.* (2011). Surface swabbing technique for the rapid screening for pesticides using ambient pressure desorption ionization with high-resolution mass spectrometry, *Rapid Commun. Mass Spectrom.*, **25**, 127–139.
18. Cajka, T., Danhelova, H., Vavrecka, A. *et al.* (2013). Evaluation of direct analysis in real time ionization–mass spectrometry (DART–MS) in fish metabolomics aimed to assess the response to dietary supplementation, *Talanta*, **115**, 263–270.
19. Vaclavik, L., Rosmus, J., Popping, B. *et al.* (2010). Rapid determination of melamine and cyanuric acid in milk powder using direct analysis in real time–time-of-flight mass spectrometry, *J. Chromatogr. A*, **1217**, 4204–4211.
20. Vaclavik, L., Cajka, T., Hrbek, V. *et al.* (2009). Ambient mass spectrometry employing direct analysis in real time (DART) ion source for olive oil quality and authenticity assessment, *Anal. Chim. Acta*, **645**, 56–63.
21. Saang'onyo, D.S. and Smith, D.L. (2012). Optimization of direct analysis in real time (DART) linear ion trap parameters for the detection and quantitation of glucose, *Rapid Commun. Mass Spectrom.*, **26**, 385–391.
22. Harris, G.A. and Fernández, F.M. (2009). Simulations and experimental investigation of atmospheric transport in an ambient metastable-induced chemical ionization source, *Anal. Chem.*, **81**, 322–329.
23. Luosujärvi, L., Arvola, V., Haapala, M. *et al.* (2008). Desorption and ionization mechanisms in desorption atmospheric pressure photoionization, *Anal. Chem.*, **80**, 7460–7466.

24. García-Reyes, J.F., Jackson, A.U., Molina-Díaz, A. *et al.* (2009). Desorption electrospray ionization–mass spectrometry for trace analysis of agrochemicals in food, *Anal. Chem.*, **81**, 820–829.
25. Berchtold, C., Müller, V., Meier, L. *et al.* (2013). Direct detection of chlorpropham on potato skin using desorption electrospray ionization, *J. Mass Spectrom.*, **48**, 587–593.
26. Lehotay, S.J. (2006). 'Quick, easy, cheap, effective, rugged, and safe approach for determining pesticide residues', in *Pesticide Protocols*, Vidal J.L.M. and Frenich A.G. (eds), Humana Press, pp 239–261. Available at: http://link.springer.com/protocol/10.1385/1-59259-929-X%3A239 [Accessed September 2, 2013].
27. Mulligan, C.C., Talaty, N. and Cooks, R.G. (2006). Desorption electrospray ionization with a portable mass spectrometer: in situ analysis of ambient surfaces, *Chem. Commun.*, 1709–1711.
28. Luosujärvi, L., Kanerva, S., Saarela, V. *et al.* (2010). Environmental and food analysis by desorption atmospheric pressure photoionization-mass spectrometry, *Rapid Commun. Mass Spectrom.*, **24**, 1343–1350.
29. Farré, M., Picó, Y. and Barceló, D. (2013). Direct peel monitoring of xenobiotics in fruit by direct analysis in real time coupled to a linear quadrupole ion trap–orbitrap mass spectrometer, *Anal. Chem.*, **85**, 2638–2644.
30. Crawford, E. and Musselman, B. (2012). Evaluating a direct swabbing method for screening pesticides on fruit and vegetable surfaces using direct analysis in real time (DART) coupled to an Exactive benchtop orbitrap mass spectrometer, *Anal. Bioanal. Chem.*, **403**, 2807–2812.
31. Schurek, J., Vaclavik, L., Hooijerink, H. (Dick). *et al.* (2008). Control of strobilurin fungicides in wheat using direct analysis in real time accurate time-of-flight and desorption electrospray ionization linear ion trap mass spectrometry, *Anal. Chem.*, **80**, 9567–9575.
32. Edison, S.E., Lin, L.A. and Parrales, L. (2011). Practical considerations for the rapid screening for pesticides using ambient pressure desorption ionisation with high-resolution mass spectrometry, *Food Addit. Contam. Part*, **28**, 1393–1404.
33. Cajka, T., Riddellova, K., Zomer, P. *et al.* (2011). Direct analysis of dithiocarbamate fungicides in fruit by ambient mass spectrometry, *Food Addit. Contam. Part*, **28**, 1372–1382.

34. Block, E., Cody, R.B., Dane, A.J. *et al.* (2010). Allium chemistry: use of new instrumental techniques to "see" reactive organosulfur species formed upon crushing garlic and onion, *Pure Appl. Chem.*, **82**, 535–539.
35. Block, E., Dane, A.J., Thomas, S. *et al.* (2010). Applications of direct analysis in real time–mass spectrometry (DART–MS) in *Allium* chemistry. 2-propenesulfenic and 2-propenesulfinic acids, diallyl trisulfane S-oxide, and other reactive sulfur compounds from crushed garlic and other *Alliums*, *J. Agric. Food Chem.*, **58**, 4617–4625.
36. Kubec, R., Cody, R.B., Dane, A.J. *et al.* (2010). Applications of direct analysis in real time–mass spectrometry (DART–MS) in *Allium* chemistry. (Z)-butanethial S-oxide and 1-butenyl thiosulfinates and their S-(E)-1-butenylcysteine S-oxide precursor from *Allium siculum*, *J. Agric. Food Chem.*, **58**, 1121–1128.
37. Zachariasova, M., Cajka, T., Godula, M. *et al.* (2010). Analysis of multiple mycotoxins in beer employing (ultra)high-resolution mass spectrometry, *Rapid Commun. Mass Spectrom.*, **24**, 3357–3367.
38. Zhou, J., Yao, S., Qian, R. *et al.* (2008). Observation of allicin-cysteine complex by reactive desorption electrospray ionization mass spectrometry for garlic, *Rapid Commun. Mass Spectrom.*, **22**, 3334–3337.
39. Li, Y. (2012). Confined direct analysis in real time ion source and its applications in analysis of volatile organic compounds of *Citrus limon* (lemon) and *Allium cepa* (onion), *Rapid Commun. Mass Spectrom.*, **26**, 1194–1202.
40. Joyce, N.I., Eady, C.C., Silcock, P. *et al.* (2013). Fast phenotyping of LFS-silenced (tearless) onions by desorption electrospray ionization–mass spectrometry (DESI–MS), *J. Agric. Food Chem.*, **61**, 1449–1456.
41. Chernetsova, E.S., Crawford, E.A., Shikov, A.N. *et al.* (2012). ID-CUBE direct analysis in real time high-resolution mass spectrometry and its capabilities in the identification of phenolic components from the green leaves of *Bergenia crassifolia L*, *Rapid Commun. Mass Spectrom.*, **26**, 1329–1337.
42. Roschek Jr., B., Fink, R.C., McMichael, M.D. *et al.* (2009). Elderberry flavonoids bind to and prevent H1N1 infection in vitro, *Phytochemistry*, **70**, 1255–1261.

43. Gerbig, S. and Takáts, Z. (2010). Analysis of triglycerides in food items by desorption electrospray ionization–mass spectrometry, *Rapid Commun. Mass Spectrom.*, **24**, 2186–2192.
44. Suni, N.M., Aalto, H., Kauppila, T.J. *et al.* (2012). Analysis of lipids with desorption atmospheric pressure photoionization–mass spectrometry (DAPPI–MS) and desorption electrospray ionization–mass spectrometry (DESI–MS), *J. Mass Spectrom.*, **47**, 611–619.
45. Hartmanova, L., Ranc, V., Papouskova, B. *et al.* (2010). Fast profiling of anthocyanins in wine by desorption nano-electrospray ionization–mass spectrometry, *J. Chromatogr. A*, **1217**, 4223–4228.
46. Li, B., Bjarnholt, N., Hansen, SH. *et al.* (2011). Characterization of barley leaf tissue using direct and indirect desorption electrospray ionization–imaging mass spectrometry, *J. Mass Spectrom.*, **46**, 1241–1246.
47. Roschek, B., Fink, R.C., Li, D. *et al.* (2009). Pro-inflammatory enzymes, cyclooxygenase 1, cyclooxygenase 2, and 5-lipooxygenase, inhibited by stabilized rice bran extracts, *J. Med. Food*, **12**, 615–623.
48. Rajchl, A., Drgová, L., Grégrová, A. *et al.* (2013). Rapid determination of 5-hydroxymethylfurfural by DART ionization with time-of-flight mass spectrometry, *Anal. Bioanal. Chem.*, **405**, 4737–4745.
49. Danhelova, H., Hradecky, J., Prinosilova, S. *et al.* (2012). Rapid analysis of caffeine in various coffee samples employing direct analysis in real-time ionization–high-resolution mass spectrometry, *Anal. Bioanal. Chem.*, **403**, 2883–2889.
50. Bai, Y., Zhang, J., Bai, Y. *et al.* (2012). Direct analysis in real time mass spectrometry combined with single-drop liquid–liquid–liquid microextraction for the rapid analysis of multiple phytohormones in fruit juice, *Anal. Bioanal. Chem.*, **403**, 2307–2314.
51. Lojza, J., Cajka, T., Schulzova, V. *et al.* (2012). Analysis of isoflavones in soybeans employing direct analysis in real-time ionization–high-resolution mass spectrometry, *J. Sep. Sci.*, **35**, 476–481.
52. Jeckelmann, N. and Haefliger, O.P. (2010). Release kinetics of actives from chewing gums into saliva monitored by direct analysis in real time mass spectrometry, *Rapid Commun. Mass Spectrom.*, **24**, 1165–1171.
53. Saka, K., Konuma, K., Asai, S. *et al.* (2009). Identification of active ingredients in dietary supplements using non-destructive mass spectrometry

and liquid chromatography–mass spectrometry, *Forensic Sci. Int.*, **191**, e5–e10.
54. Hrbek, V., Vaclavik, L., Elich, O. *et al.* (2014). Authentication of milk and milk-based foods by direct analysis in real time ionization–high resolution mass spectrometry (DART–HRMS) technique: a critical assessment, *Food Control*, **36**, 138–145.
55. Roach, P.J., Laskin, J. and Laskin, A. (2010). Nanospray desorption electrospray ionization: an ambient method for liquid-extraction surface sampling in mass spectrometry, *Analyst*, **135**, 2233.
56. Cajka, T., Riddellova, K., Tomaniova, M. *et al.* (2011). Ambient mass spectrometry employing a DART ion source for metabolomic fingerprinting/profiling: a powerful tool for beer origin recognition, *Metabolomics*, 7, 500–508.
57. Vaclavik, L., Hrbek, V., Cajka, T. *et al.* (2011). Authentication of animal fats using direct analysis in real time (DART) ionization–mass spectrometry and chemometric tools, *J. Agric. Food Chem.*, **59**, 5919–5926.
58. Novotná, H., Kmiecik, O., Gałazka, M. *et al.* (2012). Metabolomic fingerprinting employing DART–TOF-MS for authentication of tomatoes and peppers from organic and conventional farming, *Food Addit. Contam. Part*, **29**, 1335–1346.
59. Kim, H.J., Baek, W.S. and Jang, Y.P. (2011). Identification of ambiguous cubeb fruit by DART–MS-based fingerprinting combined with principal component analysis, *Food Chem.*, **129**, 1305–1310.
60. Cajka, T., Riddellova, K., Tomaniova, M. *et al.* (2010). Recognition of beer brand based on multivariate analysis of volatile fingerprint, *J. Chromatogr. A*, **1217**, 4195–4203.
61. Fraser, K., Lane, G.A., Otter, D.E. *et al.* (2013). Monitoring tea fermentation/manufacturing by direct analysis in real time (DART) mass spectrometry, *Food Chem.*, **141**, 2060–2065.
62. Chernetsova, E.S. and Morlock, G.E. (2012). Assessing the capabilities of direct analysis in real time mass spectrometry for 5-hydroxymethylfurfural quantitation in honey, *Int. J. Mass Spectrom.*, **314**, 22–32.
63. Van Berkel, G.J., Tomkins, B.A. and Kertesz, V. (2007). Thin-layer chromatography/desorption electrospray ionization mass spectrometry: investigation of goldenseal alkaloids, *Anal. Chem.*, **79**, 2778–2789.

64. Ackerman, L.K., Noonan, G.O. and Begley, T.H. (2009). Assessing direct analysis in real time–mass spectrometry (DART–MS) for the rapid identification of additives in food packaging, *Food Addit. Contam. Part*, **26**, 1611–1618.
65. Rothenbacher, T. and Schwack, W. (2010). Rapid identification of additives in poly(vinyl chloride) lid gaskets by direct analysis in real time ionisation and single-quadrupole mass spectrometry, *Rapid Commun. Mass Spectrom.*, **24**, 21–29.
66. Bentayeb, K., Ackerman, L.K., Lord, T. *et al.* (2013). Non-visible print set-off of photoinitiators in food packaging: detection by ambient ionisation mass spectrometry, *Food Addit. Contam. Part*, **30**, 750–759.
67. Mattarozzi, M., Milioli, M., Cavalieri, C. *et al.* (2012). Rapid desorption electrospray ionization–high-resolution mass spectrometry method for the analysis of melamine migration from melamine tableware, *Talanta*, **101**, 453–459.
68. Kuki, Á., Nagy, L., Zsuga, M. *et al.* (2011). Fast identification of phthalic acid esters in poly(vinyl chloride) samples by direct analysis in real time (DART)–tandem mass spectrometry, *Int. J. Mass Spectrom.*, **303**, 225–228.
69. Dane, A.J. and Cody, R.B. (2010). Selective ionization of melamine in powdered milk by using argon direct analysis in real time (DART)–mass spectrometry, *Analyst*, **135**, 696.
70. Shen, Y., van Beek, T.A., Claassen, F.W. *et al.* (2012). Rapid control of Chinese star anise fruits and teas for neurotoxic anisatin by direct analysis in real time–high-resolution mass spectrometry, *J. Chromatogr. A*, **1259**, 179–186.
71. Morlock, G. and Schwack, W. (2006). Determination of isopropylthioxanthone (ITX) in milk, yoghurt and fat by HPTLC–FLD, HPTLC–ESI/MS and HPTLC–DART/MS, *Anal. Bioanal. Chem.*, **385**, 586–595.
72. Vaclavik, L., Zachariasova, M., Hrbek, V. *et al.* (2010). Analysis of multiple mycotoxins in cereals under ambient conditions using direct analysis in real time (DART) ionization coupled to high resolution mass spectrometry, *Talanta*, **82**, 1950–1957.
73. Self, R.L. and Wu, W.-H. (2012). Rapid qualitative analysis of phthalates added to food and nutraceutical products by direct analysis in real time/orbitrap mass spectrometry, *Food Control*, **25**, 13–16.

74. D'Aloise, P. and Chen, H. (2012). Rapid determination of flunitrazepam in alcoholic beverages by desorption electrospray ionization–mass spectrometry, *Sci. Justice*, **52**, 2–8.
75. Lin, Z., Zhao, M., Zhang, S. *et al.* (2010). In situ arsenic speciation on solid surfaces by desorption electrospray ionization tandem mass spectrometry, *The Analyst*, **135**, 1268.
76. Huang, G., Xu, W., Visbal-Onufrak, M.A. *et al.* (2010). Direct analysis of melamine in complex matrices using a handheld mass spectrometer, *Analyst*, **135**, 705–711.
77. Yang, S., Ding, J., Zheng, J. *et al.* (2009). Detection of melamine in milk products by surface desorption atmospheric pressure chemical ionization mass spectrometry, *Anal. Chem.*, **81**, 2426–2436.
78. Ifa, D., Jackson, A., Paglia, G. *et al.* (2009). Forensic applications of ambient ionization mass spectrometry, *Anal. Bioanal. Chem.*, **394**, 1995–2008.
79. Curtis, M.E., Jones, P.R., Sparkman, O.D. *et al.* (2009). Determination of the presence or absence of sulfur materials in drywall using direct analysis in real time in conjunction with an accurate-mass time-of-flight mass spectrometer, *J. Am. Soc. Mass Spectrom.*, **20**, 2082–2086.
80. Grange, A.H. (2013). Semi-quantitative analysis of contaminants in soils by direct analysis in real time (DART)–mass spectrometry, *Rapid Commun. Mass Spectrom.*, **27**, 305–318.
81. Grange, A.H. (2009). Rapid semi-quantitative surface mapping of airborne-dispersed chemicals using mass spectrometry, *Environ. Forensics*, **10**, 183–195.
82. Grange, A.H. and Sovocool, G.W. (2011). Detection of illicit drugs on surfaces using direct analysis in real time (DART) time-of-flight mass spectrometry, *Rapid Commun. Mass Spectrom.*, **25**, 1271–1281.
83. Mulligan, C.C., MacMillan, D.K., Noll, R.J. *et al.* (2007). Fast analysis of high-energy compounds and agricultural chemicals in water with desorption electrospray ionization mass spectrometry, *Rapid Commun. Mass Spectrom.*, **21**, 3729–3736.
84. Strittmatter, N., Düring, R.-A. and Takáts, Z. (2012). Analysis of wastewater samples by direct combination of thin-film microextraction and desorption electrospray ionization mass spectrometry, *Analyst*, **137**, 4037–4044.

85. Campbell, I.S., Ton, A.T. and Mulligan, C.C. (2011). Direct detection of pharmaceuticals and personal care products from aqueous samples with thermally-assisted desorption electrospray ionization mass spectrometry, *J. Am. Soc. Mass Spectrom.*, **22**, 1285–1293.
86. Nizzia, J.L., O'Leary, A.E., Ton, A.T. *et al.* (2012). Screening of cosmetic ingredients from authentic formulations and environmental samples with desorption electrospray ionization mass spectrometry, *Anal. Methods*, **5**, 394–401.
87. Haunschmidt, M., Klampfl, C.W., Buchberger, W. *et al.* (2010). Determination of organic UV filters in water by stir bar sorptive extraction and direct analysis in real-time mass spectrometry, *Anal. Bioanal. Chem.*, **397**, 269–275.
88. Li, M., Chen, H., Yang, X., *et al.* (2009). Direct quantification of organic acids in aerosols by desorption electrospray ionization mass spectrometry, *Atmos. Environ.*, **43**, 2717–2720.
89. Laskin, J., Laskin, A., Roach, P.J. *et al.* (2010). High-resolution desorption electrospray ionization mass spectrometry for chemical characterization of organic aerosols, *Anal. Chem.*, **82**, 2048–2058.
90. Maleknia, S.D., Bell, T.L. and Adams, M.A. (2009). Eucalypt smoke and wildfires: temperature dependent emissions of biogenic volatile organic compounds, *Int. J. Mass Spectrom.*, **279**, 126–133.
91. Jones, R.W., Reinot, T. and McClelland, J.F. (2010). Molecular analysis of primary vapor and char products during stepwise pyrolysis of poplar biomass, *Energy Fuels*, **24**, 5199–5209.
92. Nah, T., Chan, M., Leone, S.R. *et al.* (2013). Real time in situ chemical characterization of submicrometer organic particles using direct analysis in real time–mass spectrometry, *Anal. Chem.*, **85**, 2087–2095.
93. Chan, M.N., Nah, T. and Wilson, K.R. (2013). Real time in situ chemical characterization of sub-micron organic aerosols using direct analysis in real time mass spectrometry (DART–MS): the effect of aerosol size and volatility, *Analyst*, **138**, 3749–3757.
94. Lancaster, C. and Espinoza, E. (2012). Evaluating agarwood products for 2-(2-phenylethyl)chromones using direct analysis in real time time-of-flight mass spectrometry, *Rapid Commun. Mass Spectrom.*, **26**, 2649–2656.

95. Lancaster, C. and Espinoza, E. (2012). Analysis of select Dalbergia and trade timber using direct analysis in real time and time-of-flight mass spectrometry for CITES enforcement, *Rapid Commun. Mass Spectrom. RCM*, **26**, 1147–1156.
96. Cody, R.B., Dane, A.J., Dawson-Andoh, B. *et al.* (2012). Rapid classification of white oak (*Quercus alba*) and northern red oak (*Quercus rubra*) by using pyrolysis direct analysis in real time (DART™) and time-of-flight mass spectrometry, *J. Anal. Appl. Pyrolysis*, **95**, 134–137.
97. Nyadong, L., Hohenstein, E.G., Galhena, A. *et al.* (2009). Reactive desorption electrospray ionization–mass spectrometry (DESI–MS) of natural products of a marine alga, *Anal. Bioanal. Chem.*, **394**, 245–254.
98. Lane, A.L., Nyadong, L., Galhena, A.S. *et al.* (2009). Desorption electrospray ionization mass spectrometry reveals surface-mediated antifungal chemical defense of a tropical seaweed, *Proc. Natl. Acad. Sci.*, **106**, 7314–7319.
99. Andras, T.D., Alsexander, T.S., Gahlena, A. *et al.* (2012). Seaweed allelopathy against coral: surface distribution of a seaweed secondary metabolite by imaging mass spectrometry, *J. Chem. Ecol.*, **38**, 1203–1214.
100. Vaikkinen, A., Kotiaho, T., Kostiainen, R. *et al.* (2010). Desorption atmospheric pressure photoionization with PDMS as extraction phase and sample plate material, *Anal. Chim. Acta*, **682**, 1–8.

Chapter 8

Liquid Chromatography–High-Resolution Mass Spectrometry in Environmental and Food Analysis

Paolo Lucci[a] and Claudia P. B. Martins[b]

[a] Faculty of Sciences, Pontificia Universidad Javeriana, Colombia
[b] Thermo Fisher Scientific, Villebon-sur-Yvette, France

8.1. Introduction

High-resolution mass spectrometry (HRMS) provides information concerning the exact molecular mass, elemental composition and detailed molecular structure of a compound. The full characterization of higher boiling fractions of petroleum is an early example of the analytical utility of this technique. Recently, significant instrumental advances have made it a widely used technique in various areas including, but not limited to, biological,[1] environmental,[2,3] food,[4,5] forensic and clinical applications.[6,7] The increasing interest in the use of HRMS, especially coupled to liquid chromatography (LC) is mainly due to its suitability for both targeted and non-targeted analysis; the high-resolution accurate mass full scan spectrum obtained can be successfully used to identify the presence of both targeted and non-targeted molecules. Furthermore, the use of HRMS as a detection technique also allows the simplification of sample-preparation procedures, thereby leading to faster methodologies requiring less sample manipulation.[8] This aspect is of particular importance when analyzing very complex matrices usually containing low concentrations of target compounds, such as biological fluids, food, and environmental samples. Figure 8.1 compares the different LC–HRMS workflows, from targeted to

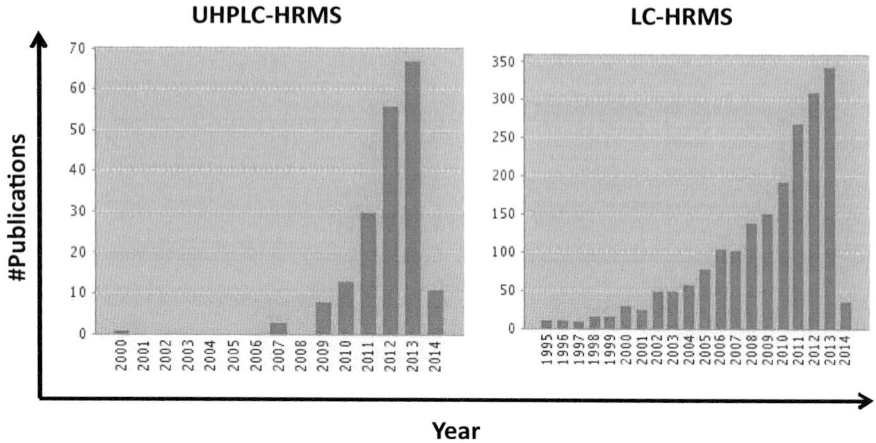

Figure 8.1. Number of publications on UHPLC/LC–HRMS extracted from 1995 to present. Web of Science (2014). Available at http://wokinfo.com/ [Accessed in February 2014].

non-targeted screening, applied to the analysis of complex environmental samples.[9] The acquisition in full-scan mode enables recording a list of ions that can be used for quantification and/or screening purposes. Other advanced scan modes can be used simultaneously, enabling confirmatory strategies by analyzing the corresponding MS/MS spectrum. In summary, the information gathered by a single injection can be used for both quantification and screening purposes, including targeted, suspect and non-targeted analysis. In addition, once the data has been recorded, it can be stored and used in a later stage: retrospective analysis. Such an approach offers many advantages to any laboratory routinely analyzing food and environmental samples due to the minimum method development necessary. This is especially relevant when performing targeted analysis, and in comparison to LC–MS/MS. Nevertheless, when the goal of the analyst is to expand from targeted analysis to both suspect and non-targeted analysis, the combination of advanced data-processing tools with a skilled operator is required. In addition, regulatory guidelines concerning the analysis of food, feed, environmental samples and other relevant matrices are not yet up-to-date on the most recent advances concerning LC–HRMS instrumentation.

The use and applicability of LC–HRMS in the field of food and environmental analysis will be discussed in this chapter. It is not our intention to summarize all the work performed in recent years, but to discuss the trends, limitations and major developments achieved in that time.

8.2. The Use and Applicability of LC–HRMS

Since the introduction of commercial orthogonal axis–time-of-flight (oa-TOF) instruments in the mid-1990s, significant improvements have been made in order to produce a high-performance instrument with modest power requirements and size, and most importantly, the ability to operate at reduced costs. From the different HRMS instrumentation available — magnetic sector, time-of-flight (TOF), Orbitrap, and Fourier transform ion cyclotron (FTICR) — TOF and Orbitrap are probably the most commonly used with LC or UHPLC systems.[10] In general, TOF instruments show a mass-resolving power of approximately 10,000–40,000 FWHM (full width at half maximum) with a mass measurement accuracy of 1–5 ppm, while the mass-resolving power of the latest Orbitrap instrument can reach up to 450,000 FWHM (at *m*/*z* 200) with <3 ppm mass accuracy in external calibration.

The flight time for each ion of a particular *m*/*z* is unique and serves as the basis for the time-of-flight analysis. This experiment takes into account the flight time experienced by a particular ion, starting when a high voltage pulse is applied to the back plate of the ion pulser and ending when the ions strike the detector. The mass (and therefore *m*/*z*) is directly related to the energy to which an ion is accelerated, the distance it has traveled and the flight time.[11] In 2005, the Orbitrap mass analyzer was introduced and its compatibility with an electrospray ionization (ESI) source described.[12] This new type of mass analyzer invented by Makarov operates by radially trapping ions about a central spindle electrode. An outer barrel-like electrode is coaxial with the inner spindle-like electrode and *m/z* values are measured from the frequency of harmonic ions oscillations, along the axis of the electric field.[12] The path experienced by a particular ion, from its generation to its detection, is therefore different in both cases. Consequently, both analyzers show different characteristics, leading to slightly different performances. Characteristics such as dynamic range, sensitivity, cycle time and mass-resolving power have been continuously improved by the manufacturers, leading to the development of different high-performance instrumentation that can be applied to a variety of applications, including food and environmental analysis. The growing attention on LC–HRMS is clearly demonstrated by the number of publications using the coupling of LC to HRMS throughout the years. Figure 8.1 illustrates the number of publications using the keywords *high-resolution mass spectrometry* combined with both *ultrahigh-pressure liquid chromatography* and *liquid chromatography* at the website 'Web of Science' made from 1995 to present.[13]

Among the possible ionization techniques in LC–HRMS analysis, ESI is probably the most widely used followed by atmospheric-pressure chemical ionization (APCI). Generally, ESI is ideal for polar to semi-polar compounds, whereas APCI provides high ionization efficiency for less polar and neutral compounds. Atmospheric-pressure photoionization (APPI) has also been recently introduced as a soft ionization technique able to broaden the group of analytes that can be analyzed to less polar compounds.[14] The sensitivity of a particular methodology can be dramatically affected by the ionization efficiency of the targeted analytes under a certain set of conditions. Therefore, it is important to select the appropriate ionization technique, independently of the mass analyzer to be used.

Fullerenes are considered to be emerging environmental contaminants and have been the focus of different monitoring studies by LC–HRMS.[14] Figure 8.2 illustrates the impact of the different ionization techniques on the ionization efficiency of a group of fullerenes, from C_{60} to C_{84}. An important improvement in response was found when using toluene-mediated APPI in negative mode when compared to other ionization techniques.[14]

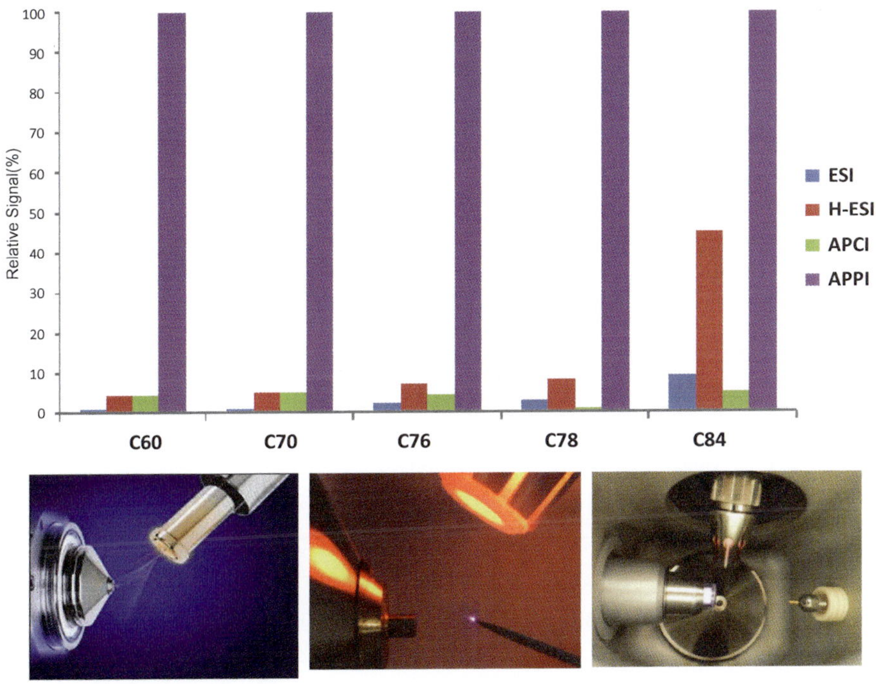

Figure 8.2. Comparison of different ionization techniques in the analysis of carbon based nanoparticles (fullerenes): electrospray ionization (ESI) and heated electrospray ionization (H-ESI), atmospheric-pressure chemical ionization (APCI) and atmospheric-pressure photoionization (APPI). Compounds selected for this study included C_{60}, C_{70}, C_{76}, C_{78}, and C_{84}.

The combination of direct analysis in real time (DART) with HRMS is an emerging tool for a diverse number of applications, not only for food quality and safety controls but also for food authentication.[15] No chromatographic separation is required, leading to a rapid, cost-effective analysis. However, a methodical validation leading to full compliance with current food regulation is still to be achieved.

Gallart-Ayala *et al.*[16] described the unequivocal identification of benzophenone (BP) by making use of an Orbitrap mass analyzer operating at mass-resolving power of 50,000 FWHM. In this experiment, the use of HRMS was crucial to successfully solving the problem of false negatives originated by the presence of an interfering compound co-eluting with the analyte of interest. The analysis by LC–MS/MS led to a considerable number of false negatives due to the direct interference of a matrix component into the ion ratio calculation following the EU Commission Decision 2002/657/EC.[17] A third transition could not be used because BP is a relatively small compound and a third product ion cannot be generated. The author has performed a comparative analysis by LC–HRMS making use of three different mass-resolving powers (10,000 FWHM, 25,000 FWHM and 50,000 FWHM). The unequivocal confirmation of BP in food samples was only achieved when using 50,000 FWHM. In addition to overcoming the confirmatory issues experienced by LC–MS/MS, the LC–HRMS method also proved to be suitable for the quantification of BP in all food matrices tested.[17]

The same approach, based on a single-stage Orbitrap mass spectrometer operating at 50,000 FWHM, was proposed by Kaufmann *et al.*[18] for the development of an improved HRMS-based multi-residue method for veterinary drugs in various food matrices, including muscle, kidney, liver, fish and honey. The reported method, which was also validated according to the EU Commission Decision 2002/657/EC, showed better analytical performances (e.g. linearity, reproducibility and detection limits) when compared to a TOF based method operating at 12,000 FWHM.[18]

Hybrid mass spectrometers can also be used by modern analytical chemists since they combine different performance characteristics (e.g. mass-resolving power, speed of analysis, and linear dynamic range) offered by the various types of analyzers in one mass spectrometer.[10] For instance, the potential of liquid chromatography–hybrid quadrupole time-of-flight mass spectrometry (QTOF) in food analysis can be easily recognized by the increasing number of screening, quantitative and confirmatory methods reported in the literature.[19,20] Hybrid QTOF-MS is a powerful tool for reducing the risk of reporting false positives thanks to the valuable information given by the product ion spectra and accurate mass measurements.[19] For example, Stolker *et al.*[3] used QTOF-MS for screening and confirmation of pharmaceuticals in surface, drinking, and ground water, based on the intensity ratios of two product ions and on their accurate masses. Besides the fact that the compounds could be confirmed by the use of both triple quadrupole (QqQ) and Q-TOF-based MS/MS methods, the latter technique had the distinct advantage of enabling the screening and confirmation of a large number of pharmaceuticals at low concentrations (1–100 ng/L) within a single run. The effectiveness of UHPLC coupled to QTOF has also been illustrated for the screening and confirmation of several antibiotics (e.g. ofloxacin, ciprofloxacin, clarythromycin, erythromycin, etc.) in water samples.[3] Furthermore, it should be noted that the high-resolution accurate-mass full-spectrum provided by TOF-MS acquisition makes this technique a reliable tool for retrospective analysis of other pharmaceuticals, or other organic contaminants in general, that could be present in the environmental samples. A QTOF-MS methodology was also used by Wang *et al.*[19,20] for the determination of more than 140 pesticide residues in several fruit and vegetable matrices as well as by Calbiani *et al.*[21] for the determination of Sudan I, II, III, and IV azo-dyes in hot chilli products with LODs ranging from 0.4 μg/g to 1.1 μg/g in matrix. A proteomic MS-based method employing a capillary LC–QTOF system has also been proven to allow qualitative and confirmative analysis of trace contamination of milk allergens in processed food matrices.[22] The LC/ESI–linear ion trap quadrupole–Orbitrap–mass spectrometry (LTQ–Orbitrap–MS) configuration represents another interesting configuration which can be successfully used in food and environmental analysis. This hybrid configuration has the advantage of combining single-stage (MS) and two-stage (MS/MS) or multiple-stage (MS^n) mass spectrometry. For instance, Vallverdú-Queralt *et al.*[23] performed the characterization of tomato polyphenols by using a LTQ–Orbitrap system with accurate mass measurements in MS, MS^2 and MS^3 modes. The analysis confirmed the presence of 38 phenolic compounds in tomato samples with a mass accuracy lower than 3 ppm. An LTQ–Orbitrap mass spectrometer operating in negative electrospray ionization mode (ESI−) has also been employed by Xu *et al.*[5] for the detection of chloramphenicol (CAP) in meat products. The high accurate mass of the molecular ion $[M-H]^-$ was recorded and the limit of quantification of the method was 0.1 μg/kg using isotope internal standard.

The use of high-resolution mass spectrometers such as Orbitrap and time-of-flight allows for excellent, accurate mass measurements, mainly due to the mass-resolving power of the new instruments on the market. However, the use of a mass-resolving power of ≥50,000 FWHM seems to be advantageous when dealing with complex matrices, and in order to avoid false negative and positive results. In addition, sensitivity and linear dynamic range have been improved considerably on the latest instrumentation introduced on the market. The hope is that combining highly sensitive full-scan with different MS/MS experiments (e.g. data-dependent, data-independent, non-selective fragmentation (all ions fragmentation, MS^E or MSALL) and targeted-MS/MS) will facilitate the identification and confirmation of certain contaminants, while providing detailed information on other possible components of the sample.

8.3. Is LC–HRMS Overtaking LC–MS/MS?

The identification and quantification of pollutants in environmental samples at low concentrations requires both sensitivity and selectivity. In addition, food safety regulations and food quality testing are becoming stricter. For a large range of compounds, selected reaction monitoring (SRM) of precursor-product ion transitions has been the method of choice. However, this technique is limited by the compromise between cycle time and the number of transitions to monitor in one run. Recently, other screening strategies taking into account full scan mode and other advanced MS/MS scan modes have been reported due to the development of a more rugged, sensitive and selective instrumentation. The benefits are quite evident: from having the possibility to look retrospectively at the sample to having an unlimited number of compounds in your methodology involving minimal optimization. Furthermore the mass-resolving power of such instrumentation is a powerful tool for identification purposes in such complex matrices, as described in the previous section.

Moschet *et al.* recently discussed the establishment of a suspect screening approach without the need of reference standards, covering nearly all Swiss-registered insecticides, fungicides and known transformation products (TPs) in surface water.[24] A careful substance selection based on water solubility and ionization of both contaminants and TPs was performed. The samples have been analysed by a hybrid Orbitrap mass analyser (Q-Exactive) in both positive and negative modes, alternating between full scan at 140,000 FWHM and Data-Dependent MS/MS (Top 5) at 17,500 FWHM. Features such as peak picking (mass accuracy below 5 ppm), automatic filters focusing on blank subtraction, peak area, peak score (chromatographic profile), signal-to-noise and isotope score were optimized and applied in order to obtain a reduced number of false positives. This methodology is cost-effective since the purchase of the standards is only necessary after the confirmation that the compound is in fact present. However, the optimization of every step of the process, particularly data processing, should be meticulously designed. The use of software is therefore becoming one of the crucial steps of the analytical methodology. The amount and complexity of the acquired spectral information demands tools that can provide a quick but also reliable result.

Screening, identification/confirmation as well as elucidation of unknown compounds are commonly targeted on a daily basis in any laboratory using high-resolution mass spectrometry for the analysis of environmental and food samples. Furthermore, with the recent hardware upgrades, it is also possible to work quantitatively with such instrumentation. The use of a database containing information on elemental composition (theoretical *m/z*, exact mass), expected retention time and possibly, exact masses of product ions, is therefore indispensable for the screening of hundreds of contaminants. The automation of such a procedure has clear benefits such as reduced analysis time, but implies the risk of having a false positive or false negative finding. A correct optimization of all parameters such as retention time windows, accurate mass tolerance, presence of adducts, isotopic fit, ion ratio and response thresholds, is essential to limit erroneous results. It is extremely important to fully understand the instrumentation and its capabilities before deciding on the optimization of such parameters.

Spectra libraries can also be used to facilitate the search and identification of both targeted and non-targeted molecules. However, the lack of a normalized interface and the different instrumentation used prevents the establishment of a comprehensive spectra library. In this context, several commercial libraries are currently available, focusing on specific instrumental conditions that the user will need to reproduce at their own laboratory. In addition, other libraries have been built by researchers and research groups, especially in life sciences, and have been made freely available to the scientific community. Examples include METLIN[25], Human Metabolome Database (HMDB),[26] MassBank[27] and LIPID Metabolites and Pathways Strategy (LIPID MAPS)[28]. Not all these options contain accurate mass spectral data. MassBank is currently being expanded with accurate mass spectral data on environmental pollutants by a network of reference laboratories: the NORMAN network.[29] The situation is quite different when using electron impact (EI) mass spectra. The current NIST reference library contains more than 240,000 EI mass spectra from 212,961 compounds,[30] whereas the Wiley Registry (ninth edition) contains 662,000 spectra from 592,000 compounds.[31]

There is a clear trend toward the use of high resolution. It has been demonstrated that the use of LC–HRMS presents clear benefits when dealing with complex samples and when on the edge of new analytical problems. However, there are still limitations such as the cost of ownership and analysis, ease-of-use of software platforms and the lack of up-to-date guidelines for compliant monitoring.

8.4. Confirmatory Strategies

The triple quadrupole (QqQ) mass analyzer running in selected reaction monitoring (SRM) mode is still the most commonly used methodology for the analysis of environmental and food samples. The selection of a precursor ion in the first quadrupole (Q), followed by its fragmentation in the collision cell (q) and the selection of a product ion in the third Q is known to be highly selective. A minimum of two precursor-product ion transitions is required to fulfill the regulatory requirements under the EU Commission Decision

2002/657/EC.[17] However, several authors have reported false-positive and negative findings, even when strictly applying the confirmatory criteria established in 2002.[16,32] The complexity of the matrix analyzed plays a crucial role in such results obtained by LC–MS/MS. Consequently the use of LC–HRMS is expected to prevent at least some of the issues by successfully resolving the targeted contaminant from the isobaric interference present in the extracted sample (Fig. 8.3). Nevertheless, it is not yet fully understood what

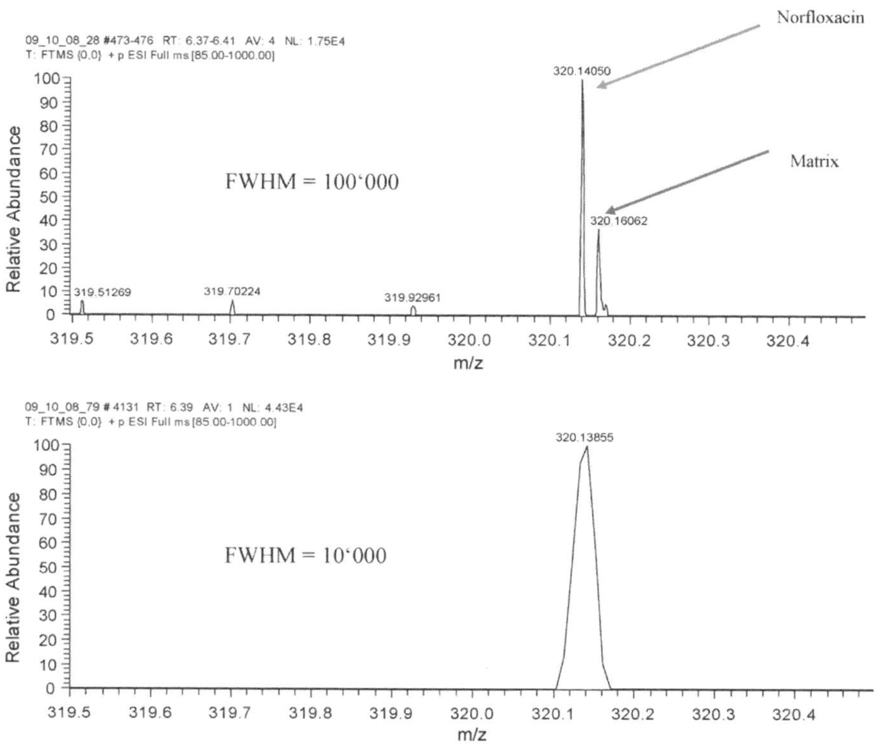

Figure 8.3. Spectrum of Norfloxacin in matrix obtained at a mass-resolving power of 100,000 FWHM (top) and 10,000 FWHM (bottom). Reprinted with permission from Kaufmann, A., Butcher, P., Maden, K. *et al.* (2010). Comprehensive comparison of liquid chromatography selectivity as provided by two types of liquid chromatography detectors (high-resolution mass spectrometry and tandem mass spectrometry): "Where is the crossover point?" *Anal. Chim. Acta*, **673**, 60–72. Copyright (2010) Elsevier.

confirmation criteria should be taken into account when using LC–HRMS. One of the difficulties is being able to find a common strategy that will take into account the different characteristics exhibited by the different mass analyzers present on the market. One example is the mass-resolving power exhibited by the Orbitrap mass analyzer and the impact of such a characteristic on the excellent mass accuracy obtained routinely.

In a recent study by Kaufmann *et al.* a direct comparison between LC–HRMS and LC–MS/MS was performed. The data gathered indicated that the selectivity of LC–HRMS exceeds that of LC–MS/MS when a mass-resolving power of 50,000 FWHM is used. [33]

The identification using HRMS is usually accomplished by the use of accurate mass measurements and retention times. The most recent applications reported for the analysis of pesticides, including metabolites and TPs, in fruit- and vegetable-based matrices using TOF and Orbitrap instruments show a mass accuracy lower than 5 ppm for different matrices and concentration levels.[34] A mass accuracy equal to or lower than 5 ppm is considered to be a good mass measurement by most analysts when performing screening of pesticides.[34]

Mol *et al.* optimized a screening methodology making use of a single-stage high-resolution Orbitrap MS working at 50,000 FWHM.[35] The study indicated that the optimization of thresholds and diagnostic ions plays a crucial role in the reduction of the number of false positives. Different pesticides will exhibit different ionization efficiencies under certain conditions related to the methodology used. This will impact the sensitivity of the methodology and therefore a relative response threshold is recommended for each pesticide. This threshold was set at half the lowest response obtained for any of the studied matrices. With this approach, the number of false positives decreased from 88 to 14 without affecting the overall percentage of pesticides detected. The other parameters adjusted were the mass tolerance window (5 ppm), retention time (RT) window (±0.5 min) and the presence of two diagnostic ions (adduct, isotope or fragment). With the two-ion approach Mol *et al.* reported 13% of false negatives at the lowest concentration studied: 0.01 mg/kg.

However, the second ion should ideally be sensitive and selective, with a consistent abundance relative to the primary diagnostic ion. Adducts were considered as the least favorable option due to the inconsistency in terms of abundance. The most abundant isotope may be favorable in certain cases. They can be predicted and ion ratios reported as stable and independent of matrix and the selectivity when compared to a low *m/z* fragment seem better. Fragments were the most sensitive secondary diagnostic ion.[35]

Multiple ions are often available, which means that different combinations can be used in order to perform confirmatory studies. Domènech *et al.* developed a multi-toxin method for quantification and confirmation of lipophilic marine biotoxins in mussels by LC–HRMS.[36] Fragments, isotope ions and ion ratios were studied and evaluated for confirmatory purposes. It was observed that both fragment ion ratio and isotope ion ratio can be used to confirm a positive result, the most favorable route being dependent on the analyte.[36] Kumar *et al.* reported the criteria established for the use of LC–HRMS for confirmatory analysis as:

- RT tolerance of ±1%;
- at least one product ion at high resolving power (>20,000 FWHM);
- a minimum resolving power of 70,000 FWHM for precursor ions;
- mass measurement accuracy of less than or equal to 5 ppm;
- monitoring of at least one ion ratio.[32]

The higher mass-resolving power used for the monitoring of the precursor ion was crucial to the differentiation between the analyte and isobaric matrix interferences. Table 8.1 compares some of the strategies discussed in this section with the main guidelines available reporting the use of HRMS. It is clear that the combination of different confirmatory ions is the best strategy. Nevertheless, parameters such as mass-resolving power, ion ratio and isotope ratio can differ between different classes of contaminants or even different matrices.

Table 8.1. Identification and confirmation criteria used in different studies involving diverse matrices and contaminants. Adapted and modified from Domènech, A., Cortes-Francisco, N., Palacios, O. *et al.* (2014). Determination of lipophilic marine toxins in mussels. Quantification and confirmation criteria using high resolution mass spectrometry, *J. Chromatogr. A*, **1328**, 16–25.[36]

	EU [17] 2002/657/EC	**SANCO [40] /12495/2011**	**EU-RL-MB SOP [41]**	**Gerssen [42] (2010)**	**Mol [35] (2012)**	**Domènech [36] (2014)**	**Kumar [32] (2014)**	**Pitarch [43] (2007)**
Analytes	—	Pesticides	Lipophilic Toxins	Lipophilic Toxins	Pesticides	Lipophilic Toxins	Ronidazole Nitroimidazoles	Priority organic micropollutants
Matrix	Food	Food and Feed	Molluscs	Shellfish	Vegetables & Fruits	Mussels	Muscle	Water
Technique	HRMS	HRMS	LC–MS/MS	LC–MS/MS	LC–HRMS/MS	LC–HRMS/MS	LC–HRMS/MS	GC–MS/MS
Mass Accuracy	—	<5ppm	—	—	<5ppm	<5ppm	<5ppm	—
Resolution (FWHM)	—	≥20,000	—	—	≥20,000	≥20,000	≥70,000	—
Retention Time (RT) Tolerance	2.5%	2.5%	Does not exceed 3%	5%	1%	Mean ±3SD (not relative to time)	±1%	Agreement: RT samples & standards

(Continued)

Table 8.1. *(Continued)*

	EU[17] 2002/657/EC	SANCO[40] /12495/2011	EU-RL-MB SOP[41]	Gerssen[42] (2010)	Mol[35] (2012)	Domènech[36] (2014)	Kumar[32] (2014)	Pitarch[43] (2007)
Diagnostic Ions	≥ 2	≥ 2	1 precursor	1 precursor	≥ 2	1	2	1 or 2 precursors
Fragment Ions	—	At least one	At least 2 precursor-product transitions	2 products	At least one	1	At least 1, >20,000 FWHM	At least 2 precursor-product transitions
Isotope Ions	—	—	—	—	M+1 M+2	M+1	—	—
Ion Ratio	Relative intensity (% at base peak)	Relative intensity (% at base peak)	Must be recorded	As described 2002/657/EC	Fragment ion ratio; diagnostic/fragment isotope ion ratio; diagnostic/M+1 (M+2)	Fragment ion ratio; diagnostic/fragment isotope ion ratio; diagnostic/M+1	At least one	Ratio between quantitative and confirmation transition
Fragment-Isotope Ion Ratio Tolerance	2 IPs for precursor ion; 2.5 IPs for a product	As described 2002/657/EC	—	As described 2002/657/EC	Independent of relative intensity between ions: ±50%	As described 2002/657/EC	—	As described 2002/657/EC

8.5. The Identification of Unknowns

Despite recent advances in hardware as well as software tools, the identification of unknowns remains challenging. In general, the process leading to the identification of a new entity not present on a database requires several methodical steps, from measurement of data to compound identification. The overall process consists of the use of automated peak-detection by exact mass-filtering from the chromatographic run, followed by the assignment of an elemental composition and database search for possible structures. The findings obtained when using large compound databases (e.g. PubChem and ChemSpider) require further ranking using MS/MS data. An important consideration to take into account when identifying unknown compounds is the mass accuracy of the measurements. As the number of elements increases (C, H, O, N, S, F...) less uncertainty in the measurement should be allowed, because the possible number of candidates will increase exponentially with the mass error of the measurement.[37] In silico fragmentation is also discussed as a valuable tool.[38] However, the number of experimentally observed fragments is normally much lower than the ones measured by computational fragmentation strategies.

The prediction of retention times may also aid in identifying an unknown or confirming the presence of a contaminant where reference materials were not originally included in the analytical screen. Miller *et al.* showed that artificial neural networks (ANNs) could be used to predict chromatographic retention times.[39]

The study involved the selection of 86 compounds included in the London 2012 Olympic and Paralympic Games. After training, verification, and testing of a range of computational models, the work showed that ANNs could be used to predict chromatographic retention times for 93% of the compounds selected within 0.5 min of their true value and within 1 min for all other compounds. Furthermore, the model showed the same level of accuracy when applied to the prediction of unknowns alone.[39]

8.6. Conclusions and Future Perspectives

The impact of LC–HRMS in environmental and food analysis is expected to increase in the coming years. Despite the apparent higher investment cost, this technique is making a clear transition from research to routine analysis due to the development of simplified hardware and software. The development of sensitive instruments, compliance with regulatory guidelines, and user-friendly software packages seem to play a decisive role in this transition. Powerful strategies combining the targeted, suspect, and non-targeted analysis in the same run have been reported and are in use, offering clear benefits over LC–MS/MS analysis performed by a triple quadrupole mass spectrometer. Furthermore, the possibility to perform retrospective analysis enables searching for a 'new' contaminant years after data recording. This option is very helpful when analyzing food and environmental samples, especially when taking into account the complexity concerning the degradation and transformation of contaminants in the environment or the introduction of new contaminants into food products.

References

1. Zhang, H., Heinig, K. and Henion, J. (2000). Atmospheric pressure ionization time-of-flight mass spectrometry coupled with fast liquid chromatography for quantitation and accurate mass measurement of five pharmaceutical drugs in human plasma, *J. Mass Spectrom.*, **35**, 423–31.
2. Petrovic, M., Gros, M. and Barcelo, D. (2006). Multi-residue analysis of pharmaceuticals in wastewater by ultra-performance liquid chromatography–quadrupole–time-of-flight mass spectrometry, *J. Chromatogr. A.*, **1124**, 68–81.
3. Stolker, A.A., Niesing, W., Hogendoorn, E.A. *et al.* (2004). Liquid chromatography with triple-quadrupole or quadrupole-time-of-flight mass spectrometry for screening and confirmation of residues of pharmaceuticals in water, *Anal. Bioanal. Chem.*, **378**, 955–63.

4. Peters, R.J., Bolck, Y.J., Rutgers, P. *et al.* (2009). Multi-residue screening of veterinary drugs in egg, fish and meat using high-resolution liquid chromatography–accurate mass time-of-flight mass spectrometry. *J. Chromatogr. A.*, **1216**, 8206–16.
5. Xu, H., Zhang, J., He, J. *et al.* (2011). Rapid detection of chloramphenicol in animal products without clean-up using LC–high resolution mass spectrometry, *Food Addit. Contam. Part A*, **28**, 1364–1371.
6. Ojanperä, L., Pelander, A., Laks, S. *et al.* (2005). Application of accurate mass measurement to urine drug screening. *J. Anal. Toxicol.*, **29**, 34–40.
7. Ojanperä, I., Kolmonen, M. and Pelander, A. (2012). Current use of high-resolution mass spectrometry in drug screening relevant to clinical and forensic toxicology and doping control, *Anal. Bioanal Chem.*, **403**, 1203–20.
8. Gallart-Ayala, H., Lucci, P. and Núñez, O. (2012). Recent advances in LC–MS analysis of food-packaging contaminants, *Trends Anal. Chem.*, **42**, 99–124.
9. Krauss, M., Singer, H. and Hollender, J. (2010). LC–high-resolution MS in environmental analysis: from target screening to the identification of unknowns, *Anal. Bioanal. Chem.*, **397**, 943–951.
10. Glish, G.L. and Burinsky, D.J. (2008). Hybrid mass spectrometers for tandem mass spectrometry. *J. Am. Soc. Mass Spectrom.*, **19**, 161–172.
11. Fjeldsted, J. (2009). 'Accurate mass measurements with orthogonal axis time-of-flight mass spectrometry', in Ferrer, I. and Thurman, E.M. (eds), *Liquid Chromatography Time-of-Flight Mass Spectrometry: Principles, Tools and Applications for Accurate Mass Analysis*, John Wiley & Sons, New Jersey, pp. 3–15.
12. Hu, Q., Noll, R.J., Li, H. *et al.* (2005). The Orbitrap: a new mass spectrometer, *J. Mass Spectrom.*, **40**, 430–443.
13. Web of Science (2014). Available at http://wokinfo.com/ [Accessed in February 2014].
14. Núñez, O., Gallart-Ayala, H., Martins, C.P.B. *et al.* (2012). Atmospheric pressure photoionization mass spectrometry of fullerenes, *Anal. Chem.*, **84**, 5316–5326.
15. Hajslova, J., Cajka, T. and Vaclavik, L. (2011). Challenging applications offered by direct analysis in real time (DART) in food-quality and safety analysis, *Trends Anal. Chem.*, **30**, 204–218.

16. Gallart-Ayala, H., Nuñez, O., Moyano, E. *et al.* (2011). Preventing false negatives with high-resolution mass spectrometry: the benzophenone case, *Rapid Commun. Mass Spectrom.*, **25**, 3161–3166.
17. Commission Decision 2002/657/EC, OFF., *J. Euro. Commun.*, L221, p. 8.
18. Kaufmann, A., Butcher, P., Maden, K. *et al.* (2011). Development of an improved high-resolution mass spectrometry-based multi-residue method for veterinary drugs in various food matrices, *Anal. Chim. Acta*, **700**, 86–94.
19. Wang, J., Leung, D. and Chow, W. (2010). Applications of LC–ESI–MS/MS and UHPLC–Qq–TOF-MS for the determination of 148 pesticides in berries, *J. Agric. Food Chem.*, **58**, 5904–5925.
20. Wang, J., Chow, W. and Leung, D. (2010). Applications of LC–ESI–MS/MS and UHPLC–Qq–TOF-MS for the determination of 148 pesticides in fruits and vegetables. *Anal. Bioanal. Chem.*, **396**, 1513–1538.
21. Calbiani, F., Careri, M., Elviri, L. *et al.* (2004). Accurate mass measurements for the confirmation of Sudan azo-dyes in hot chilli products by capillary liquid chromatography–electrospray tandem quadru-pole orthogonal-acceleration time-of-flight mass spectrometry, *J. Chromatogr. A*, **1058**, 127–135.
22. Chassaigne, H., Nørgaard, J.V. and Hengel, A.J. (2007). Proteomics-based approach to detect and identify major allergens in processed peanuts by capillary LC–Q–TOF-(MS/MS), *J. Agric. Food Chem.*, **55**, 4461–73.
23. Vallverdú-Queralt, A., Jáuregui, O., Medina-Remón, A. *et al.* (2010). Improved characterization of tomato polyphenols using liquid chromatography/electrospray ionization linear ion trap quadrupole Orbitrap mass spectrometry and liquid chromatography/electrospray ionization tandem mass spectrometry, *Rapid Commun. Mass Spectrom.*, **24**, 2986–2992.
24. Moschet, C., Piazzoli, A., Singer, H., *et al.* (2013). Alleviating the reference standard dilemma using a systematic exact mass suspect screening approach with liquid chromatography–high-resolution mass spectrometry, *Anal. Chem.*, **85**, 10312–10320.
25. Smith, C.A., O'Maille, G., Want, E.J. *et al.* (2005). METLIN: a metabolite mass spectral database, *Ther. Drug Monit.*, **27**, 747–751.

26. Human Metabolome Database, version 3.5. (2014). Available at http://www.hmdb.ca [Accessed in February 2014].
27. Horai, H., Arita, M., Kanaya, S. *et al.* (2010). MassBank: a public repository for sharing mass spectral data for life sciences, *J. Mass Spectrom.*, **45**: 703–714.
28. LIPID Metabolites and Pathways Strategy (LIPID MAPS). (2014). Available at http://www.lipidmaps.org/ [Accessed in February 2014].
29. Network of reference laboratories for monitoring of emerging environmental pollutants (Norman). (2014). Available at http://www.norman-network.net/.
30. NIST Standard Reference Data. (2014). Available at http://www.nist.gov/srd/nist1a.cfm [Accessed in February 2014].
31. *Wiley Registry of Mass Spectral Data,* 9th Edition (2011). John Wiley & Sons.
32. Kumar, P., Rubies, A., Centrich, F. *et al.* (2014). A false-positive case due to matrix interference in the analysis of ronidazole residues in muscle tissue using LC–MS/MS, *Meat Sci.*, **97**, 214–219.
33. Kaufmann, A., Butcher, P., Maden, K. *et al.* (2010). Comprehensive comparison of liquid chromatography selectivity as provided by two types of liquid chromatography detectors (high-resolution mass spectrometry and tandem mass spectrometry): "where is the crossover point?" *Anal. Chim. Acta.*, **12**, 60–72.
34. Gómez-Ramos, M.M., Ferrer, C., Malato, O. *et al.* (2013). Liquid chromatography–high-resolution mass spectrometry for pesticide residue analysis in fruit and vegetables: screening and quantitative studies, *J. Chromatogr. A*, **1287**, 24–37.
35. Mol, H.G.J., Zomer, P. and Koning, M. (2012). Qualitative aspects and validation of a screening method for pesticides and fruits based on liquid chromatography coupled to full scan high-resolution (Orbitrap) mass spectrometry, *Anal. Bioanal. Chem.*, **403**, 2891–2909.
36. Domènech, A., Cortes-Francisco, N., Palacios, O. *et al.* (2014). Determination of lipophilic marine toxins in mussels. Quantification and confirmation criteria using high resolution mass spectrometry, *J. Chromatogr. A*, **1328**, 16–25.
37. Cortes-Francisco, N., Flores, C., Moyano, E. *et al.* (2011). Accurate mass measurements and ultrahigh-resolution: evaluation of different

mass spectrometers for daily routine analysis of small molecules in negative electrospray ionization mode, *Anal. Bioanal. Chem.*, **400**, 3595–3606.

38. Hill, D.W., Kertesz, T.M., Fontaine, D. *et al.* (2008). Mass spectral metabonomics beyond elemental formula: chemical database querying by matching experimental with computational fragmentation spectra, *Anal. Chem.*, **80**, 5574–5582.
39. Miller, T.H., Musenga, A., Cowan, D.A. *et al.* (2013). Prediction of chromatographic retention time in high-resolution anti-doping screening data using artificial neural networks, *Anal. Chem.*, **85**, 10330–10337.
40. European Commission, Directorate General Health and Consumer Protection, SANCO/12495/2011.
41. European Union Reference Laboratory for Marine Biotoxins, EU-RL-MB SOP, Version 4, 2011.
42. Gerssen, A., van Olst, E.H.W., Mulder, P.P.J. *et al.* (2010). In-house validation of a liquid chromatography–tandem mass spectrometry method for the analysis of lipophilic marine toxins in selfish using matrix-matched calibration, *Anal. Bioanal. Chem.*, **397**, 3079–3088.
43. Pitarch, E., Medina, C., Portoles, T. *et al.* (2007). Determination of priority organic micro-pollutants in water by gas chromatography coupled to triple quadrupole mass spectrometry, *Anal Chim Acta*, **583**, 246–258.

Chapter 9

Liquid Chromatography–Mass Spectrometry: Quantification and Confirmation Aspects

Jaume Aceña,[a] Daniel Rivas,[a] Bozo Zonja,[a] Sandra Pérez[a] and Damià Barceló[a, b]

[a] Water and Soil Quality Research Group, Department of Environmental Chemistry, IDAEA-CSIC, Spain
[b] Catalan Institute for Water Research (ICRA), Spain

9.1. Introduction

This chapter presents a discussion of the development of validated methods for the quantification and confirmation of small organic compounds in biological, environmental and food samples using low-resolution (LR) and high-resolution (HR) mass spectrometers. Together with the increase in resolution of the mass spectrometers, the limits of detection (LODs) have steadily improved over the years (Fig. 9.1). This has allowed the development of new reliable analytical procedures, making use of LR and HR mass spectrometry (MS) for the quantification of small molecules in a variety of matrices.

Nowadays, two approaches are the most frequently used for the quantification of organic compounds in biological, environmental and food samples: tandem mass spectrometry (MS/MS) on a triple quadrupole mass spectrometer (QqQ) and HRMS. Since the 1990s, when liquid chromatography (LC) coupled to QqQ began to be marketed (Fig. 9.1), it was rapidly considered the best strategy for routine quantitative analysis of small molecules in biological, food and environmental samples. In contrast, typical applications of HRMS focused on accurate mass measurements aimed at determining elemental compositions of molecular ions and fragment ions. These

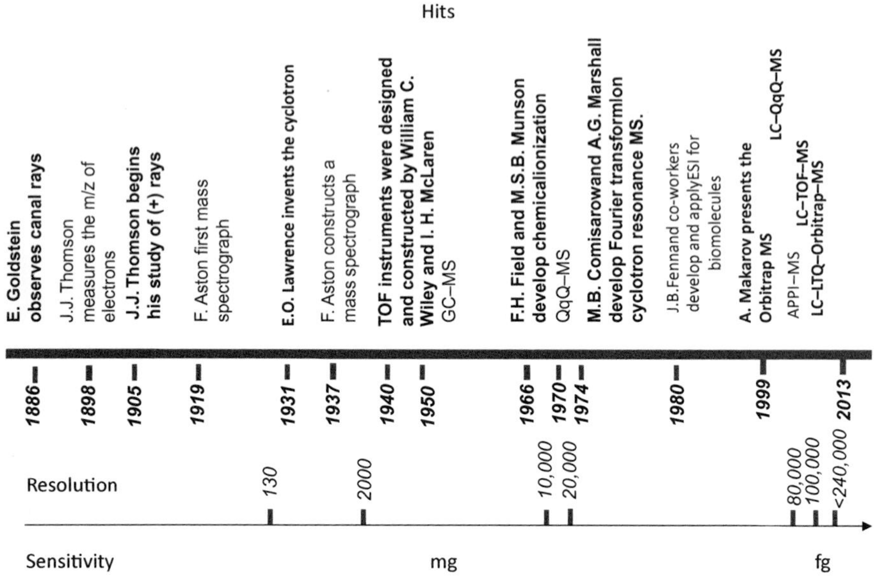

Figure 9.1. History of mass spectrometry.

conventional HRMS systems were designed for structure elucidation of unknown compounds such as transformation products or metabolites, but not for quantitative purposes, with their requirements on wide linear range and very low LODs. However, during the last 10 years, technological developments have produced less expensive HRMS instruments (both TOF and Orbitrap MS systems) with improved capabilities for quantitative analysis.

Therefore, here we review the requirements for the development and validation of methods for the quantification and confirmation of small polar organic compounds in environmental samples, foodstuff and biofluids using LC coupled to low and high resolution mass spectrometers. We also describe and discuss some LC/LR/HRMS applications that have been published over the past five to ten years.

9.2. Increasing Chromatographic Resolution: from HPLC to UHPLC

LC is a separation technique used in many different areas to achieve the separation of substances for their further identification and quantification in MS systems. Currently, two major LC modes are used: high-performance liquid chromatography (HPLC) and ultra-high-performance liquid chromatography (UHPLC). UHPLC is a special variant of HPLC with the main physical difference being particle size. This has a strong influence on separation efficiency and back-pressure. In UHPLC, particle sizes are lower than 2 μm, while in HPLC particle sizes are generally between 3 μm and 5 μm. The effect of using smaller particles as the stationary phase is described by the well-known Van Deemter equation, which establishes the relationship between linear velocity (flow rate) and plate height (column efficiency). Particle diameter is one of the variables in this equation, and as particle size decreases mass transfer improves and diffusion is reduced. At the same time the separation efficiency is relatively independent of the flow rate for a wide range of linear velocities, which allows working at different flow rates with minimum loss of separation efficiency. The higher performance of UHPLC produces narrower peaks at reduced analysis time. Typically, HPLC columns can be operated at a maximum pressure of 600 bar whereas UHPLC systems can reach up to 1200 bar. In UHPLC, internal column diameters typically range from 1 mm to 2.2 mm in order to keep the temperature at acceptable levels.

The sensitivity of UHPLC is more than two to three times higher than that of HPLC separations.[1,2] Churwell *et al.*[3] analyzed isoflavones and their metabolites, tamoxifen metabolites, β-agonists and ephedra alkaloids in environmental samples demonstrating that the use of UHPLC generally produced an improvement in sensitivity and total analysis time. It was also noted that the hybrid particles used in UHPLC often showed unique selectivity when compared to conventional HPLC packings and that extraordinary separations of geometric isomers could be achieved through UHPLC without additional effort.[3] Another study reported by Leandro *et al.*[4] identified 16 priority pesticides and their transformation products specified in the EU Directive 2013/13/EC in baby food comparing HPLC and UHPLC coupled to MS/MS. The UHPLC–MS/MS allowed confirmation of the pesticide disulfoton in the tested baby foods due to the enhancement of peak response in the analyses. Another significant advantage of the use of UHPLC was the speed of the chromatographic separation, with run time being 2.5 times shorter than HPLC.[4] For the determination of procyanidins and alkaloids in cocoa samples, Ortega *et al.*[5] compared UHPLC–MS/MS and HPLC–MS/MS. They observed a reduction in analysis time by a factor of 7 in UHPLC compared to HPLC, as well as greater sensitivity and selectivity for the quantification of the target compounds.[5]

In order to not lose the achieved chromatographic resolution, the detector cell of any in-line optical detection system, such as ultraviolet, photodiode array or fluorescence, should have minimal dispersion volume and a sufficiently high sampling rate for recording the elution of very narrow peaks. UHPLC systems are coupled mainly to MS due to its high acquisition rate. The coupling of UHPLC to MS/MS presents some important advantages including reduction of matrix effects, improvement in sensitivity, higher throughput and better resolution.[6]

9.3. From Low- to High-Resolution Mass Spectrometry Instruments

Mass resolution is defined as the minimum mass difference, $m_2 - m_1$, between two mass spectral peaks and it measures the ability to distinguish two peaks of slightly different mass-to-charge-ratios ΔM.[7] The history of mass spectrometry is the history of resolution. The first example of mass spectrometry (LR) was performed by Joseph John Thomson, who was able to separate neon isotopes by their masses in 1913 (Fig. 9.1). In the 20th century many new developments in MS were achieved (Fig. 9.1), for instance increasing the resolution of the mass spectrometer from 130 to 80,000. Another great achievement in the 20th century was the ability to couple a separation system with a mass spectrometer such as the coupling of gas chromatography to MS (GC–MS) in the 1950s, with commercial instruments becoming available in the 1970s. However, GC–MS analysis is used mainly for volatile compounds and a new technique suitable for polar compounds trace analysis was always desirable. Back then, the relative incompatibility of existing MS ion sources with a continuous LC stream delayed the coupling with MS

(LC–MS). In the mid-20th century several interfaces for LC–MS were developed but they were not reliable, therefore their success on the market was very limited. However, the situation changed with the development of atmospheric pressure ionization (API) techniques, which enabled the coupling of LC to MS. The electrospray ionization (ESI) source, which is currently the most widely used interface, was developed by Fenn in the 1980s.[8,9] Manufacturers rapidly developed instruments equipped with ESI sources providing a simple and robust interface, allowing LC–ESI–MS to become a routine technique in the determination of a wide range of mid-polar to polar analytes. However, it is known that ESI is the interface most sensitive to matrix effects. Nowadays, LC–MS instruments almost exclusively rely on API interfaces such as ESI, atmospheric pressure chemical ionization (APCI), and the more recently developed atmospheric pressure photoionization (APPI) for the detection of organic compounds in virtually any kind of sample (extract) that can be loaded onto an LC column. The two latter interfaces are capable of ionizing compounds of low to medium polarity and thus are complementary to ESI.

In the last two decades, different LR analyzers have been coupled to LC; however, only few of them afford the sensitivity and selectivity required for the reliable quantification of small polar compounds in different types of samples. LC interfaced with single quadrupoles (Q) offers good sensitivity, but when very complex matrices like raw sewage are investigated, insufficient selectivity often impairs the unequivocal identification of the analytes. Tandem MS, which was developed in 1970s,[10] offers superior performance in terms of sensitivity and selectivity in comparison with Q instruments, and the ability to isolate in a first stage the molecular ion of the compound of interest. LC techniques coupled to QqQ or the hybrid mass spectrometer quadrupole linear ion traps (QqLIT) are the most widely used LR mass spectrometers for the quantification of organic compounds analysis. Due to their sensitivity and selectivity, QqQ analyzers are work-horses in the quantitative trace analysis in all sorts of matrices. However, special concern is devoted to the detection of false positives and false negatives on

QqQ instruments, which can be avoided by employing HR mass spectrometers. Over the past decade, the MS landscape has been transformed by access to instruments of increasingly high mass-resolving power. There are two fundamental designs of HRMS instruments for analytical, chemical and biochemical applications: time-of-flight (TOF) and Fourier transform (FT), which includes Orbitrap and ion cyclotron resonance MS. Orbitrap and TOF-based platforms are the two most widely used ones for polar organic compound analysis, while Fourier transform ion cyclotron resonance (FTICR) mass spectrometers are barely found in routine analysis due to their high cost. The determination of the mass-to-charge ratio in TOF analyzers relies on the measurement of the flight time of accelerated ions in a field-free drift tube. In contrast, ion detection in the Orbitrap analyzer, which was invented by Makarov in 1999[11] (Fig. 9.1), is based on orbital trapping of ions around a central electrode and recording of an image current followed by Fourier transformation to ultimately generate a mass spectrum.[12]

9.4. Method Validation

Several guidelines for the validation of quantitative methods for the determination of organic compounds in food and biological samples have been established. These deal with the analysis of residues of drugs and pesticides in foodstuff[13] and the analysis of pharmaceuticals in biosamples of human origin.[14–16] These guidelines describe in detail how a quantitative method has to be validated in order to guarantee that the obtained analytical results for the targeted compounds are accurate and reliable. In this section, key aspects of selected guidelines, including those from the Food and Drug Administration (FDA), International Conference on Harmonisation (ICH), European Medicines Agency (EMA) and the European Commission, on evaluation of quantitative parameters of the analytical methods are reviewed with respect to the validation parameters: accuracy, precision, recovery, linearity, sensitivity, and selectivity.

9.4.1. *Accuracy, precision and recovery*

The accuracy of an analytical method describes the closeness of the determined value obtained by the analytical method to a reference value.[13] It is determined by replicate analysis of the standard (minimum five replicates) containing known amounts of the analyte.[15] It is related to the precision, which describes the closeness of repeated individual measures of analyte. The precision is usually expressed as variance, relative standard deviation (RSD) or coefficient of variation (CV) of a series of measurements. Precision is further subdivided into within-run, intra-batch precision or repeatability, which assesses precision during a single analytical run, and between-run, inter-batch precision or repeatability, which measures precision over time, and may involve different analysts, equipment, reagents, and laboratories.[14] The FDA guidance document on bioanalytical Method Validation proposes that precision should be measured using a minimum of five determinations per concentration, and recommends a minimum of three concentrations in the range of expected sample concentrations.[14] The precision determined at each concentration level should not exceed 15% of the CV except for the limit of quantification (LOQ), where it must not exceed 20% of the CV.[14] Another important parameter in assessing the performance of a methodology is the recovery, which can be defined as the fraction of the analyte determined after addition of a known amount of the analyte to a sample. It is used for the measurement of effectiveness of the analytical method. The FDA guideline[14] for bioanalytical method validation proposes that the recovery experiments should be performed by comparing the analytical results for extracted samples at three concentrations (low, medium, and high) with unextracted standards that represent 100% recovery. While in bioanalytical and food analysis three levels of spiking are used, generally two levels of spiking are used in environmental analysis.

9.4.2. *Linearity, sensitivity and stability*

The linearity of an analytical method is its ability to obtain test results that are directly proportional to the concentration of analyte in the sample. For the establishment of linearity at least five concentrations of the standards are recommended.[15] The standards are part of the calibration curve, which is the relationship between instrument response and known concentrations of the analyte.[14] For bioanalytical analysis the FDA guideline proposes that the calibration curve has to be prepared in the same matrix as the samples and at least at six different concentrations.[14] If it is not possible to prepare the standards in matrix, surrogate matrices can be used instead.[14]

The sensitivity of a method is defined as the lowest concentration that can be measured with an acceptable limit of accuracy and precision.[17] It is expressed as LOD and LOQ. The LOD of an individual analytical procedure is the lowest amount of analyte in a sample which can be detected but not necessarily quantified with sufficiently precision.[15] The LOQ of an individual analytical procedure is the lowest amount of analyte in a sample which can be quantitatively determined with suitable precision and accuracy.[15] Stability testing should evaluate the stability of the analytes during sample collection, handling and storage, and after freeze-thaw cycles and the analytical process.[14]

9.4.3. *Selectivity of the method*

Selectivity is the ability of an analytical method to differentiate and quantify the analyte in the presence of other components in the sample.[14] For applications of GC–MS/MS and LC–MS/MS to the analysis of residues in products from animal origin the EU Commission proposes a minimum of four identification points (IPs) per compound. Each MS technique is associated with IPs (see Table 9.1 and Table 9.2). However, in order to qualify for the identification points required for confirmation and the sum of identification points to be calculated:

- a minimum of at least one ion ratio shall be measured;
- all relevant measured ion ratios shall meet the criteria described above;
- a maximum of three independent techniques can be combined to achieve the maximum number of identification points.[13]

Table 9.1. IP for low- and high-resolution MS techniques.

MS technique	Identification point earned per ion
Low-resolution MS	1.0
Low-resolution MS precursor ion	1.0
Low-resolution MS transition products	1.5
High-resolution MS	2.0
High-resolution precursor ion	2.0
High-resolution transition products	2.5

Extracted from European Commission Office (2002) *J. Eur. Comm. L*, **221**, 8.

Table 9.2. Examples of number of IP for low and high MS techniques earned for combinations of LC–MS (n = an integer). Extracted from European Commission Office (2002) *J. Eur. Comm. L*, **221**, 8.

Technique(s)	Number of ions	Identification points
LC–MS	N	n
LC–MS/MS	1 precursor and 2 daughters	4
LC–MS/MS	2 precursors each with 1 daughter	5
LC–MS/MS/MS	1 precursor, 1 daughter and 2 granddaughters	5.5
HRMS	N	$2n$
LC–MS	2 + 2	4

In more detail, the specificity of a LC–MS method is related to the selectivity of the method.[18] One of the main drawbacks of LC–API–MS methods is matrix effects, which have to be carefully addressed by the analyst. Matrix effects are defined as suppression or enhancement effects on the analyte signal that are caused by matrix components of the sample that ultimately compromise reproducibility, linearity, and accuracy of the analytical method.[19,20] Matrix compounds compete with analyte ions in the interface during the ionization process and therefore may influence the ionization efficiency of the analyte.[21] In order to minimize matrix effects, or at least to compensate for them, several approaches can be applied to improve the various stages of the analytical procedures including sample-preparation protocols, LC separation, and MS detection. One of the first thoughts is to improve sample-preparation procedures in order to reduce the amount of, or ideally to completely eliminate, potentially interfering matrix components from the sample.[22] Using liquid–liquid extraction (LLE) the amount of matrix in the final extract can be largely reduced, but the major disadvantages of LLE are the considerable solvent consumption and the difficulty for efficient automation of the process.

For liquid samples, solid-phase extraction (SPE) is considered one of the best options in terms of the versatility of available sorbent chemistries, the ability to add wash steps prior to elution of the analytes from the cartridge, the low requirements of organic solvents, and the potential for on-line extractions with direct introduction of the analytes into LC–MS.[22] In fact, SPE is nowadays the most widespread method for extraction and clean-up of biological, food and environmental samples. The parameters of the SPE for liquid samples or the clean-up of extracts from solid samples have to be carefully optimized in order to minimize the presence of unwanted matrix components in the final extract. While this is relatively straightforward in the case of single-analyte protocols, multi-analyte methodologies targeting several tens or even more than a hundred compounds are far more difficult to optimize in this respect. Optimization parameters include sample pH, sorbent type, and composition of wash and elution solvents.[22] Another approach is the use of on-line sample treatment such as SPE (which offers the same advantages as off-line SPE but less solvent is used for the sample preparation). A promising alternative is turbulent-flow chromatography (TFC) which utilizes columns with particle and pore sizes of ~40–150 μm and ~60 Å, respectively; it pumps the solvent at high flow rates through the column thereby causing it to behave in the turbulent manner. TFC also simultaneously combines size exclusion and analyte selectivity because at higher flow rates larger molecules diffuse much more slowly into the particles than do smaller ones, which in turn are retained in the turboflow column.[6] This approach is particularly attractive for bioanalytical analysis because the matrix contains large biomolecules that are easy to get rid of; however, it has also been applied recently and with considerable success to food and environmental matrices.[23,24] A very simple approach to reduce the load of matrix introduced into MS is the direct injection of (diluted) sample into the LC system without any sample preconcentration step. Omitting any extraction procedure offers substantial savings of time and material but its applicability is limited to samples with sufficiently high analyte concentrations.[20] A gain in sensitivity of the aforementioned approach is achieved by the so-called large-volume injection (LVI) technique in which a sample volume of as high as 10% of the analytical column void volume is injected onto the HPLC column. The efficiency of this approach with respect to matrix effects was recently evaluated in a comparison with conventional SPE.[25] For a series of environmental contaminants, including four estrogens, eight perfluoroalkyl carboxylates and five perfluoroalkyl sulfonates, the authors demonstrated that the analytical signals were similar for both methodologies, with comparable degree of matrix effects.[25]

Modifications of the chromatographic conditions (selection of stationary phase, mobile phase composition, and gradient profiles) can also help diminish matrix effects by adjusting retention times of the target analytes such that co-elution with (known) interferences is avoided. As the last stage of the overall analytical methodology, operational parameters of the mass spectrometer may also be considered to address matrix effects. Some authors have reported that the change of the ionization mode from positive to negative ESI can lead to a more selective ionization process with reduced matrix effects.[22] It has to be borne in mind, however, that even with appropriate changes in the mobile phase pH, many analytes, including most small-molecule drugs, yield responses in the (–)-mode that are far below those produced in the (+)-mode. The use of ion sources other than the very popular ESI has been proposed in several studies. Although for many polar compounds ESI usually leads to higher peak intensities as compared to APCI or APPI, it is much more prone to matrix-dependent ionization interferences.[19] Among the three interface types, APPI is the most recently developed technique which is based on photoionization of the analyte ionizing many species more efficiently than ESI and APCI (these are based on charge affinity), rendering ion suppression generally less prominent than in APCI or even ESI.[19] Yang *et al.*[9] compared APCI and APPI performance in the analysis of the non-steroidal estrogen antagonist, idoxifene, and its two metabolites in human plasma. They observed that chemical noise in APPI was much lower than in APCI, which resulted in improved selectivity for the analytes. In particular, APPI proved to give better sensitivity than APCI for the neutral metabolite.[9]

A further approach to account for matrix effects relies on the use of distinct calibration methods. One of the least expensive approaches to compensate for matrix effects is the standard addition method, in which each sample is split into several aliquots and spiked with small, identical volumes of standard solutions containing increasing analyte concentrations. Graphical extrapolation then allows estimating the concentration of the compound of interest. Despite its simplicity and robustness, it is rarely employed in routine analysis because of the need to prepare a calibration series for every single sample.[26] The most widely used approach to address matrix effects in LC–MS relies on the use of isotope dilution. For this purpose internal standards (IS) are used such as stable isotope-labeled internal standard (SIL-IS). The addition of an IS to the final sample extract can correct the variability of analyte signal in the MS. If IS is added prior to the sample process, the IS will compensate for matrix effects and the recovery of the analyte. The IS has to present similar ionization properties and elute closely to the target analytes. The common practice is to use a deuterium-labeled IS, however some recent works have demonstrated that deuterated ISs are not an appropriate standard because they do not exactly co-elute with the target analyte, thereby suffering different matrix effects.[27] Therefore, ^{13}C, ^{15}N or ^{18}O ISs have been recommended in several papers.[28–30] Several of these papers point out a combination of two or more of the approaches described in this section to take measures against matrix effects. For instance, some works[26,31] propose the use of SIL-IS in combination with UHPLC–MS/MS, because the overall methodology can almost completely account for matrix effects. Using an SIL-IS approach, Van de Steene compared the matrix effects in surface water samples for nine basic pharmaceuticals obtained with HPLC against those in UHPLC. In the HPLC–MS/MS analysis the matrix effects were strong even when SIL-IS was used in the analysis; however, in the analysis with UHPLC–MS/MS the matrix effects were minor or eliminated. Nevertheless, in another work[32] they conclude that the use of a single approach in order to compensate for matrix effects using SIL-IS in calibration standards in water or in mobile phase instead of the combination of SIL-IS with matrix match standard approach is the simplest option and therefore the best approach.

9.5. Quantification and Confirmation in LC–MS

Since 1953, when the basic principles of Q were published[22] (Fig. 9.1), thousands of studies have reported on the analysis of organic compounds in the environment and food samples.[34–36] In order to overcome the limitations of LC–Q–MS, HPLC coupled to MS/MS replaced methods on Q instruments, affording enhanced selectivity and sensitivity. More recently, HRMS has been used for quantitative measurements, while previously it was almost exclusively used for qualitative analysis. In the following section, several examples are described to illustrate the advantages and limitations of both analytical detection concepts.

9.5.1. *Quantification with LC/LR–MS/MS*

LC–QqQ–MS/MS platforms include two sequential mass analyzers. In the first stage (pseudo) molecular ions are selected; they are then fragmented in the collision cell and the second mass filter analyzes all or just specific formed fragment ions. In the last decade, the use of LC–QqQ–MS/MS based on selected reaction monitoring (SRM), has become the cornerstone of quantitative analysis for environmental and food contaminants working in accordance to international guidelines.[37,38] In these studies QqQ working in SRM mode showed excellent sensitivity, selectivity and robustness. Over the last two decades, due to the capability of QqQ instruments to analyze over 300 compounds in one run, a large number of multi-residue methods based on MS/MS for the quantitative analysis of organic contaminants in food and environmental samples have been published.[37–41] Ortelli *et al.*[40] developed a multiresidue method for the analysis of 74 pesticides in fruits and vegetables by LC–QqQ–MS/MS. Good sensitivity and selectivity were obtained with LOQs of 0.01 mg/kg in almost all cases. Recoveries, RSD and accuracy values fulfilled the criteria of the European Commission. Also using a LC–QqQ–MS/MS, Gros *et al.*[41] developed a multi-residue method for the determination of 28 pharmaceuticals in surface and wastewaters. Recoveries obtained for most target compounds were higher than 60% and LODs in the low ng/L range for both river and wastewater. For four pharmaceuticals the number of IPs required for the European commission were not achieved due to their low fragmentation.

A closely related QqQ variant, which also works in tandem MS, is the so-called QqLIT, in which the second mass filter is replaced by a linear ion trap (LIT). It retains the classical QqQ acquisition modes, but offers additional functions involving trap experiments in the LIT. Hopfgartner *et al.* discussed the different QqLIT operation modes and their applications and concluded that QqLIT has provided satisfactory results for quantitative studies and also new analytical approaches at a confirmatory level.[8] In Section 9.5.2, some examples of employing LIT-based modes for confirmation of target analytes for compounds which only form a single abundant fragment ion are reviewed.

9.5.2. *Confirmation aspects in LC/LR–MS/MS*

According to EU Commission Decision 2002/657/EC for the confirmation of compounds in food samples by means of LC–MS/MS, at least four (IPs) are required for a target compound analysis when mass fragments are included in the method using techniques other than full-scan MS. In QqQ–MS the confirmation of a positive finding has to be done by evaluating the intensity ratio between different

SRM transitions. In SRM mode, this implies that four IPs can be collected by obtaining two transition precursor ions: product ion (1.5 IP each) and one point for the precursor. Also, for confirmation of a compound, the LC retention time of the standard has to match the sample.[41] In MS/MS, the SRM transitions are optimized to achieve high fragment ion intensities. As this optimization process has to be carried out for each individual analyte on at least two transitions, it may constitute a very laborious and time-consuming task when multi-residue methods are developed. Since the optimum collision energies for a given compound are instrument-specific, simple method transfer to MS platforms from different vendors are generally not feasible. Despite the high selectivity of the SRM mode there is still a small chance of false positives if matrix compounds produce the same fragment ions as the analyte of interest. Another problematic case may emerge if a co-eluting isobaric compound generates one of the two fragment ions monitored because this implies a change in ion ratio. In the literature, several studies reported false positive results.[42–44] Co-eluting interferences producing a signal at the same SRM transition can lead to ion ratio errors beyond the 20% acceptance criterion. In this instance, the suspected positive finding must be flagged according to the EU Commission decision.[13] In the case of ambiguous results, the directive recommends monitoring a third transition, which is not possible for certain analytes due to insufficient abundance or little fragmentation. For the confirmation of benzophenone in foodstuffs Gallart-Ayala *et al.*[42] proposed the use of HRMS because benzophenone yielded only two product ions. With respect to QqLIT systems, they afford a powerful feature for confirmation purposes by taking advantage of the high scan sensitivity of the LIT analyzer. While running a quantitative method monitoring a number of SRM transitions, the detection of a signal for a given SRM channel above a defined threshold triggers an information-dependent acquisition (IDA). The preferred acquisition mode in the IDA experiment is an enhanced product ion (EPI) scan, in which fragment ions produced in the collision cell are accumulated in the LIT and subsequently analyzed by the detector. This combination of a sensitive survey scan with an EPI scan providing structural information is very useful as it allows the comparison of product ion profiles between the sample and authentic standards (Fig. 9.2). The applicability of the IDA approach was demonstrated in several multi-residue methods developed for the analysis of pharmaceuticals in environmental waters[45–49] and proved very useful for analytes for which only a single SRM transition was monitored.

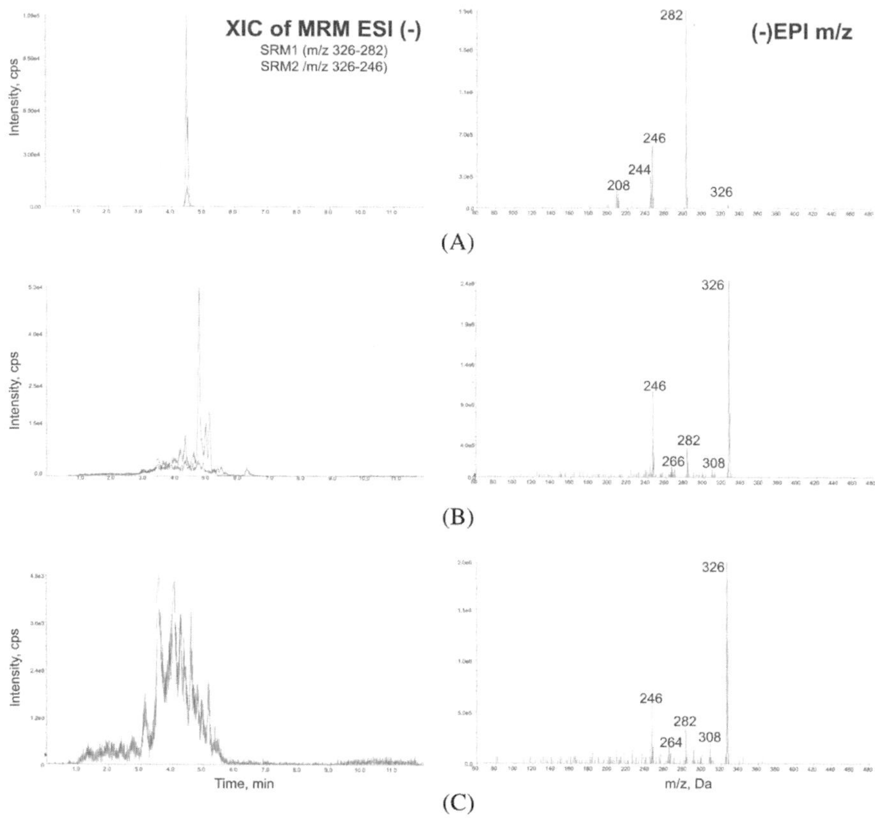

Figure 9.2. Example of an IDA experiment performed for the determination of the transformation product 4',5-diOH-diclofenac in (A) a standard, (B) an urban effluent and (C) an urban influent. Reproduced with permission from Osorio, M.I.V., Zonja, B., Abad, J.L. *et al.* (2014). Simultaneous determination of diclofenac, its human metabolites and microbial nitration/nitrosation transformation products in wastewaters by liquid chromatography/quadrupolelinear ion trap mass spectrometry, *J. Chromatogr. A*, **1347**, 63–71, Elsevier.

Even if the criteria of the IPs according to the EU Commission decision are applied in a consistent manner, the detection of false-positives can never be completely ruled out.[13] Concerning the use of SRM mode for analysis of organic contaminants, the Commission decision recommends specific neutral losses in order to confirm findings instead of unspecific transitions, such as the loss of water or ammonia. In addition, co-eluting interfering matrix compounds may produce two common transitions and the obtained results are consequently considered as correct. As an example, Schurmann *et al.*[43] reported false positive results for sebuthylazine residues using MS/MS according to the EU directive. By monitoring a third SRM transition, the authors were able to track down false positive detection through mismatching fragment ion ratios. In response to the inherent shortcomings of LR quadrupole-based instruments in which mass resolution is set to unit resolution, modification of the shape of the quadrupole rods has allowed for a substantial increase in mass resolution while almost maintaining sensitivity. Using instruments equipped with hyperbolic quadrupoles, resolutions of as low as 0.02 Da could be achieved. Under these conditions, the QqQ was able to deliver accurate mass measurements with errors below 5 mDa using internal lock mass corrections.[7] Gallart-Ayala *et al.*[50] reported a quantitative method for a hyperbolic rod-equipped QqQ for the determination of isopropylthioxanthone in packaged food. This study confirmed higher selectivity without sensitivity loss. Martinez-Villalba *et al.*[51] used this strategy to develop a method for the analysis of toltrazuril and its metabolites in food. When comparing hyperbolic to round-shaped quadrupoles, especially for complex matrixes, the first can be seen as more selective and able to improve signal to noise ratio, allowing the avoidance of false positive findings.[51]

In a comparison of LR–MS with HRMS, Kaufmann *et al.*[47] detected positive findings of nitroimidazole in honey using the former technique while the latter clearly demonstrated the absence of the antibiotic in the food samples. These studies proved that the EU Commission decision[13] is a very useful document but not infallible for complex matrixes. As a complementary tool, or even as a replacement for LR–MS-based methodologies, exact mass analysis on HRMS platforms can provide valuable information to minimize the risk of interferences from matrix components.

9.5.3. *Quantification with LC–HRMS*

Over the past couple of years, HRMS has emerged as a promising technique for quantitative methods thereby challenging established LR–MS-based protocols. It is known that LR–MS methods require the optimization of tandem MS/MS parameters for each analyte and therefore potential issues related to cross-talk, adduct formation, and unspecific SRM transitions have to be taken into account. As a consequence, the setup of the SRM methods is a tedious but essential part of the method development. Moreover, the LR of QqQ does not allow the identification of the co-eluting matrix ions with a very close *m/z* value to the target analyte. Therefore, one possible solution to overcome this drawback is to work with HRMS.[52] For instance, Henry *et al.*[52] developed a HRMS method in full-scan mode activating the so-called All Ion Fragmentation (AIF) function. Here the instrument alternates at a frequency of about 2 Hz between a conventional full scan, in which mostly intact molecular ions generated in the ion-source are sent as packets into the Orbitrap, and a high-energy collision-induced dissociation (HCD) step, where the ions being sampled into the mass spectrometer are first directed towards the collision cell for unselective fragmentation without any precursor ion selection and then analyzed in the Orbitrap detector. While the first scan mode produces relatively simple total ion chromatograms (TICs), from which specific ion masses can be extracted with high selectivity by setting very narrow mass extraction windows, the far more complex TIC of the AIF requires deconvolution to identify all those ions belonging to the same molecular entity. The application of this approach is particularly valuable for retrospective qualitative analysis because in a single chromatographic run a great deal of structural information is gathered. Other Orbitrap–MS configurations, such as the LTQ–Orbitrap, proved to be a suitable tool for quantitation of small molecules and simultaneous metabolite identification.[53]

There are a growing number of reports focusing on the comparison of LR–MS/MS with HRMS but only few of them offer a more detailed examination of selectivity. Henry *et al.*[52] compared the capabilities of HR and tandem QqQ for quantitative analysis of drugs in plasma samples.[52] Three instruments, namely the high-resolution Orbitrap–MS (Exactive) and the two QqQ instruments (TSQ Quantum Discovery and TSQ Quantum Ultra featuring the H-SRM mode), were benchmarked by evaluating the analysis of 17 therapeutic drugs. Similar detection specificity, assay precision, accuracy, linearity, and sensitivity were obtained for the three instruments. In 2010, Kaufmann *et al.*[54] evaluated the selectivity provided by Orbitrap–MS (Exactive model) and a conventional QqQ. High selectivity with similar sensitivity was obtained for HRMS, suggesting that HRMS is likely to become a dominating technique in the field of multi-residue quantification in the near future. Kaufmann *et al.* also compared three different platforms, including two HRMS instruments such as TOF–MS, Orbitrap–MS and one LR mass spectrometer QqQ.[55] This study concluded that TOF analysis afforded poorer analytical performance than the analysis by QqQ and Orbitrap–MS but the employed TOF instrument was not from the latest generation; it was equipped with a time-to-digital detector (TDC) and not with the newest analog-to-digital detector (ADC). The most recent generation of TOF–MS has improved capabilities, especially in terms of dynamic range (by ADC detector or dynamic range enhancement functions[56]) and resolution (from 15,000 FWHM to 40,000 FWHM), thereby reducing the weaknesses to sensitivity and precision. Nowadays, several methods have been successfully developed according to different guidelines with satisfactory quality parameters.[57,58] Several authors compared QqQ versus TOF–MS and/or QTOF HRMS systems for the determination of pesticides in fruits and vegetables.[59,60] These studies confirm that MS/MS is the most sensitive and robust approach but QTOF–MS offers good performance, the most powerful confirmation, and additional sample information. However, the most recent models of QqQ systems are still more sensitive than HRMS instruments, particularly when only a small number of compounds, and therefore SRM transitions, are to be monitored. But as the number of target analytes within a single method increases, a turning point is reached at which the unselective data recording in full-scan mode on the HRMS exceeds the sensitivity of the LRMS detection. As mentioned before, HRMS measurements largely resolve the issue of interfering matrix compounds and are of great help for unmasking false positive MS/MS results. An example of the benefits of HRMS analysis involving complex matrices is the separation of isobaric interferences from the contaminant signals of interest. In the benzophenone case, Gallart-Ayala *et al.* showed that a resolution of 50,000 FWHM allowed the distinguishing of benzophenone from an interfering compound (Fig. 9.3).[42] Ferrer *et al.*[61] used these capabilities in a QTOF–MS, working at 30,000 FWHM, to develop a quantitative method for 100 drugs and their degradates, among which were a number isobaric compounds. A further benefit described in many publications is the possibility of performing quantitative and qualitative analysis simultaneously. This advantage allows the development of methods for the simultaneous screening of suspect or non-target compounds and the quantification of target analytes. This powerful tool is currently being used for environmental[62–64] and dietary supplement[65] analysis, providing information on the concentration of known compounds using calibration curves of authentic standards but also on the presence of other compounds.

In summary, HRMS offers several benefits: the possibility to store a high volume of full-scan information data; data-dependent MS/MS with high mass accuracy, which allows for retrospective analysis; and the ease of method setup. The robustness of any HR full-scan approach depends to a large extent on the stability of mass accuracy, the precision, and the resolving power as a means of distinguishing between very similar ion masses. Nowadays, many laboratories are considering replacing QqQ-based methods with those relying on the use of the latest generation of HRMS.

9.5.4. *Confirmation aspects in LC–HRMS*

HRMS can provide a great deal of information, allowing the confirmation of the identity of each analyte; accurate mass and isotope cluster of the molecular ion are directly available from the full-scan

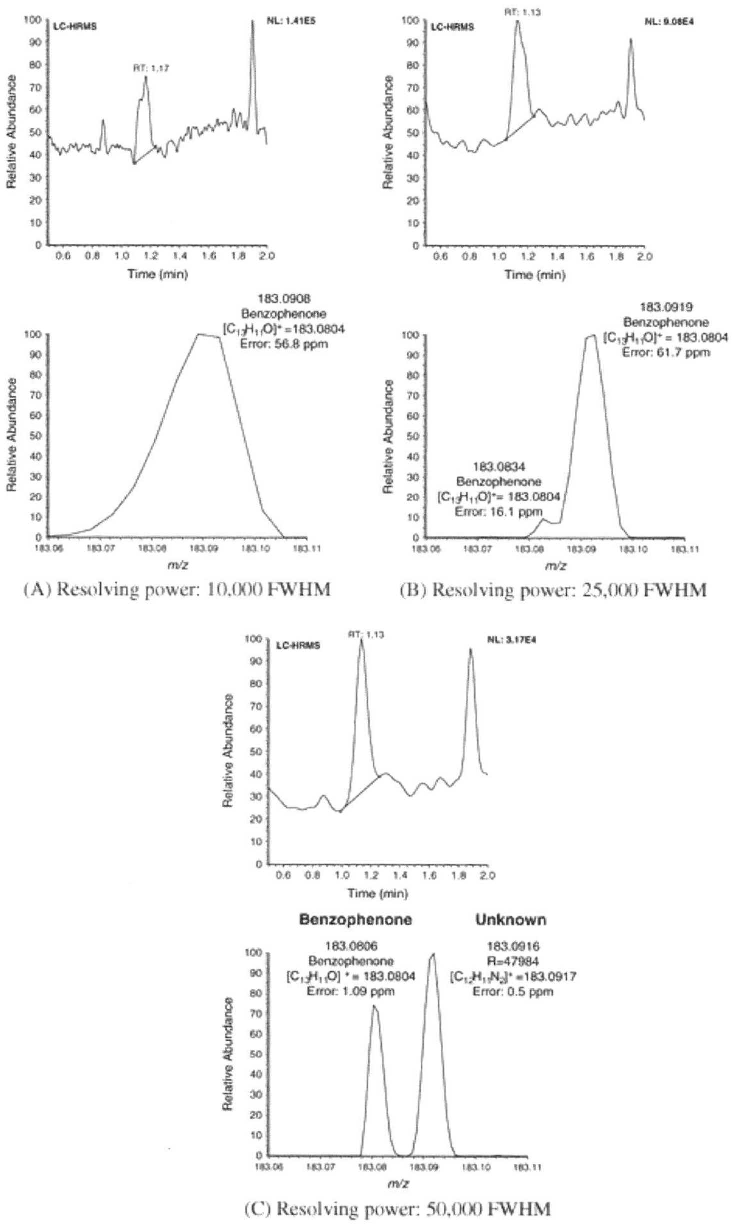

Figure 9.3. Preventing false negatives with HRMS: the benzophenone case. Reproduced from Gallart-Ayala, H., Núñez, O., Moyano, E. *et al.* (2011) Preventing false negatives with high-resolution mass spectrometry: the benzophenone case, *Rapid Commun. Mass Spectrom.* **25**, 3161–3166. With permission from the authors.

acquisition. Of particular value, however, is mass spectral information on fragment ions, which can be generated using platform-specific functions. The two most popular approaches are non-selective fragmentation (All Ion Fragmentation or MS^E modes) on HRMS instruments and data-dependent acquisitions to produce high-quality MS/MS spectra. The latter mode is only feasible on hybrid instruments (QTOF, Q–Exactive, LTQ–Orbitrap) in which precursor ions can be isolated in a first stage of mass filtering and then selectively fragmented. Regardless of whether AIF or data-dependent MS/MS switching modes are run, the use of specific software to facilitate the data-mining process, and therefore the confirmation, is considered essential. In particular, automated matching of MS/MS-product ion spectra against spectral libraries is of great help. One of the limitations of the two above-described acquisition modes is related to the ion abundance of the molecular ions in the full-scan mode; very weak signals are unlikely to generate any useful fragment ion above the noise level. In this instance, the level of confidence in compound identity can be higher in modern LR–MS instruments programmed to acquire at least two selective SRM transitions. Regarding the EU Commission decision,[13] an ion measured at a resolution of 10,000 produces 0.5 IPs more than a one MS/MS transition (from 1.5 to 2 identification points).

In HRMS false positives can be revealed using narrow mass extraction windows. A resolving power of 20,000 is generally adequate to obtain accurate quantitation but some authors have requested the establishment of additional guidelines for appropriate mass extraction windows and mass accuracy criteria.[66,67] False negatives can also occur, when too narrow mass windows and insufficient MS resolution are applied.[43,55] Figure 9.4 shows how narrow extraction windows can help eliminate background ions.

In some cases, full-scan HRMS with fragment ion confirmation or HRMS in SRM mode are not enough to prevent false positive results. Ramanathan *et al.*[58] reported a case when a co-eluting endogenous isobaric compound caused interferences in prednisolone analysis. Prednisolone and cortisone are isomers and show several common product ions. To quantify this compound it is necessary to choose specific product ions.

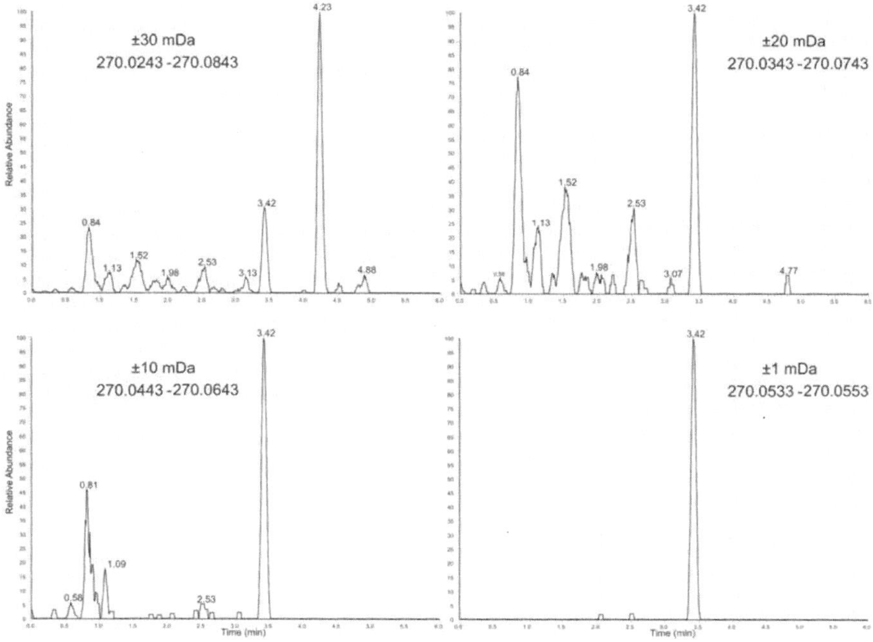

Figure 9.4. Different mass extraction windows applied for the quantification of OH-sulfamethoxazole by Orbitrap–MS (Q–Exactive) in full scan mode.

9.6. Conclusions and Future Needs

In comparison to HPLC, UHPLC coupled to MS provides faster chromatographic separation methods, improves the sensitivity of the quantitative analytical method two- to three-fold, and reduces matrix effects. Tandem MS based on LRMS constitutes the most frequently used technique for the quantitative determination of organic trace contaminants in environmental and food samples. Technological advances have led to the development of a new generation of HRMS instruments, which can compete with LRMS in terms of sensitivity and cost. That means that the practice of using HRMS exclusively for qualitative purposes (e.g. in identification of metabolites and transformation products) needs to be revised. Comparing the performance of LR–MS/MS and HRMS for quantitative analysis, both approaches can now afford comparable results in terms of specificity, accuracy, reproducibility and robustness. As far as sensitivity is concerned, acquisitions done in SRM mode on LR–MS/MS generally provide lower LODs and LOQs. The sensitivity, however, steadily decreases as the number of compounds, and therefore MRM transitions, increases. In contrast, the sensitivity in HRMS instruments is independent of the number of target analytes because data acquisition is usually done in full-scan mode; the acquisition of data over a wide *m/z* range and detection of compounds of interest relies on the use of very narrow extraction windows around the *m/z* value of the analyte. A further advantage of HRMS is the fact that no compound-specific parameters, such as selection of the most adequate fragment ion and optimization of the collision energy, need to be tuned. Apart from this timesaving aspect, the risk of false positive findings is lower on HRMS, particularly in complex matrices with co-eluting isobaric compounds. Moreover, HRMS is the most powerful tool for the confirmation of compound identity.

In summary, LC–HRMS is a powerful tool for simultaneous quantitative and qualitative analysis, allowing the quantitation of target compounds and the search for suspected metabolites and transformation products. LC–HRMS provides more information about each sample, allowing retrospective analysis. To this end, new commercial software as well as user-built libraries help to mine the data in an efficient and comprehensive fashion. It is necessary to keep in mind that HRMS data file sizes are typically 10–100 times larger than MS/MS files. As of today, HRMS holds great promise to become in the near future a routine tool for the quantitative analysis of organic contaminants. In light of this growing popularity and acceptance, current guidelines need to be revised.

Acknowledgements

This work has been supported by the Spanish Ministry of Science and Innovation [project Consolider-Ingenio 2010 Scarce CSD2009-00065]. Bozo Zonja acknowledges the Marie Curie Actions ITN CSI:Environment PITN-GA-2010-264329 for the Early Stage Researcher contract and funding. Sandra Pérez acknowledges the contract from the Ramón y Cajal Program of the Spanish Ministry of

Economy and Competitiveness. Prof. Damia Barceló acknowledges King Saud University (Riyadh, Saudi Arabia) for his contract position as Visiting Professor. This work was partly supported by the Generalitat de Catalunya (Consolidated Research Group: Water and Soil Quality Unit 2009-SGR-965). This work reflects only the authors' views and the European Community is not liable for any use that may be made of the information contained therein.

References

1. Chen, S. and Kord, A. (2009). Theoretical and experimental comparison of mobile phase consumption between ultrahigh-performance liquid chromatography and high performance liquid chromatography, *J. Chromatogr. A*, **1216**, 6204–6209.
2. Gumustas, M., Kurbanoglu, S., Uslu, B. *et al.* (2013). UPLC versus HPLC on drug analysis: advantages, applications and their validation parameters, *Chromatographia*, **76**, 1365–1427.
3. Churchwell, M.I., Twaddle, N.C., Meeker, L.R. *et al.* (2005). Improving LC–MS sensitivity through increases in chromatographic performance: comparisons of UPLC–ES/MS/MS to HPLC–ES/MS/MS, *J. Chromatogr. B*, **825**, 134–143.
4. Leandro, C.C., Hancock, P., Fussell, R.J. *et al.* (2006). Comparison of ultra-performance liquid chromatography and high-performance liquid chromatography for the determination of priority pesticides in baby foods by tandem quadrupole mass spectrometry, *J. Chromatogr. A*, **1103**, 94–101.
5. Ortega, N., Romero, M.-P., Macià. A. *et al.* (2010). Comparative study of UPLC–MS/MS and HPLC–MS/MS to determine procyanidins and alkaloids in cocoa samples, *J. Food Comp. Anal.*, **23**, 298–305.
6. Núñez, O., Gallart-Ayala, H., Martins, C.P.B. *et al.* (2012). New trends in fast liquid chromatography for food and environmental analysis, *J. Chromatogr. A*, **1228**, 298–323.
7. Marshall, A.G. and Hendrickson, C.L. (2008). High-resolution mass spectrometers, *Annual Rev. Anal. Chem.*, **1**, 579–599.
8. Hopfgartner, G., Varesio, E., Tschäppät, V. *et al.* (2004). Triple quadrupole linear ion trap mass spectrometer for the analysis of small molecules and macromolecules, *J. Mass Spectrom.*, **39**, 45–855.

9. Yang, C. and Henion, J. (2002). Atmospheric pressure photoionization liquid chromatographic–mass spectrometric determination of idoxifene and its metabolites in human plasma, *J. Chromatogr. A*, **970**, 155–165.
10. Yost, R. and Enke, C. (1978). Selected ion fragmentation with a tandem quadrupole mass spectrometer, *J. Amer. Chem. Soc.*, **100**, 2274–2275.
11. Makarov, A.A. (1999). Mass spectrometer. Google Patents.
12. Eichhorn, P., Pérez, S. and Barceló D. (2012) Time-of-flight mass spectrometry versus orbitrap-based mass spectrometry for the screening and identification of drugs and metabolites: is there a winner? *Comprehensive Anal. Chem.*, **58**, 217–272.
13. European Commission (2002). Commission Decision 2002/657/EC of 12 August 2002 implementing Council Directive 96/23/EC concerning the performance of analytical methods and the interpretation of results. *Off. J. Eur. Comm.*, **L.221**, 8–36.
14. Administration FaD. (2013). 'Guidance for industry, bioanalytical method validation' in *Services*, USDoHaH (ed.).
15. ICH (2005). ICH harmonised tripartite guideline vlaidation of analyticial procedures: text and methodology.
16. EMEA/CHMP/EWP/192217/2009. (2011). Guideline on bioanalytical method validation.
17. Bansal, S. DeStefano, A. (2007). Key elements of bioanalytical method validation for small molecules, *AAPS J.*, **9**, E109–E114.
18. SANCO/12495/2011. (2012). Method validation and quality control procedures for pesticide residues analysis in food and feed.
19. Kloepfer, A., Quintana, J.B. and Reemtsma, T. (2005). Operational options to reduce matrix effects in liquid chromatography–electrospray ionisation-mass spectrometry analysis of aqueous environmental samples, *J. Chromatogr. A*, **1067**, 153–160.
20. Gosetti, F., Mazzucco, E., Zampieri, D. *et al.* (2010). Signal suppression/enhancement in high-performance liquid chromatography tandem mass spectrometry, *J. Chromatogr. A*, **1217**, 3929–3937.
21. Taylor, P.J. (2005). Matrix effects: the Achilles heel of quantitative high-performance liquid chromatography–electrospray–tandem mass spectrometry, *Clin. Biochem.*, **38**, 328–334.
22. Trufelli, H., Palma, P. Famiglini, G. *et al.* (2011). An overview of matrix effects in liquid chromatography–mass spectrometry, *Mass Spectrom. Rev.*, **30**, 491–509.

23. Mottier, P., Hammel, Y.-A., Gremaud, E. *et al.* (2007). Quantitative high-throughput analysis of 16 (fluoro)quinolones in honey using automated extraction by turbulent flow chromatography coupled to liquid chromatography–tandem mass spectrometry, *J. Agric. Food Chem.*, **56**, 35–43.
24. Krebber, R., Hoffend, F.-J. and Ruttmann, F. (2009). Simple and rapid determination of enrofloxacin and ciprofloxacin in edible tissues by turbulent flow chromatography/tandem mass spectrometry (TFC–MS/MS), *Anal. Chim. Acta*, **637**, 208–213.
25. Backe, W.J. and Field, J.A. (2012). Is SPE necessary for environmental analysis? A quantitative comparison of matrix effects from large-volume injection and solid-phase extraction based methods, *Environ. Sci. Technol.*, **46**, 6750–6758.
26. Van de Steene, J.C. and Lambert, W.E. (2008). Comparison of matrix effects in HPLC–MS/MS and UPLC-MS/MS: analysis of nine basic pharmaceuticals in surface waters, *J. Amer. Soc. Mass Spectrom.*, **19**, 713–718.
27. Wang, S., Cyronak, M. and Yang, E. (2007). Does a stable isotopically labeled internal standard always correct analyte response? A matrix effect study on a LC/MS/MS method for the determination of carvedilol enantiomers in human plasma, *J. Pharm. Biomed. Anal.*, **43**, 701–707.
28. Couchman, L. and Morgan, P.E. (2011). LC–MS in analytical toxicology: some practical considerations, *Biomed. Chromatogr.*, **25**, 100–123.
29. Van Eeckhaut, A., Lanckmans, K., Sarre, S. *et al.* (2009). Validation of bioanalytical LC–MS/MS assays: evaluation of matrix effects, *J. Chromatogr. B*, **877**, 2198–2207.
30. Stokvis, E., Rosing, H. and Beijnen, J.H. (2005). Stable isotopically labeled internal standards in quantitative bioanalysis using liquid chromatography/mass spectrometry: necessity or not? *Rapid Commun. Mass Spectrom.*, **19**, 401–407.
31. Berg, T. and Strand, D.H. (2011). 13C labelled internal standards — A solution to minimize ion suppression effects in liquid chromatography–tandem mass spectrometry analyses of drugs in biological samples?, *J. Chromatogr. A*, **1218**, 9366–9374.
32. Hewavitharana, A.K. (2011). Matrix matching in liquid chromatography–mass spectrometry with stable isotope labelled internal standards — Is it necessary?, *J. Chromatogr. A*, **1218**, 359–361.

33. Paul, W. and Steinwedel, H. (1953). Quadrupole mass filter, *Z Naturforforsch A.*, **8**, 448.
34. Lindsey, M.E., Meyer, M. and Thurman, E.M. (2001). Analysis of trace levels of sulfonamide and tetracycline antimicrobials in groundwater and surface water using solid-phase extraction and liquid chromatography/mass spectrometry, *Anal. Chem.*, **73**, 4640–4646.
35. Wang, Y., Catana, F., Yang, Y. *et al.* (2002). An LC–MS method for analyzing total resveratrol in grape juice, cranberry juice, and in wine, *J. Agric. and Food Chem.*, **50**, 431–435.
36. López de Alda, M.J. and Barceló D. (2000). Determination of steroid sex hormones and related synthetic compounds considered as endocrine disrupters in water by liquid chromatography–diode array detection–mass spectrometry, *J. Chromatogr. A*, **892**, 391–406.
37. Picó, Y., Blasco, C. and Font, G. (2004). Environmental and food applications of LC–tandem mass spectrometry in pesticide-residue analysis: an overview, *Mass Spectrom. Reviews*, **23**, 45–85.
38. Petrović, M., Hernando, M.D., Díaz-Cruz, M.S. *et al.* (2005). Liquid chromatography–tandem mass spectrometry for the analysis of pharmaceutical residues in environmental samples: a review, *J. Chromatogr. A*, **1067**, 1–14.
39. Gago-Ferrero, P., Diaz-Cruz, M.S. and Barcelo, D. (2013). Liquid chromatography–tandem mass spectrometry for the multi-residue analysis of organic UV filters and their transformation products in the aquatic environment, *Anal. Methods*, **5**, 355–366.
40. Ortelli, D., Edder, P. and Corvi, C. (2004). Multiresidue analysis of 74 pesticides in fruits and vegetables by liquid chromatography–electrospray–tandem mass spectrometry, *Anal. Chim. Acta.*, **520**, 33–45.
41. Gros, M., Petrović, M. and Barceló, D. (2006) Development of a multiresidue analytical methodology based on liquid chromatography–tandem mass spectrometry (LC–MS/MS) for screening and trace level determination of pharmaceuticals in surface and wastewaters, *Talanta*, **70**, 678–690.
42. Gallart-Ayala, H., Núñez, O., Moyano, E. *et al.* (2011). Preventing false negatives with high-resolution mass spectrometry: the benzophenone case. *Rapid Commun. Mass Spectrom.*, **25**, 3161–3166.

43. Schürmann, A., Dvorak, V., Crüzer, C. *et al.* (2009). False-positive liquid chromatography/tandem mass spectrometric confirmation of sebuthylazine residues using the identification points system according to EU directive 2002/657/EC due to a biogenic insecticide in tarragon, *Rapid Commun. Mass Spectrom.*, **23**, 1196–1200.
44. Kaufmann, A., Butcher, P., Maden, K. *et al.* (2009) Are liquid chromatography/electrospray tandem quadrupole fragmentation ratios unequivocal confirmation criteria?, *Rapid Commun. Mass Spectrom.*, **23**, 985–998.
45. Jelić, A., Petrović, M. and Barceló, D. (2009). Multi-residue method for trace level determination of pharmaceuticals in solid samples using pressurized liquid extraction followed by liquid chromatography/quadrupole-linear ion trap mass spectrometry, *Talanta*, **80**, 363–371.
46. Gros, M., Pizzolato, T.-M., Petrović, M. *et al.* (2008). Trace level determination of β-blockers in waste waters by highly selective molecularly imprint ed polymers extraction followed by liquid chromatography–quadrupole-linear ion trap mass spectrometry, *J. Chromatogr. A*, **1189**, 374–384.
47. Gros, M., Rodríguez-Mozaz, S. and Barceló, D. (2012). Fast and comprehensive multi-residue analysis of a broad range of human and veterinary pharmaceuticals and some of their metabolites in surface and treated waters by ultra-high-performance liquid chromatography coupled to quadrupole-linear ion trap tandem mass spectrometry, *J. Chromatogr. A*, **1248**, 104–121.
48. Gros, M., Petrović,M. and Barceló, D. (2008). Tracing pharmaceutical residues of different therapeutic classes in environmental waters by using liquid chromatography/quadrupole-linear ion trap mass spectrometry and automated library searching, *Anal. Chem.*, **81**, 898–912.
49. Osorio, M.I.V, Zonja, B., Abad, J.L. *et* al. (2014). Simultaneous determination of diclofenac, its human metabolites and microbial nitration/nitrosation transformation products in wastewaters by liquid chromatography/quadrupolelinear ion trap mass spectrometry, *J. Chromatogr. A*, *submitted.*
50. Gallart-Ayala, H., Moyano, E. and Galceran, M.T. (2008). Liquid chromatography/tandem mass spectrometry (highly selective selected

reaction monitoring) for the analysis of isopropylthioxanthone in packaged food, *J. Chromatogr. A*, **1208**, 182–188.

51. Martínez-Villalba, A., Moyano, E., Martins, C.B. *et al.* (2010). Fast liquid chromatography/tandem mass spectrometry (highly selective selected reaction monitoring) for the determination of toltrazuril and its metabolites in food, *Anal. Bioanal. Chem.*, **397**, 2893–2901.
52. Henry, H., Sobhi, H.R., Scheibner, O. *et al.* (2012). Comparison between a high-resolution single-stage Orbitrap and a triple quadrupole mass spectrometer for quantitative analyses of drugs, *Rapid Commun. Mass Spectrom.*, **26**, 499–509.
53. Zhang, N.R., Yu, S., Tiller, P. *et al.* (2009). Quantitation of small molecules using high-resolution accurate mass spectrometers – a different approach for analysis of biological samples, *Rapid Commun. Mass Spectrom.*, **23**, 1085–1094.
54. Kaufmann, A., Butcher, P., Maden, K. *et al.* (2010). Comprehensive comparison of liquid chromatography selectivity as provided by two types of liquid chromatography detectors (high resolution mass spectrometry and tandem mass spectrometry): "Where is the crossover point?", *Anal. Chim. Acta*, **673**, 60–72.
55. Kaufmann, A., Butcher, P., Maden, K. *et al.* (2011). Quantitative and confirmative performance of liquid chromatography coupled to high-resolution mass spectrometry compared to tandem mass spectrometry, *Rapid Commun. Mass Spectrom.*, **25**, 979–992.
56. Ibáñez, M., Guerrero, C., Sancho, J.V. *et al.* (2009). Screening of antibiotics in surface and wastewater samples by ultra-high-pressure liquid chromatography coupled to hybrid quadrupole time-of-flight mass spectrometry, *J. Chromatogr. A*, **1216**, 2529–2539.
57. Hermo, M.P., Barrón, D. and Barbosa, J. (2008). Determination of multiresidue quinolones regulated by the European Union in pig liver samples: high-resolution time-of-flight mass spectrometry versus tandem mass spectrometry detection, *J. Chromatogr. A*, **1201**, 1–14.
58. Ramanathan, R., Jemal, M., Ramagiri, S. *et al.* (2011). It is time for a paradigm shift in drug discovery bioanalysis: from SRM to HRMS, *J.Mass Spectrom.*, **46**, 595–601.
59. Grimalt, S., Sancho, J.V., Pozo, Ó.J. *et al.* (2010). Quantification, confirmation and screening capability of UHPLC coupled to triple quadrupole

and hybrid quadrupole time-of-flight mass spectrometry in pesticide residue analysis. *J. Mass Spectrom.*, **45**, 421–436.

60. Wang, J., Leung, D. and Chow, W. (2009). Applications of LC/ESI–MS/MS and UHPLC QqTOF MS for the determination of 148 pesticides in berries†, *J. Agric. Food Chem.*, **58**, 5904–5925.
61. Ferrer, I. and Thurman, E.M. (2012). Analysis of 100 pharmaceuticals and their degradates in water samples by liquid chromatography/quadrupole time-of-flight mass spectrometry, *J. Chromatogr. A.*, **1259**, 148–157.
62. Chiaia-Hernandez, A.C., Krauss, M. and Hollender, J. (2012). Screening of lake sediments for emerging contaminants by liquid chromatography atmospheric pressure photoionization and electrospray ionization coupled to high resolution mass spectrometry. *Environ. Sci. Technol.*, **47**, 976–986.
63. Krauss, M., Singer, H. and Hollender, J. (2010). LC–high resolution MS in environmental analysis: from target screening to the identification of unknowns, *Anal. Bioanal. Chem.*, **397**, 943–951.
64. Moschet, C., Piazzoli, A., Singer, H. *et al.* (2013). Alleviating the reference standard dilemma using a systematic exact mass suspect screening approach with liquid chromatography–high-resolution mass spectrometry, *Anal. Chem.*, **85**, 10312–10320.
65. Shi, F., Guo, C., Gong, L. *et al.* (2014). Application of a high resolution benchtop quadrupole-Orbitrap mass spectrometry for the rapid screening, confirmation and quantification of illegal adulterated phosphodiesterase-5 inhibitors in herbal medicines and dietary supplements, *J. Chromatogr. A*, in press. Available online, 17 December.2013.
66. Xia, Y.-Q., Lau, J., Olah, T. *et al.* (2011). Targeted quantitative bioanalysis in plasma using liquid chromatography/high-resolution accurate mass spectrometry: an evaluation of global selectivity as a function of mass resolving power and extraction window, with comparison of centroid and profile modes, *Rapid Comm. in Mass Spectrom.*, **25**, 2863–2878.
67. Rochat, B., Kottelat, E. and McMullen, J. (2012). The future key role of LC–high-resolution-MS analyses in clinical laboratories: a focus on quantification, *Bioanal.*, **4**, 2939–2958.

Part 3

Relevant LC-MS Applications in Food and Environmental Analysis

Chapter 10

Multiresidue Analysis of Pesticides: LC–MS/MS versus LC–HRMS

Juan V. Sancho and María Ibáñez

Research Institute for Pesticides and Water, University Jaume I, Spain

10.1. Introduction

Pesticide residue analysis (PRA) is one of the most widely studied topics in analytical laboratories working in the field of food safety, not only for the protection of human health but also to comply with regulatory controls. Moreover, due to the worldwide application of intensive agricultural methods during the last decades, the scientific community has shown increasing concern about the possible adverse effects associated with the presence of pesticides in the environment.

Due to the high number of pesticides that might be present in food or in different environmental compartments (more than 1,000 active substances are currently used around the world), the most useful approach is the application of multiresidue methods (MRMs) able to simultaneously determine (i.e. reliable identification and/or accurate quantification) as many compounds as possible in one analysis. However, the development of pesticide MRMs is a challenging task due to the very different physicochemical characteristics of pesticides, the low levels typically present in samples, the strict regulations, and the variety of matrices of different composition (food, water, soil, air etc.)

This fact forces the analytical chemist to develop highly sensitive and selective methods based on the use of powerful techniques. There has been a clear trend in the analytical methodology applied, from gas chromatography–mass spectrometry (GC–MS) with single quadrupole and liquid chromatography (LC) with UV, photodiode array, or fluorescence detectors in the 1970s and 1980s, towards GC–MS/MS and LC–MS/MS, which are the main techniques applied at present.

Over the last few years, LC–MS/MS with electrospray ionization (ESI) has been gaining importance as a consequence of its very high multiclass pesticide detection capability along with an increase in LC-amenable compounds new to the market compared to those amenable to GC.[1–2] With some exceptions (such as organochlorine pesticides or pyretroids), LC–MS/MS gives an appropriate analytical answer for pesticides determination at the residue level. The advantages of triple quadrupole (QqQ) working on selected reaction monitoring (SRM) mode (i.e. acquisition of MS/MS transitions) are widely recognized.[3–6] LC–MS/MS QqQ is the workhorse at PRA laboratories due to its robustness, excellent sensitivity, selectivity and confirmation capabilities. This analyzer has a notorious predominance over other MS analyzers, like ion trap or single quadrupole, in LC–MS methods.

Most methods developed until now include several tens of analytes. Although in the last years the number of analytes has considerably increased with the development of faster and more sensitive instruments, not many methods have been reported for more than 200 pesticides.[4–8] These LC–MS/MS methods are highly useful, and are developed, optimized and validated specifically for those target analytes included in the scope of the method (pre-target methods). Typically, they are focused on quantification and the information acquired is analyte-specific, which means that other pesticides that might be also present in the samples would be ignored in analyses, as only those transitions selected for the target analytes are acquired.

In the last decade, high-resolution (HR) mass spectrometers (such as time-of-flight (TOF) and Orbitrap) are gaining popularity for multiresidue analysis due to the continuous instrumental improvements and the decrease in price. The excellent features of HRMS come from its full-spectrum acquisition at accurate mass with satisfactory sensitivity, which enables the screening of the samples for a large number of pesticides; the data file of each sample can be examined for a theoretically unlimited number of pesticides (even without reference standards) and/or even reprocessed a posteriori for additional compounds not included in the initial analysis.[9–10]

Typically, extracted ion chromatograms (XICs) at selected analyte *m/z* values are reconstructed from full-scan accurate-mass data with reasonable sensitivity, using narrow *m/z* windows (e.g. ± 0.01 Da). Identification of the compounds detected is based on accurate masses of the (de)protonated molecule and of fragment ions when present, the compatibility of the suggested structures of fragment ions with the chemical structure of the compound detected, the isotopic distribution, and the retention time.

Tandem mass configurations (typically hybrid quadrupole TOF (QTOF) and linear ion trap quadrupole (LTQ)–Orbitrap) provide relevant additional information by obtaining product ion accurate-mass spectra. QTOF–MS/MS experiments are an excellent way to confirm potential positives revealed by, for example, TOF–MS or QqQ analysis, and are highly useful for elucidating the structures of (unknown) non-target compounds or in metabolism/degradation studies.[11]

The screening of a large number of pesticides in food/environment using HRMS will surely be one of the most relevant subjects of research in the next few years due to the benefits of having available as much information as possible on pesticides' presence in one single analysis.

This chapter presents the evolution and current state-of-the-art of both LC–MS/MS and LC–HRMS analysis for multiresidue pesticide analysis in food and environmental samples. On one hand, LC–MS/MS seems to be the most attractive and efficient technique nowadays for developing quantitative MRMs; on the other hand, the use of HRMS in combination with LC will be also treated, as this is one of the fields of major interest at present, especially in the qualitative field. So, the main advantages and drawbacks of each methodology will be briefly discussed.

10.2. LC–MS/MS (QqQ)

As commented on in Section 10.1, more than 1,000 pesticides and their metabolites, belonging to more than 100 chemical classes, are used during food production and may be present in the environment. Thus, fast and powerful analytical methods are needed to be able to detect very low concentrations of pesticides. Recent regulations and guidelines on food and environmental analysis applicable in Europe, USA and Japan require that these methods be based on MS detection.[12–14]

Despite the use of MS detection, sample preparation in different sample matrices normally involves extraction, clean-up and concentration steps before LC–MS analysis. Only clean samples, such as ground or spring water, are usually directly injected into the LC–MS system without pre-treatment.[15–16] The recent trends for multiresidue multiclass determination, thanks to the advanced LC–MS/MS systems, make the simultaneous extraction of a large number of pesticides with a broad range of polarity an extremely challenging task.

Solvent extraction in the case of solid samples, and solid-phase extraction (SPE) for liquid samples are the most widely used extractions methods in PRA.[17–23] Different solvents of varying polarity have been used for an effective extraction of pesticides, but the use of acetonitrile has recently gained popularity through the QuEChERS (quick, easy, cheap, effective, rugged and safe) method, which combines acetonitrile pesticide extraction with dispersive SPE clean-up (primary-secondary amine, PSA) in a convenient way.[24–27]

The most widely used LC separation technique for pesticides continues to be reversed-phase LC using C18 material on silica-based stationary phases. The mobile phase is chosen as a compromise between good chromatographic separation and overall MS performance. In this sense, methanol is preferred over acetonitrile due to its enhanced ionization efficiency in positive mode. Moreover, the benefits of using volatile additives (such as formic acid or ammonium formate, among others) in the mobile phase should be evaluated.

The most important development in LC during the last years is the expanding use of ultraperformance liquid chromatography (UPLC). UPLC uses columns packed with particles with diameters below 2 μm, which significantly increase efficiency even at high mobile phase flow rate, achieving faster separations.[28–31] More than a hundred pesticides can be easily separated by UPLC coupled to MS/MS in less than 10 minutes.[7,32]

The second development is the routine introduction of hydrophilic interaction liquid chromatography (HILIC) in PRA. In HILIC, the stationary phase is polar and the mobile phase tends to be a water-organic solvent gradient with increasing water content. The compounds are eluted in order of increasing polar character, and can be an effective alternative to ion-pairing chromatography in order to avoid ionization suppression in the source.[12]

In this sense, a critical aspect of LC–MS is the so-called 'matrix effect'. It appears as a consequence of the influence of co-extracted matrix components on analyte ionization in the atmospheric pressure ionization (API) interfaces.[33] Matrix effect typically results in an important loss of sensitivity due the ionization suppression from the co-extracted components, hampering the analyte quantification,[29] although, less frequently, enhancement of the analyte signal can also occur. This undesirable effect causes a loss in method accuracy, precision and sensitivity, leading to incorrect quantification and also to problems in the confirmation process. Matrix effect depends on each particular analyte-sample matrix combination, and also on the API source of the instrument.

Different approaches are used to remove or minimize matrix effects, like applying an efficient sample clean-up, the standard additions method or a simple sample extract dilution.[34] The first two approaches are laborious, time-consuming and can introduce analytical errors associated to sample manipulation. The dilution, on the other hand, is a fast and simple way to minimize matrix interferences, but only if the sensitivity is enough. Undoubtedly, however, the most widely applied strategies to compensate for matrix effect are the use of matrix-matched standards calibration or isotope-labeled internal standards (ILIS). Normally, the use of analyte-ILIS is the preferred way to solve matrix effects, but it is limited because of the commercial unavailability of reference standards and their high cost, making this approach difficult to apply in MRMs.

Regarding the ionization interfaces, as stated above, the API ones are the most widely used, despite the matrix effect observed. ESI, atmospheric pressure chemical ionization (APCI) and atmospheric pressure photoionization (APPI), in this order, are the most used, applied from very polar pesticides, ESI, to non-polar ones, APPI.

Methods based on QqQ analyzers are the most widely used for PRA in food and water samples due to the high sensitivity and selectivity, wide linear dynamic range, and high precision achieved by these systems.[35–37] The use of SRM acquisitions allows an excellent sensitivity together with reduced spectral interferences offering identification/confirmation of the detected pesticide by monitoring at least two SRM transitions per target pesticide.

Although most MS-based methods have satisfactory selectivity, some matrix components can share the same MS properties as the analyte, especially in complex-matrix samples. Confirmation of potential positives is a matter of concern due to the undesirable effects associated with erroneous identifications (i.e. reporting of either false positives or false negatives).[38] Reliable analytical methods that lead to accurate quantification of analytes are obviously necessary but also it is even more important that the compound detected is properly identified, especially when the maximum permitted levels are exceeded in the samples analyzed.[39]

Another relevant aspect is the capability to rapidly switch the ionization polarity during the chromatographic run. This fact allows the simultaneous analysis of pesticides from different chemical classes in a single run. Moreover, the short dwell times offered by modern triple quads allow an easy coupling to UPLC where very narrow chromatographic peaks are obtained.

However, the total number of SRM transitions that can be measured per run is usually limited by the time involved in monitoring a transition (dwell time), the chromatographic peak width, the number of points/peak required, and the level of analytes' co-elution. Therefore, the number of pesticides that can be satisfactorily analyzed in a single run usually is around 250–300.

Although the majority of methodologies proposed focus on development of multi-class MRMs, several deal with the determination of highly polar pesticides not amenable for these multi-analyte approaches.[40]

Another disadvantage of QqQ geometry is its restriction to pre-target analysis: a priori knowledge of the pesticides of interest is mandatory and reference standards must be used for deriving the analytes SRM transitions. This is an important drawback of this approach, as possible non-selected pesticides present in the samples are not actually detected. Additionally, if the SRM transition's specificity is not evaluated, false positives may be reported due to the presence of co-eluted compounds sharing common transitions.

These pitfalls of QqQ analyzers have driven the recent growing interest on the use of HRMS coupled to LC. LC–HRMS would allow a wide-scope screening for the presence of any ionizable pesticide in the samples. This topic is covered in depth in the following section.

Figure 10.1 shows, as a graphical summary, different aspects (matrix effect, confirmation, and UPLC separation) discussed when using QqQ mass analyzers for PRA. We can see the different SRM chromatograms for the insecticide dimethoate and its ILIS (dimethoate-d6) for a reference standard and different environmental water samples (groundwater (GW), effluent waste water (EWW), and urban solid waste leachate (LW)). As can be seen, slight ionization suppression is observed for GW and EWW, but a significant effect for LW, where only one quarter of the signal was obtained. In order to perform a correct quantification of the analyte, a matrix-matched calibration plot should be obtained if blank samples are available. A more robust solution would be to add ILIS to standards and samples to correct the matrix effect, if available. In this case, dimethoate-d6 was used and a satisfactory matrix effect correction was possible for all cases, even LW, as can be seen in Fig. 10.1.

Regarding the confirmation of the identity of the compound responsible for the peak at 4.47 min, apart from the quantification (Q) transition, two additional transitions (q_1 and q_2) were also acquired in order to check the identity through ion-ratio deviations. As can be seen, deviations observed for both transitions fall below the maximum tolerances acceptable (20%) in the applicable guidelines. Moreover, the benefits of using UPLC separations in this

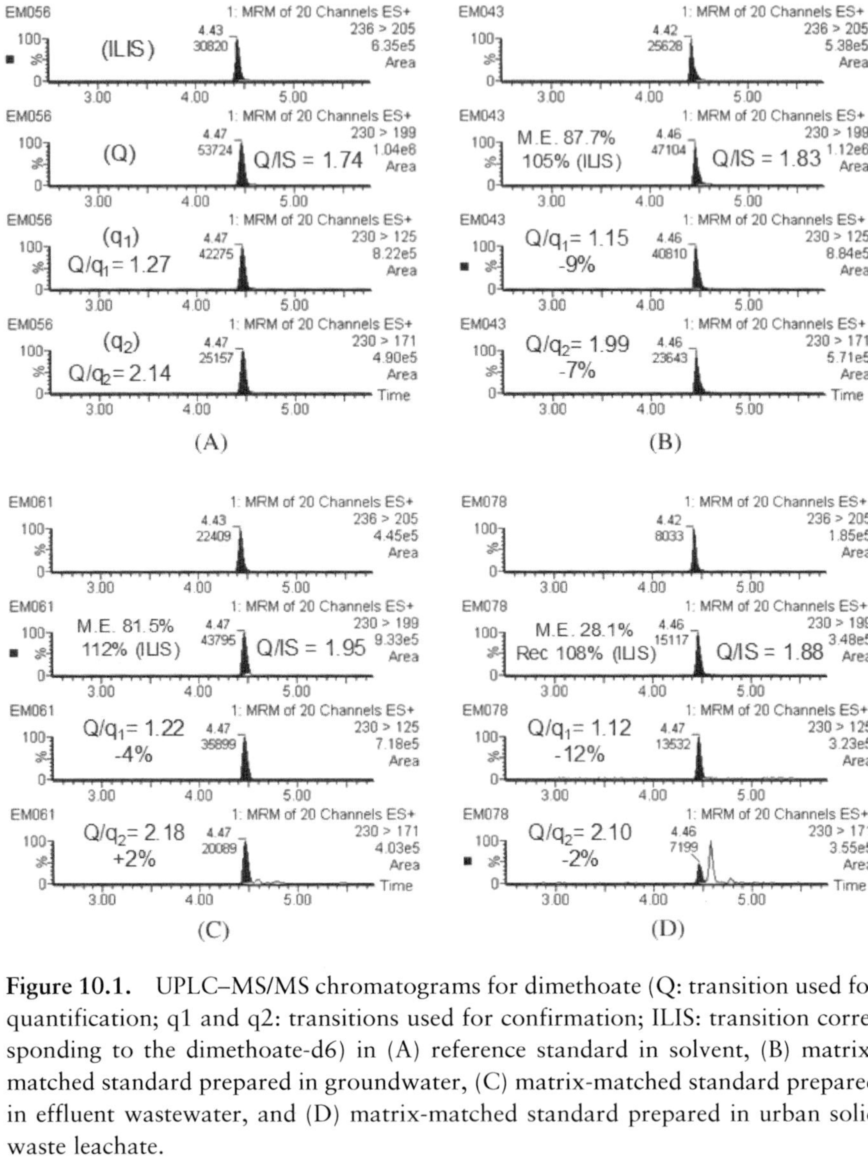

Figure 10.1. UPLC–MS/MS chromatograms for dimethoate (Q: transition used for quantification; q1 and q2: transitions used for confirmation; ILIS: transition corresponding to the dimethoate-d6) in (A) reference standard in solvent, (B) matrix-matched standard prepared in groundwater, (C) matrix-matched standard prepared in effluent wastewater, and (D) matrix-matched standard prepared in urban solid waste leachate.

confirmation step are noticed in the case of LW sample (Fig. 10.1D), where the q_2 transition is shared by other compounds present in the sample, which would have affected to the ion-ratio calculation, rendering a potential false negative.

10.3. High-Resolution Mass Spectrometry

Both water pollution monitoring and food quality control typically make use of target analysis (normally based on usage of the pesticides, their occurrence or their inclusion in priority lists), the scope of which rarely exceeds several tens of analytes, being quite unusual to find analytical methods applied to more than 150 pesticides. Nowadays, LC–MS/MS in SRM mode has become one of the most powerful techniques in (multi)residue pesticide analysis. However, working in MS/MS mode makes any MS system blind to all other compounds. Under these circumstances, only a limited number of analytes is recorded and therefore other potentially harmful, non-targeted analytes might be also present in the samples and would not be detected in a target analysis.

As an alternative to MS/MS instruments, HRMS analyzers provide the selectivity and sensitivity required for an efficient and wide-scope screening, as they combine high full-spectral sensitivity with both elevated mass accuracy and mass resolution allowing any ionizable component in the sample to be accurately mass-measured. The advantage of HRMS analyzers for screening comes from their ability to examine a data file for a theoretically unlimited number of pesticides and their metabolites.

In the last ten years a growing interest in HRMS, commanded by TOF and Orbitrap mass analyzers, has been observed, which has been more evident in LC–MS analysis. The increasing interest of using HRMS in environmental sciences relies on its capability to perform both targeted and untargeted analysis, based on its sensitivity in full scan acquisition mode and high mass accuracy. The LC–HRMS approach enables screening for hundreds (and even thousands) of compounds with high sensitivity within one run.

High-resolution mass spectrometry (HRMS) has gained wider acceptance in the last few years. This development is based on the availability of more rugged, sensitive and selective instrumentation. The benefits provided by HRMS (TOF and Orbitrap instruments) over classical unit-mass-resolution tandem mass spectrometry are considerable. These benefits include the collection of full-scan spectra, and consequently, the freedom for the analyst to measure compounds without previous compound-specific tuning; the possibility of retrospective data analysis; and the capability to perform structural elucidation of unknown or suspected compounds. HRMS strongly competes with classical tandem mass spectrometry in the field of quantitative multiresidue methods.[41]

With the same instrument one can perform pre- and post-target analysis, retrospective analysis, metabolite and transformation products discovering, and non-target analysis. All these possibilities are relevant to environmental sciences where the analyst has to face thousands of potential organic contaminants of quite different chemical composition. Thus, wide-scope screening of environmental samples is one of the main applications of HRMS.

In this section, a review of the present role of HRMS in multiresidue methods is made. Needless to say, it is not the intention of the authors to summarize all contributions of HRMS in this field, as is done in classical descriptive reviews, but rather to give an overview of the main characteristics of HRMS, its strong potential in environmental mass spectrometry and the trends observed over the last few years. Literature has been taken since 2000, coinciding with the beginning of HRMS in this field.

Interesting reviews/trends articles have been published in the last few years dealing with the current role of high-resolution mass spectrometry including TOF, QTOF and Orbitrap analyzers in food and water analysis; a reading of them is recommended for additional information on this subject.[1,11,12,38,41–51] The monograph 'Liquid chromatography time-of-flight mass spectrometry' also gives relevant information on the use and applications of this technique in different fields.[52]

Different tandem mass configurations have been reported in the literature, the most popular being QTOF and LTQ–Orbitrap. These hybrid tandem mass instruments provide relevant structural information by obtaining full product ion spectra at accurate mass. MS/MS experiments are an excellent way to confirm potential positives revealed by, for example, TOF-MS or QqQ analysis, and are highly useful for elucidating the structures of (unknown) non-target compounds. However, the pre-selection of the precursor ion is required for MS/MS product ion generation, and therefore a second injection is needed. In order to overcome these limitations, product ion spectra can be collected using data-dependent acquisition (DDA), in which the instrument switches automatically from MS to MS/MS acquisitions based on predefined selection criteria.

Another possibility is the fragmentation of compounds without a previous precursor ion selection. This implies the fragmentation of all ions in the collision cell (QTOF) or in the HCD cell (Orbitrap).[10,53–54] These experiments involve the simultaneous acquisition of accurate mass data at low and high collision energy. By applying low energy (LE), fragmentation is minimized and the information obtained corresponds normally to the parent molecule (adducts in some cases). However, at high collision energy (HE), fragmentation of the molecule is favored. So, both (de)protonated molecule data and fragment ion data are enabled in a single acquisition without the need of selecting the precursor ion.[10,53]

Throughout the years, HR mass analyzers have undergone several technical advances that have transformed these instruments into valuable tools for pesticide residue analysis. In the following section, we will review the use of LC–HRMS in multiresidue pesticide analysis, in both multiresidue target pesticide analysis and non-target analysis. Although it is out of the scope of this article, we also discuss briefly the different approaches to illustrate the extraordinary potential of HRMS analyzers for structural analysis, basically in identification/elucidation of pesticide metabolites/degradation products.

10.3.1. *Target screening of pesticides*

TOF instruments, which accurately measure mass-to-charge ratios on the basis of the mass-dependent flight time differences from the entrance of the analyzer to the detector, afforded fast full-spectral acquisition rates, full-spectral sensitivity, at high mass-resolution (10,000–40,000, depending on the instrument) with high mass accuracy (typically lower than 2 mDa) to pesticide analysis.

Therefore, XICs at selected exact masses can be performed from full-scan data with high sensitivity and selectivity using narrow mass windows (e.g. ±0.01 Da). Identification of the compounds detected is then based on accurate masses of the (de)protonated molecule and fragment ions when present, the isotopic distribution and the retention time. In contrast to GC, LC–TOF-MS with the widely used ESI source typically generates the (de)protonated molecule as the base peak of the spectrum. Monitoring this expected ion is usually the first step of the detection and identification strategy.

The first work dealing with LC–TOF-MS application in the environmental field, to our knowledge, was published in 1999. Hogenboom *et al.* explored the potential of LC–TOF-MS to routinely perform accurate mass determination of polar microcontaminants in surface water as well as to unambiguously identify pesticides in mixtures of isobaric and/or isomeric compounds.[55] Two years later, Maizels and Budde also showed the usefulness of LC–TOF-MS for studies in environmental samples, by analyzing 10 test pesticides in standard solutions.[56] Analyte confirmation was achieved by exact mass measurement, with errors of 0–5.4 ppm. In 2002, Thurman *et al.* reported the analysis of ethanesulfonic acid degradates of acetochlor and alachlor using LC–TOF-MS.[57]

Application of LC–TOF-MS has been successful in fruit and vegetables,[58–63] fruit-based baby food,[64] water,[65–66] olive oil and olives,[67] palm oil,[68–69] soybean,[70] fruit-based soft drinks,[71–73] wine[74–75] and milk.[76] In most of these applications, accuracy values (RSD) were usually better than 10%. In addition, the linearity — over two or three orders of magnitude, depending on the instrument used — approached that attainable by QqQ instruments.

The capability of (Q)TOF-MS for screening, quantification and/or confirmation of pesticides has been continuously compared to LC–MS/MS, typically with QqQ but also with ion-trap analyzers.[38,77–81] In these works, both systems were demonstrated to be complementary. On the one hand, LC–MS/MS proved to be the first choice for quantification of pre-target analysis due its good sensitivity and good repeatability. On the other, QTOF provided accurate mass measurements, being therefore ideal for post-target screening and confirmation. So, the most suitable strategy seemed to be an automated pesticide screening and identification by LC–(Q)TOF instruments, followed by quantification by LC–MS/MS.[38,82]

So, although historically the use of (Q)TOF instruments for quantification of target analytes in environmental samples has been quite limited, in the last few years, there has been an increased use of this method, now that the limitations of lower sensitivity and linear dynamic range have been largely solved. This has contributed to extending the usefulness of LC–(Q)TOF-MS for quantitation of target pesticides in water and food.[83–87]

But undoubtedly, it is in the qualitative field where HRMS can take full advantage of its excellent capabilities based on sensitive accurate-mass full-spectrum acquisition. This offers the possibility to investigate the presence of compounds once the sample analysis has been performed and the MS data acquired (i.e. without pre-selection of analytes), and is especially suited for screening/identification of contaminants and confirmation of presumptive positive samples reported by other techniques.

As a consequence of the intrinsic characteristics of HRMS analyzers, their use enables screening for a huge number of contaminants with high sensitivity within one run, with the obvious restrictions derived from the chromatographic and ionization processes in the LC–MS as well as from sample pre-treatment.

In 2007, we found the first large-scale multiresidue method for pesticides in food. García-Reyes *et al.* combined automated screening for 100 target pesticides by LC–TOF-MS in crops, and subsequent confirmation and quantitation by LC–MS/MS using a hybrid quadrupole linear ion trap (QqLIT) mass spectrometer.[88] Afterwards, different authors have dealt with the comprehensive multi-residue method for the accurate mass identification covering around 100 pesticides and their degradation products using LC–(Q)TOF-MS in water,[87] fruits and vegetables,[80,87,89–92] fruit- and vegetable-based infant foods[93] and fruit-based soft drinks.[94] In all cases, the identification of the targeted species was performed by retention time matching combined with accurate mass measurements of the targeted (de)protonated molecule, and their main fragment/product ions.

Very recently, LC–(Q)TOF instruments have started to be applied for wide-scope screening of large numbers of pollutants of different chemical families, including pesticides. Thus, now it is quite common to find articles dealing with the automated screening of hundreds of pesticides in water,[95] fruit and vegetables,[96–98] fish[99] or animal feed.[99,100]

The easy reviewing step and the relevant information obtained, such as accurate mass spectra of the peak, mass error for the protonated molecule and the most abundant fragments, and isotopic distribution, gives high confidence to the confirmation of potential positives even without reference standards being available. In these cases, the presence of the (de)protonated molecule measured at its accurate mass is evaluated in the samples. When a peak was detected, collision-induced dissociation (CID) fragments (or characteristic isotopic ions) are evaluated and justified taking into account the structure of the molecule.[53] As an example, Fig. 10.2 illustrates the detection and identification of the fungicide prochloraz in cucumber by UPLC–QTOF-MS. The protonated molecule of prochloraz was detected in the LE function (Fig. 10.2A, bottom), with a mass error of 1.7 mDa. Moreover, the combined spectrum of this chromatographic peak showed a typical three-chlorine-atoms isotopic pattern, being therefore in accordance with the chemical structure of prochloraz ($C_{15}H_{16}Cl_3N_3O_2$). As the reference standard was not available at our laboratory, the accurate mass of the fragment ions was justified using the MassFragment software (Waters). In order to avoid spectrum interferences that would complicate the identification process, recognizing which ions are fragments and which are not becomes mandatory. UPLC proved valuable for choosing perfectly co-eluting ions (see chromatographic peak at 12.17 min, Fig. 10.2B). Reliable elemental composition for the three fragments detected in the HE function (*m/z* 308.0027, 267.9516 and 70.0302) was calculated, obtaining errors below 2 mDa in relation to the theoretical exact masses predicted (Fig. 10.2A, top). The tentative identification of prochloraz was supported by the MS/MS product ions reported in the literature. After this careful evaluation process, the reference standard was finally acquired and injected, allowing the ultimate confirmation of this compound in the sample.

This makes it possible to enlarge drastically the number of contaminants included in the databases, even up to thousands.[53,99]

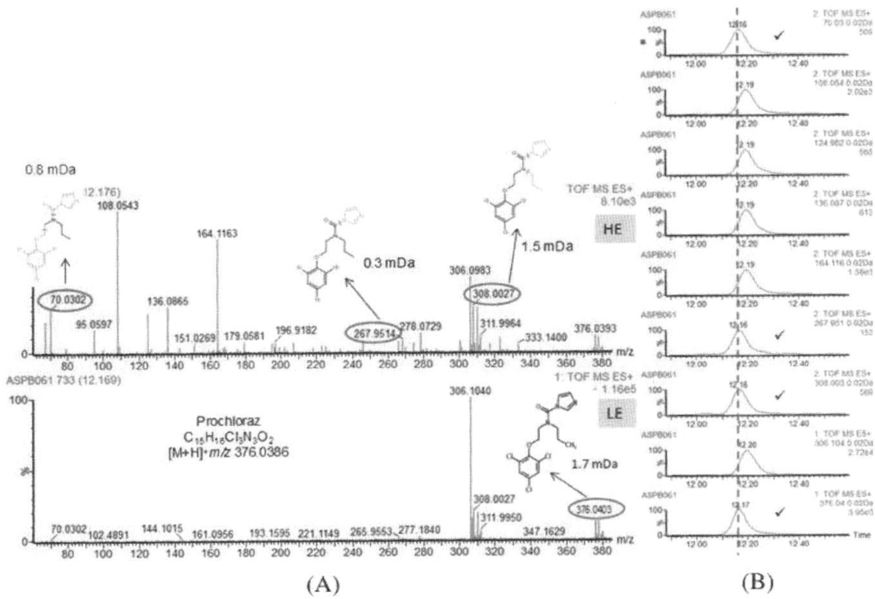

Figure 10.2. Detection and identification of the fungicide prochloraz in cucumber by MS^E (fragmentation of compounds in the collision cell, without previous precursor ion selection). (A) Spectra of the sample after applying collision energy of 4 eV (LE function, bottom) and ramp 15–40 eV (HE function, top). MassFragment justification of observed fragments using UPLC–QTOF-MS (B) XICs at 20 mDa mass window for $[M+H]^+$ in LE function and different ions observed in HE function. Those corresponding to $[M+H]^+$ as well as to main fragments are marked with ü.

This opens a new scenario in screening, favoring a more realistic overview when investigating organic contaminants in different applied fields.

A few years ago, a new type of mass analyzer was invented by Alexander Makarov.[101–104] Orbitrap operates by radially trapping ions about a central electrode and measures mass/charge values from the frequency of ion oscillations. This device shows high mass resolution (>100,000 FWHM), high mass accuracy (<5 ppm) and acceptable dynamic range (10^3). The main drawback is its scanning speed, inverse to mass resolution. One can only acquire one scan per second when selecting 100,000 resolution, affecting to the correct chromatographic peak shape when coupled to UHPLC where the peak widths are only a few seconds. When faster scanning is selected (10 scans/s), resolution decreases dramatically (e.g. 10,000 FWHM). Thus, a compromise between achievable resolution and adequate chromatography should be found.[105–106]

In 2009, Kellmann *et al.* published the first application of Orbitrap in multiresidue pesticide analysis, where different parameters affecting the accuracy of mass assignment (such as analyte concentration, complexity of the matrix and resolving power) were evaluated.[105] The results showed that for an accurate quantitation of analytes at low levels in complex animal feed matrices, a resolving power ≥50,000 FWHM was required. At lower resolving-power settings, the error in the mass increased due to the co-elution of analytes with isobaric interferences at very similar accurate-mass. In the case of the less complex matrix honey, a resolving power of 25,000 was generally sufficient for mass accuracy ≤2 ppm down to low concentration levels of 10 ng/g.

An example of the benefits of using 100,000 instead of 10,000 FWHM resolution is illustrated in Fig. 10.3, which shows the co-elution of two analytes imazalil and flunixin, differing by 30 mDa in their exact masses.[105] The mass spectra at three time points across the imazalil elution profile shows a mass accuracy better than 2 ppm for all measurements of the high resolving experiment, but mass deviations up to 95 ppm were encountered for the measurement at a resolving power set at 10,000. This is due to flunixin presence, which is partially coeluting with imazalil (dashed line), and could not be resolved at the 10,000 resolving power setting. A resolving power setting of 100,000 provides more than enough mass-resolution to detect both compounds, independently of each other, with correctly masses assigned.

Different authors have explored the Orbitrap characteristics, usually coupled to LC, for accurate mass-screening and identification of multi-class pesticides in fruits and vegetables,[107–111] bakery ingredients,[112] honey,[113] agricultural soils[114] and lake sediments,[115] but also by direct analysis in real time (DART).[116–118]

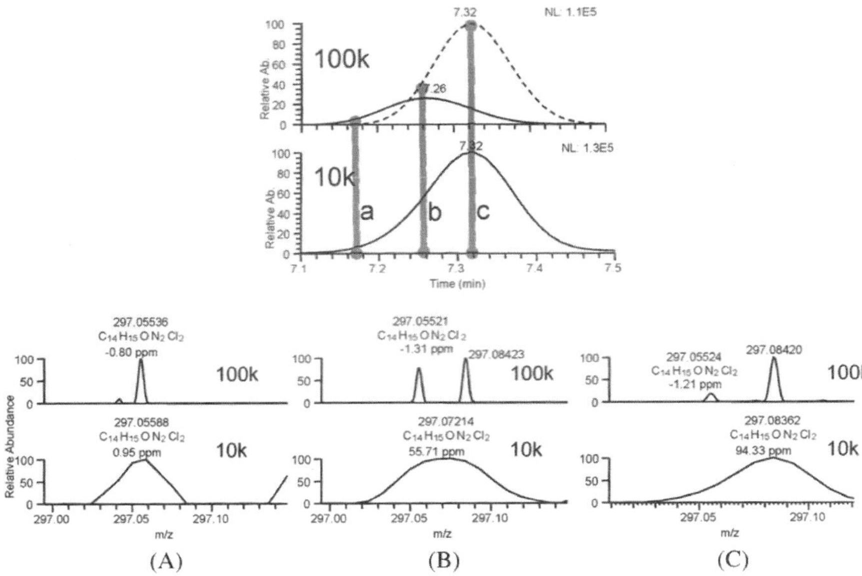

Figure 10.3. Effect of the resolving power on assigned mass accuracy of two co-eluting analytes, imazalil (MH^+= 297.05560, $C_{14}H_{14}Cl_2N_2O$, RT = 7.26 min) and flunixin (MH^+ = 297.08454, $C_{14}H_{11}F_3N_2O_2$, RT = 7.32 min). Upper figure: extracted ion chromatograms (±5 ppm at 100 k, and ±100 ppm at 10 k). Bottom figures: mass profiles at two resolving power settings 10,000 (10k) and 100,000 (100 k) for (A) RT = 7.17 min, (B) RT = 7.26 min and (C) RT = 7.32 min. Reprinted with permission from Kellmann, M., Muenster, H., Zomer, P. *et al.* (2009). Full scan MS in comprehensive qualitative and quantitative Residue analysis in food and feed matrices: how much resolving power is required?, *J. Am. Soc. Mass Spectrom.*, 20, 1464–1476. Copyright (2009) Springer.

Until now, there have been few published studies regarding the use of Orbitrap instruments for the quantitation of pharmaceuticals and pesticides in marine environment[119] and honey.[113] The performances of these analyzers have also been compared versus other MS instruments.[108,120–121]

10.3.2. *Non-target screening*

Full-spectrum acquisitions at accurate masses provided by HRMS also open the possibility to investigate non-target compounds in food and environmental samples. Different criteria are used in the bibliography for the definition of target and non-target analysis. Thus, in the field of PRA, 'non-target' has been used for non-expected compounds or for pesticides not included in routine monitoring protocols although specifically searched in analysis. However, other authors use the term 'non-target screening' when searching for unknowns, which in a strict sense starts without any information on the compounds to be detected. It must be noticed that the term 'unknown' does not mean that the compound discovered in the analysis is new or an unreported compound.[122] In a recent article, Krauss *et al.* perform an overview of the state-of-the-art and future trends of LC–high-resolution MS applied to the environmental analysis of polar micropollutants.[48] Figure 10.4 shows an illustrative systematic workflow for quantitative target analysis with reference standards, suspects screening without reference standards, and non-target screening of unknowns.

A genuine non-target analysis involves the automated component detection from the total ion chromatogram (TIC) and the mass spectra deconvolution for a subsequent comparison with mass spectral libraries. As is well known, commercial, standardized ESI mass spectra library are not available as ESI is not an ion source as stable and reproducible as electron ionization (EI). Instead, theoretical mass spectra libraries based on molecular formula databases can be built, which facilitates increasing the number of compounds that can be searched. Homemade empirical libraries can also be used, but these normally include much fewer compounds due to the need to inject standards. These experimental libraries offer fragmentation and retention time information as well, providing more confidence in the compound-identification process.[123] However, the possibility of detecting and identifying the sample contaminants, using both library approaches in a non-target analysis, depends on the success of the deconvolution process (i.e. the capability of the software to find the component peaks and to obtain mass spectra as free as possible of sample interferents). Obviously, the more complex the matrix, the more difficult the deconvolution will be.

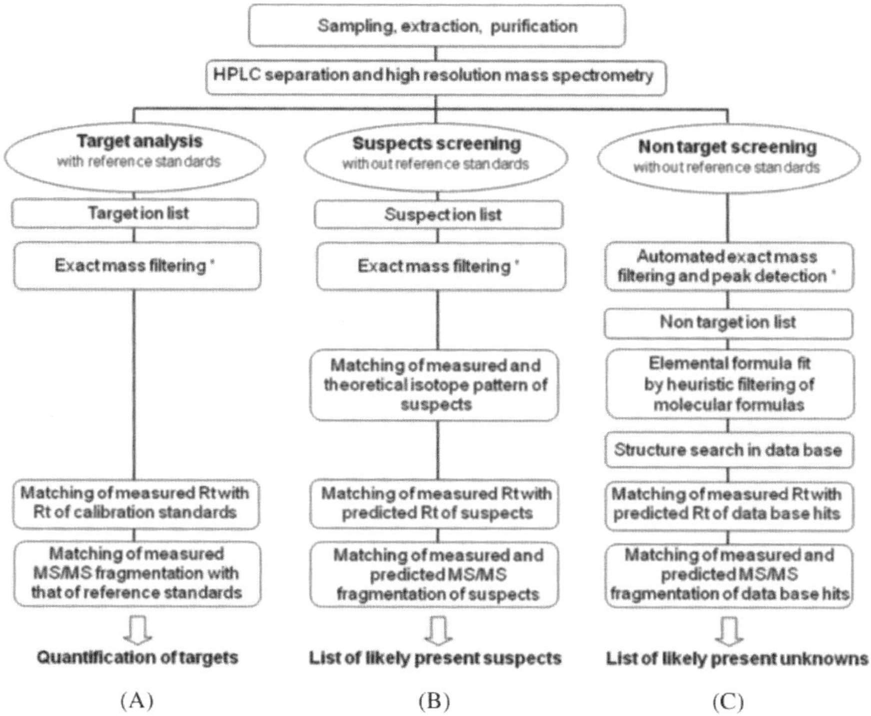

Figure 10.4. Comparison of systematic workflows for (A) quantitative target analysis with reference standards, (B) suspects screening without reference standards, and (C) non-target screening of unknowns in environmental samples by using LC–high-resolution (tandem) mass spectrometry. *Note that the m/z range of the extraction window for the exact mass filtering depends on the mass accuracy and the resolving power of the mass spectrometer used. Reprinted with permission from Krauss, M., Singer, H. and Hollender, J. (2010). LC–high-resolution MS in environmental analysis: from target screening to the identification of unknowns, *Anal. Bioanal. Chem.*, **397**, 943–951. Copyright (2010) Springer.

The first application of LC–QTOF-MS in the non-target screening dates from 2001. Bobeldijk *et al.* explored the possibilities of QTOF-MS for the identification of unknown compounds in water samples.[124] For this purpose, they developed a model based on known pesticides. Based on the accurate mass, the elemental compositions for the precursor and product ions were calculated, and these were searched against available databases as well as a CID library. Applying the developed procedure to a surface water extract, the structures of three unknown compounds were elucidated.

A non-target analysis would require a visual inspection of the TIC chromatogram to find potential components present in the sample. However, low abundant compounds might not be apparent by visual inspection. A genuine non-target analysis requires deconvolution software able to detect the components in the sample. In a previous work we explored the potential of LC–QTOF-MS in the elucidation of non-target unknown compounds in different water samples.[125] It was concluded that the success of this procedure depends in some way on the availability of compound databases where the search can be performed. Further restrictions came from the need to pre-select 'relevant' peaks normally based on peak intensity, as a compromise between time, invested effort, and information gathered has to be reached.

Therefore, rather than performing a general screening for unknowns, it seems reasonable to focus the identification only on 'relevant' compounds for man or environment, and not on all unknown peaks present in the sample.[126] This can be facilitated by making use of vast spectral libraries. Following this idea (a semi-automated 'molecular-feature' and database-searching consisting of the exact monoisotopic mass of 100 compounds), at least one exact mass in-source fragment for each compound and chromatographic retention time were used by Ferrer *et al.* to identify pesticides in food and water samples.[61]

Recently, UPLC–TOF-MS in combination with specialized software (ChromaLynx™) has been applied for non-target screening of environmental and waste water[123] and also in honeybee samples, after different poisoning cases.[127] After chromatographic peak deconvolution, the software discriminated ions coming from organic compounds present in the sample from background ions. In a second step, a library search was performed to match the experimental deconvoluted mass spectra with the existing entries in available libraries. Finally, the formula from the library hit was submitted to an elemental composition calculator and the most intense ions were either confirmed or rejected based on accurate mass criteria and isotopic pattern.

The main difficulty associated to this approach is the non-availability of LC–MS-standardized libraries as a consequence of the well-known differences in the ionization efficiency between the existing interfaces together with the variability in the results depending on the mobile phase composition or on the cone voltage applied. Under these circumstances, the building of home-made libraries to facilitate the searching is necessary. Obviously, the higher the number of compounds included in the library, the wider the possibilities to detect as many contaminants as possible in positive samples. Luckily, nowadays it is possible to build both empirical and/or theoretical library databases within the available softwares.[123] The advantages and limitations of using theoretical or empirical libraries have been already commented in previous works.[47,123]

10.3.3. *Elucidation of metabolites and degradation products*

Metabolites and breakdown products of pesticides may occur in many environmental compartments, in animal feed, or in food for human consumption. Nowadays, there is a growing interest in this subject because:

- some metabolites can be as hazardous as parent compounds;
- present regulations include relevant metabolites to be investigated;
- data show their high occurrence in environment.

So, before incorporating the most relevant metabolites into current analytical methods, it is necessary to secure their reliable identification.

In this sense, TOF mass spectrometry, especially QTOF-MS, plays an important role in the performance this task. Some interesting review/trend articles have been published that deal with the investigation of pesticide metabolites/degradation products in food and water by HRMS.[128–129] Different situations might be considered for discovering pesticide metabolites. One of the most extended approaches is looking for (un)expected metabolites in parent-positive samples. If metabolites are reported in the literature, it is a simple operation to plot a narrow-window XIC at its theoretical exact mass. However, other unreported metabolites might be present. To overcome this limitation, the manual selection of each abundant peak in the total ion chromatogram (TIC) and examination of its accurate mass spectrum has been used when investigating non-target chlorinated pesticides in citrus.[130–132] Another strategy, based on the use of 'fragmentation-degradation' relationships, has been proposed to avoid this manual search and inspection.[133] Thus, the accurate mass of ions of interest is used to establish relationships between fragmentation of the parent pesticides (in-source CID fragmentation) and possible degradation products.[132,134] A similar approach, based on the selection of one diagnostic ion characteristic, has been proposed by Ferrer and Thurman[87] as a potential tool to identify pesticides and metabolites from the same family. An extension of this approach has been applied, based not only on the fragmentation pathway of the parent compound but also on those identified metabolites present in the samples.[132,134]

The second approach consists of performing laboratory or field experiments under controlled conditions in order to investigate pesticide metabolism/degradation. In the recent literature, several examples can be found dealing with degradation of pesticides and elucidation of the products formed. LC–(Q)TOF-MS has been applied to study the photodegradation in water of triazine herbicides,[135] pesticides belonging to different chemical classes,[44] diazinon,[136–137] bentazone[138] and fenhexamid[139] or imazaquin in soil,[140] among others.

The methodology is based on comparison between blank and treated samples, using usually specialized software (e.g. MetaboLynx™), assuming that differential chromatographic peaks corresponded to potential metabolites. The molecular formula for each potential metabolite is then obtained from the accurate masses and the isotopic pattern observed in the mass spectrum, both necessary for differentiation between isobaric compounds. MS/MS experiments with accurate mass measurements were necessary to elucidate their structures. Other similar studies have been performed to characterize photodegradation products of alachlor,[141] carbofuran[142] and bentazone.[138]

In the case of vegetables, an interesting possibility is analysis of samples from field residue trials, where blank specimens can be collected.[132,143] A similar study for the organophosphorous pesticide fenthion[144] and amitraz[145] has been reported in fruit samples by using QTOF-MS. This strategy was also applied to the elucidation of metabolites of the insecticide coumaphos in honeybee samples, after a case of poisoning.[127]

Regarding Orbitrap instruments, a hybrid LTQ–Orbitrap used for the degradation of two triketone herbicides using four different advanced oxidation processes[146] has been recently investigated. The use of HRMS allowed a more comprehensive interpretation of the TPs generated.

10.4. Conclusions and Future Trends

Currently, QqQ is the first choice for routine quantitative pesticide residue analysis. Such systems allow target residue identification and quantification in complex matrices at extremely low levels. Nevertheless, they present limitations when high numbers of compounds (several hundreds) have to be analyzed in a single run.

HRMS systems provide high-resolution, accurate-mass and high full-scan sensitivity and selectivity, making them attractive for both target and non-target screening. Other advantages are the possibility of performing retrospective data examination or their usefulness in identifying unknown compounds.

An aspect that still must be improved for any LC–MS instrument is the heavy matrix effect suffered in the ionization source. New designs or ionization concepts that overcome this limitation, would be very welcome.

Regarding QqQ mass analyzers, the expected improvements should be related to higher ionization or transport yields, rendering even more sensitive instruments. In the case of HRMS, better sensitivity and linearity is expected in order to be a true alternative to QqQ in quantitative applications. Improvements in the scan speed for Orbitraps and mass resolving power for TOFs are also expected in order to become the golden standards in HRMS.

Finally, new hybrid analyzers incorporating ion mobility spectrometry (IMS), either before or between mass analyzers, are appearing in the market, adding an orthogonal separation mechanism to classical mass-to-charge ratio. IMS could improve identification of co-eluting isomers or confirmation in complex samples, allowing us to obtain extra identification points when following SANCO guidelines.

References

1. Gómez-Ramos, M.M., Ferrer, C., Malato, O. *et al.* (2013). Liquid chromatography–high-resolution mass spectrometry for pesticide residue analysis in fruit and vegetables: screening and quantitative studies, *J. Chromatogr. A*, **1287**, 24–37.
2. Alder, L., Greulich, K., Kempe, G. *et al.* (2006). Residue analysis of 500 high priority pesticides: better by GC–MS or LC–MS/MS?, *Mass Spectrom. Rev.*, **25**, 838–865.
3. Hernández, F., Pozo, O.J., Sancho, J.V. *et al.* (2006). Multiresidue liquid chromatography tandem mass spectrometry determination of 52 non-gas-chromatography-amenable pesticides and metabolites in different food commodities, *J. Chromatogr. A*, **1109**, 242–252.
4. Kmellár, B., Abrankó, L., Fodora, P. *et al.* (2010). Routine approach to qualitatively screening 300 pesticides and quantification of those frequently detected in fruit and vegetables using liquid chromatography tandem mass spectrometry (LC–MS/MS), *Food Addit. Contam. A*, **27**, 1415–1430.
5. Pihlström, T., Blomkvist, G., Friman, P. *et al.* (2007). Analysis of pesticide residues in fruit and vegetables with ethyl acetate extraction using gas and liquid chromatography with tandem mass spectrometric detection, *Anal. Bioanal. Chem.*, **389**, 1773–1789.
6. Pizzutti, I.R., de Kok, A., Hiemstra, M. *et al.* (2009). Method validation and comparison of acetonitrile and acetone extraction for the analysis of 169 pesticides in soya grain by liquid chromatography–tandem mass spectrometry, *J. Chromatogr. A*, **1216**, 4539–4552.
7. Lehotay, S.J., de Kok, A., Hiemstra, M. *et al.* (2005). Validation of a fast and easy method for the determination of residues from 229

pesticides in fruits and vegetables using gas and liquid chromatography and mass spectrometric detection, *J. AOAC Int.*, **90**, 595–614.

8. Hiemstra, M. and de Kok, A. (2007). Comprehensive multi-residue method for the target analysis of pesticides in crops using liquid chromatography-tandem mass spectrometry, *J. Chromatogr. A*, **1154**, 3–25.
9. Ibáñez, M., Guerrero, C., Sancho, J.V. *et al.* (2009). Screening of antibiotics in surface and wastewater samples by ultrahigh-pressure liquid chromatography coupled to hybrid quadrupole time-of-flight mass spectrometry, *J. Chromatogr. A*, **1216**, 2529–2539.
10. Hérnández, F., Bijlsma, L., Sancho, J.V. *et al.* (2011). Rapid wide-scope screening of drugs of abuse, prescription drugs with potential for abuse and their metabolites in influent and effluent urban wastewater by ultrahigh-pressure liquid chromatography–quadrupole–time-of-flight mass spectrometry, *Anal. Chim. Acta*, **684**, 87–97.
11. Sancho, J.V., Pozo, Ó.J., Ibáñez, M. *et al.* (2006). Potential of liquid chromatography/time-of-flight mass spectrometry for the determination of pesticides and transformation products in water, *Anal. Bioanal. Chem.*, **386**, 987–997.
12. Botitsi, H.V., Garbis, S.D., Economu, A. *et al.* (2011). Current mass spectrometry strategies for the analysis of pesticides and their metabolites in food and water matrices, *Mass Spectrom. Rev.*, **30**, 907–939.
13. Picó, Y., Font, G., Ruiz, MJ. *et al.* (2006). Control of pesticide residues by liquid chromatography–mass spectrometry to ensure food safety, *Mass Spectrom. Rev.*, **25**, 917–960.
14. Hernández, F., Sancho, J.V. and Pozo, O.J. (2005). Critical review of the application of liquid chromatography–mass spectrometry to the determination of pesticide residues in biological samples, *Anal. Bioanal. Chem.*, **382**, 934–946.
15. Pareja, L., Martínez-Bueno, M.J., Cesio, V., *et al.* (2011). Trace analysis of pesticides in paddy field water by direct injection using liquid chromatography–quadrupole-linear ion trap-mass spectrometry, *J. Chromatogr. A*, **1218**, 4790–4798.
16. Pozo, O.J., Pitarch, E., Sancho, JV. *et al.* (2001). Determination of the herbicide 4-chloro-2-methylphenoxyacetic acid and its main metabolite, 4-chloro-2-methylphenol in water and soil by liquid

chromatography–electrospray tandem mass spectrometry, *J. Chromatogr. A*, **923**, 75–85.

17. Shinomiya, M. (2013). Determination of pesticides originating from golf courses by solid-phase extraction and liquid chromatography–tandem mass spectrometry, *Int. J. Environ. Anal. Chem.*, **93**, 858–871.
18. Loos, R., Tavazzi, S., Paracchini, B. *et al.* (2013). Analysis of polar organic contaminants in surface water of the northern Adriatic Sea by solid-phase extraction followed by ultrahigh-pressure liquid chromatography–QTRAP® MS using a hybrid triple-quadrupole linear ion trap instrument, *Anal. Bioanal. Chem.*, **405**, 5875–5885.
19. Taylor, M.J., Melton, L.M., Sharp, E.A. *et al.* (2013). A liquid chromatography–electrospray tandem mass spectrometry method for the determination of multiple pesticide residues involved in suspected poisoning of non-target vertebrate wildlife, livestock and pets, *Anal. Methods*, **5**, 248–259.
20. Botero-Coy, A.M., Marín, J.M., Ibáñez, M. *et al.* (2012). Multi-residue determination of pesticides in tropical fruits using liquid chromatography–tandem mass spectrometry, *Anal. Bioanal. Chem.*, **402**, 2287–2300.
21. Cazorla-Reyes, R., Fernández-Moreno, J.L., Romero-González, R. *et al.* (2011). Single solid phase extraction method for the simultaneous analysis of polar and non-polar pesticides in urine samples by gas chromatography and ultrahigh-pressure liquid chromatography coupled to tandem mass spectrometry, *Talanta*, **85**, 183–196.
22. Cahill, M.G., Capriolo, G., Stack, M. *et al.* (2011). Semi-automated liquid chromatography–mass spectrometry (LC–MS/MS) method for basic pesticides in wastewater effluents, *Anal. Bioanal. Chem.*, **400**, 587–594.
23. Aguilera-Ruiz, M.M., Plaza-Bolaños, P., Romero-González, R. *et al.* (2011). Comparison of the efficiency of different extraction methods for the simultaneous determination of mycotoxins and pesticides in milk samples by ultrahigh-performance liquid chromatography–tandem mass spectrometry, *Anal. Bioanal. Chem.*, **399**, 2863–2875.
24. Liu, S.Y., Jin, Q., Huang, X.H. *et al.* (2013). Determination of residues of seven carbamate pesticides in milk using ultraperformance liquid chromatography–tandem mass spectrometry and QuEChERS methods, *J. AOAC Int.*, **96**, 657–662.

25. Rajski, L., Lozano, A., Uclés, A. *et al.* (2013). Determination of pesticide residues in high oil vegetal commodities by using various multi-residue methods and clean-ups followed by liquid chromatography tandem mass spectrometry, *J. Chromatogr. A*, **1304**, 109–120.
26. Chamkasem, N., Ollis, L.W., Harmon, T. *et al.* (2013). Analysis of 136 pesticides in avocado using a modified QuEChERS method with LC–MS/MS and GC–MS/MS, *J. Agric. Food Chem.*, **61**, 2315–2329.
27. Núñez, O., Gallart-Ayala, H., Ferrer, I. *et al.* (2012). Strategies for the multi-residue analysis of 100 pesticides by liquid chromatography–triple quadrupole mass spectrometry, *J. Chromatogr. A*, **1249**, 164–180.
28. Benvenuto, F., Marín, J.M., Sancho, J.V. *et al.* (2010). Simultaneous determination of triazines and their main transformation products in surface and urban wastewater by ultrahigh-pressure liquid chromatography–tandem mass spectrometry, *Anal. Bioanal. Chem.*, **397**, 2791–2805.
29. Marín, J.M., Gracia-Lor, E., Sancho, J.V. *et al.* (2009). Application of ultrahigh-pressure liquid chromatography–tandem mass spectrometry to the determination of multi-class pesticides in environmental and wastewater samples. Study of matrix effects, *J. Chromatogr. A*, **1216**, 1410–1420.
30. Garrido-Frenich, A., Martínez-Vidal, J.L., Pastor-Montoro, E. *et al.* (2008). High-throughput determination of pesticide residues in food commodities by use of ultraperformance liquid chromatography–tandem mass spectrometry, *Anal. Bioanal. Chem.*, **390**, 947–959.
31. Mezcua, M., Agüera, A., Lliberia, J.L. *et al.* (2006). Application of ultraperformance liquid chromatography–tandem mass spectrometry to the analysis of priority pesticides in groundwater, *J. Chromatogr. A*, **1109**, 222–227.
32. Mastovska, K., Dorweiler, K.J., Lehotay, S.J. *et al.* (2010). Pesticide multiresidue analysis in cereal grains using modified QuEChERS method combined with automated direct sample introduction GC–TOFMS and UPLC–MS/MS techniques, *J. Agric. Food Chem.*, **58**, 5959–5972.
33. Dijkman, E., Mooibroek, D., Hoogerbrugge, R. *et al.* (2001). Study of matrix effects on the direct trace analysis of acidic pesticides in water using various liquid chromatographic modes coupled to tandem mass spectrometric detection, *J. Chromatogr. A*, **926**, 113–125.

34. Sancho, J.V., Pozo, O.J., López, F.J. *et al.* (2002). Different quantitation approaches for xenobiotics in human urine samples by liquid chromatography–electrospray tandem mass spectrometry, *Rapid Commun. Mass Spectrom.*, **16**, 639–645.
35. Masiá, A., Ibáñez, M., Blasco, *et al.* (2013). Combined use of liquid chromatography–triple quadrupole–mass spectrometry and liquid chromatography–quadrupole–time-of-flight mass spectrometry in systematic screening of pesticides and other contaminants in water samples, *Anal. Chim. Acta*, **761**, 117–127.
36. Dagnac, T., García-Chao, M., Fernández-Alvarez, M. *et al.* (2013). Study of the presence of priority pesticides in surface water of river basins located in two areas of intensive dairy farming in the NW Spain (Galicia), *Int. J. Environ. Anal. Chem.*, **92**, 995–1011.
37. Fenoll, J., Hellín, P., Martínez, C.M. *et al.* (2011). Determination of 48 pesticides and their main metabolites in water samples by employing sonication and liquid chromatography–tandem mass spectrometry, *Talanta*, **85**, 975–982.
38. Hernández, F., Pozo, O.J., Sancho, J.V. *et al.* (2005). Strategies for quantification and confirmation of multi-class polar pesticides and transformation products in water by LC–MS2 using triple quadrupole and hybrid quadrupole time-of-flight analyzers, *Trends Anal. Chem.*, **24**, 596–612.
39. Pozo, O.J., Sancho, J.V., Ibáñez, M. *et al.* (2006). Confirmation of organic micropollutants detected in environmental samples by liquid chromatography–tandem mass spectrometry: achievements and pitfalls, *Trends Anal. Chem.*, **25**, 1030–1042.
40. Botero-Coy, A.M., Ibáñez, M., Sancho, J.V. *et al.* (2013). Improvements in the analytical methodology for the residue determination of the herbicide glyphosate in soils by liquid chromatography coupled to mass spectrometry, *J. Chromatogr. A*, **1292**, 132–141.
41. Kaufmann A. (2012). The current role of high-resolution mass spectrometry in food analysis, *Anal. Bioanal. Chem.*, **403**, 1233–1249.
42. Hernández, F., Sancho, J.V., Ibáñez, M. *et al.* (2012). Current use of high-resolution mass spectrometry in the environmental sciences, *Anal. Bioanal. Chem.*, **403**, 1251–1264.
43. García-Reyes, J.F., Hernando, M.D., Molina-Díaz A. *et al.* (2007). Comprehensive screening of target, non-target and unknown pesticides in food by LC–TOF-MS, *Trends Anal. Chem.*, **26**, 828–841.

44. Hernández, F., Ibáñez, M., Pozo, O.J. *et al.* (2008). Investigating the presence of pesticide transformation products in water by using liquid chromatography–mass spectrometry with different mass analyzers, *J. Mass Spectrom.*, **43**, 173–184.
45. Lacorte S. and Fernández-Alba, A.R. (2006). Time-of-flight mass spectrometry applied to the liquid chromatographic analysis of pesticides in water and food, *Mass Spectrom. Rev.*, **25**, 866–880.
46. Martos, P. and McCormick, H. (2012). Selected applications of accurate mass determinations using LC–TOF for toxic compounds in products of animal origin, *Compr. Anal Chem.*, **58**, 111–167.
47. Ibáñez, M., Sancho, J.V., Hernández, F. *et al.* (2008). Rapid non-target screening of organic pollutants in water by ultraperformance liquid chromatography coupled to time-of-light mass spectrometry, *Trends Anal. Chem.*, **27**, 481–489.
48. Krauss, M., Singer, H. and Hollender, J. (2010). LC–high-resolution MS in environmental analysis: from target screening to the identification of unknowns, *Anal. Bioanal. Chem.*, **397**, 943–951.
49. Petrovic, M. and Barceló, D. (2006). Application of liquid chromatography–quadrupole time-of-flight mass spectrometry (LC–QqTOF-MS) in the environmental analysis, *J. Mass Spectrom.*, **41**, 1259–1267.
50. Croley, T.R., White, K.D., Callahan, J.H. *et al.* (2012). The chromatographic role in high-resolution mass spectrometry for non-targeted analysis, *J. Am. Soc. Mass Spectrom.*, **23**, 1569–1578.
51. Soler, C., Mañes J. and Picó, Y. (2008). The role of the liquid chromatography–mass spectrometry in pesticide residue determination in food, *Crit. Rev. Anal. Chem.*, **38**, 93–117.
52. Ferrer, I. and Thurman, E.M. (2009). *Liquid Chromatography Time-of-Flight Mass Spectrometry. Principles, Tools and Applications for Accurate Mass Analysis*, John Wiley & Sons, Inc., Volume 173.
53. Díaz, R., Ibáñez, M., Sancho, J.V. *et al.* (2012). Target and non-target screening strategies for organic contaminants, residues and illicit substances in food, environmental and human biological samples by UHPLC–QTOF–MS, *Anal. Methods.*, **4**, 196–209.
54. Plumb, R.S., Johnson, K.A., Rainville, P. *et al.* (2006). High-temperature ultraperformance liquid chromatography coupled to hybrid quadrupole time-of-flight mass spectrometry applied to ibuprofen metabolites in human urine, *Rapid Commun. Mass Spectrom.*, **20**, 1989–1994.

55. Hogenboom, A.C., Niessen, W.M.A., Little, D. *et al.* (1999). Accurate mass determinations for the confirmation and identification of organic microcontaminants in surface water using on-line solid-phase extraction liquid chromatography electrospray orthogonal-acceleration time-of-flight mass spectrometry, *Rapid Commun. Mass Spectrom.*, **13**, 125–133.
56. Maizels, M. and Budde, W.L. (2001). Exact mass measurements for confirmation of pesticides and herbicides determined by liquid chromatography–time-of-flight mass spectrometry, *Anal. Chem.*, **73**, 5436–5440.
57. Thurman, E.M., Ferrer, I. and Parry, R. (2002). Accurate mass analysis of ethanesulfonic acid degradates of acetochlor and alachlor using high-performance liquid chromatography and time-of-flight mass spectrometry, *J. Chromatogr. A*, **957**, 3–9.
58. Ferrer, I., Thurman, E.M. and Fernández-Alba, A.R. (2005). Quantitation and accurate mass analysis of pesticides in vegetables by LC–TOF-MS, *Anal Chem.*, **77**, 2818–2825.
59. Ferrer, I., García-Reyes, J.F., Mezcua, M. *et al.* (2005). Multi-residue pesticide analysis in fruits and vegetables by liquid chromatography–time-of-flight mass spectrometry, *J. Chromatogr. A*, **1082**, 81–90.
60. Ferrer, I., García-Reyes, J.F. and Fernández-Alba, A.R. (2005). Identification and quantitation of pesticides in vegetables by liquid chromatography–time-of-flight mass spectrometry, *Trends Anal. Chem.*, **24**, 671–682.
61. Ferrer, I., Fernández-Alba, A., Zweigenbaum, J.A. *et al.* (2006). Screening and confirmation of 100 pesticides in food samples by liquid chromatography–tandem mass spectrometry, *Rapid Commun. Mass Spectrom.*, **20**, 3659–3668.
62. Soler, C., Hamilton, B., Furey, A. *et al.* (2007). Liquid chromatography–quadrupole–time-of-flight mass spectrometry analysis of carbosulfan, carbofuran, 3-hydroxycarbofuran, and other metabolites in food, *Anal. Chem.*, **79**, 1492–1501.
63. Anagnostopoulos, C.J., Liapis, K., Haroutounian, S. *et al.* (2013). Simultaneous determination of different classes of plant growth regulator in high water content agricultural products by liquid chromatography–tandem mass spectrometry and time-of-flight mass spectrometry, *J. Liq. Cromatogr. R. T.*, **36**, 315–335.
64. Gilbert-López, B., García-Reyes, J.F., Ortega-Barrales, P. *et al.* (2007). Analyses of pesticide residues in fruit-based baby food by liquid

chromatography–electrospray ionization–time-of-flight mass spectrometry, *Rapid Commun. Mass Spectrom.*, **21**, 2059–2071.

65. Holm, A., Molander, P., Lundanes, E. *et al.* (2003). Determination of rotenone in river water utilizing packed capillary column switching liquid chromatography with UV and time-of-flight mass spectrometric detection, *J. Chromatogr. A*, **983**, 43–50.
66. Sasaki, H., Yonekubo, J. and Hayakawa, K. (2006). A new on-line sample preparation system for the liquid chromatography–time-of-flight mass spectrometry simultaneous analysis of pesticides in river water, *Anal. Sciences*, **22**, 835–840.
67. García-Reyes, J.F., Ferrer, C.M., Thurman, E.M. *et al.* (2006). Analysis of herbicides in olive oil by liquid chromatography–time-of-flight mass spectrometry, *J. Agr. Food Chem.*, **54**, 6493–6500.
68. Sobhanzadeh, E., Kabar, N.K.A., Abas, M.R.B. *et al.* (2011). Low temperature followed by matrix solid-phase dispersion-sonication procedure for the determination of multiclass pesticides in palm oil using LC–TOF-MS, *J. Haz. Mat.*, **186**, 1308–1313.
69. Sobhanzadeh, E., Bakar, N.K.A., Abas, M.R.B. *et al.* (2012). A simple and efficient multi-residue method based on QuEChERS for pesticides determination in palm oil by liquid chromatography–time-of-flight mass spectrometry, *Environ. Monit. Assess.*, **184**, 5821–5828.
70. Xinyi, C., Xiwu, G., Xiaogang, C. *et al.* (2007). Application of liquid chromatography–time-of-flight-mass spectrometry for the analysis of urea herbicides residues in soybean, *Anal. Letters*, **40**, 1117–1130.
71. Gilbert-López, B., García-Reyes, J.F., Mezcua, M. *et al.* (2007). Determination of postharvest fungicides in fruit juices by solid-phase extraction followed by liquid chromatography–electrospray–time-of-flight mass spectrometry, *J. Agric. Food Chem.*, **55**, 10548–10556.
72. Gilbert-López, B., Jaén-Martos, L., García-Reyes, J.F. *et al.* (2012). Study on the occurrence of pesticide residues in fruit-based soft drinks from the EU market and Morocco using liquid chromatography–mass spectrometry, *Food Control*, **26**, 341–346.
73. Gilbert-López, B., García-Reyes, J.F., Mezcua, *et al.* (2010). Multiresidue determination of pesticides in fruit-based soft drinks by fast liquid chromatography–time-of-flight mass spectrometry, *Talanta*, **81**, 1310–1321.

74. Fontana, A.R., Rodríguez, I., Ramil, M. *et al.* (2011). Liquid chromatography–time-of-flight mass spectrometry following sorptive microextraction for the determination of fungicide residues in wine, *Anal Bioanal. Chem.*, **401**, 767–775.
75. Pérez-Ortega, P., Gilbert-López, B., García-Reyes, J.F. *et al.* (2012). Generic sample treatment method for simultaneous determination of multiclass pesticides and mycotoxins in wines by liquid chromatography–mass spectrometry, *J. Chromatogr.*, **1249**, 32–40.
76. Gao, F., Zhao, Y., Shao, B. *et al.* (2012). Determination of residues of pesticides and veterinary drugs in milk by ultraperformance liquid chromatography coupled with quadrupole–time-of-flight mass spectrometry, *Chinese J. Chromatogr.*, **30**, 560–567.
77. Hernández, F., Ibáñez, M., Sancho, J.V. *et al.* (2004). Comparison of different mass spectrometric techniques combined with liquid chromatography for confirmation of pesticides in environmental water based on the use of identification points, *Anal. Chem.*, **76**, 4349–4357.
78. Grimalt, S., Sancho, J.V., Pozo, O.J. *et al.* (2010). Quantification, confirmation and screening capability of UHPLC coupled to triple quadrupole- and hybrid quadrupole–time-of-flight mass spectrometry in pesticide residue analysis, *J. Mass Spectrom.*, **45**, 421–436.
79. Wang J., Chow, W. and Leung, D. (2010). Applications of LC–ESI–MS/MS and UHPLC–Qq–TOF-MS for the determination of 148 pesticides in fruits and vegetables, *Anal. Bioanal. Chem.*, **396**, 1513–1538.
80. Wang, J., Leung, D. and Chow, W. (2010). Applications of LC–ESI–MS/MS and UHPLC–Qq–TOF-MS for the determination of 148 pesticides in berries, *J. Agric. Food Chem.*, **58**, 5904–5925.
81. Picó, Y., Blasco, C., Farré, M. *et al.* (2009). Analytical utility of quadrupole time-of-flight mass spectrometry for the determination of pesticide residues in comparison with an optimized column high-performance liquid chromatography–tandem mass spectrometry method, *J. AOAC Int.*, **82**, 734–744.
82. Fernández-Alba, A.R. and García-Reyes, J.F. (2008). Large-scale multiresidue methods for pesticides and their degradation products in food by advanced LC–MS, *Trends Anal. Chem.*, **27**, 973–990.

83. Nurmi, J. and Pellinen, J. (2011). Multiresidue method for the analysis of emerging contaminants in wastewater by ultraperformance liquid chromatography–time-of-flight mass spectrometry, *J. Chromatogr. A*, **1218**, 6712– 6719.
84. Wille, K., Vanden Bussche, J., Noppe, H. *et al.* (2010). A validated analytical method for the determination of perfluorinated compounds in surface-, sea- and sewage-water using liquid chromatography coupled to time-of-flight mass spectrometry, *J. Chromatogr. A*, **1217**, 6616–6622.
85. Lara-Martín, P.A., González-Mazo, E. and Brownawell. B.J. (2011). Multi-residue method for the analysis of synthetic surfactants and their degradation metabolites in aquatic systems by liquid chromatography–time-of-flight mass spectrometry, *J. Chromatogr. A*, **1218**, 4799–4807.
86. Farré, M., Gros, M., Hernández, B. *et al.* (2008). Analysis of biologically active compounds in water by ultraperformance liquid chromatography–quadrupole–time-of-flight mass spectrometry, *Rapid Commun. Mass Spectrom.*, **22**, 41–51.
87. Ferrer, I. and Thurman, E.M. (2007). Multi-residue method for the analysis of 101 pesticides and their degradates in food and water samples by liquid chromatography–time-of-flight mass spectrometry, *J. Chromatogr. A*, **1175**, 24–37.
88. García-Reyes, J.F., Hernando, M.D., Ferrer, C. *et al.* (2007). Large-scale pesticide multiresidue methods in food combining liquid chromatography–time-of-flight mass spectrometry and tandem mass spectrometry, *Anal. Chem.*, **79**, 7308–7323.
89. Malato, O., Lozano, A., Mezcua, M. *et al.* (2011). Benefits and pitfalls of the application of screening methods for the analysis of pesticide residues in fruits and vegetables, *J. Chromatogr. A*, **1218**, 7615–7626.
90. Akiyama, Y., Matsuoka, T. and Mitsuhashi, T. (2009). Multiresidue screening method of acidic pesticides in agricultural products by liquid chromatography–time-of-flight mass spectrometry, *J. Pest. Sci.*, **34**, 265–272.
91. Lacina, O., Urbanova, J., Poustka, J. *et al.* (2010). Identification/quantification of multiple pesticide residues in food plants by ultrahigh-performance liquid chromatography–time-of-flight mass spectrometry, *J. Chromatogr. A*, **1217**, 648–659.

92. Taylor, M.J., Keenan, G.A., Reid, K.B. *et al.* (2008). The utility of ultraperformance liquid chromatography–electrospray ionization–time-of-flight mass spectrometry for multi-residue determination of pesticides in strawberry, *Rapid Commun. Mass Spectrom.*, **22**, 2731–2746.
93. Wang, J. and Leung, D. (2009). Applications of ultraperformance liquid chromatography–electrospray ionization–quadrupole–time-of-flight mass spectrometry on analysis of 138 pesticides in fruit- and vegetable-based infant foods, *J Agric. Food Chem.*, **57**, 2162–2173.
94. García-Reyes, J.F., Gilbert-López, B., Molina-Díaz, A. *et al.* (2008). Determination of pesticide residues in fruit-based soft drinks, *Anal. Chem.*, **80**, 8966–8974.
95. Gómez, M.J., Gómez-Ramos, M.M., Malato, O. *et al.* (2010). Rapid automated screening, identification and quantification of organic micro-contaminants and their main transformation products in wastewater and river waters using liquid chromatography–quadrupole–time-of-flight mass spectrometry with an accurate-mass database, *J. Chromatogr. A*, **1217**, 7038–7054.
96. Mezcua, M., Malato, O., García-Reyes, J.F. *et al.* (2009). Accurate-mass databases for comprehensive screening of pesticide residues in food by fast liquid chromatography–time-of-flight mass spectrometry, *Anal. Chem.*, **81**, 913–929.
97. Zhao, Z., Shi Zhihong, S., Kang, J. *et al.* (2013). Rapid screening and confirmation of 281 pesticide residues in apples, tomatos and cabbages by liquid chromatography–quadrupole–time-of-flight mass spectrometry, *Chinese J. Chromatogr.*, **31**, 372–379.
98. Mezcua, M., Malato, O., Martínez-Uroz, M.A. *et al.* (2011). Evaluation of relevant time-of-flight MS parameters used in HPLC–MS full-scan screening methods for pesticide residues, *J. AOAC Int.*, **94**, 1674–1684.
99. Nácher-Mestre, J., Ibáñez, M., Serrano, R. *et al.* (2013). Qualitative screening of undesirable compounds from feeds to fish by liquid chromatography coupled to mass spectrometry *J. Agric. Food Chem.*, **61**, 2077–2087.
100. Aguilera-Ruiz, M.M., Romero-González, R., Plaza-Bolaños, P. *et al.* (2013). Wide-scope analysis of veterinary drug and pesticide residues

in animal feed by liquid chromatography coupled to quadrupole–time-of-flight mass spectrometry *Anal. Bioanal. Chem.*, **405**, 6543–6553.

101. Hardman, M., Mol, H.J. and Makarov, A. (2003). Interfacing the orbitrap mass analyzer to an electrospray ion source, *Anal. Chem.*, **75**, 1699–1705.
102. Hu, Q., Noll, R.J., Li. H. *et al.* (2005). The Orbitrap: a new mass spectrometer, *J. Mass Spectrom.*, **40**, 430–443.
103. Makarov, A., Denisov, E., Kholomeev, A. *et al.* (2006). Performance evaluation of a hybrid linear ion trap–Orbitrap mass spectrometer, *Anal. Chem.*, **78**, 2113–2120.
104. Perry, R.H., Cooks, R.G. and Noll, R.J. (2008). Orbitrap mass spectrometry: instrumentation, ion motion and applications, *Mass Spectrom. Rev.*, **27**, 661–699.
105. Kellmann, M., Muenster, H. and Zomer, P. (2009). Full scan MS in comprehensive qualitative and quantitative residue analysis in food and feed matrices: how much resolving power is required?, *J. Am. Soc. Mass Spectrom.*, **20**, 1464–1476.
106. Makarov, A. and Scigelova, M. (2010). Coupling liquid chromatography to Orbitrap mass spectrometry, *J Chromatogr. A*, **1217**, 3938–3945.
107. Zhang, A., Chang, J.S., Gu, C. *et al.* (2010). Non-targeted screening and accurate mass confirmation of 510 pesticides on a high-resolution LC–MS Orbitrap system, *Brazilian J. Anal. Chem.*, **1**, 60–74.
108. Hayward, D.G., Wong, J.W., Zhang, K. *et al.* (2011). Multiresidue pesticide analysis in ginseng and spinach by non-targeted and targeted screening procedures, *J. AOAC Int.*, **94**, 1741–1751.
109. Shi, Y., Chang, J.S., Esposito, C.L. *et al.* (2011). Rapid screening for pesticides using automated on-line sample preparation with a high-resolution benchtop Orbitrap mass spectrometer, *Food Addit. Contam.*, **28**, 1383–1392.
110. Wang, J., Chow, W., Leung, D. *et al.* (2012). Application of ultrahigh-performance liquid chromatography and electrospray ionization quadrupole Orbitrap high-resolution mass spectrometry for determination of 166 pesticides in fruits and , *J. Agric. Food Chem.*, **60**, 12088–12104.
111. Mol, H.G.J., Zomer, P. and de Koning, M. (2012). Qualitative aspects and validation of a screening method for pesticides in vegetables and

fruits based on liquid chromatography coupled to full scan high-resolution (Orbitrap) mass spectrometry, *Anal Bioanal. Chem.*, **403**, 2891–2908.

112. De Dominicis, E., Commissati, I. and Suman, M. (2012). Targeted screening of pesticides, veterinary drugs and mycotoxins in bakery ingredients and food commodities by liquid chromatography–high-resolution single-stage Orbitrap mass spectrometry, *J. Mass Spectrom.*, **47**, 1232–1241.
113. Gómez-Pérez, M.L., Plaza-Bolaños, P., Romero-González, R. *et al.* (2012). Comprehensive qualitative and quantitative determination of pesticides and veterinary drugs in honey using liquid chromatography–Orbitrap–high-resolution mass spectrometry, *J. Chromatogr. A*, **1248**, 130–138.
114. Padilla-Sánchez, J.A., Plaza-Bolaños, P., Romero-González, R. *et al.* (2012). Innovative determination of polar organophosphonate pesticides based on high-resolution Orbitrap mass spectrometry, *J. Mass Spectrom.*, **47**, 1458–1465.
115. Chiaia-Hernández, A.C., Krauss, M. and Hollander, J. (2013). Screening of lake sediments for emerging contaminants by liquid chromatography atmospheric pressure photoionization and electrospray ionization coupled to high-resolution mass spectrometry, *Environ. Sci. Technol.*, **47**, 976–986.
116. Edison, S.E., Lin, L.A., Gamble, B.M. *et al.* (2011). Surface swabbing technique for the rapid screening for pesticides using ambient pressure desorption ionization with high-resolution mass spectrometry, *Rapid Commun Mass Spectrom.*, **25**, 127–139.
117. Edison, S.E., Lin, L.A. and Parrales, L. (2011). Practical considerations for the rapid screening for pesticides using ambient pressure desorption ionisation with high-resolution mass spectrometry, *Food Addit & Contaminants*, **28**, 1393–1404.
118. García-Reyes, J.F., Gilbert-López, B., Agüera, A. *et al.* (2012). The potential of ambient desorption ionization methods combined with high-resolution mass spectrometry for pesticide testing in food, *Comprehens. Anal. Chem.*, **58**, 339–366.
119. Wille, K., Claessens, M., Rappe, K. *et al.* (2011). Rapid quantification of pharmaceuticals and pesticides in passive samplers using

ultrahigh-performance liquid chromatography coupled to high-resolution mass spectrometry, *J. Chromatogr. A*, **1218**, 9162–9173.

120. Alder, L., Steinborn, A. and Bergelt, S. (2011). Suitability of an Orbitrap mass spectrometer for the screening of pesticide residues in extracts of fruits and vegetables, *J. AOAC Int.*, **94**, 1661–1673.
121. Kaufmann, A., Dvorak, V., Crüzer, C. *et al.* (2012). Study of high-resolution mass spectrometry technology as a replacement for tandem mass spectrometry in the field of quantitative pesticide residue analysis, *J. AOAC Int.*, **95**, 528–48.
122. Hernández, F., Sancho, J.V., Ibáñez, M. *et al.* (2011). 'TOF- and QTOF-MS for identifying unknown contaminants and degradation products in the environment' in Meyers, R.A. (ed), *Encyclopedia of Analytical Chemistry*, John Wiley and Sons, Inc., Chichester.
123. Díaz, R., Ibáñez, M., Sancho, J.V. *et al.* (2011). Building an empirical mass spectra library for screening of organic pollutants by ultrahigh-pressure liquid chromatography–hybrid quadrupole–time-of-flight mass spectrometry, *Rapid Commun. Mass Spectrom.*, **25**, 355–369.
124. Bobeldijk, I., Vissers, J.P.C., Kearney, G. *et al.* (2001). Screening and identification of unknown contaminants in water with liquid chromatography and quadrupole–orthogonal acceleration time-of-flight tandem mass spectrometry, *J. Chromatogr. A*, **929**, 63–74.
125. Ibáñez, M., Sancho, J.V, Pozo, O.J. *et al.* (2005). Use of quadrupole time-of-flight mass spectrometry in the elucidation of unknown compounds present in environmental water, *Rapid Commun. Mass Spectrom.*, **19**, 169–178.
126. Bobeldijk, I., Stoks, P.G.M., Vissers, J.P.C. *et al.* (2002). Surface and wastewater quality monitoring: combination of liquid chromatography with (geno)toxicity detection, diode array detection and tandem mass spectrometry for identification of pollutants, *J. Chromatogr. A*, **97**, 167–181.
127. Portolés, T., Ibáñez, M., Sancho, J.V. *et al.* (2009). Combined use of GC–TOF-MS and UHPLC–(Q)TOF-MS to investigate the presence of nontarget pollutants and their metabolites in a case of honeybee poisoning, *J. Agric. Food Chem.*, **57**, 4079–4090.
128. Hernández, F., Sancho, J.V., Ibáñez, M. *et al.* (2008). Investigation of pesticide metabolites in food and water by LC–TOF-MS, *Trends Anal. Chem.*, **27**, 862–872.

129. Picó, Y. and Barceló, D. (2008). The expanding role of LC–MS in analyzing metabolites and degradation products of food contaminants, *Trends Anal. Chem.*, 27, 821–835.
130. Thurman, E.M., Ferrer, I., Zweigenbaum, J.A., *et al.* (2005). Discovering metabolites of post-harvest fungicides in citrus with liquid chromatography–time-of-flight mass spectrometry and ion trap–tandem mass spectrometry, *J. Chromatogr. A*, **1082**, 71–80.
131. García-Reyes, J.F., Ferrer, I., Thurman, E.M. *et al.* (2005). Searching for non-target chlorinated pesticides in food by liquid chromatography–time-of-flight mass spectrometry, *Rapid Commun. Mass Spectrom.*, **19**, 2780–2788.
132. Thurman, E.M., Ferrer, I., Zavitsanos, P. *et al.* (2013). Identification of imidacloprid metabolites in onion (*Allium cepa L.*) using high-resolution mass spectrometry and accurate mass tools, *Rapid Commun. Mass Spectrom.*, **27**, 1891–1903.
133. García-Reyes, J.F., Molina-Díaz, A. and Fernández-Alba A.R. (2007). Identification of pesticide transformation products in food by liquid chromatography–time-of-flight mass spectrometry via "fragmentation-degradation" relationships, *Anal. Chem.*, **79**, 307–321.
134. Hernández, F., Grimalt, S., Pozo, O.J. *et al.* (2009). Use of ultrahigh-pressure liquid chromatography–quadrupole–time-of-flight MS to discover the presence of pesticide metabolites in food samples, *J. Sep. Science*, **32**, 2245–2261.
135. Ibáñez, M., Sancho, J.V, Pozo, O.J. *et al.* (2004). Use of quadrupole–time-of-flight mass spectrometry in environmental analysis: elucidation of transformation products of triazine herbicides in water after UV exposure, *Anal. Chem.*, **76**, 1328–1335.
136. Ibáñez, M., Sancho, J.V, Pozo, O.J. *et al.* (2006). Use of liquid chromatography–quadrupole–time-of-flight mass spectrometry in the elucidation of transformation products and metabolites of pesticides. Diazinon as a case study, *Anal. Bioanal. Chem.*, **384**, 448–457.
137. Kouloumbos, V.N. and Tsipi, D.F. (2003). Identification of photocatalytic degradation products of diazinon in TiO2 aqueous suspensions using GC–MS/MS and LC–MS with quadrupole–time-of-flight mass spectrometry, *J. Am. Soc. Mass Spectrom.*, **14**, 803–817.

138. Peschka, M., Petrovic, M., Knepper, T.P. *et al.* (2007). Determination of two phototransformation products of bentazone using quadrupole–time-of-flight mass spectrometry, *Anal. Bioanal. Chem.*, **388**, 1227–1234.
139. Lambropoulou, D.A., Konstantinou, I.K., Albanis, T.A. *et al.* (2011). Photocatalytic degradation of the fungicide Fenhexamid in aqueous TiO2 suspensions: identification of intermediates products and reaction pathways, *Chemosphere*, **83**, 367–378.
140. Yao, H.B., Han, G.J., Liu, G.X., *et al.* (2010). Determination of imazaquin and its metabolite by liquid chromatography–quadrupole–time-of-flight tandem mass, *Bull. Environ. Contam. Toxicol.*, **85**, 142–146.
141. Hogenboom, A.C., Niessen, W.M.A. and Brinkman, U.A.Th. (2000). Characterization of photodegradation products of alachlor in water by on-line solid-phase extraction liquid chromatography combined with tandem mass spectrometry and orthogonal-acceleration time-of-flight mass spectrometry, *Rapid Commun. Mass Spectrom.*, **14**, 1914–1924.
142. Detomaso, A., Mascolo, G. and Lopez, A. (2005). Characterization of carbofuran photodegradation by-products by liquid chromatography–hybrid quadrupole–time-of-flight mass spectrometry, *Rapid Commun. Mass Spectrom.*, **19**, 2193–2102.
143. Grimalt, S., Pozo, O.J., Sancho, J.V. *et al.* (2007). Use of liquid chromatography coupled to quadrupole–time-of-flight mass spectrometry to investigate pesticide residues in fruits, *Anal. Chem.*, **79**, 2833–2843.
144. Picó, Y., Farré, M., Soler, C. *et al.* (2007). Confirmation of fenthion metabolites in oranges by IT–MS and QqTOF-MS, *Anal. Chem.*, **79**, 9350–9363.
145. Picó, Y., Farré, M., Tokman, N. *et al.* (2008). Rapid and sensitive ultrahigh-pressure liquid chromatography–quadrupole–time-of-flight mass spectrometry for the quantification of amitraz and identification of its degradation products in fruits, *J. Chromatogr. A*, **1203**, 36–46.
146. Jović, M., Manojlović, D., Stanković, D. *et al.* (2013). Degradation of triketone herbicides, mesotrione and sulcotrione, using advanced oxidation processes, *J. Haz. Mat.*, **260**, 1092–1099.

Chapter 11

Food-Packaging Contaminants

Silvia Lacorte,[a] Montse Cortina,[b] Albert Guart[c] and Antonio Borrell[c]

[a]Department of Environmental Chemistry, IDAEA-CSIC, Barcelona, Catalonia, Spain
[b]Escola Universitària Salesiana de Sarrià, Barcelona, Catalonia, Spain
[c]Laboratorio Dr. Oliver-Rodés, Moreres S.A. El Prat de Llobregat, Spain

11.1. Introduction

Packaging has become an indispensable element in the food manufacturing process. Food packaging is an important way to store food at different temperatures, to extend the shelf-life of the product, and to enable foods to travel safely for long distances from their point of origin and still be wholesome at the time of consumption. Food packaging safeguards foods from natural agents, such as air, and can retard product deterioration, retain the beneficial effects of processing, and maintain or increase the quality, organoleptic properties and safety of food.[1] The design and manufacture of food packaging is in continuous development to cover the different and ever increasing types of foods available on the market. Materials that have traditionally been used in food packaging include glass, metals (aluminum foils and laminates, tinplate, and tin-free steel), paper and paper boards, and plastics.[1] Among plastics, thermoplastics have emerged as an ideal material for food packaging due to their low cost and functional advantages (chemically resistance, light weight, physical and optical properties, thermosealability, microwavability). Thermoplastics are polymers that soften upon exposure to heat and can then be shaped and molded into sheets, shapes and structures, offering considerable design flexibility. In fact, many plastics are heat-sealable, easy to print, and can be integrated into food production processes where the package is formed, filled, and sealed in the same production line. Moreover, almost all thermoplastics are recyclable (by melting and reuse as raw materials for production of new products). The major disadvantage of plastics is their variable permeability to light, gases, vapours, and to low molecular weight molecules.[1]

There are several types of plastic, each with unique properties and application in the food sector. The most common plastic materials used in food packaging are polyethylene terephthalate (PET), high-density polyethylene (HDPE), polycarbonate (PC), low-density polyethylene (LDPE), polyvinyl chloride (PVC) and polystyrene (PS) (Table 11.1). Each polymeric material has been given a code, the so-called 'resin identification coding system' developed by the Society of the Plastic Industry (SPI). There are other new plastics (resin code 7) that have appeared in recent years, such as TritanTM copolyester, which is used to replace PC. Plastics are manufactured from various polymeric materials. A polymeric material is composed of one or more polymers and additives. A polymer molecule is a repetition of small and simple chemical units that are connected with covalent bonds. These small units may be repeated, as in 2,2-bis(4-hydroxiphenyl)propane (Bisphenol A or BPA) in PC, or may be different, as in ethylene glycol and dimethyl terephthalate (DMTP) in PET. In addition, plastic material used to manufacture food packaging containers contains several additives which are incorporated into the polymer to give added value to the manufactured product, such as colour, strength, flexibility or mouldability. Additives can be accelerators, catalysers, stabilisers, lubricants, anti-static and anti-blocking agents, coupling agents, plasticisers and antioxidants which have been developed to improve the performance of polymeric packaging materials.[2] Often, additives added as processing aids during the manufacturing of polymeric materials are not declared in the final product.[3] The main additives are:

- Accelerators, also known as promotors, which are often added to resins to aid in polymerisation. They react only when a catalyst

Table 11.1. Typical plastic packaging used for foods.

Polymer	Characteristics, use and structure
Polyethylene terephthalate (PET or PETE) 1 PET	Used in the manufacture of soft drink and water bottles, salad domes, biscuit trays, salad dressing and peanut butter containers. PET has a density of 1.37 g/cm^3 and is manufactured from terephthalic acid and etilenglycol. It has a good gas barrier and it is light and recyclable. 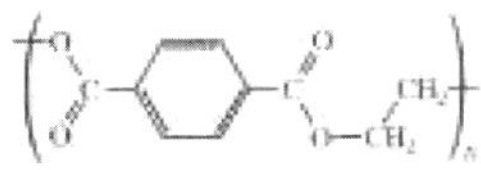
High density polyethylene (HDPE) 2 HDPE	Used in the manufacture of caps for PET bottles, crinkly shopping bags, freezer bags, milk bottles, ice cream containers, juice bottles and milk crates. HDPE has a density between 0.945 g/cm^3 and 0.964 g/cm^3. It is manufactured from ethylene. It is harder than PET but with a worse gas barrier. 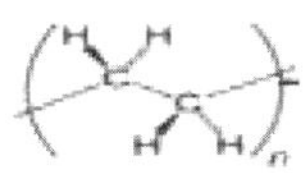
Polyvinyl chloride (PVC) 03 PVC	Used in the manufacture of tableware as well as reusable bottles and food storage containers that can be conveniently used in the refrigerator and microwave. Rigid PVC has a density between 1.3 g/cm^3 and 1.45 g/cm^3, wheras flexible PVC has a density between 1.1 g/cm^3 and 1.35 g/cm^3. It is manufactured from vinyl chloride. 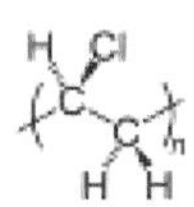

(*Continued*)

Table 11.1. (*Continued*)

Polymer	Characteristics, use and structure
Low density polyethylene (LDPE)	Used in the manufacture of caps for PC carboys, squeeze bottles and food storage containers. LDPE has a density between 0.915 g/cm^3 and 0.940 g/cm^3. It has good hardness and flexibility. 
Polypropylene (PP)	Used in the manufacture of 8 L bottles, ice cream tubs, potato chip bags, straws, microwave dishes, kettles and lunch boxes. LDPE has a density between 0.90 g/cm^3 and 0.91 g/cm^3. It is a good water vapour barrier and has fat-resistant properties. 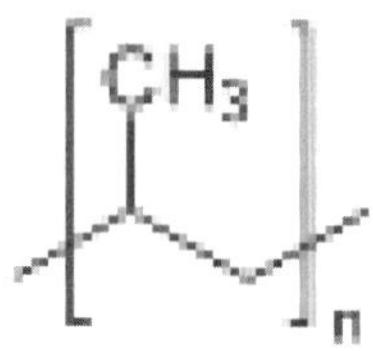
Polystyrene (PS)	Used in the manufacture of liners for LDPE caps and plastic cutlery. PS has a density between 1.04 g/cm^3 and 1.12 g/cm^3. PS may be copolymerised with other monomers and it is often substituted for silicones in LDPE caps. 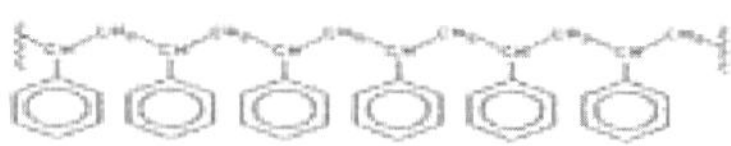

(*Continued*)

Table 11.1. (*Continued*)

Polymer	Characteristics, use and structure
Polycarbonate (PC)	Used in the manufacture of reusable bottles and 13 L, 18.9 L and 20 L carboys. These carboys are coupled to the point-of-use (PoU) called 'coolers', which are commonly placed in offices, hospitals, etc. PC has a density between 1.20 g/cm^3 and 1.24 g/cm^3. It is manufactured from BPA and it is reused after cleaning with detergents and water. 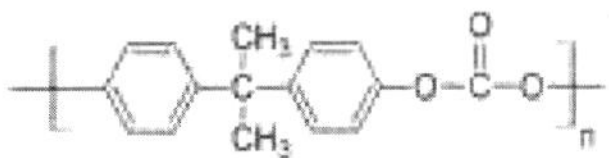
Tritan^{TM} (registered by Eastman Chemical Company)	Used in the manufacture of reusable bottles and sport bottles. Nowadays, it is considered as a possible substitute for PC carboys. It is manufactured from dimethyl terephthalate (DMTP), 2,2,4,4-tetramethyl-1,3-cyclobutanediol and 1,4-cyclohexanedimethanol as principal monomers (Eastman, 2010). In spite of knowing the principal monomers, final structure is unknown.

is added and the reaction that causes polymerisation produces heat. A common accelerator used with the catalyst methyl ethyl ketone peroxide is cobalt naphtenate.

- Catalysts, sometimes called hardeners (more correctly initiators), which are chemicals that help the monomers to join and/or cross-link. Organic peroxides are used to polymerise and cross-link thermoplastics (PVC, PS, LDPE, EVA and HDPE) in addition to thermosetting polyesters. The most widely used catalysts are unstable peroxides called azo compounds. Benzoyl peroxide and methyl ethyl ketone peroxide are widely used as organic catalysts.

- Stabilisers corresponding to a spread range of compounds, which are antioxidants, antiacids, antimicrobials, dehydrating agents and heat and light stabilisers. The main group is antioxidants, which give protection to the plastic against oxygen-triggered degradation and extend the plastic service life. Due to their function, their characteristic requirements are low concentration of use, compatibility with the substrate, stability, low toxicity, ease of use, and cost.[2,3] Organophosphites, phenols, thiosters and amines are the most commonly used antioxidants. They are used to protect the color and molecular weight of the polymer during processing, decompose peroxides, as well as chelate and react with metals.
- Coupling agents, which are especially important in processing composites. They are used as surface treatments to improve the interfacial bond between the matrix and the reinforcements, fillers or laminates. Without this treatment, many resins and polymers will not adhere to reinforcements or other substrates. Good adhesion is essential for the polymer matrix to transfer stress from one fibre, particle or laminar substrate to the next. Commonly used coupling agents are silane and titanate.
- Plasticisers, which give the plastic flexibility, resistance, mechanical strength, durability, and stretchability, and improve its processibility by decreasing the melt viscosity, temperature and elasticity of the final product without the alteration of the chemical character of the polymer. Normally, a combination of plasticisers is used in a single plastic to achieve the desired plastic characteristics. Due to their function, their characteristic requirements are compatibility with the polymer, low extractability by water and solvents, stability to heat and light, good resistance to low-temperature properties, ease of processing, and low odour, taste and toxicity. The most commonly used plasticisers are phthalates.

Other types of additives are[3]:

- Antifogging agents, which are used in packaging containing food with high water content.
- Antistatic agents, which reduce the chargeability of plastics, so-called 'static electricity'. They can be applied either from solutions on the plastic surfaces or mixed into the plastic masses during processing.
- Blowing agents, which are used in expanded PVC or PS packaging materials of the catering industry.
- Colourants, which may be either mixed into the polymer mass or applied as printing inks on plastic surfaces. Some processing aids, such as dispersants, binding agents (acrylic, alkyd, polyester, or melamine resins), or solvents, have to be used together with colourants.
- Fillers and reinforcing agents, which are usually powdered inorganic additives (e.g. calcium carbonate, talc, kaolin or silica), and are used to increase bulk and improve mechanical and physical properties of plastics. Other additives, such as glass, carbon, and polyester fibers, are used as specific reinforcing additives in the manufacture of large rigid containers.
- Lubricants, which can be classified as 'outer' lubricants, which reduce the external friction between plastics and processing equipment interfaces, and protect from sticking to the mould of the machinery, and 'inner' lubricants, which reduce the internal friction on macromolecule and macromolecule interfaces, improving the movement of polymeric chains.
- Nucleating agents, which assist in achieving consistent properties and morphology of semicrystalline plastics.
- Optical brighteners such as fluorescent agents, which improve the whiteness or brilliant white appearance of finished articles.

All food-packaging materials have to be manufactured with materials approved for food contact and these materials must be inert and must not alter food composition. Moreover, materials and devices must not react with food and their design has to allow good hygienic maintenance. However, in recent years concern has increased due to the migration of low molecular weight substances such as stabilisers, plasticisers, antioxidants, monomers, and oligomers from plastic packaging materials into food[4] by thermal and mechanical stress or excessive contact time. Chemicals that leach from food packaging into food are usually intentionally added compounds, like additives, processing aids and un-reacted monomers. However, non-intentionally added substances (NIAS) (side products like oligomers and impurities) can contribute to overall leaching. The presence of these compounds in food, if not properly controlled, can affect the organoleptic properties of food and produce a risk for human health if the levels exceed the legislated or toxicological values. For safety reasons, polymers used for packaging which are in contact with food must be analysed before use to prevent migration of any of the components to the food at concentrations that may cause health effects.[5]

11.2. Hazardous Compounds in Food-Packaging Materials

Regulation 10/2011 on plastic materials and articles intended to come into contact with food lists 885 authorised monomers, other starting substances, macromolecules obtained from microbial fermentation, additives, and polymer production aids that can be used in the manufacture of plastic layers in plastic materials and articles. Among several chemical families, this chapter will be focussed on those compounds that are used in large quantities in the plastic industry and those of general scientific and social interest due to their high migration potential and potential toxic, carcinogenic or endocrine-disrupting effects. These compounds are alkylphenols, phthtalates, BPA and related compounds, ultraviolet (UV)-ink photoinitiators, perfluorochemicals and NIAS (Table 11.2). A brief description of each chemical family will be made to stress their uses in the food packaging sector, and their migration potential and effects.

Table 11.2. Food packaging contaminants.

Compounds	CAS	Uses	SML (mg/kg)
Alkylphenols and phenols			—
n-nonylphenol (mixed isomers) nonylphenol nonylphenol-industrial	25154-52-3 104-40-5 84852-15-3	— Monomer in the production of phenolic resins — Intermediate in the production of TNPP — Catalyst in the cure of epoxy resins	0.01
Octylphenol	27193-28-8	— Coating industries, rubbers	—
2,4-di-tert-butylphenol	96-76-4	— Synthesis of triarylphosphates — Antioxidant in plastics	—
2,6-di-tert-butyl-4-methylphenol	128-37-0	— Phenolic antioxidant	3
4-tert-butylphenol	98-54-4		0.05
4-cumylphenol	599-64-4		0.05
Phthalates			
Di-ethylhexyl phthalate	117-81-7	— Additive, plasticiser in repeated use materials/articles for non-fatty foods — Technical support agent (up to 0.1% of final product)	1.5
Benzyl butylphthalate	85-68-7	— Additive in repeated-use articles — Additive in single-use articles for non-fatty foods (infant food excluded) — Technical support agent (up to 0.1% of final product)	30

(Continued)

Table 11.2. (*Continued*)

Compounds	CAS	Uses	SML (mg/kg)
Dibutylphthalate	84-74-2	— Additive in repeated-use materials contacting non-fatty foods — *Technical support agent in PE, PP up to 0.05% of the final product	0.3
Diallyl phthalate	131-17-9	— Monomer	—
Ortho-phthalic acid	88-99-3	— Monomer — Additive	—
Diisononyl phthalate (DINP): substance mixture	28553-12-0 68515-48-0	— Additive in repeated-use articles — Additive in single-use articles for non-fatty foods (infant foods excluded) — Technical support agent (up to 0.1% in the final product)	9 (applies to group of substances)
Diisodecyl phthalate (DIDP): substance mixture	26761-40-0 68515-49-1	— Additive in repeated-use articles — Additive in single-use articles for non-fatty foods (infant foods excluded) — Technical support agent (up to 0.1% in the final product)	9 (applies to group of substances)
Bisphenol A and related compounds			
Bisphenol A	80-05-7	— Starting substance for polycarbonate plastics and epoxy resins — Coating of thermal papers	0.6

(*Continued*)

Table 11.2. *(Continued)*

Compounds	CAS	Uses	SML (mg/kg)
Bisphenol-F-diglycidyl ether	39817-09-9	— Additive of organosols	—
BADGE	1675-54-3	— Derivative of BPA that is used in epoxy resins for its cross-linking properties	—
NOGE	28064-14-4 9003-36-5	— Additive of organosols	—
UV filters			
Benzophenone	119-61-9		0.6
2-hydroxy-4-methoxybenzophenone	131-57-7	— To protect the food packaging from degradation	—
2-(2H-benzotriazol-2-yl)-4-methyl-phenol	2440-22-4	— To protect food contained from harmful UV light	—
Bumetriziole	3896-11-5		—
2,4-bis(1,1-dimethylethyl)-phenol, phosphite (3:1)	31570-04-4		—
Perfluoro chemicals			
Perfluorooctanoic acid, ammonium salt	3825-26-1	— Use in plastic food contact materials for repeated use, e.g. for non-slip surfaces on metal (Teflon™) pans	—

(Continued)

Table 11.2. (*Continued*)

Compounds	CAS	Uses	SML (mg/kg)
Primary Aromatic Amines			
Aniline	62-53-3	— Used in polyamide kitchenware	—
4,4′-Methylenedianiline	101-77-9		—
2,4-Toluenediamine	95-80-7		—

SML = legal limits for different phthalates in food contact material plastics (plastic layer in direct contact with food) as listed in EU 2011/10.
* = fatty foods in Annex V: no testing requirements for simulants D1 or D2. '—' = no SML.

11.2.1. *Alkylphenols and phenols*

Long-chain alkylphenols have been used extensively for over 40 years as additives for lubricants, as polymers, and as components in phenolic resins. These compounds are also used as building block chemicals that are employed in making thermoplastic elastomers, antioxidants and fire retardant materials. Through their downstream use in making alkylphenolic resins, alkylphenols are also found in adhesives and coatings. Alkylphenols (e.g. nonylphenol (NP), octylphenol (OP) and 2,4-di-tert-butylphenol (2,4-DTBP) can migrate from packaging into food and as a consequence have been detected in many types of foods.

NP is an alkylphenol whose name can apply to a large number of isomeric compounds of general formula $C_6H_4(OH)C_9H_{19}$. NPs may vary in two ways: the substitution position of the nonyl group on the phenol molecule, and the degree of branching of the nonyl group. The main use of NP in the plastics industry is as a monomer in the production of phenol/formaldehyde resins. It is also an intermediate in the production of tri-(4-nonylphenyl) phosphite (TNPP) and as a catalyst in the curing of epoxy resins. NP is not used as a free additive in resins, plastics or stabilisers and it may be present in detergents. There is a potential for consumer exposure due to the use of epoxy resins. NP has a specific migration limit (SML) of 0.010 mg/kg, which was assessed for risks to human health under the existing substances regulation (ESR) 793/93/EEC[6] and is currently the subject of marketing and use restrictions under Council Directive 76/769/EEC.[7]

OP is a high-production-volume chemical and is the most likely immediate replacement for NP. It is used in paints and printing inks, in the rubber and coatings industries, in adhesive formulation, and in the production of polymers.

The alkylphenol 2,4-DTBP is primarily used for the synthesis of triaryl phosphites and as an antioxidant in plastics. It is also used to produce primary phenolic antioxidants and can be converted to benzotriazole derivates or to the ester 3,5-di-tert-butyl-4-hydroxybenzoic acid, both of which are used as UV stabilisers. Another common phenol is 2,6-di-tert-butyl-4-methylphenol, which is used as a monomer with a SML of 3 mg/kg[8] or as a phenolic antioxidant used for polyolefin applications such as food products and packaging. The FDA approved it for food contact as adhesives (no limitations), pressure sensitive adhesives (0.1% max) and rubber articles intended for repeated use (5% max).[2] Most alkylphenols are endocrine disruptors[9] and therefore their uptake through food should be minimised.

11.2.2. *Phthalates*

Phthalates are additives usually added to PVC for softening, therefore they are also known as plasticisers.[10] Phthalate-based catalysts are also used in the production of polypropylene plastics. Phthalates are present in foods as a result of migration from food packaging. Exposure to phthalates is of concern, because these substances are linked to reduced fertility, reproductive toxicity and testicular toxicity in animal studies. An overview of the economic and social interest in the control of phthalate esters in food analyses is reviewed by Gómez-Hens.[11] The following phthalates are of concern since they have been repeateadly detected in food and aqueous matrices due to migration from containers.

Benzyl butyl phthalate (BBP) is a plasticiser added to polymers to give flexibility and softness. It is used in flexographic inks for food packaging applications[2] and it is considered to be a very toxic compound due to its mutagenic properties, acute oral and reproductive toxicity, and carcinogenicity.[2,12] Regulation No. 10/2011[8] indicates that BBP can be used as a plasticiser in repeated-use materials and articles; as a plasticiser in single-use materials and articles contacting non-fatty foods except for infant formulae or processed cereal-based foods and baby foods for infants and young children; and as a technical support agent in concentrations up to 0.1% in the final product. Its SML is 30 mg/kg and its specific migration limit as a sum of substances (SML(T)) is 60 mg/kg.

Dibutyl phthalate (DBP) can only be used as a plasticiser in repeated use materials and in articles in contact with non-fatty foods and as a technical support agent in polyolefins in concentrations up to 0.05% in the final product[8]. SML and SML(T) are 0.3 mg/kg and 60 mg/kg, respectively. The European Chemical Agency (ECHA)[12] established DMP use as a polymer and industrial plasticiser.

Di-(ethylhexyl) phthalate (DEHP) can be used in repeated-use materials and articles in contact with non-fatty foods and as a technical support agent in concentrations up to 0.1% in the final product.[8] SML and SML(T) are 1.5 mg/kg and 60 mg/kg, respectively.

Diethyl phthalate (DEP) is used as a plasticiser to improve plastic flexibility and is commonly used in products such as food packaging. DEP is not legislated in Commission Regulation 10/2011.[8]

11.2.3. *Bisphenol A and related compounds*

BPA was synthesised in 1891 and it is nowadays used in the manufacture of PC plastics and epoxy resins. PC is used in food-contact plastics such as reusable beverage bottles, infant feeding bottles, and storage containers, whereas epoxy resins are used in protective liners for food and beverage cans. Heat (sterilisation process) and contact with acidic or basic foods increase the hydrolysis of the ester bond linking BPA monomers to one another to form a polymer, releasing BPA into materials with which it comes into contact, for example food or water.[13] On the other hand, epoxy-based lacquers or vinylic organosol (PVC) materials are used for coating the inside of food cans, in big storage vessels, and in food containers to reduce food spoilage or degradation. These lacquers are epoxy phenolic resins synthetised from the polymerisation of BPA-diglycidyl ether (BADGE) and novolac glycidyl ether (NOGE). During thermal-coating treatments and storage or due to contact with food (especially aqueous or acidic), BADGE, bisphenol F-diglycidyl ether, and their chlorinated and hydrolysed derivatives can be released into the packaged foods.[14] From some years, there have existed concerns related to BPA ingestion from food and beverages. The ingestion can occur from the migration in plastic bottles.

BPA is considered to be an endocrine-disrupting chemical (EDC) with estrogenic activity. In 2006, the EFSA set a tolerable daily intake (TDI) for BPA of 50 μg BPA/kg body weight/day whereas Health Canada established a provisional TDI of 25µg/kg bw/day. The TDI was based on a no observed adverse effect level (NOAEL) of 5 mg/kg bw/day, identified in two multi-generational reproductive toxicity studies in rodents, where the critical effects were changes in body and organ weights in adult and offspring rats and liver effects in adult mice, respectively. This TDI was set to protect the human population for life-time exposure, including sensitive groups such as pregnant and lactating women, infants (1–12 months) and young children (12–36 months). In 2008, the EFSA reaffirmed this TDI, concluding that age-dependent toxicokinetic differences of BPA in animals and humans would have no implication for the default uncertainty factor of 100 for the TDI. For BADGE, a TDI of 150 µg/kg bw/day was established considering a NOAEL of 15 mg/kg bw/day derived from chronic toxicity/carcinogenicity study in rats. Hydrolyzed BADGE derivatives are included in the TDI, whereas for BADGE chlorohydrins, a restriction of 1 mg/kg food is appropriate.[15]

Regulation 10/2011 authorises the use of BPA in food contact materials with a SML of 0.6 mg/kg or 100 µg/dm^2.[8] However, Commission Directive of 28 January 2011 amending Directive 2002/72/EC restricted the use of BPA in plastic infant feeding bottles. In relation to BADGEs, the SML is of 9 mg/kg for the sum of BADGE and hydrolysed derivatives and 1 mg/kg for the sum of BADGE·HCl, BADGE·2HCl and BADGE·HCl·H_2O.[16]

11.2.4. *UV filters*

The alert of UV filters arose in 2005 when the Italian Food Control Authority detected that 2-isopropylthioxanthone migrated into baby milk at concentrations of 120–300 µg/L.[14] UV filters are added to food packaging in order to protect the food packaging from degradation, as well as the food contained within from harmful UV light. UV filters may migrate into foodstuffs. They have been associated with endocrine activity, cancer and contact sensitisation. Apart from being used in polymer-based materials, UV filters are added to printing inks. Here they act as photo initiators, which start the reaction that eventually dries the ink rapidly and prevents set-off effects of other substances contained in ink into the food.[17] Leaching of UV-screens both from printing inks as well as polymers has been observed.[18–20] UV screens produced in quantities larger than 1,000 tonnes/year and detected in food stuffs are 2-(2H-benzotriazol-2-yl)-4-methyl-phenol (CAS 2440-22-4), bumetriziole (CAS 3896-11-5), 2,4-bis(1,1-dimethylethyl)-phenol, phosphite (3:1) (CAS 31570-04-4) and benzophenone (CAS 119-61-9), all of which show endocrine-disrupting effects.[21–23]

Benzophenones have aroused interest as they have been detected in food. Benzophenones are non-substituted diphenylketones used as UV stabilisers to prevent discoloration, cracking and loss of physical properties due to sunlight.[24] Benzophenone can be used as an additive or polymer production aid[8] and as photoinitiator catalysers for inks and lacquers that are cured with UV light.

11.2.5. *Perfluorochemicals*

The use of perfluorinated compounds (PFCs) as grease and stain repellents in consumer products started in the 1950. Their 'non-stick' characteristic is due to the fully fluorinated carbon chain, which gives them unique molecular properties that make PFCs particularly unreactive. Perfluorochemicals are used in the manufacturing of food-contact substances such as non-stick coatings (polytetrafluoroethylene (PTFE) for cookware and also in paper coatings for oil and moisture resistance). Perfluorooctane sulfonic acid (PFOS) is additionally a residual impurity in some paper coatings used for food contact and perfluorooctanoic acid (PFOA) is a processing aid in the manufacture of PTFE used for many purposes including non-stick cookware.

11.2.6. *Primary aromatic amines*

Another group of compounds not intended to migrate from food-contact materials are primary aromatic amines (PAAs), which are considered as carcinogens. PAAs are formed as a degradation product from the hydrolysis of residual isocyanates in polyurethane adhesives.[3] These adhesives are the most common adhesives used in laminated materials. European legislation[8] established that PAAs shall not release in a detectable quantity (0.01 mg as Σ PAAs per kg of food).

11.2.7. *NIAS*

NIAS are chemical compounds that are present in a material but have not been intentionally added for technical reasons during the production process. NIAS originate from breakdown products of food-contact materials, impurities of starting materials, unwanted side-products and various contaminants from recycling processes. Breakdown processes may occur during manufacture processing, storage and/or contact with the food itself. Generally, it is accepted that only compounds < 1000 Da are considered NIAS, because substances with a higher molecular weight are regarded as inert towards migration due to their larger size.

Leaching of NIAS is generally defined as migration, but polymers can be degraded under the influence of acidic or alkaline foodstuffs, UV light or heat, and as a consequence monomers can release into food.[24] Their presence in food contact materials is generally not known by the consumer and often is a challenge for the food-contact materials producer.

11.3. Sample Preparation

Compounds used in the packaging sector have been found to represent a source of contamination through the migration of substances from the packaging into food. Various analytical methods have been developed to analyse the migrants in the foodstuff, and migration evaluation procedures based on theoretical prediction of migration from plastic food contact material were also introduced recently. The regulatory control, analytical methodology, factors affecting the migration and migration evaluation will be reviewed in this section.

Analysis of food-packaging contaminants is intrincate due to the low concentrations of contaminants and the complexity of the food matrix. Sensitive and selective analytical methods are needed to meet the EU requirements. These methods should:

- Selectively extract target compounds from a food commodity and eliminate matrix interferences that may affect detection and quantification.

- Unequivocally identify plastic monomers and additives in a food commodity.
- Quantify target compounds with high precision and accuracy at levels imposed by current legislation.
- Be robust enough to permit a high throughput analysis.
- Be cost effective and quick enough for decision-making actions.

The methods used for the analysis of food-packaging contaminants are countless and depend on the compounds to be analysed and the type of food. Physico chemical properties of contaminants such as the octanol-water partitioning coefficient (Kow), solubility and vapour pressure together with the characteristics of the food commodity as regards to water content, lipid content, sugar and proteins, solid or liquid state, and presence of pigments, among others, will determine both the extraction and the analytical procedure to be used. The main steps to be used on the analytical procedures to determine food-packaging contaminants are: (i) sampling, (ii) homogenization, (iii) extraction, (iv) clean-up of the analytes from the sample, (v) concentration of the analytes, and (vi) detection. The sampling step has to be performed so that the sample is a representative part of a whole sample and has to be properly homogenised to avoid differences between sample aliquots. Regardless of the type of food, the most critical step in the analysis of contaminants is the extraction procedure, which must be selective and efficient and must remove the potential interferents that may affect the analytical determination. When analysing fruits and vegetables, cooked food, meat, and fish, among other foodstuffs, extraction must be followed by a clean-up step. The aim of the clean-up step is the removal of any substance(s) (protein, carbohydrates or fats) from the food that could interfere or obscure the signal of the analytes investigated. This step needs to be performed with caution to avoid losses in the recovery of target compounds and to avoid external contamination during manipulation. The type of extraction and sample clean-up used depends on how much substance is expected to be present and the characteristics of the substance and the matrix from which it is being extracted.

11.3.1. *Solid food matrices*

For the extraction of food contact materials from solid matrices, Soxhlet extraction, solvent extraction under reflux, pressurised liquid extraction (PLE), ultrasonic extraction or more recently, quick, easy, cheap, effective, rugged and safe (QuEChERS)[25] have been reported for the analysis of food-packaging contaminants. Camela[26] reviews recent extraction techniques for solid matrices. Table 11.3 provides the advantages and disadvantages of each method, and their applicability.

Soxhlet extraction is a classical technique for the extraction of food packaging contaminants from food commodities and provides the main reference for evaluating the performance of other extraction methods. The technique is based on the choice of a solvent coupled with the use of heat and/or agitation to extract the contaminants. Different solvents will yield different extracts and extract compositions.[27] In recent years, improvements in the Soxhlet extraction method have been proposed that entail reducing the extraction time, solvent or energy consumption.[28] Eight phthalates were analysed in twenty-five kinds of plastic products for food use, including packaging bags, packaging film, containers, boxes for microwave oven use, sucking tubes, spoons, cups, and plates, by gas chromatography–mass spectrometry (GC–MS) and permitted to detect concentrations <10 µg/kg with good reproducibility.[29] Soxhlet extraction was used to determine the migration of dioctylphthalate and dioctyladipate plasticisers in PVC films into ground meat and levels of 2 mg/kg to 80 mg/kg of meat (0.12–4.8 mg/dm^2) were detected after 8 days at 4 °C.[30] Other studies report the use of Soxhlet for the analysis of alkylphenols[31] and other types of contaminants.[32] The advantages of Soxhlet are sample-fresh solvent contact during the whole process, no need for filtration procedure, good reproducibility and efficiency, and little extract manipulation. However, compared with the novel fast extraction techniques such as ultrasound-assisted or accelerated solvent extractions, Soxhlet is a time- and solvent-consuming extraction technique.

Ultrasonic extraction is a common extraction technique, which provides good extraction yields for organic contaminants. Sound

Table 11.3. Advantages and disadvantages of food extraction methods.

Method solid matrices	Advantages	Disadvantages
Soxhlet	Standardised method	Time- and solvent-consuming technique
	Little manipulation	Blank contamination
	High extraction yields	No automatisation
	Good reproducibility	
Ultrasounds	Inexpensive	No automatisation
	Simple and efficient	
	High extraction yields	
	Recovers thermolabile compounds	
	Miniaturisation of extraction	
	Low solvent consumption	
	Versatile and robust	
PLE	Automated	Abrasive extraction
	Effective in time and labour	Need of clean-up
	High sample throughput	Expensive equipment
	Low solvent consumption	
	Versatile	
QuEChERS	Effective in time and labour	No automatisation
	Low solvent consumption	
	High extraction yields	
	Clean extracts	
Liquid matrices		
LLE	Standardised method	Restricted volume of sample
	Cost-effective	Time and solvent consumption
	Versatile	Formation of emulsions
		Clean-up often necessary
		High sample manipulation
		Evaporation losses

(*Continued*)

Table 11.3. (*Continued*)

Method solid matrices	Advantages	Disadvantages
		Inaccurate and poor reproducibility
SPE	Standardised	Clogging of cartridges
	High preconcentration volumes	Breakthrough for polar compounds
	Low sample manipulation	
	Effective in time and labour	
	Automated Versatile (off-and-on-line)	
SPME	Effective in time and labour	Low capacity
	Low cost	
	High sensitivity	
	Low sample manipulation	
	Versatile	
SBSE	Effective in time and labour	Loss of sample if problem occurs
	Low cost	No simultaneous extraction of polar/apolar compounds
	High sensitivity	
	Low sample manipulation	
	Versatile	
	High capacity	
	Automated	

waves having frequencies higher than 20 kHz are mechanical vibrations in a liquid that produce liquid jets that have strong impact on the extracted matrix. Ultrasound in extraction can disrupt biological cell walls, facilitating the release of contaminants, leading to the enhancement of extraction with ultrasonic power.[33] Sample properties such as moisture and fat content and particle size, together with the solvent used for the extraction and the frequency, pressure, temperature and sonication time, will define the efficieny of ultrasound-assisted extraction.[34] Ultrasonic solvent extraction combined with solid-phase microextraction (SPME) with calyx arene/hydroxy-terminated silicone oil coated fiber and gas chromatography was used to extract eight phthalate acid esters (PAEs) in blood bags, transfusion tubing, food packaging bags, and mineral water bottles. Both the extraction parameters (i.e. extraction time, extraction temperature, ionic strength) and the conditions of the thermal desorption in a GC injector were optimised and good extraction yields and sensitivity were obtained.[35] Ultrasound-assisted extraction is an inexpensive, simple and efficient alternative to conventional extraction techniques. The main benefits of use of ultrasound in solid–liquid extraction include the increase of extraction yield, faster kinetics and the possibility to extract thermolabile compounds by reducing the operating temperature.

The PLE employs liquid solvents at elevated temperatures and pressures to effectively extract organic contaminants from a solid matrix, in this case, food. The advantage of this system is that the extraction time is very much reduced in comparison to other methods such as Soxhlet or ultrasonic extraction because it is automated. PLE is quite an abrasive technique and is capable of disrupting cell membranes and releasing contaminants from food matrices and thus, extraction yields are improved. However, this highly abrasive extraction may also extract matrix compounds, which may interfere with the analysis of target analytes. Therefore, a clean-up step is mandatory to obtain a clean extract to be analysed by GC or LC. The clean-up can either be performed off-line using SPE cartridges of Florisil, alumina or silica, or by adding these sorbents in the PLE extraction cell and performing the clean-up in parallel to the extraction procedure. This system is effective both in time and labour, and is highly adopted for the analysis of food contaminants. Shao *et al.*[36] obtained excellent extraction efficiency and sensitivity in the determination of BPA and alkylphenols in meat by using PLE with subsequent clean-up step using amino-propyl solid phase extraction cartridges. With PLE using ethyl acetate at 70 °C, reversed-phase silica C18 as the dispersing agent, and three cycles of extraction, NP, OP and BPA were recovered at 84–92% and detected in powdered milk at the low mg/kg level.[37]

Some of the techniques indicated above are complicated, time-consuming and require expensive equipment. The QuEChERS extraction method, developed by Anastassiades *et al.*[25] consists of an acetonitrile extraction and then a liquid–liquid partitioning by anhydrous magnesium sulphate. Afterwards, an aliquot of organic layer is cleaned up by SPE with primary secondary amine (PSA) to remove organic acids and anhydrous magnesium sulphate to remove water from extract. Other sorbents, such as graphitised carbon black (GCB) can be used to remove pigments such as chlorophyll and carotenoids, and sterol or C18 for lipids.[38,39] This method needs only a short time-of-application to process a large number of samples through a few actions, low quantities of organic solvent, and it shows high recoveries for detection of some chemicals. The QuEChERS method is highly effective and selective as it eliminates chromatographic interferences and increases the sensibility of the detection by either GC or LC coupled to MS.[40] In the last few years, QuEChERS concepts have been adapted for the simultaneous extraction of polar or non-polar pesticides from fruits and vegetables[41,42] and for acrylamide, clinical and veterinary drug residues, perfluorinated compounds, polycyclic aromatic hydrocarbons, alkaloids and mycotoxin detection.[40] QuEChERS has been applied to determine 2-isopropylthioxanthone (2-ITX) in food and food packaging-materials on the German market and obtained recoveries between 80 LOD and 105 and LOD between 2 µg/L and 5 µg/L.[43] Biedermann utilised the method to determine the migration of mineral oil, photoinitiators and plasticisers from recycled paperboard.[44]

11.3.2. *Liquid matrices*

For the analysis of food-packaging contaminants in liquid matrices, the most used extraction techniques developed are liquid–liquid extraction (LLE), SPE, solid phase microextraction (SPME) and more recently, stir bar sorptive extraction (SBSE).

LLE has traditionally been used for the analysis of contaminants in liquid matrices such as water, milk or juices. LLE is based on the partitioning of different components between the aqueous solution and an immiscible organic solvent. Most of the current methodologies for the analysis of food-packaging contaminants in water samples involve extraction with various solvents such as dichloromethane, butyl acetate, chloroform or hexane, and can be done either in acid conditions to avoid hydrolysis of compounds or at neutral pH. Other approaches have been described using ion-pair-forming agents by adjusting the pH to a basic value and additioning quaternary ammonium salts. The selectivity of LLE depends on the used solvent, extraction pH, ionic strength, water-to-solvent ratio, number of extractions and on the chemical nature and concentration of the analytes. LLE was used in combination with automated large volume injection (LVI) and GC–MS analysis for the determination of phthalates in water samples.[45] Amiridou and Voutsa[46] determined alkylphenols, phthalates and BPA in bottled water by LLE–GC–MS with derivatisation. Detection limits were calculated from the standard deviation of seven replicates and ranged from 0.0022 μg/L for OP up to 0.030 μg/L for DMP. Phthalates were determined in milk using LLE with tert-butyl methyl ether and hexane, obtaining good extraction yields.[47] However, LLE has some drawbacks such as a requirement for a large amount of generally toxic and inflammable organic solvent, the generation of emulsions and the frequent necessity of further clean-up. LLE procedures are tedious and time-consuming, and they involve extensive sample manipulation with risk of contamination. In addition, evaporative losses may occur during removal of the large solvent volumes, thus making the analysis inaccurate and scarcely reproducible. Moreover, its automation requires the use of expensive robots. Hence, there is a general trend to change current LLE procedures to other analytical protocols, such as miniaturised LLE, dispersive liquid–liquid microextraction[48] or SPE.

SPE has been successfully applied to analyse phthalates, alkylphenols and BPA in water,[49] BPA in beverages[50] and as a clean-up procedure for food samples.[51] The main advantage of using cartridges is that manipulation, solvent disposal and analysis time are reduced in comparison to conventional LLE techniques. The procedure is based on the use of an adsorbent material, which is capable of selectively retaining the target compounds from the matrix. Briefly, the mode of action consists in percolating the sample through the sorbent where the contaminants are retained; afterwards, they are selectively eluted with an organic solvent. Nowadays there exist a wide variety of SPE methods, which may be divided into off-line and on-line (or precolumn technology) methods. The chemistry and principles of these two approaches are identical. The method involves different steps: (i) washing the sorbent with an organic solvent, often the same solvent which will be used for the elution step; (ii) activation of the sorbent (wetting) with a water-miscible solvent (normally methanol or acetonitrile); (iii) removal of the excess of solvent with pure water; (iv) percolation of the water sample; (v) removal of interferences (clean-up) by flowing water or solvent through the sorbent bed at a high flow rate; and (vi) elution of trapped analytes with a selective solvent, normally compatible with further chromatographic detection system. With off-line SPE methods, the aqueous sample is percolated through the sorbent bed, which is packed in disposable cartridges or enmeshed in a membrane extraction disk. Samples can be percolated by gravity through the sorbent bed or by positive (syringe) or negative (vacuum manifold) pressure. In general, off-line practices are well-accepted due to the large variety of packing materials available (C8, C18, polymeric phases and graphitisied carbon black being the most common ones), with different sizes and volumes, which allows a great operational flexibility. In the cartridge format, 30 mg to 2 g or more of sorbent (normally of 40–150 μm particle size) is usually packed in a polypropylene, Teflon or glass syringe, the latter used to avoid sample contamination due to contact with the plastic SPE support. When cartridges are used, the sample is normally percolated at flow rates of 5–10 mL/min to avoid the formation of preferential channels in the sorbent bed that could reduce adsorption of the analytes in the sorbent. Percolation of water samples can be done manually, but nowadays automated devices such as the ASPEC XL from Gilson (France) or the Zymark system (USA) permit the automatation of the technique, making it useful for routine analysis. In food analysis applications, it is important to get high enrichment factors so that one can achieve high analytical sensitivity. Prior to SPE, samples have to be filtered through 0.45 μm filters to remove gross material and to avoid clogging of the SPE cartridges. In SPE, the extraction efficiency depends on the type of matrix to be analysed. In general, better recoveries and detection limits are obtained when analysing water, since the amount of interferences are very low. Contrarily, when preconcentrating beverages, soups, and milk, among others, impurities are also preconcentrated, so that worse quality parameters are achieved.

Using SPE and GC–MS, Casajuana and Lacorte[52] analysed BPA, BADGE, alkylphenols and phthalates in bottled water obtaining recoveries between 72% and 80%, except for DMP, DEP, DEHP and BADGE, which were 56%, 42%, 21% and 120%, respectively. Li *et al.*[53] determined 4-NP, BPA and triclosan in tap water, bottled water and baby bottles by SPE–GC–MS with a prior derivatisation step. Three spiking levels of 5 ng, 100 ng, and 200 ng were added to 1 L of surface water, obtaining recoveries from 74% to 118% and the limit of detection (LOD) values for 4-NP and BPA were of 0.002 μg/L and 0.0007 μg/L, respectively.

An alternative to off-line SPE with cartridges is the use of Empore Extraction disks, where the sorbent particles are meshed in Teflon fibrils to form strong sheets or membranes that can be used in standard filtration devices. Since the channeling phenomenon does not occur in the disk format, higher sample-flow rates can be applied in comparison to cartridges. The main drawback is that automation is not straightforward. In the Speedisk format, the active sorbent is placed below a mesh that acts as a filter and therefore, filtration and extraction are performed in a single step. With these disks, sample flow rate can be increased to 200 mL/min. The overall conditioning and extraction time using speedisks is 10 min. Moreover, racks of six positions are available, increasing sample throughput.

A different approach for SPE is the use of on-line systems connected with LC or sometimes with GC. In this case, the sorbent is packed in a stainless steel or Teflon cylindrical precolumn, which is generally 2 mm long and of 2–3 mm internal diameter, and is packed with 10–60 μm sorbent material. The sample is pumped through the precolumn, which is placed at the loop position of a six–port valve. After preconcentration of the sample, the valve is electrically switched from the preconcentration position to the elution position and the contaminants are desorbed by the mobile phase and are directed to the analytical column where separation will take place. The main advantage of on-line systems is that they can be fully automated, which makes them especially suitable for routine monitoring programs. The Prospekt (Spark Holland, The Netherlands) and the OSP-2 (Merck, Germany) are the two systems commercially available for automated on-line preconcentration with their own precolumn designs, which cover a large variety of packing materials. Parameters that can be optimised for on-line SPE are the volume of water to be preconcentrated, the flow rate, normal or backflush elution and type of sorbent used. This is an engaging approach for the routine monitoring of food packaging contaminants in situations where low concentrations are expected in complicated liquid matrices such as beverages or canned liquids. In relation to these techniques, the use of small precolumns permits the miniaturisation and total automation of the preconcentration technique. LODs at levels of ng/L and excellent reproducibility can be obtained by using on-line approach because the entire sample is transferred to the analytical column and losses during sample manipulation are minimised. An automated on-line SPE coupled to fast LC–tandem mass spectrometry (MS/MS) was developed for the simultaneous analysis of BPA, bisphenol F, bisphenol E, bisphenol B and bisphenol S in canned soft drinks without any previous sample treatment. A C18 (12 μm particle size) loading column was used for the SPE on-line preconcentration to obtain accurate and reproducible quantification at ng/L level using matrix-matched calibration.[54] New trends related to on-line preconcentration methods have been recently reviewed and compared to other extraction procedures.[55]

SPME is a technique developed in 1990, which uses a fibre coated with an extracting phase where compounds are attached for the posterior injection. SPME reduces the time necessary for sample preparation, decreases purchase and disposal costs of solvents and can improve detection limits.[56] In the last decade, SPME has been applied to the analysis of phthalates in water matrices due to its simplicity.[57–59] This technique requires a contact period to extract analytes, which affects the extraction efficiency. For example, long periods and the use of salts that change the ionic strength, increase the amount of analyte extracted.[60] Dévier *et al.*[61] determined phthalates by SPME–GC–MS in samples and in blanks at similar levels and demonstrated that the few detected compounds originated from the background laboratory contamination. Therefore, this study showed the complexity of reaching a reliable measure to qualify the contamination of a sample at ultra-trace level. SPME followed by GC–MS was also used for the determination of BPA and its derivate BADGE in different food simulants (distilled water, 3% acid acetic and 10% ethanol).[60] Nerín *et al.*[62] used SPME for the analysis of BPA, bisphenol F and derivates in aqueous foodstuffs with a previous derivatisation followed by high-performance liquid chromatography (HPLC). Luks-Betlej *et al.*[57] analysed phthalates in drinking waters by SPME–GC–MS and obtained LODs between 0.005 μg/L and 0.04 μg/L. BP, naphthalene, BHT and 2,4-DTBP were analysed by SPME–GC–MS in recycled PET and recycled HDPE multilayer[63] obtaining values of BHT and 2,4-DTBP above 320 ng/g HDPE. SPME with an 85 μm polyacrylate fiber, coupled to GC–MS was used to determine six phthalate esters and bis(2-ethylhexyl) adipate in water samples to evaluate the material of the recipients on the concentration of phthalates (Peñalver *et al.*, 2000).[58] Montouri *et al.*[64] also demonstrated the applicability of SPME to determine phthalate migration in bottled water.

SBSE has been developed to extract volatile and semi-volatile compounds from water.[65] This technique is based on the sorption of apolar solutes present in aqueous samples onto a polydimethylsiloxane (PDMS) stir bar, and is based on the principles of SPME, thus partitioning of the analytes between the sample and an extracting phase, proportional to the Kow. In short, a stir bar coated with PDMS is introduced in the sample and after a stirring time period, it is then removed, rinsed, dried and placed in an "injector tray". The compounds are thermo desorbed and cryofocused in a programmable temperature vaporization (PTV) injector and finally analysed by GC–MS.[65,66] SBSE presents advantages over LLE and SPE since no solvent is used (Krüger et al., 2011) and can handle small sample volumes since all the preconcentrated analytes are desorbed. With respect to SPME, lower LODs (sub-ng/l to ng/l), higher capacity and better recoveries can be achieved since extraction is performed with a larger amount of PDMS. SBSE is also characterised by its high reproducibility due to minimal sample preparation and manipulation.[65] However, the SBSE technique has difficulty in extracting efficiently both hydrophobic and hydrophilic compounds.[67,68] Both SPME and SBSE provide enhanced sensibility compared to traditional extraction procedures because the devices are introduced directly into the thermal desorption port without any losses.[68] The application to a variety of water samples makes this technique very useful.[69] Recently Guart *et al.*[70] used SBSE to determine a large number of contaminants in bottled water from around the world, and plasticisers and additives were among the most detected compounds.

11.4. Analytical Methodologies

Present trends for food packaging contaminants are focused on the use of GC and LC coupled to MS, MS/MS or high resolution mass spectrometry (HRMS) that provide very high sensitivity to obtain LODs of parts per trillion.

GC has been one of the most popular techniques due to its high separation capabilities and sensitivity. Within the wide range of volatile compounds that are normally analysed by GC–MS, phthalates have been largely studied in different water samples such as bottled waters.[49,52] Although GC is a good technique for the identification and quantification of plastic components, it may be necessary to derivatise phenolic compounds to improve GC–MS sensitivity.[71,72] Nerín *et al.*[73] indicate that GC–MS is limited for some BPA derivates due to their low volatility. Other plastic compounds have been characterised using GC. Monteiro *et al.*[74] analysed UV stabilisers in PET samples performing ultrasonic bath extraction with dichloromethane followed by GC–MS.

On the other hand, LC is used for the analysis of non-volatile and more-polar compounds without the requirement of a derivatisation step. Avoiding this additional manipulation of the sample allows saving time and increases reproducibility.[71] Gallart-Ayala reviews the recent trends in LC–MS analysis of food packaging contaminants.[14] Bentayeb *et al.*[75] used LC–MS to analyse terephthalic acid and ethylenglycol that were extracted from recycled PET samples and indicated disadvantages of LC–MS for identifying unknown compounds due to the lack of mass spectra libraries and its relatively limited sensitivity, especially compared to GC–MS. Most food packaging contaminants are analysed with reverse-phase stationary phases, which permit the retention of lipophilic contaminants, such as alkylphenols, bisphenol A, phthtalates, and other substances. However, the more polar compounds are difficult to retain and therefore cannot be identified.

The advent of hydrophilic interaction liquid chromatography (HILIC) has represented a valuable complementary approach to reversed-phase liquid chromatography for the analysis of highly hydrophilic and polar compounds. The stationary phase is based in pure silica materials where silanols provide retention of polar-ionic compounds which otherwise would elute at or near the dead volume using reverse-phase chromatography. The surface of the silica can be modified with cyano (–CN), amino ($-NH_2$), or diol ($-CH(OH)-CH_2-OH$) groups, to increase the ability to retain certain hydrophilic compounds. HILIC separation relies on the use of mobile phases with a high percentage of organic solvent, which is beneficial for MS detection as enhanced ionization results in an increased sensitivity. At low percentages of organics in the eluent (high water content) there is no retention because the hydrophilic compound prefers to be in the mobile phase. When the mobile phase is sufficiently nonpolar (significant amounts of organic solvent), then retention is observed for many polar compounds. However, typical hydrophobic compounds will have little or no retention throughout the entire mobile phase composition range. Thus, while a separation of some mixtures of polar compounds is achievable, those samples that contain one or more nonpolar compounds cannot be analysed effectively because the hydrophobic solutes will be eluted at or near the dead volume over the entire mobile phase composition range. To date, few applications of HILIC chromatography for the analysis of food packaging contaminants have been published, and most of them refer to the analysis of melamine. Melamine was analysed in 41 types of retail melamine-ware products in Malaysia. Analysis was carried out using LC–MS/MS with a HILIC column and mobile phase consisting of ammonium acetate/formic acid (0.05%) in water and ammonium acetate/formic acid (0.05%) in acetonitrile (95:5, v/v) and limits of quantification of 5 ng/ml were obtained. Melamine migration was detected from all samples and it was found that excessive heat and acidity affected melamine migration, although at levels below the SML of 30 mg/kg set out in European Commission Directive 2002/72/EC.[76] In another study, melaware articles were tested for the migration of melamine into the food simulant 3% w/v acetic acid as a benchmark, and into other food simulants, beverages and foods for comparison. An NH_2 column was used for the determination of melamine in dairy products, using acetonitrile-H_3PO_4 (15 mmol/L, pH = 6.5) (v/v = 88:12) as mobile phase and UV detection at 215 nm, obtaining LOD of 4 μg/L.[77] Bradley *et al.*[78] used a HILIC column to carry out a comprehensive assessment of phthalic acid phthalate monoesters (mono-isopropyl phthalate, mono-n-butyl phthalate, mono-isobutyl phthalate, mono-benzyl phthalate, mono-cyclohexyl phthalate, mono-n-pentyl phthalate, mono-(2-ethylhexyl) phthalate, mono-n-octyl phthalate and mono-isononyl phthalate) in foods after extraction using acidified acetonitrile and dichloromethane, and permitted to detect concentrations of 5–100 μg/kg. The use of HILIC will expand in the near future to determine food-packaging contaminants, especially those not amenable to reverse-phase columns.

When investigating the NIAS of a food-contact material, not all compounds of interest can usually be detected with state-of-the-art analytical techniques. The analysis of NIAS is complicated by the lack of knowledge concerning the chemical properties of unidentified substances. Extracted samples can be separated chromatographically and analysed by MS. As in many chemical analyses, sample extraction is one of the most crucial steps, because incomplete transfer hinders full analysis. Furthermore, new mass spectra might not be assigned to a known chemical structure. Applying direct thermal desorption techniques prevents the extraction step, but the results are even more difficult to interpret due to complicated fragmentation patterns. Finally, knowledge about the concentrations of NIAS is needed to carry out risk assessment.

Both GC- and LC-based techniques allow the detection of food-packaging contaminants at very low concentrations. To ensure the correct analysis of target compounds in several food commodities, quality parameters such as LOD, recoveries and relative standard deviation (RSD) have to be calculated. In addition, blank analyses need to be assessed to control the possible external contamination and to avoid false positives. False positives in plastic-components analysis are often the result of phthalate contamination. Phthalates are widely used in the manufacture of many items, and are therefore present in air, water, organic solvents, adsorbed on glass and, of course, in plastic materials that could contaminate the samples.[79,80] Therefore, it is important to minimise the risk of external contamination by using clean material, high-quality water and organic solvents, and a clean atmosphere. The use of HPLC or Milli-Q water blanks in parallel with the samples permit the ascertainment of the possible contamination sources so they can be controlled or avoided. Nerín *et al.*[81] observed contamination of DEHP attributed to the contribution of plastic syringes, fittings, glass wool, and other common materials used in the laboratory. In general, it is practically impossible to obtain blank samples without any contribution of phthalates or BPA. Two options can be used to control the blank contribution. The first one relies on the analysis of a large number of blanks (n = 50) and calculate the LOD as three times the standard deviation of their contribution or as the mean value plus three times the standard contribution. Another option is to substract the blank contribution in each sample, but this has to be done only when the blank contribution is well controlled in each extraction/analytical blank and when blanks are analysed in each extraction batch.

11.5. Migration: Overall and Specific Migration

Food-contact materials must not transfer their components into the foods in unacceptable quantities. The safety in the use of polymeric materials is a subject of concern due to the transfer of plastic material constituents to the food by a diffusion process called migration.[82–84] Migration is a term used to describe the transfer of components from a certain material to the foodstuff in contact with this material. A compound placed in plastic material may migrate either because it did not react during manufacturing or it was released as a consequence of degradation by the contact with foodstuff or environment such as food acidity or UV light. Migration is of relevance for smaller size compounds (below 1000 Da). The extent to which migration occurs depends on various factors:

- The physico chemical properties of the migrant, of the packaging material, and the food (e.g. fat content).
- The temperature.
- The storage time.
- The size of the packaging in proportion to the foodstuff volume (smaller size packaging has a larger surface to volume ratio).

The types of chemicals that can migrate from packaging into food are highly diverse and depend on the type of packaging material. According to the Regulation 10/2011, there are two classes of migration:

- Overall migration, which is defined as the total amount of non-volatile substance released from a material or object into the food or food simulant. Overall-migration assessment serves to determine the whole chemical transfer from packaging into food without necessarily knowing their chemical identity.
- Specific migration, which is defined as the amount of a specific substance released from a material or article into the food or food simulant.

Different migration tests or assays, mainly for plastics, have been developed to test migration within the frame of the European Committee for Standardisation. These migration assays consist of the preparation of the plastic or object to favour migration and they are important because they allow recreating a real long-period contact between material and food by performing short-period contact in laboratory-controlled conditions (time, temperature and food simulant). That means the migration assays are standardised and may be recreated in different laboratories at the same conditions.

To determine the extent of chemical transfer from packaging into food, migrants are measured in food simulants, not actual foodstuffs. Food simulants are used in the migration assays as substitutes for food due to the simplification of chemical analysis. Food simulants vary according to the kind of food that is in contact with plastic material, thus representing several different food types: hydrophilic (water-based), lipophilic (fatty foods) or amphiphillic (foods with both watery and fatty properties). According to the Commission Regulation (UE) No 10/2011,[85] food simulants used for specific migration are:

- Distilled water. It is used for migration assays of plastic intended to contain aqueous food with pH>4.5. According to current legislation[85] it can be used until 31th December 2015. After this date, only the Simulants A to E can be used.
- Simulant A — Ethanol 10% (v/v). It is used for molasses, sugar syrups, honey, nut paste or cream, fresh vegetables peeled or cut, preserved vegetables in oily medium and animal products such as fish or meat.
- Simulant B — Acetic acid 3% (w/v). It is used for migration assays of plastic intended to contain clear drinks such as water, ciders, clear fruit or vegetable juices, infusions, coffee, soft drinks, energy drinks, cloudy drinks such as juices with fruit pulp and liquid chocolate, fruit and vegetables in the form of purée or preserves, animal products in aqueous medium, fermented milk such as yoghurt, cream and sour cream, processed cheese and

preserved cheese in aqueous medium, vinegar, sauces, mustard, and concentrated extracts of an alcoholic strength ≥6%.

- Simulant C — Ethanol 20% (v/v). It is used for migration assays of plastic intended to contain clear drinks, alcoholic beverages of an alcoholic strength of between 6% and 20%, confectionary chocolate products in paste form, fruit and vegetables in the form of purée or preserves, animal products in an aqueous medium and ice-creams.
- Simulant D1 — Ethanol 50% (v/v). It is used for migration assays of plastic intended to contain cloudy drinks, alcoholic beverages of an alcoholic strength >20% and all cream liquors, fruit and vegetables preserved in alcoholic medium, preserved meat in aqueous medium, liquid and cooked eggs, milk, fermented milk, cream and sour cream, preserved cheese in aqueous medium, and concentrated extracts of an alcoholic strength ≥6%.
- Simulant D2 — Vegetable oil. It is used for migration assays of plastic intended to contain pastry, biscuits, cakes, bread, etc. with fatty substances, chocolate and its confectionary products in solid or paste form with fatty substances on the surface, fruit and vegetables preserved in oily medium, nuts in paste or cream form, fats and oils, fish, crustaceans and molluscs preserved in oily medium, meat of all zoological species, preserved meat in a fatty or oily medium, natural cheese without rind or with edible rind and melting cheese, preserved cheese in oily medium, fried or roasted foods, cocoa paste, spices and seasoning in oily medium such as pesto or curry paste and other products with fatty character.
- Simulant E — Poly(2,6-diphenyl-p-phenylene oxide). It is used for migration assays of plastic intended to contain all kind of cereals and chocolate not contemplated before, sugar and its products in crystal or powder form, dried or dehydrated fruits and vegetables, nuts not in paste or cream form, powdered, dried or frozen eggs, powdered milk, cheese (whole with inedible rind) and other foods not contemplated before such as cocoa powder, frozen foods, pepper and salt.
- Ethanol 95% (v/v). It is used as a substitute for simulant D2 for migration assays of plastic intended to contain undenaturated ethyl alcohol beverages.

- Isooctane. It is used as a substitute for simulant D2 for migration assays when simulant D2 is considered to be unstable.

Table 11.4 summarises the general assignment of food simulants depending on the characteristics of each food. To test the overall migration, the food simulant must be chosen according to each type of food in compliance with Annex III,[85] as indicated in Table 11.5.

To determine the specific migration, materials and articles intended for contact with all types of food will be tested with food simulants A, B and D2. When materials are intended only for specific

Table 11.4. General assignment of food simulants to foods.

Food simulant	Abbreviation	Food assigned	Special uses
Ethanol 10%	Simulant A	Foods with hydrophilic character	
Acetic acid 3%	Simulant B	Foods with hydrophilic character	For foods with pH below 4.5
Ethanol 20%	Simulant C	Foods with hydrophilic character	For alcoholic foods with alcohol up to 20%; more lipophilic
Ethanol 50%	Simulant D1	Foods with lipophilic character	For alcoholic foods with alcohol above 20% and for oil in water emulsions
Vegetable oil	Simulant D2	Foods with lipophilic character	For foods which contain free fats at the surface
poly (2,6-diphenyl -p-phenylene oxide)	Simulant E	Specific migration for dry foods	

Table 11.5. Food simulant assignment for overall migration.

Food	Simulants assigned
All types of food	Distilled water or Simulant A and B and D2
All types of food except acidic foods	Distilled water or Simulant A and D2
All aqueous and alcoholic foods and milk products	Simulant D1
All aqueous, acidic and alcoholic foods and milk products	Simulants B and D1
All aqueous and alcoholic foods up to an alcohol content of 20%	Simulant C
All aqueous, acidic and alcoholic foods up to an alcohol content of 20%	Simulants B and C

types of foods, they have to be tested with food simulants indicated in Annex III of Commission Regulation 10/2011.

Test migration conditions such as contact time and temperature (Table 11.6) have to be set according to the conditions of use of Regulation 10/2011 (Annex V, 2.1.3 — Conditions of contact when using food simulants). Migration tests correspond to the worst foreseeable conditions of contact between the plastic material and the foodstuff. The standardised conditions of time and temperature are given in Table 11.6. If the plastic material is intended for a food-contact application covered by a combination of two or more times and temperatures, the migration test shall be carried out subjecting the test specimen successively to all the applicable worst foreseeable conditions appropriate to the sample, using the same portion of food simulant. For articles that are used repeatedly, the assay must be done in triplicate, changing the simulant each time. The compliance with migration limits will be evaluated on the basis of the level of the migration found in the third test. When the materials and articles are labelled for use at room temperature or below or when the materials and articles by their nature are clearly intended for use at room temperature and below, the incubation of the product (food) or simulant shall be carried out for 10 days at 20°C , 40°C , 50°C or 60°C, depending on each type of food. This is the most-used test condition

Table 11.6. Temperature and test time for specific migration assays in worst foreseeable use.

Conditions of worst foreseeable use contact time	**Test conditions test time**
$t \leq 5$ min	5 min
5 min $< t \leq$ 0.5 hour	0.5 hours
0.5 hours $< t \leq$ 1 hour	1 hour
1 hour $< t \leq$ 2 hours	2 hours
2 hours $< t \leq$ 6 hours	6 hours
6 hours $< t \leq$ 24 hours	24 hours
1 day $< t \leq$ 3 days	3 days
3 days $< t \leq$ 30 days	10 days
$t \geq 30$ days	Test conditions which are recognised to be the most severe on the basis of scientific evidence (e.g. 10 days at 40 °C or 60 °C)
Contact temperature	
$T \leq$ 5°C	5 °C
5°C $< T \leq$ 20°C	20 °C
20°C $< T \leq$ 40°C	40 °C
40°C $< T \leq$ 70°C	70 °C
70°C $< T \leq$ 100°C	100 °C or reflux temperature
100°C $< T \leq$ 121°C	121 °C
121°C $< T \leq$ 130°C	130 °C
130°C $< T \leq$ 150°C	150 °C
150°C $< T \leq$175°C	175 °C
$T >$ 175°C	Adjust the temperature to a real temperature at the interface with the food

for contact times above 30 days at room temperature or below. For materials and articles intended for use in microwave ovens, migration testing may use either a conventional or a microwave oven provided the appropriate time and temperature conditions are selected.

Migration tests may be performed in four ways:

- Single-sided testing using a migration cell. It is particularly important for multi-layer materials. Only one surface of the material is in contact with the simulant.
- Single-sided testing using a pouch. It is preferred for flat articles, which have sufficient seal strength to form durable pouches, as this does not require specialised apparatus and allows more efficient use of oven space. Like a migration cell, only one surface is in contact with the food simulant. The surface to volume ratio in a pouch is conventionally 2 dm^2 of material to 100 mL of food simulant.
- Testing by total immersion. Samples are cut out in order to obtain specimens of 1 dm^2 and are immersed in the simulant. With an immersion test, both faces of the sample are in contact with the simulant.
- Single-sided testing by filling. For articles in container form it is usually most convenient to test them by filling with the food simulant. For very large containers testing by filling may not be practicable and it may be necessary to fabricate smaller test specimens representing the article to be tested.

As already mentioned, migration can be assessed in two ways: by determining the overall migration or by the specific migration. In an overall-migration test the total quantity of all of the substances that have migrated from the test specimen of a food-contact material to a food simulant are determined gravimetrically using the procedures outlined in European Commission Directive 2002/72/EC,[5] as amended, and the detailed methods in the EN 1186 series of standards (UNE-EN 1186:2002). In a specific-migration test, the quantity of an individual substance (monomer, additive, etc.) or group of substances is determined in a food simulant following the exposure of a food-contact specimen to the food simulant for the prescribed period of time at the prescribed temperature using an appropriate analytical method (UNE-EN 13130:2004). Both of these migration tests consist of two parts: (1) the exposure of the test specimen to the food simulant (or food for specific migration) and (2) the analytical determination. The outcome of the exposure part (and therefore the test result) is furthermore dependent on the material tested (e.g. degree of homogeneity and interaction with the food or food simulant and the test conditions applied).

The methods for determining the overall migration into the aqueous simulants (A, B and C) are as follows:

- Bring the FCM in contact with the simulant for a selected time and temperature.
- Separate the sample from the simulant and evaporate the simulant.
- Determine the weight of the residue and calculate the overall migration. As a consequence volatile chemicals, which migrate, are not included in the overall migration value for the aqueous simulants.

The methods for determining the overall migration into the fatty food simulant olive oil are more complicated since it cannot be simply evaporated. However, the use of olive oil is preferred over the use of alternative simulants, because in most cases 95% ethanol or isooctane is a more stringent simulant resulting in a much higher value of overall migration than the value that would be obtained when olive oil is used. In this case, the value of overall migration is measured by determining weight loss from the sample, but because the sample might have absorbed components of the fatty simulant during contact, the weight loss of the sample must be corrected for the amount of absorbed fat. The procedure of determining the migration into fat is:

- Determine the weight before contact (W_1).
- Bring the FCM in contact with the simulant for a specific time and at a defined temperature.
- Separate the sample from the simulant and remove as much simulant as possible.
- Determine the weight after contact (W_2).

- Determine the amount of fat absorbed in the sample using a suitable method (F).
- Calculate the migration (migration is $W_1 - W_2 + F$).

Overall migration methods are described in detail in official CEN methods (UNE-EN 1186:2002) and are listed in Table 11.7.

While the determination of the overall migration requires weighing as the only quantitative analytical technique, the determination of specific migration requires additional analytical testing following the migration step to identify and quantify each compound. In this case, the analytical approach will be dependent on:

- the volatility of the substance(s),
- the polarity of the substance(s),
- the nature of the food or food simulant (e.g. aqueous or fatty),
- the level of determination (e.g. high or low) and
- the functional groups of the substance(s) (considered to define the detection method).

Analytical procedures are described in detail in official CEN methods (UNE-EN 13130:2004) and are listed in Table 11.8. However, any analytical method described in Sections 11.3 and 11.4 can be used to identify and quantify compounds able to migrate.

The EU limit for the overall migration is 10 mg/dm^2 of the food contact surface for all substances that can migrate from food contact material to foods or 60 mg/kg otherwise. On the other hand, the SML for individually authorised substances is fixed on the basis of a toxicological evaluation. The SML is set according to the Acceptable Daily Intake or the Tolerable Daily Intake established by the Scientific Committee on Food. The limit is set on the assumption that every day throughout their lifetime, a person weighing 60 kg eats 1 kg of food packaged in plastics containing the substance in the maximum permitted quantity. Specific migration limits for some substances are defined as not detectable using a method with a detection limit of 10 µg substances/kg food (or food simulant). Furthermore, it should be borne in mind that analytical error in the

Table 11.7. CEN standards related to overall migration in plastics.

Reference Plastics	Title **Materials and articles in contact with foodstuffs – Plastics –**
EN 1186-1:2002	Part 1: Guide to the selection of conditions and test methods for overall migration
EN 1186-2:2002	Part 2: Test methods for overall migration into olive oil by total immersion
EN 1186-3:2002	Part 3: Test methods for overall migration into aqueous food simulants by total immersion
EN 1186-4:2002	Part 4: Test methods for overall migration into olive oil by cell
EN 1186-5:2002	Part 5: Test methods for overall migration into aqueous food simulants by cell
EN 1186-6:2002	Part 6: Test methods for overall migration into olive oil using a pouch
EN 1186-7:2002	Part 7: Test methods for overall migration into aqueous food simulants using a pouch
EN 1186-8:2002	Part 8: Test methods for overall migration into olive oil by article filling
EN 1186-9:2002	Part 9: Test methods for overall migration into aqueous food simulants by article filling
EN 1186-10:2002	Part 10: Test methods for overall migration into olive oil (modified method for use in cases where incomplete extraction of olive oil occurs)
EN 1186-11:2002	Part 11: Test methods for overall migration into mixtures of C-labelled synthetic triglycerides
EN 1186-12:2002	Part 12: Test methods for overall migration at low temperatures
EN 1186-13:2002	Part 13: Test methods for overall migration at high temperatures
EN 1186-14:2002	Part 14: Test methods for 'substitute tests' for overall migration from plastics intended to come into contact with fatty foodstuffs using test media isooctane and 95% ethanol
EN 1186-15:2002	Part 15: Alternative test methods to migration into fatty food simulants by rapid extraction into isooctane and/or 95% ethanol

Table 11.8. Examples of CEN standards/technical specifications on specific migration for plastics.

Reference Plastics	Title Materials and articles in contact with foodstuffs – Plastics –
EN 13130-1:2004	Part 1: Guide to test methods for the specific migration of substances from plastics to foods and food simulants and the determination of substances in plastics and the selection of conditions of exposure to food simulants
EN 13130-2:2004	Part 2: Determination of terephthalic acid in food simulants
EN 13130-3:2004	Part 3: Determination of acrylonitrile in food and food simulants
EN 13130-4:2004	Part 4: Determination of 1,3-butadiene in plastics
EN 13130-5:2004	Part 5: Determination of vinylidene chloride in food simulants
EN 13130-6:2004	Part 6: Determination of vinylidene chloride in plastics
EN 13130-7:2004	Part 7: Determination of monoethylene glycol and diethylene glycol in food simulants
EN 13130-8:2004	Part 8: Determination of isocyanates in plastics
CEN/TS 13130-9:2005	Part 9: Determination of acetic acid, vinyl ester in food simulants
CEN/TS 13130-10:2005	Part 10: Determination of acrylamide in food simulants
CEN/TS 13130-11:2005	Part 11: Determination of 11-aminoundecanoic acid in food simulants
CEN/TS 13130-12:2005	Part 12: Determination of 1,3-benzenedimethanamine in food simulants
CEN/TS 13130-13:2005	Part 13: Determination of 2,2-bis (4-hydroxyphenyl)propane (bisphenol A) in food simulants
CEN/TS 13130-14:2005	Part 14: Determination of 3,3-bis(3-methyl-4-hydroxyphenyl)-2-indoline in food simulants
CEN/TS 13130-15:2005	Part 15: Determination of 1,3-butadiene in food simulants

(*Continued*)

Table 11.8. (*Continued*)

Reference Plastics	Title Materials and articles in contact with foodstuffs – Plastics –
CEN/TS 13130-16:2005	Part 16: Determination of caprolactam and caprolactam salt in food simulants
CEN/TS 13130-17:2005	Part 17: Determination of carbonyl chloride in plastics
CEN/TS 13130-18:2005	Part 18: Determination of 1,2-dihydroxybenzene, 1,3-dihydroxybenzene, 1,4-dihydroxybenzene, 4,4'-dihydroxybenzophenone and 4,4'dihydroxybiphenyl in food simulants
CEN/TS 13130-19:2005	Part 19: Determination of dimethylaminoethanol in food simulants
CEN/TS 13130-20:2005	Part 20: Determination of epichlorohydrin in plastics
CEN/TS 13130-21:2005	Part 21: Determination of ethylenediamine and hexamethylenediamine in food simulants
CEN/TS 13130-22:2005	Part 22: Determination of ethylene oxide and propylene oxide in plastics
CEN/TS 13130-23:2005	Part 23: Determination of formaldehyde and hexamethylenetetramine in food simulants
CEN/TS 13130-24:2005	Part 24: Determination of maleic acid and maleic anhydride in food simulants
CEN/TS 13130-25:2005	Part 25: Determination of 4-methyl-1-pentene in food simulants
CEN/TS 13130-26:2005	Part 26: Determination of 1-octene and tetrahydrofuran in food simulants
CEN/TS 13130-27:2005	Part 27: Determination of 2,4,6-triamino-1,3,5-triazine in food simulants
CEN/TS 13130-28:2005	Part 28: Determination of 1,1,1-trimethylolpropane in food simulants

determination of the overall migration was determined by the EU as 2 mg/dm^2 or 12 mg/kg for the aqueous food simulants (A, B and C), whereas the error is 3 mg/dm^2 or 20 mg/kg for the fatty food simulant (D).

11.6. Food Packaging Migration Studies

Several authors studied the specific migration in different polymeric materials using food simulants or the water inside the plastic bottles. Bentayeb *et al.*[75] used PET to perform migration assays at 70 °C using several food simulants (water, 3% acetic acid, 10% ethanol and 95% ethanol). These conditions recreated the normal use of PET bottles for soft drinks, where diethylenglycol and terephthalic acid were detected at the highest concentrations of 1.060 µg/kg and 0.841 µg/kg, respectively. Votavová *et al.*[86] used the food simulants distilled water, 3% acid acetic and 95% ethanol for 10 days at 40 °C to determine the migration of NP in PVC films. They found that in 95% ethanol, NP release was up to 0.449 mg/g polymer, for distilled water up to 0.091 mg/g polymer and for 3% acetic acid up to 0.079 mg/g polymer. It was concluded that although NP was not used as a direct additive into polymers, it might be originated as a component of a more complex additive preparation (e.g. stabiliser). Li *et al.*[53] determined the migration of BPA from baby bottles filled with Milli-Q grade water for 24 h at different temperatures (24 °C, 40 °C and 100 °C), obtaining the highest value of 4.500 µg/L at 100 °C . Amiridou and Voutsa[46] analysed the migration from 5 brands of 1 L PET bottles and from a 18.9 L PC reusable container. In this case, each brand's own bottled water was used to perform the assays for each respective brand. The assay was performed analysing 3 samples first and then storing 2 more samples outdoors and directly exposed to sunlight for 15 days and 30 days. This assay showed a BPA increase with time from 0.112 µg/L to 0.170 µg/L in the PC container. In PET bottles, BPA, NP, di-(2-ethylhexyl) phthalate (DEHP), dibutyl phthalate (DBP) and diethyl phthalate (DEP) were detected at concentrations up to 0.350 µg/L. Casajuana and Lacorte[52] determined the migration of phthalates, NP, BPA and BADGE using each brand's own bottled water contained in PE, PET and glass containers which were analysed initially and after 10 weeks outdoor storage at temperatures up to 30 °C. In this study, there was also an increase of the detected compounds for all three kinds of samples obtaining the highest value of 0.196 µg DEHP/L for PE samples after the 10 days storage. Le *et al.*[87] performed assays for new and used PC and for HDPE water bottles with a 7 day incubation at room temperature to test the BPA migration. Along the 7 days, there was an increase of BPA migration. For new PC bottles the increase was from 0.36 µg/L to 1.33 µg/L, for used PC it was from 0.29 µg/L to 0.93 µg/L and for new HDPE it was from 0.08 µg/L to 0.19 µg/L. Furthermore, when PC migration was tested at 100 °C , the BPA value was detected up to 7.67 µg/L. Gallart-Ayala *et al.*[54] detected BPA in 11 canned soft drinks including soda, beer, cola beverages, tea and energy drinks at concentrations ranging from 0.044 µg/L to 0.607 µg/L.

On the other hand, other studies used a solvent such as dichloromethane to dissolve the plastic material and then identify their components. Monteiro *et al.*[74] dissolved PET samples with dichloromethane, let the samples macerate for 6 h and sonicated 1 h prior to injection in the GC–MS. Nerín *et al.*[81] identified and quantified the compounds present in a commercially available PC container used for microwave applications. A total dissolution of the polymer was performed with dichloromethane and after reprecipitation of the polymer with methanol, compounds were analysed by HPLC with both UV and fluorescence detection. GC–MS was used for compound confirmation. This procedure showed BPA concentrations of 30 µg/g PC and 2,4-DTBP of 76 µg/g PC at room temperature in the PC container used in a microwave. Votavová *et al.*[86] studied the migration of NP in PVC films, performing an extraction with methanol under reflux for 2 h followed by GC–MS, after using several simulants, and found NP at a concentration up to 0.449 mg/g polymer. Biles *et al.*[88] dissolved PC materials from baby bottles and cups with dichloromethane and also detected BPA ranging from 7 µg/g to 58 µg/g PC.

Benzophenone has been reported to migrate to foodstuffs by mass transference,[32,43] which can occur by set-off (as a result of the contact of the external printed face of the packaging with the inner non-printed face) or by a transfer through the substrate. Benzophenone was found in all tested beverage samples from Italy which where packaged in multilayer laminated carton bricks[89] and in 32 out of 77 bottled waters from all over the world.[14] There is a TDI for benzophenone and 4-hydroxybenzophenone of 10 μg/kg body weight/day and the SML is set at 0.6 mg/kg.

Other authors detected contaminants in plastic material without using any food simulant or solvent. Dutra *et al.*[63] placed pellets of recycled PET and recycled HDPE multilayer into a 20 mL vial and after 10 min the SPME fibre was exposed to the vapours. The results of this study showed that the presence of high levels of some contaminants such as 2,4-DTBP and BHT could be attributed to the misuse of post-consumer PET material and a lack of control in the collection of this material, or due to recontamination in the recycling system or even by external contamination. Sanches-Silva *et al.*[32] used HPLC-UV to perform mathematical models for the prediction of the migration of photoinitiators (e.g. benzophenone), which are used as catalysers for inks and lacquers that are cured with UV light and then can contaminate foodstuffs by mass transference.

Vera *et al.*[90,91] analysed 12 market samples of multilayer materials (laminates) for packaging dry food (tomatoes, cakes, cookies, breadcrumbs, flour or salt) or fresh food (pizza and pastry) that were produced with 5 different adhesives. A total of 25 different compounds from adhesives were detected, including butyric acid, acetic acid, methyl butyrate, 1-butanol and nonanal, which are odorous compounds. The highest concentration was acetic acid at 200 μg/dm^2 for an adhesive.

Plasticisers and additives have also been detected in paper packages. Gartner *et al.*[92] analysed 20 infant food samples (such as sugar, rice, and maize flour) packed in recycled paperboard containers and detected phthalates (mainly diisobutyl phthalate, DiBP) and diisopropyl naphthalenes (DIPN), known incorporated substances in recycled paper. Brauer and Funke[93] detected di-n-butylphthalate, diisopropylnaphthaline, benzophenone and 2-phenylphenol at levels up to 3,000–5,000 μg/kg, observed mainly in finely ground foods like icing sugar or flour. Benzophenone was found in printed paperboard used for cake packaging at a range of 232.7–580.9 mg/m^2. In this study, polypropylene plastic sleeves were found not to be an effective barrier against benzophenone diffusion because it was found in cakes protected with this film after 48 h at 70 °C , where 3,800 μg/kg of benzophenone was found. A derivative of benzophenone, 4-methylbenzophenone, was found in different European breakfast cereal samples packaged in paperboard at levels of up to 3.7 mg/kg in the food.[15] This compound was also detected at a concentration of 798 μg/kg in chocolate crunch muesli (a breakfast cereal) produced in Belgium and packaged in a polyethylene bag inside a printed carton board outer package[94] and in seven cereal samples at concentrations from 384 μg/kg up to 3,729 μg/kg.[95] In a following study, 4-methylbenzophenone was followed in several products on the Belgian market (32 food samples (17 breakfast cereals, 4 croquettes, 4 warm ready meals, 3 biscuits, 2 coffees, 1 ice cream and 1 milk based drink)). This derivate was found in 8 breakfast cereals, 2 croquettes and 1 biscuit at concentrations up to 5,400 μg/kg food.[95]

Droz and Grob[96] investigated the migration of volatile mineral oils present in inks used for printing paper and cardboard and found that the inks can contaminate food, such as cereals and dry baby-food products, at concentrations between 10 mg/kg and 150 mg/kg. Mineral oil migration was also investigated by Biedermann and Grob[44] in a study where 70% (19 ppm) of the volatile mineral oil fraction was transferred from a rice packaging into the rice after a storage time of 8 months.

Little information is available on the types of perfluorochemicals that have the potential to migrate from perfluoro coatings into food. Results from migration tests show mg/kg amounts of perfluoro paper additives/coatings transfer to food oil. Analysis of PTFE cookware showed residual amounts of PFOA in the low μg/kg range. PFOA is present in microwave popcorn bag paper at amounts as high as 300 μg/kg.[19] Another study analysed several polymer coatings (polyethersulphone, PTFE, BPA/epichlorohydrin) intended for stovetop use. The overall migration was measured after heating at 250 °C for 30 min to simulate actual conditions and none of the products exceeded an overall migration limit of 10 mg/dm^2 and perfluorochemicals were not detected.[97]

Mortensen *et al.*[98] developed and applied a multi-analyte method for the determination of 20 PAAs associated with polyurethane (PU) products or azo-colours. The method was validated in-house for water and 3% acetic acid food simulants using spiked migrates from plastic laminates. PAA migration from plastic laminates and black nylon cooking utensils were determined with high levels of 4,4'-methylenedianiline and aniline in 50% of the tested cooking utensils. Pezo *et al.*[99] evaluated the presence of PAAs in eighteen different laminates with PU adhesives. It was showed that 40 NIASs were identified in the 18 different laminates, all of them with similar chemical behaviour to that of PAAs. All the samples complied with the legislation because the total concentration of PAAs was lower than 10 ng/g.

The risk assessment of FCS may be done using the classical approach, which entails the screening of migrates, the identification of unknowns, and their subsequent toxicological evaluation. This procedure is very complex and expensive. However, it is simply not applicable for substances that cannot be identified by current analytical techniques, such as NIAS, or produced in sufficient amounts for toxicological testing. Therefore, much effort is being given to develop analytical methodologies to identify those compounds that can be produced in food packaging materials and to determine the toxicological endpoints to ascertain the risk for humans.

Although there are many studies on the toxicity of food-packaging contaminants and their response effects, it is difficult to establish a relationship between ingestion and adverse health effects such as cancer (exposure-response relationship). Some compounds, such as phthalates, BPA and alkylphenols, are not persistent or they do not bioaccumulate in organisms, but they are constantly ingested throughout life and can produce long-term endocrine disrupting effects well after ingestion (lag time). Hence, some authors consider them as pseudo-persistent. Furthermore, these compounds can act at low doses. Traditionally, chemical testing focuses on doses ranging from 1 mg/kg bw and upwards, but EDs could act at µg/kg bw or ng/kg bw.

11.7. General Legislation

In 1982, Council Directive 82/711/EEC,[100] laying down the basic rules necessary for testing migration of the constituents of plastic materials and articles intended to come into contact with foodstuffs, was published to describe the simulants and test conditions (times and temperatures). These simulants and test conditions were carried out to select the conditions that correspond most closely to the normal or foreseeable conditions of contact for the plastic materials or articles being studied. Some years later, in 1993 and 1997, the Commission Directives 93/8/EEC[6] and 97/48/EC[101] amending Council Directive 82/711/EEC[100] were published to describe in more detail the migration test conditions and describing the basic rules for testing migration of constituents of plastic materials.

Next, Commission Directive 2002/72/EC[5] was published indicating a list of positive substances related to plastic materials intended to come into contact with foodstuffs. In 1987 the European list of the Scientific Committee on Food (SCF) established a positive list of permitted substances for the manufacture of polymeric materials. Moreover, other legislations were published in relation with plastic contact, as Regulation (EC) 1935/2004,[102] Council Directive 85/572/EEC[103], and Commission Directives 2004/19/EC[104] and 2007/19/EC.[105] In addition, in relation to these regulations, two series of European standards were published, EN 1186 (UNE-EN 1186:2002) and EN 13130 (UNE-EN 13130:2005) for overall and specific migration assays, respectively. Both European standards are interpretations of the regulations described before and indicate that the plastic material could be cut into pieces to perform the migration assays.

Nowadays, the main legislation for the safety of materials in contact with food is Regulation (EC) 1935/2004, on materials and articles intended to come into contact with food[102] and Regulation (EC) 2023/2006 which describes the good manufacturing practice of food-contact materials.[106] Both regulations ensure that any molecule transferred to food does not cause changes on organoleptic properties or raise safety concerns. Specifically, Regulation 1935/2004 indicates in its Annex I a list of materials which may be covered by specific measures: active and intelligent materials and articles, adhesives, ceramics, cork, rubbers, glass, ion-exchange resins, metals and alloys, paper and board, plastics, printing inks, regenerated cellulose, silicones, textiles, varnishes and coatings, waxes and wood.[104] In Article 6, it indicates that in the absence of specific measures for the different materials, this Regulation shall not prevent Member States from maintaining or adopting national provisions.

In addition to Regulation 1935/2004, other regulations for specific materials have been published. For plastic materials, taking into account all the regulations and directives set until that moment, European countries decided to unify all of them into the Commission Regulation (EU) No 10/2011 on plastic materials and articles intended to come in contact with food.[85] Regulation 10/2011 covers a list of authorised substances to be used in plastic manufacture, including monomers, additives, polymer production aids, and macromolecules; a description of migration tests; the overall and specific migration; and food simulants and testing conditions for the different uses of plastics in contact with food. In fact, there are some changes concerning the use of different simulants in comparison with the previous regulations. Nowadays there are two regulations amending Commission Regulation 10/2011:[85] Commission Regulations 321/2011,[107] regarding the use of BPA in plastic infant feeding bottles, and 1282/2011,[8] where some new substances are included. According to Regulation 10/2011, to evaluate the compliance with migration limits, the results of specific migration testing obtained with food prevail over the results obtained with food simulants.

Other European legislation on specific materials are Regulation 450/2009 on active and intelligent materials and articles,[94] Regulation 282/2008 on recycled plastic materials,[108] Directive 2007/42/EC on regenerated cellulose film[105] and Directive 84/500/EEC on ceramic articles.[109]

11.8. Conclusions

Plastic components, such as monomers or additives, may migrate into food during processing or storage. Food characteristics such as fat, protein, pigment, and water content, and the plastic type and properties affect the potential migration of contaminants. Compounds that have generated alarm in the food-packaging sector are phthalates, alkylphenols, perfluorinated compounds, BPA and derivatives, primary aromatic amines, and NIAS. These compounds have been detected in food at concentrations that may cause health effects. To control and reduce their presence, the European Legislation has set up procedures and limits for a large number of compounds used in food packaging. To control their presence, overall-migration tests are used to qualitatively determine the migration of plastic components, whereas specific-migration tests are used to identify specific compounds able to migrate. Such tests are generally performed using food simulants, which permit the standardisation of the tests and are less laborious and more precise than the analysis of food directly. For such purposes, several analytical methodologies have emerged for the identification of food-packaging contaminants, quantification of their levels and evaluation of their risk. Such methodologies are based on a selective extraction and analysis by GC or LC coupled to MS. Overall, these activities are intended to control human exposure to a set of compounds that have endocrine-disrupting activities or are potentially carcinogenic or toxic.

Acknowledgements

This chapter was financed by the Ministry of Education, Science and Innovation in Spain (INNPACTO, IPT-2011-0709-060000).

References

1. Marsh, K. and Bugusu, B. (2007). Food packaging: roles, materials, and environmental issues, *J. Food Sci.*, 72, R39–R55.
2. Bolgar, M., Hubball, J., Groeger, J. *et al.* (2008). *Handbook for the Chemical Analysis of Plastic and Polymer Additives*, CRC Press, Boca Raton, FL, USA.
3. Piringer, O.G. and Baner, A.L. (2008). *Plastic Packaging. Interactions with Food and Pharmaceuticals*, Wiley-VCH, Weinheim, Germany.

4. Arvanitoyannis, I.S. and Bosnea, L. (2004). Migration of substances from food-packaging materials to foods, *Crit. Rev. Food Sci. Nutr.*, **44**, 63–76.
5. EU (2002). Commission Regulation 2002/72/EC of 6 August 2002 relating to plastic materials and articles intended to come into contact with foodstuffs, *Off. J. Eur. Communities*.
6. EU (1993). Council Regulation (EEC) No 793/93 of 23 March 1993 on the evaluation and control of the risks of existing substances, *Off. J. Eur. Communities*.
7. EU (1976). Council Directive 76/769/EEC of 27 July 1976 on the approximation of the laws, regulations and administrative provisions of the Member States realting to restrictions on the marketing and use of certain dangerous substances and preparations, *Off. J. Eur. Communities*.
8. EU (2011). Commission Regulation (EU) No 1282/2011 of 28 Nov-ember 2011 amending and correcting Commission Regulation (EU) No 10/2011 on plastic materials and articles intended to come into contact with food, *Off. J. Eur. Communities*.
9. Céspedes, R., Petrovic, M., Raldúa, D. *et al.* (2004). Integrated procedure for determination of endocrine-disrupting activity in surface waters and sediments by use of the biological technique recombinant yeast assay and chemical analysis by LC–ESI-MS, *Anal. Bioanal. Chem.*, **378**, 697–708.
10. Bach, C., Dauchy, X., Severin, I. *et al.* (2013). Effect of temperature on the release of intentionally and non-intentionally added substances from polyethylene terephthalate (PET) bottles into water: chemical analysis and potential toxicity, *Food Chem.*, **139**, 672–680.
11. Gómez-Hens, A. and Aguilar-Caballos, M.P. (2003). Social and economic interest in the control of phthalic acid esters, *TrAC, Trends Anal. Chem.*, **22**, 847–857.
12. ECHA, (2013). *European Chemical Agency website.* Available at: http://echa.europa.eu/ [Accessed March 18, 2013].
13. Voom Saal, F.S. and Hughes, C. (2005). An extensive new literature concerning low-dose effects of bisphenol A shows the need for a new risk assessment, *Environ. Health Perspect.*, **113**, 926–933.
14. Gallart-Ayala, H., Núñez, O. and Lucci, P. (2013). Recent advances in LC–MS analysis of food-packaging contaminants, *TrAC, Trends Anal. Chem.*, **42**, 99–124.

15. EFSA (2004). Opinion of the scientific panel on food additives, flavourings, processing aids and materials in contact with food (AFC) on a request from the commission related to 2,2-bis(4-hydroxyphenyl)propane-bis-(2,3-epoxypropyl) ether (bisphenol A diglycidyl ether, BADGE), *EFSA J.*, **86**, 1–40.
16. EU (2005). Commission Regulation (EC) No 1895/2005 of 18 November 2005 on the restriction of use of certain derivatives in materials and articles intended to come into contact with food, *Off. J. Eur. Commun.*
17. Aurela, B., Science, K. and Söderhjelm, L. (2007). 'Food packaging inks and varnishes and chemical migration into food', in K. Barnes, R. Sinclair and D. Watson (eds), *Chemical Migration and Food Contact Materials*, CRC Press, Boca Raton, FL, USA.
18. Bradley, E.L., Stratton, J.S., Leak, J. *et al.* (2012). Printing ink compounds in foods: UK survey results, *Food Addit. Contam., Part B*, **6**, 73–83.
19. Begley, T.H., Biles, J.E., Cunningham, C. *et al.* (2004). Migration of a UV stabilizer from polyethylene terephthalate (PET) into food simulants, *Food Addit. Contam.*, **21**, 1007–1014.
20. Monteiro, M., Nerín, C. and Reyes, F.G.R. (1996). Determination of UV stabilizers in pet bottles by high performance-size exclusion chromatography, *Food Addit. Contam.*, **13**, 575–586.
21. Ogawa, Y., Kawamura, Y., Wakui, C. *et al.* (2006). Estrogenic activities of chemicals related to food contact plastics and rubbers tested by the yeast two-hybrid assay, *Food Addit. Contam.*, **23**, 422–430.
22. Morohoshi, K., Yamamoto, H., Kamata, R. *et al.* (2005). Estrogenic activity of 37 components of commercial sunscreen lotions evaluated by in vitro assays, *Toxicol. In Vitro*,**19**, 457–469.
23. Suzuki, T., Kitamura, S., Khota, R. *et al.* (2005). Estrogenic and anti-androgenic activities of 17 benzophenone derivatives used as UV stabilizers and sunscreens, *Toxicol. Appl. Pharmacol.*, **203**, 9–17.
24. Muncke, J. (2011). Endocrine disrupting chemicals and other substances of concern in food contact materials: an updated review of exposure, effect and risk assessment, *J. Steroid. Biochem. Mol. Biol.*, **127**, 118–127.
25. Anastassiades, M., Lehotay, S.J., Štajnbaher, D. *et al.* (2003). Fast and easy multiresidue method employment acetonitrile extraction/partitioning and 'dispersive solid-phase extraction' for the determination of pesticide residues in produce, *J. AOAC Int.*, **86**, 412–431.

26. Camel, V. (2001). Recent extraction techniques for solid matrices–supercritical fluid extraction, pressurized fluid extraction and microwave-assisted extraction: their potential and pitfalls, *Analyst*, **126**, 1182–1193.
27. Zarnowski, R. and Suzuki, Y. (2004). Expedient Soxhlet extraction of resorcinolic lipids from wheat grains, *J. Food Compos. Anal.*, **17**, 649–663.
28. Luque de Castro, M.D. and Priego-Capote, F. (2010). Soxhlet extraction: past and present panacea, *J. Chromatogr. A*, **1217**, 2383–2389.
29. Shen, H.-Y. (2005). Simultaneous screening and determination eight phthalates in plastic products for food use by sonication-assisted extraction/GC–MS methods, *Talanta*, **66**, 734–739.
30. Kondyli, E., Demertzis, P.G. and Kontominas, M.G. (1992). Migration of dioctylphthalate and dioctyladipate plasticizers from food-grade PVC films into ground-meat products, *Food Chem.*, **45**, 163–168.
31. Fernandes, A.R., Rose, M. and Charlton, C. (2008). 4-Nonylphenol (NP) in food-contact materials: analytical methodology and occurrence, *Food Addit. Contam., Part A*, **25**, 364–372.
32. Sanches-Silva, A., Andre, C., Castanheira, I. *et al.* (2009). Study of the migration of photoinitiators used in printed food-packaging materials into food simulants, *J. Agric. Food Chem.*, **57**, 9516–9523.
33. Mason, T.J., Paniwnyk, L. and Lorimer, J.P. (1996). The uses of ultrasound in food technology, *Ultrason. Sonochem.*, **3**, S253–S260.
34. Luque-Garcı´a, J.L. and Luque de Castro, M.D. (2003). Ultrasound: a powerful tool for leaching, *TrAC, Trends Anal. Chem.*, **22**, 41–47.
35. Li, X., Zeng, Z., Chen, Y. *et al.* (2004). Determination of phthalate acid esters plasticizers in plastic by ultrasonic solvent extraction combined with solid-phase microextraction using calix[4]arene fiber, *Talanta*, **63**, 1013–1019.
36. Shao, B., Han, H., Li, D. *et al.* (2007). Analysis of alkylphenol and bisphenol A in meat by accelerated solvent extraction and liquid chromatography with tandem mass spectrometry, *Food Chem.*, **105**, 1236–1241.
37. Ferrer, E., Santoni, E., Vittori, S. *et al.* (2011). Simultaneous determination of bisphenol A, octylphenol, and nonylphenol by pressurised liquid extraction and liquid chromatography–tandem mass spectrometry in powdered milk and infant formulas, *Food Chem.*, **126**, 360–367.

38. Cunha, S.C., Lehotay, S.J., Mastovska, K. *et al.* (2007). Evaluation of the QuEChERS sample preparation approach for the analysis of pesticide residues in olives, *J. Sep. Sci.*, **30**, 620–632.
39. Mastovska, K., Dorweiler, K.J., Lehotay, S.J. *et al.* (2009). Pesticide multiresidue analysis in cereal grains using modified QuEChERS method combined with automated direct sample introduction GC–TOFMS and UPLC–MS/MS techniques, *J. Agric. Food Chem.*, **58**, 5959–5972.
40. Lehotay, S.J., Son, K.A., Kwon, H. *et al.* (2010). Comparison of QuEChERS sample preparation methods for the analysis of pesticide residues in fruits and vegetables, *J. Chromatogr. A*, **1217**, 2548–2560.
41. Koesukwiwat, U., Lehotay, S.J., Miao, S. *et al.* (2010). High throughput analysis of 150 pesticides in fruits and vegetables using QuEChERS and low-pressure gas chromatography–time-of-flight mass spectrometry, *J. Chromatogr. A*, **1217**, 6692–6703.
42. Padilla-Sánchez, J.A., Plaza-Bolaños, P., Romero-González, R. *et al.* (2010). Application of a quick, easy, cheap, effective, rugged and safe-based method for the simultaneous extraction of chlorophenols, alkylphenols, nitrophenols and cresols in agricultural soils, analyzed by using gas chromatography–triple quadrupole-mass spectrometry/mass spectrometry, *J. Chromatogr. A*, **1217**, 5724–5731.
43. Rothenbacher, T., Baumann, M. and Fügel, D. (2007). 2-Isopropylthioxanthone (2-ITX) in food and food packaging materials on the German market, *Food Addit. Contam.*, **24**, 438–444.
44. Biedermann, M., Ingenhoff, J.-E., Zurfluh, M. *et al.* (2013). Migration of mineral oil, photoinitiators and plasticisers from recycled paperboard into dry foods: a study under controlled conditions, *Food Addit. Contam., Part A*, **30**, 885–898.
45. Tienpont, B., David, F., Dewulf, E. *et al.* (2005). Pitfalls and solutions for the trace determination of phthalates in water samples, *Chromatographia*, **61**, 365–370.
46. Amiridou, D. and Voutsa, D. (2011). Alkylphenols and phthalates in bottled waters, *J. Hazard. Mater.*, **185**, 281–286.
47. Sørensen, L.K. (2006). Determination of phthalates in milk and milk products by liquid chromatography/tandem mass spectrometry, *Rapid Commun. Mass Spectrom.*, **20**, 1135–1143.

48. Cunha, S.C., Almeida, C., Mendes, E. *et al.* (2011). Simultaneous determination of bisphenol A and bisphenol B in beverages and powdered infant formula by dispersive liquid–liquid micro-extraction and heart-cutting multidimensional gas chromatography–mass spectrometry, *Food Addit. Contam., Part A*, **28**, 513–526.
49. Bono-Blay, F., Guart, A., Fuente, B. *et al.* (2012). Survey of phthalates, alkylphenols, bisphenol A and herbicides in Spanish source waters intended for bottling, *Environ. Sci. Pollut. Res.*, **19**, 3339–3349.
50. Cao, X.-L., Corriveau, J. and Popovic, S. (2009). Levels of bisphenol A in canned soft drink products in Canadian markets, *J. Agric. Food Chem.*, **57**, 1307–1311.
51. Casajuana, N. and Lacorte, S. (2004). New methodology for the determination of phthalate esters, bisphenol A, bisphenol A diglycidyl ether and nonylphenol in commercial whole milk samples, *J. Agric. Food Chem.*, **52**, 3702–3707.
52. Casajuana, N. and Lacorte, S. (2003). Presence and release of phthalic esters and other endocrine disrupting compounds in drinking water, *Chromatographia*, **57**, 649–655.
53. Li, X., Ying, G.-G., Su, H.-C. *et al.* (2010). Simultaneous determination and assessment of 4-nonylphenol, bisphenol A and triclosan in tap water, bottled water and baby bottles, *Environ. Int.*, **36**, 557–562.
54. Gallart-Ayala, H., Moyano, E. and Galceran, M.T. (2011). Analysis of bisphenols in soft drinks by on-line solid phase extraction fast liquid chromatography–tandem mass spectrometry, *Anal. Chim. Acta*, **683**, 227–233.
55. Núñez, O., Gallart-Ayala, H., Martins, C.P.B. *et al.* (2012). New trends in fast liquid chromatography for food and environmental analysis, *J. Chromatogr. A*, **1228**, 298–323.
56. Coelho, E., Ferreira, C. and Almeida, C.M.M. (2008). Analysis of polynuclear aromatic hydrocarbons by SPME-GC-FID in environmental and tap waters, *J. Brazil. Chem. Soc.*, **19**, 1084–1097.
57. Luks-Betlej, K., Popp, P., Janoszka, B. *et al.* (2001). Solid-phase microextraction of phthalates from water, *J. Chromatogr. A*, **938**, 93–101.
58. Peñalver, A., Pocurull, E., Borrull, F. *et al.* (2000). Determination of phthalate esters in water samples by solid-phase microextraction and

gas chromatography with mass spectrometric detection, *J. Chromatogr. A*, **872**, 191–201.
59. Polo, M., Llompart, M., Garcia-Jares, C. *et al.* (2005). Multivariate optimization of a solid-phase microextraction method for the analysis of phthalate esters in environmental waters, *J. Chromatogr. A*, **1072**, 63–72.
60. Salafranca, J., Batlle, R. and Nerı´n, C. (1999). Use of solid-phase microextraction for the analysis of bisphenol A and bisphenol A diglycidyl ether in food simulants, *J. Chromatogr. A*, **864**, 137–144.
61. Dévier, M.-H., Le Menach, K., Viglino, L. *et al.* (2013). Ultra-trace analysis of hormones, pharmaceutical substances, alkylphenols and phthalates in two French natural mineral waters, *Sci. Total Environ.*, **443**, 621–632.
62. Nerı´n, C., Philo, M.R., Salafranca, J. *et al.* (2002). Determination of bisphenol-type contaminants from food packaging materials in aqueous foods by solid-phase microextraction–high-performance liquid chromatography, *J. Chromatogr. A*, **963**, 375–380.
63. Dutra, C., Pezo, D., Freire, M.T.d.A. *et al.* (2011). Determination of volatile organic compounds in recycled polyethylene terephthalate and high-density polyethylene by headspace solid phase microextraction gas chromatography mass spectrometry to evaluate the efficiency of recycling processes, *J. Chromatogr. A*, **1218**, 1319–1330.
64. Montuori, P., Jover, E., Morgantini, M. *et al.* (2008). Assessing human exposure to phthalic acid and phthalate esters from mineral water stored in polyethylene terephthalate and glass bottles, *Food Addit. Contam., Part A*, **25**, 511–518.
65. David, F. and Sandra, P. (2007). Stir bar sorptive extraction for trace analysis, *J. Chromatogr. A*, **1152**, 54–69.
66. Prieto, A., Zuloaga, O., Usobiaga, A. *et al.* (2007). Development of a stir bar sorptive extraction and thermal desorption–gas chromatography–mass spectrometry method for the simultaneous determination of several persistent organic pollutants in water samples, *J. Chromatogr. A*, **1174**, 40–49.
67. León, V.M., Álvarez, B., Cobollo, M.A. *et al.* (2003). Analysis of 35 priority semivolatile compounds in water by stir bar sorptive extraction–thermal desorption–gas chromatography–mass spectrometry: I. Method optimisation, *J. Chromatogr. A*, **999**, 91–101.

68. Sampedro, M.C., Goicolea, M.A., Unceta, N. *et al.* (2009). Sequential stir bar extraction, thermal desorption and retention time locked GC–MS for determination of pesticides in water, *J. Sep. Sci.*, **32**, 3449–3456.
69. Serôdio, P. and Nogueira, J.M.F. (2004). Multi-residue screening of endocrine disrupters chemicals in water samples by stir bar sorptive extraction-liquid desorption-capillary gas chromatography–mass spectrometry detection, *Anal. Chim. Acta*, **517**, 21–32.
70. Guart, A., Calabuig, I., Lacorte, S. *et al.* (2014). Continental bottled water assessment by stir bar sorptive extraction followed by gas chromatography–tandem mass spectrometry (SBSE–GC–MS/MS), *Environ. Sci. Pollut. Res.*,**21**, 2846–2855.
71. Gallart-Ayala, H., Moyano, E. and Galceran, M.T. (2010). Recent advances in mass spectrometry analysis of phenolic endocrine disruptors and related compounds, *Mass Spectrom. Rev.*, **29**, 776–805.
72. Stuart, J.D., Capulong, C.P., Launer, K.D. *et al.* (2005). Analyses of phenolic endocrine disrupting chemicals in marine samples by both gas and liquid chromatography–mass spectrometry, *J. Chromatogr. A*, **1079**, 136–145.
73. Nerín, C. and Acosta, D. (2002). Behavior of some solid food simulants in contact with several plastics used in microwave ovens, *J. Agric. Food Chem.*, **50**, 7488–7492.
74. Monteiro, M., Nerín, C., Rubio, C. *et al.* (1998). A GC–MS method for determining UV stabilizers in polyethyleneterephthalate bottles, *J. High Resolut. Chromatogr.*, **21**, 317–320.
75. Bentayeb, K., Batlle, R., Romero, J. *et al.* (2007). UPLC–MS as a powerful technique for screening the nonvolatile contaminants in recycled PET, *Anal. Bioanal. Chem.*, **388**, 1031–1038.
76. Chik, Z., Haron, D.E.M., Ahmad, E.D. *et al.* (2011). Analysis of melamine migration from melamine food contact articles, *Food Addit. Contam., Part A*, **28**, 967–973.
77. Zheng, X.-L., Yu, B.-S., Li, K.-X. *et al.* (2012). Determination of melamine in dairy products by HILIC-UV with NH_2 column, *Food Control*, **23**, 245–250.
78. Bradley, E.L., Burden, R.A., Bentayeb, K. *et al.* (2013). Exposure to phthalic acid, phthalate diesters and phthalate monoesters from

foodstuffs: UK total diet study results, *Food Addit. Contam., Part A*, **30**, 735–742.
79. Capdeville, M.J. and Budzinski, H. (2011). Trace-level analysis of organic contaminants in drinking waters and groundwaters, *TrAC, Trends Anal. Chem.*, **30**, 586–606.
80. Munch, J.W. (1995). EPA Method 525.2: determination of organic compounds in drinking water by liquid–solid extraction and capillary column gas chromatography/mass spectrometry -Revision 2.0. (EPA/600/R-95-131).
81. Nerín, C., Fernández, C., Domeño, C. *et al.* (2003). Determination of potential migrants in polycarbonate containers used for microwave ovens by high-performance liquid chromatography with ultraviolet and fluorescence detection, *J. Agric. Food Chem.*, **51**, 5647–5653.
82. Brocca, D., Arvin, E. and Mosbæk, H. (2002). Identification of organic compounds migrating from polyethylene pipelines into drinking water, *Water Res.*, **36**, 3675–3680.
83. Bruchet, A. and Janex-Habibi, M. (2004). Endocrine disruptors in environmental and drinking waters: a review, *Tech., Sci., Methodes*, **4**, 81–90.
84. Nerín, C. and Asensio, E. (2007). Migration of organic compounds from a multilayer plastic–paper material intended for food packaging, *Anal. Bioanal. Chem.*, **389**, 589–596.
85. EU (2011). Commission Regulation (EU) No 10/2011 of 14 January 2011 on plastic materials and articles intended to come in contact with food, *Off. J. Eur. Communities*.
86. Votavova, L., Dobiás, J., Voldrich, M. *et al.* (2009). Migration of nonylphenols from polymer packaging materials into food simulants, *Czech J. Food Sci.*, **27**, 293–299.
87. Le, H.H., Carlson, E.M., Chua, J.P. *et al.* (2008). Bisphenol A is released from polycarbonate drinking bottles and mimics the neurotoxic actions of estrogen in developing cerebellar neurons, *Toxicol. Lett.*, **176**, 149–156.
88. Biles, J.E., McNeal, T.P., Begley, T.H. *et al.* (1997). Determination of bisphenol-A in reusable polycarbonate food-contact plastics and migration to food-simulating liquids, *J. Agric. Food Chem.*, **45**, 3541–3544.

89. Sagratini, G., Caprioli, G., Cristalli, G. *et al.* (2008). Determination of ink photoinitiators in packaged beverages by gas chromatography–mass spectrometry and liquid chromatography–mass spectrometry, *J. Chromatogr. A*, **1194**, 213–220.
90. Vera, P., Canellas, E. and Nerín, C. (2014). Migration of odorous compounds from adhesives used in market samples of food packaging materials by chromatography olfactometry and mass spectrometry (GC–O–MS), *Food Chem.*, **145**, 237–244.
91. Vera, P., Uliaque, B., Canellas, E. *et al.* (2012). Identification and quantification of odorous compounds from adhesives used in food packaging materials by headspace solid phase extraction and headspace solid phase microextraction coupled to gas chromatography–olfactometry–mass spectrometry, *Anal. Chim. Acta*, **745**, 53–63.
92. Gärtner, S., Balski, M., Koch, M. *et al.* (2009). Analysis and migration of phthalates in infant food packed in recycled paperboard, *J. Agric. Food Chem.*, **57**, 10675–10681.
93. Brauer, B. and Funke, T. (2008). Bestimmung von Kontaminanten - Papier aus recycelten Fasern und verpackten Lebensmitteln [Detection of contaminants. Paper from recycled fibre and packaged food], *Dtsch. Lebensm.-Rundsch.*, **104**, 330-335.
94. EU (2009). Commission Regulation (EU) No 450/2009 of 29 May 2009 on active and intelligent materials and articles intended to come into contact with food, *Off. J. Eur. Communities.*
95. EFSA (2009). EFSA statement on the presence of 4-methylbenzophenone found in breakfast cereals, *EFSA J.*, **RN-243**, 1–19.
96. Droz, C. and Grob, K. (1997). Determination of food contamination by mineral oil material from printed cardboard using on-line coupled LC–GC–FID, *Z. Lebensm. Unters. Forsch.*, **205**, 239–241.
97. Bradley, E.L., Read, W.A. and Castle, L. (2007). Investigation into the migration potential of coating materials from cookware products, *Food Addit. Contam.*, **24**, 326–335.
98. Mortensen, S.K., Trier, X.T., Foverskov, A. *et al.* (2005). Specific determination of 20 primary aromatic amines in aqueous food simulants by liquid chromatography–electrospray ionization-tandem mass spectrometry, *J. Chromatogr. A*, **1091**, 40–50.

99. Pezo, D., Fedeli, M., Bosetti, O. *et al.* (2012). Aromatic amines from polyurethane adhesives in food packaging: the challenge of identification and pattern recognition using quadrupole–time-of-flight–Mass SpectrometryE, *Anal. Chim. Acta*, 756, 49–59.
100. EU (1982). Council Directive 82/711/EEC of 18 October 1982 laying down the basic rules necessary for testing migration of the constituents of plastic materials and articles intended to come into contact with foodstuffs, *Off. J. Eur. Communities.*
101. EU (1997). Commission Directive 97/48/EC of 29 July 1997 amending for the second time Council Directive 82/711/EEC laying down the basic rules necessary for testing migration of the constituents of plastic materials and articles intended to come into contact with foodstuffs, *Off. J. Eur. Communities.*
102. EU (2004). Regulation (EC) No 1935/2004 of the European Parliament and of the Council of 27 October 2004 on materials and articles intended to come into contact with food and repealing Directives 80/590/EEC and 89/109/EEC, *Off. J. Eur. Communities.*
103. EU (1985). Council Directive 85/572/EEC of 19 December 1985 laying down the list of simulants to be used for testing migration of constituents of plastic materials and articles intended to come into contact with foodstuffs, *Off. J. Eur. Communities.*
104. EU (2004). Commission Directive 2004/19/EC of 1 March 2004 amending Directive 2002/72/EC relating to plastic materials and articles intended to come into contact with foodstuffs, *Off. J. Eur. Communities.*
105. EU (2007). Commission Directive 2007/42/EC of 29 June 2007 relating to materials and articles made of regenerated cellulose film intended to come into contact with foodstuffs, *Off. J. Eur. Communities.*
106. EU (2006). Commission Regulation (EC) No 2023/2006 of 22 December 2006 on good manufacturing practice for materials and articles intended to come into contact with food, *Off. J. Eur. Communities.*
107. EU (2011). Commission Implementing Regulation (EU) No 321/2011 of 1 April 2011 amending Regulation (EU) No 10/2011 as regards the restriction of use of Bisphenol A in plastic infant feeding bottles, *Off. J. Eur. Communities.*

108. EU (2008). Commission Regulation (EU) No 282/2008 of 27 March 2008 on recycled plastic materials and articles intended to come into contact with foods and amending Regulation (EC) No 2023/2006, *Off. J. Eur. Communities*.
109. EU (1984). Council Directive 84/500/EEC of 15 October 2004 on the approximation of the laws of the Member States relating to ceramic articles intended to come in contact with foodstuffs, *Off. J. Eur. Communities*.

Chapter 12

Liquid Chromatography–Mass Spectrometry for the Analysis of Perfluorinated Compounds in Water Samples

Marianna Rusconi, Stefano Polesello and Sara Valsecchi

CNR–IRSA, Water Research Institute, Brugherio, Italy

12.1. Introduction

In the last decade concern about perfluoroalkyl substances (PFAS) has rapidly grown in the scientific community because of their worldwide distribution in different environmental compartments.[1-4] This class of chemicals has been used in a wide range of industrial and consumer products for the past six decades mainly to repel dirt, water and oil.[5,6] PFAS include thousands of chemicals but the environmental studies have been concentrated mainly on perfluoroalkylsulfonic acids (PFSA), such as perfluorooctanesulfonic acid (PFOS), perfluorosalkylsulfonamides (PFASA) and perfluoroalkylcarboxylic acids (PFCA), which include perfluorooctanoic acid (PFOA). PFSA and PFCA are low-molecular-weight surfactants in which all carbons are bonded to fluorine atoms, and which consist of a homologous series of molecules that differ in carbon chain length. PFOS and PFOA have been demonstrated to be persistent in the environment and bioaccumulative in the trophic chain. The accumulation in the aquatic trophic chain poses concerns about the risks for the end consumers, including humans. After a risk-assessment study, the European Commission very recently included PFOS in the list of priority hazardous substances which must be monitored in the EU water bodies, setting an Environmental Quality Standard (EQS) of 0.65 ng/L for freshwater,[7] while the US Environmental Protection

Agency (EPA) proposed Provisional Health Advisories of 400 ng/L and 200 ng/L respectively for PFOA and PFOS in drinking waters[9]. The introduction of regulatory restrictions in the use of PFOS and PFOA[9,10] induced the major PFAS producers to find substitutes for these compounds, especially among the congeners with different chain lengths. These series of homologues, usually from 4 to 14 for PFCA and from 6 to 10 for PFSA, show very different physico-chemical behaviours, which presents a serious challenge for the simultaneous determination of these compounds in water samples.

12.2. Analytical Challenges in the Analysis of Perfluorinated Compounds in Water

The main sources of uncertainties in the determination of PFAS in environmental matrices have been discussed by several authors.[11–13] A fundamental paper on analytical challenges in PFAS analysis[12] listed the main analytical gaps that negatively influence the data accuracy, including availability and purity of native and labelled standards, different isomer profiles of the available commercial standards,[14] matrix effects, isobaric mass interferences, and ion suppression or enhancement. All the recent analytical methods employed for the analysis of various PFAS use stable isotope-labelled internal standards (IS) to monitor recoveries of analytes and to mitigate the effects of matrix during instrumental analysis and quantitation. A certain number of stable isotope-labelled PFCA, PFSA, and PFASA are currently commercially available.

A further problem in PFAS determination is the contamination from lab equipments and instrumentation. It is common to have leakage of PFAS (especially PFOA) from the HPLC systems since perfluorinated polymers are often used in the manufacturing of tubing, septa and seals. A simple solution is to add an extra column (generally a reversed-phase HPLC column) before the injector to delay the elution of PFAS coming from the pump and the system in order to separate sample PFAS from those due to the system contamination.

Another practical analytical challenge of PFAS determination is to find methods that have acceptable limits of quantification, but also reduce the time needed for sample preparation and analysis, notwithstanding the different physicochemical properties of the compounds, in order to achieve the simultaneous determination of different congeners in the various classes of PFAS (PFCA, PFSA, perfluoroalkylphosphonates (PFPA) and PFASA).

As an example of the variability of the physicochemical characteristics of PFCA, it is known that solubility strongly decreases by increasing the chain length (e.g. from 100 g/L for PFHpA to 0.1 g/L for PFUnDA[15,16]), while acidity decreases as the chain length increases (pK_as vary from 0.1 to 3.8 in the range PFBA–PFDoDA).[17] The increase of number of CF_2 moieties also leads to a significant increase in lipophilicity expressed as pK_{ow} (e.g. from PFHxA to PFDoDA, pK_{ow}s increase from 3.68 to 9.21[17]), sulfonates being generally more lipophilic than carboxylates for a certain chain length. The complexity of the physicochemical characteristics of these classes of compounds and the need to develop extraction and separation methods, which should be able to determine this large set of compounds in water in a single run, induced researchers to explore new advancements in chromatographic science; the most significant achievements in this field are reviewed in the present chapter.

12.3. Novel Approaches for High-Throughput Sample Extraction Procedures

The sample extraction step is usually the time-consuming bottleneck of the whole analytical procedure, but in the case of the PFAS analysis it is also a significant source of contamination.[12] Some attempts have been made to overcome the extraction step by direct injection.[18] Centrifugation followed by large-volume injection (LVI) (500 µL) of the supernatant has been used for the determination of 11 PFAS in wastewater with excellent recoveries from WWTP raw influents and effluents in the ranges of about 80–100%.[19] LVI has been demonstrated to be comparable with reversed phase (HLB and C18) SPE as regards the recovery of PFAS with more than 6 carbon atoms with a significant reduction (5–9%) in matrix effects and the possibility of determining short chain PFAS such as PFBA.[20]

12.3.1. *Off-line extraction methods*

Notwithstanding the attempts to avoid the extraction step, liquid–liquid extraction (LLE)[21] or solid-phase extraction (SPE) are still the usual methods for enrichment and isolation of trace levels of PFAS in water samples.[22,23] Innovative versions of the former traditional method, such as vortex-assisted liquid–liquid micro-extraction (VALLME)[24], allow the extraction of lower sample volumes. According to the review of Jahnke and Berger,[25] LLE showed better recoveries for PFAS with carbon chain lengths >7, whereas SPE was best suited for PFAS with <10 carbon atoms. Major drawbacks of the SPE approach in PFAS analysis are the sample contamination and possible losses of the surface-active PFAS to container walls and other materials (tubing, connections), besides the problems inherent in SPE such as breakthrough and clogging of the column.[25] Moreover, it requires the use of large sample volumes followed by solvent evaporation, which translates into the extension of the analysis time to hours.

SPE extraction has usually been carried out with various types of cartridges in the off-line mode. A short review on SPE applications to PFAS analysis has been published in Petrovic *et al.*[23] Even if there is still a limited use of the classical C18 phase,[26] most of the application in recent years employed the functionalized water-wettable, polymer-based SPE sorbent, such as the OASIS family (Waters Corp., Milford, Massachusetts, USA) with different functionalization: hydrophilic-lipophilic-balanced (HLB) reversed-phase[27,28] and weak anion exchange (WAX).[29–32] According to Taniyasu *et al.*[29] HLB reversed-phase sorbent and WAX cartridges showed comparable results for most compounds, but short-chain compounds were only efficiently trapped by WAX cartridges. This is particularly useful for application in deposition chemistry where trifluoroacetic acid (TFA) and several short-chain PFAS have to been monitored.[29] Use of WAX allowed the determination, in a single analytical run, of 40 PFAS including 16 PFCA, 7 PFSA, 6:2 fluorotelomer sulfonate (6:2 FTSA), 3 perfluoroalkyl sulfinates (PFSiA), 4 PFASA, 3 perfluoroalkyl sulfonamidoethanols (PFASE), 3 fluorotelomer carboxylic acids (FTCA) and 3 unsaturated fluorotelomer carboxylic acids (FTUCA) in Rhine river water.[30] In another application of WAX extraction, neutral compounds and telomer alcohols were separated from other poly- and perfluorinated acids by sequential elution with sodium acetate buffer, methanol, and 0.1% NH_4OH in methanol.[31] This method is selective, simple, and robust, with overall recoveries of the target analytes ranging from 75% to 132%.[29] The method based on WAX SPE and LC–MS/MS determination has been adopted as the ISO Standard for water analysis[32].

Alternative sorbents have been tested for PFAS preconcentration from waters. Mixed hemi-micelles-based phases such as cetyltrimethylammonium bromide (CTAB)-coated silica and sodium dodecyl sulfate (SDS)-coated alumina were compared[33], showing better recoveries for CTAB for a narrow range of PFAS (PFBA, PFHpA, PFOA, PFNA and PFDA and PFOS). Efficient extraction and sub-ng/L detection of PFOA in water samples have also been achieved by employing bamboo charcoal as an SPE sorbent.[34] Simultaneous determination of PFPA (carbon chain lengths C6, C8, C10), PFCA (C5–C12), and PFSA (C4, C6, C8, C10) in drinking water has been achieved by a mixed-mode co-polymeric sorbent (C8+quaternary amine),[35] but the whole-method recoveries in HPLC grade water were really satisfactory mainly for PFCA (56–97%), being only 40–56%, and 55–77% for PFPA and PFSA, respectively.

An innovative approach to speed up the extraction step is to adapt solid-phase microextraction (SPME), designed for direct introduction in a GC injector, with LC–MS methods after solvent extraction with very small volumes (generally hundreds of μL). An off-line SPME method for the determination of PFOS and PFOA in environmental samples was developed by sol–gel deposition of a mixed-mode coating to an anodized Ti wire support.[36] The mixed-mode coating is composed of 3-(trimethoxysilyl)-1-propanamine and dimethyloctadecyl [3-(trimethoxysilyl)propyl] ammonium chloride, which provide a synergistic effect of hydrophobic interactions and electrostatic interactions, enhancing the selectivity and the extraction capability toward PFOS and PFOA. The comparison of extraction performance of the proposed SPME fiber with those of 100 μm polydimethylsiloxane (PDMS) and 85 μm polyacrylate (PA) commercial fibers showed that recoveries of proposed mixed mode coating SPME fiber, ranging from

88% to 102%, were 4 times those of PDMS and 55 times those of PA, with detection limits of a few ng/L.[36]

Current research efforts have been devoted to developing affinity media selective for some hazardous fluorous compounds, such as PFOA, which can be used as both a removal medium from contaminated drinking water and a sorbent for analytical purpose. Molecular imprinting, a synthesis methodology for obtaining polymeric artificial receptors, was expected to be suitable for this purpose because the methodology can locate plural functional moieties around a given template molecule to construct a selective binding site as follows: (i) a template molecule is mixed with monomers to form polymerizable complexes, (ii) the complexes are polymerized in the presence of cross-linkers to produce a network polymer with the complexes immobilized and (iii) the template molecule is extracted from the network polymer, which results in a binding site complementary to the template molecule. To date, the molecular recognition ability exhibited by MIPs has been utilized in many analytical applications, such as chromatography, SPE and sensors. MIPs selective for specific fluorous compounds (e.g. PFOA) were synthesized using a fluorous monomer and a fluorous cross-linker, which were expected to show fluorine–fluorine interaction with PFOA[37]. The fluorous MIP selective for PFOA would be potentially useful as a solid-phase extraction sorbent and a sensor chip membrane, although more detailed assessment of selectivity would be required before it is routinely applied. Furthermore, the molecular imprinting with the fluorous monomer and cross-linker would also be useful for the synthesis of a MIP selective for other environmentally concerned fluorous compounds such as PFOS. A monomer with a cationic functionality such as 2-(dimethylamino)ethyl methacrylate, which provides electrostatic interaction with the sulfonate group, would be suitable for the synthesis of a PFOS-selective MIP, whereas methacrylic acid (MAA) was adopted to exploit the formation of a double hydrogen bond between the carboxylic group of MAA and that of PFOA.[37]

Alternatively, the sorption properties of tunable urethane-based copolymer materials containing beta-cyclodextrin (beta-CD) were evaluated with PFOA anions in aqueous solutions. The copolymer materials can be considered as a MIP since their design strategy incorporates a porogen macromolecule, beta-CD, within a cross-linked hexamethylene diisocyanate (HDI) framework.[38] A similar approach has been used to efficiently remove and recover PFOA and PFOS from water, by using surface-tethered beta-CDs on the surface of polystyrene (PS) particles (PS-beta-CDs). The PS-beta-CDs with a 36% beta-CD content showed high adsorption capability against these PFAS from water via inclusion complex formation. The PFCA adsorbed onto PS-beta-CDs were easily and quantitatively recovered by washing with acetone.[39]

Newly developed monolithic capillary column was also tested for selectively enriching PFAS from water. The organic–inorganic hybrid fluorous monolithic capillary column was synthesized by a 'one-pot' approach via the polycondensation of γ-methacryloxy-propyltrimethoxy-silane, then in situ copolymerization of 1H,1H,7H-dodecafluoroheptyl methacrylate and vinyl group on the precondensed siloxanes. The results demonstrated that the optimal column possessed good mechanical stability and high permeability, and PFOA and PFOS in water samples were successfully concentrated about 160 times by this monolithic column, which showed high capacity and selectivity and allowed higher flow rate during the extraction step than usual polymer or silica-based adsorbents.[40]

Other recently developed phases, tested only in research work and not in routine monitoring, are the interior-walls decyl-perfluorinated functionalized magnetic mesoporous microspheres (F17–Fe_3O_4@$mSiO_2$), applied as adsorbents to extract and concentrate PFAS from water samples. The fluorous functionalized interior pore-walls contributed to the high-selective preconcentration of PFAS due to fluorous affinity, and abundant silanol groups on the exterior surface of microspheres contributed to the good dispersibility in water sample. The optimized procedure took only 10 min to extract analytes with high recoveries and merely 800 μL acetonitrile to elute analytes from the magnetic adsorbents. Validation experiments showed good precision (2.6–7.6%) and high recovery (93.4–105.7%)

of the proposed method, but the limits of detection were still too high for environmental applications, being from 8 ng/L to 125 ng/L.[41]

12.3.2. *Automation in extraction procedures*

Commercial stations for automated SPE have been used in several monitoring campaigns in order to process a large number of samples.[27,42] These systems allowed reaching higher concentration factors (1:1000) in order to measure PFAS in low contaminated samples such as sea water at pg/L level.[42] Reduction in sample-loading time can be achieved by using an automated SPE disk extractor; higher flow rate of the SPE disk shortens the extraction time and enables a larger volume of drinking water to be processed, increasing the concentration factor. Ten PFCA, PFSA and PFASA were determined in 500 mL of non-filtered drinking water, adjusted to pH 3.5, by using 47 mm HLB disks at a sample-loading rate of 70–86 mL/min. Analytes were eluted from disks with methanol containing 0.1% NH_4OH (v/v) and the eluates were filtered on nylon syringe filters (0.22 µm) and concentrated to 1 mL at 50 °C under nitrogen.[43] Though the use of automated extractors for disk speeds up the sample loading steps, the remaining preparation steps are labour-intensive and time consuming and the risk of contamination from manipulation is still present.

Only the implementation of on-line SPE made possible the effective development of faster methods by reducing the analysis time and thus increasing the analytical productivity.[22,44] The combination of on-line SPE cleanup and preconcentration and LC–MS/MS was shown to be effective in the analysis of PFAS in natural and wastewater samples.[45–49] The method is labour-saving and cost effective, as the SPE column can be reused more than 300 times (depending on the characteristics of the samples). The on-line SPE method would be particularly suitable in studies where only limited sample material is available, as the sample volume can be as low as 1 mL. Also, compared to more conventional methods, the on-line SPE method uses very small amounts of solvent for sample cleanup, and the method is less labour-intensive and reduces the risk of sample contamination because the only sample preparation necessary for water samples is sedimentation by centrifugation.[47]

The challenge of the on-line SPE methods is still to optimize pre-concentration and elution procedures to achieve a satisfactory accuracy in a single run for a wide numbers of PFAS, which include classes of compounds with different physicochemical variables. In addition, the coupling of on-line SPE with UHPLC is not straightforward because of the high back-pressure generated by the combination of a high flow rate (loading step) with a low particle size (2 μm) column. The use of on-line SPE coupled with UHPLC using sub-2 μm particle size columns has made possible the development of faster methodology, by reducing the analysis time and thus increasing the sample throughput. However, if a multi-residual analysis of a series of analytes with a broad polarity range is carried out, it may be difficult to achieve a satisfying analysis for all target compounds because of the variability in SPE recovery and the loss of chromatographic efficiency. Ideally, by the time the extraction column is switched into the analytical flow path, the trapped analytes should be eluted and re-focused onto the analytical column by the analytical elution gradient. However in multi-residual analysis, the gradient elution for reversed-phase separations usually starts with a high percentage of water in the mobile phase, and the slow elution from the SPE pre-concentration column results in peak broadening, which may cause a decrease in sensitivity.[50–51]

The process of optimization of an on-line SPE method can be long and troublesome because of the effects of the elution gradient; sample volume as well as matrix modifications should be investigated. Understanding how these factors impact an on-line SPE methodology helps to develop a rapid and reliable method for the simultaneous determination of PFAS in water. The optimization work, presented in the following sections (12.3.2.1–12.3.2.3), were carried out in our laboratory for PFCA and PFSA ranging respectively from 4 to 12 and from 4 to 8 carbon atoms (Valsecchi *et al.*, unpublished results). The on-line SPE–UHPLC has been carried out by a Thermo Fisher Scientific EQuan system, which consists of two LC pumps (conventional and UHPLC) with a polar endcapped C18 pre-concentration column (Thermo Hypersil GOLD aQ 12 µm, 20 mm × 2.1 mm), an analytical column (Thermo Hypersil GOLD PFP 1.9 µm, 50 mm × 2.1 mm), and an autosampler equipped with three 6-way VICI valves (Fig. 12.1). Samples were injected into a high volume loop (Fig. 12.1a) and then transferred onto the pre-concentration column by the loading pump using 2 mM ammonium acetate (NH_4OAc) with 5% methanol (MeOH) eluent at 1200 µL/min (Fig. 12.1b). When the enrichment step is completed (260 s), the 6-way valve switches and the elution UHPLC pump starts the elution gradient, composed of two eluents ((A) 2 mM NH_4OAc–5% MeOH and (B) MeOH) at 300 µL/min, through the pre-concentration column to the analytical column (Fig. 12.1c). The loading and the elution gradients are illustrated in Table 12.1. In order to delay the interfering background peaks of PFAS, which are present in solvents or are released from the analytical system, a C18 trap column has been placed between the analytical pump and the injection valve.

12.3.2.1. *Optimisation of the on-line SPE parameters: the elution gradient*

In the on-line SPE method, during the sample-loading step analytes are trapped by the stationary phase of the pre-concentration column. Then, elution of analytes is achieved in back-flush mode by putting in-line the pre-concentration column with the eluting mobile phase.

The first approach is to transfer the separation gradient elution program, optimised by direct injection of analytes, to the on-line procedure. In this case (Table 12.1: 'unchanged gradient'), the gradient program started at 4.34 min, after the end of the loading of the sample into the pre-concentration column. Throughout the loading time, the chromatographic column is maintained in an isocratic flow with the starting gradient conditions (5% methanol). By using these settings, peak broadening and distortion has been observed for the shorter chain and the more polar homologues (PFBA and PFPeA), because these compounds, which show lower affinity for the pre-concentration column stationary phase, are poorly focused on the pre-concentration column and are thereby

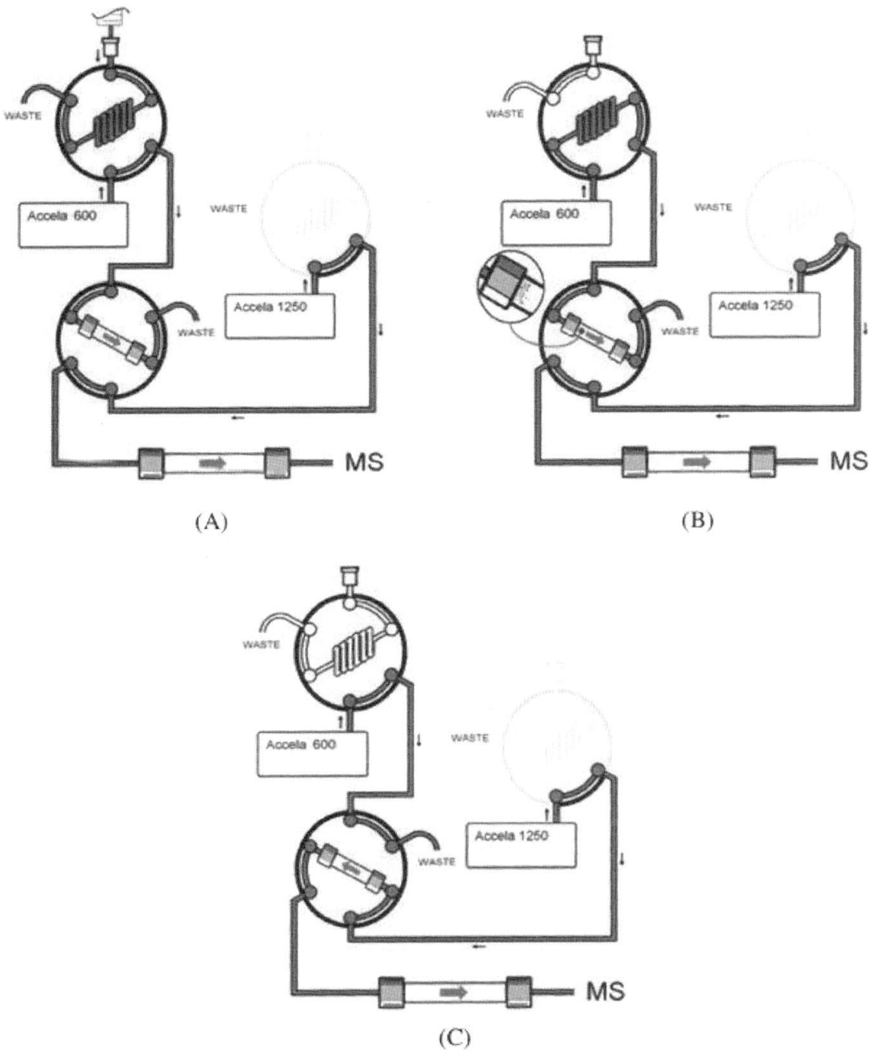

Figure 12.1. Schematic representation of the on-line SPE system used. A) Loading of the sample into the high volume loop. B) Transfer of the sample from the injection loop to the preconcentration column. C) Transfer of the analytes retained in the preconcentration column to the chromatographic column. Reprinted with permission from Valsecchi, S., Mazzoni, M. and Polesello, S. (2013). Analisi multiresiduale LC–MS mediante arricchimento in linea del campione (on-line SPE/UHPLC–ESI–MS/MS) per la determinazione di acidi perfluoroalchilcarbossilati e perfluoroalchilsolfonati nelle acque dolci naturali, *Notiziario dei Metodi Analitici*, **1**, 2–12.[55]

Table 12.1. Elution gradients used by the loading pump and the elution pump. Elution pump flowed at 300 μL/min. Loading time was 260 s. Sample volume was 5 mL.

	Elution pump (*unchanged gradient*)		**Elution pump** (*plug gradient*)		**Elution pump** (*early gradient*)		**Loading pump**		
Time (min)	**A%**	**B%**	**A%**	**B%**	**A%**	**B%**	**Flow (μL/min)**	**A%**	**B%**
0	95	5	95	5	95	5	1200	100	0
3.00					95	5			
3.99			95	5					
4.00			20	80					
4.34	95	5							
4.50			20	80			1200	100	0
4.51			65	35					
4.75			55	45					
5.00			30	70	30	70			
6.00			30	70					
6.34	30	70							
6.50							200	10	90
10.00					0	100			
11.00			0	100					
11.34	0	100							
11.50							200	10	90
13.50					0	100			
14.50			0	100	95	5	1200	100	0
14.84	0	100							
15.50			95	5	95	5			
15.84	95	5							
16.50	95	5	95	5			1200	100	0

(A) 2 mM ammonium acetate with 5% methanol

(B) methanol

transferred already as a broadened band into the chromatographic column. To overcome this problem it is better to implement the 'solvent-plug injection technique'.[50] This technique provides elution bands of only a few seconds width with high percentage solvent in order to obtain a rapid transfer of the analytes from the pre-concentration column to the chromatographic column as well as to keep them focused. Gode *et al.*[50] generated short plugs of high-elution-strength solvent by means of an external loop and an additional LC-pump. We achieved the same high-elution strength solvent plugs by the UHPLC pump, used for LC separation, which has a very low dead volume. A high percentage solvent step (80% B) was inserted in the elution gradient at the switching time, in order to provide a narrow high elution strength eluent band containing all the eluted analytes (Table 12.1: 'plug gradient'). The following separation gradient was adjusted in order to optimize the chromatographic separation (Table 12.1).

The effects of the different gradients used for the elution of pre-concentrated analytes are shown in Fig. 12.2a, where the ratios of the analyte peaks obtained by the plug gradient and those achieved by the unchanged gradient were plotted. Both early- and late-eluting PFCA benefit from 'plug gradient' because of the improvement of the peak shapes, whereas no significant effects are evidenced for PFSA. The peak heights improved up to 3.7 times for the least retained PFCA. Nevertheless, the optimization of the solvent-plug injection technique is laborious and time-consuming because it needs several tests to find out the best gradients both for SPE elution and chromatographic separation, in order to avoid the loss of retention along with severe peak distortion and artefact formation.[50]

A further and simpler approach can be tested: the start of the gradient program on the separation column is anticipated when the loading on the pre-concentration column is still ongoing, and in this way, when the valve is switched on the pre-concentration column, it is the mobile phase, with a higher solvent percentage, which makes it possible to quickly transfer analytes on the separation column with a focusing effect (Table 12.1: 'early gradient'). The optimum mobile-phase composition at the switching time is determined by calculating the composition when the first peak (PFBA) elutes in direct-injection analysis. Improvements in the PFAS peak heights, compared with 'unchanged gradient', are analogous to those of the 'injection plug' and even better for the longer chain PFCA (Fig. 12.2a).

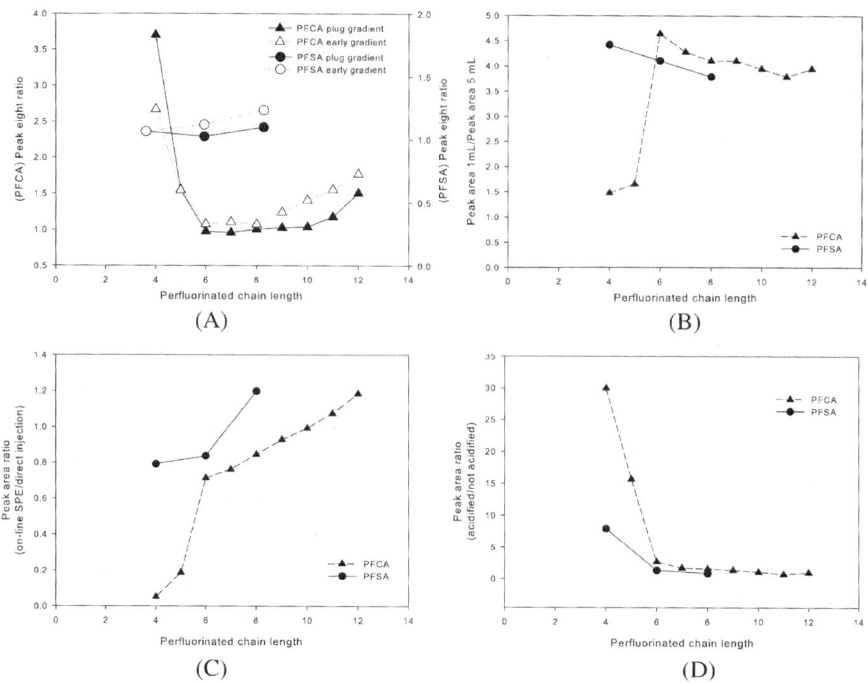

Figure 12.2. On-line SPE method development. A) Effect of different elution gradients, 'plug gradient' or 'early gradient' compared with 'unchanged gradient', on the analyte peak height; injection of 5 mL of acidified aqueous standard at 200 ng/L. B) Effect of the sample volume on the peak area; injection in 'early gradient' mode of aqueous standard at 200 ng/L. C) Extraction efficiency of the 5 mL on-line SPE injection volume; on-line SPE injection in 'early gradient' mode of 5 mL of aqueous standard at 200 ng/L and direct injection of 25 µL of aqueous standard at 40 µg/L. D) Effect of acidification of the sample on the peak area; injection in 'early gradient' mode of aqueous standard at 200 ng/L. Data from (Valsecchi *et al.*, unpublished results)

12.3.2.2. *Optimisation of the on-line SPE parameters: the sample volume*

The effect of sample volume has been evaluated by comparing the peak areas obtained after the injection of 1 mL and 5 mL of aqueous standard (200 ng/L). Proportionality is satisfactory (Fig. 12.2b), with an average response ratio of 4.1, for PFSA and PFCA from 6 to 12 carbon atoms, despite a slight peak broadening. Only PFBA and PFPeA do not show proportional improvement in the response when the sample volume is increased, because the shorter homologues in the PFCA series have smaller breakthrough volumes on the concentration column.

In order to evaluate the actual recovery of the on-line SPE method, the same mass (100 pg) of each analyte has been directly injected into the chromatographic column using a 25 μL loop or has been diluted in 5 mL and pre-concentrated on the on-line SPE. In this way it is possible compare directly peak areas obtained by the two methods: as shown in Fig. 12.2c, ratios of peak areas close to 1 (0.75–1.2) have been obtained for all PFSA homologues and the PFCA with carbon chain length greater than 6, indicating good recovery for these compounds. On the other hand, responses of the more soluble PFCA (PFBA and PFPeA) to on-line SPE methods are significantly lower than the response obtained by direct injection (ratios: 0.1–0.2) confirming that these compounds have poorer affinity for the pre-concentration phase and smaller breakthrough volumes.

12.3.2.3. *Optimisation of the on-line SPE parameters: chemical modification of the sample*

Since all target analytes are acids, samples should be acidified to pH 3 by adding 50 μl of concentrated formic acid before the extraction in order to improve their retention on the endcapped C18 phase of the pre-concentration column. Acidification significantly improves the peak areas of the less retained and more soluble homologues (PFBA, PFPeA, PFHxA and PFBS) whereas no effects are pointed out for the homologues with longer carbon chain (Fig. 12.2d).

The effect of the addition of organic solvent (acetonitrile and methanol) to the water sample before the injection has also been explored. Addition of organic solvent (10%) caused a decrease in the response of shorter and longer-chain PFCA homologues (peak area ratios ranging from 0.65 to 0.75). However, the same decrease in response is not observed when injecting 25 μL of the aqueous standards directly onto the analytical column without a pre-concentration step; in this case peak areas and symmetry increase with perfluorinated carbon chains when standards are prepared in water and methanol. The addition of solvent before the preconcentration step probably causes a loss of retention on the pre-concentration column, because the organic solvent plug competes with the active site of the pre-concentration phase.

12.3.2.4. *Optimisation of the on-line SPE parameters: choice of the preconcentration column*

The optimisation work described in the above paragraphs has been carried out by using a polar embedded reverse-phase column, such as Hypersil GOLD aQ column (2.1 mm × 20 mm, 12 μm particle size) which has been demonstrated to be able to retain a wide range of PFAS, including PFCA (from 4 to 18 carbon atoms), PFSA (from 4 to 10), PFPA (6, 8 and 10) with good limits of detection (LOD) and quantification (LOQ) ranging, in general, from 0.8–10 ng/L to 3–50 ng/L.[46] This method has been validated for different types of matrices (ultrapure water, tapwater and treated wastewater). As an alternative, endcapped, ultrapure, silica-based C18 column has been also used to preconcentrate a narrower range of PFAS (PFHpA, PFOA, PFNA, PFDA, PFBS and PFOS).[47]

Furthermore, polymeric packings have been tested as pre-concentrators in on-line SPE–UHPLC–MS/MS analysis. While the Dionex HRGP (2 mm × 10 mm, 20 μm) gave a poor retention, the POROS HQ column (2.1 mm × 30 mm, 10 μm), working in perfusion mode, permitted good performances in terms of both sorption and chromatographic separation.[48] POROS HQ columns are polymeric packings designed for anion exchange chromatography of peptides, proteins, polynucleotides and other biomolecules in the perfusion flow-through particle chromatography mode. They consist of crosslinked poly(styrene-divinylbenzene) flow-through particles with a bimodal pore size distribution for very rapid mass transport. POROS HQ media is surface-coated with fully quaternized polyethyleneimine. It is a strong anion exchanger with complete surface ionization over a pH range of 1 to 14. The composition of the UHPLC mobile phase was optimized in order to reach the complete elution of the analytes from the sorbent together with a good chromatographic separation on a C18 column for UHPLC. Different buffers (ammonium acetate, ammonium formate and ammonium carbonate) with different concentrations (1 mM, 5 mM and 10 mM) and different organic solvents (methanol, acetonitrile) were tested and compared. The use of 5 mM ammonium acetate solution at pH 8.2 favoured the desorption of PFAS from the SPE cartridge and gave the best separation[48].

C18 microbore columns were also used as enrichment column in an on-line SPE with nano-LC/nano-ESI–MS method for rapid and sensitive determination of PFOA and PFOS in river water.[49] In this case 1 mL of sample was loaded onto a microbore Kromasil C18 enrichment column (5 mm × 1 mm I.D, 5 μm) by a carrier solution consisting of 10 mM NH_4Ac in acetonitrile-water (10:90, v/v) at a flow rate of 250 μL/min, providing on-line analyte enrichment and sample clean-up.

Saito *et al.*[52] developed an automated in-tube SPME method to determine PFOS and PFOA by using an open tubular fused-silica capillary with an inner surface coating as the SPME device. A GC capillary column (60 cm × 0.32 mm I.D.) was used as the in-tube SPME device, and placed between the injection loop and injection needle of the autosampler. CP-Sil 5CB (100% polydimethylsiloxane, film thickness 5 μm), CP-Sil 19CB (14% cyanopropyl phenyl methylsilicone, film thickness 1.2 μm), CP-Wax 52CB (polyethylene glycol, film thickness 1.2 μm), CP-Pora PLOT amine (basic modified styrene divinylbenzene polymer, film thickness 10 μm), Supel-Q PLOT (divinylbenzene polymer, film thickness 17 μm), and Carboxen 1006 PLOT (Carboxen molecular sieves, film thickness 15 μm) were tested for comparison of extraction efficiency. The optimum in-tube SPME conditions were 20 repeated draw/inject cycles of 40 μL of sample using a CP-Pora PLOT amine capillary column as the extraction device. The extracted compounds could be desorbed easily from the capillary by the mobile phase, and no carryover was observed. However, in-tube SPME method has some limitations in the extraction of 'dirty' environmental samples because the capillary used for the extraction is prone to clogging.

12.4. Advanced Chromatographic Separation for the Determination of PFAS in Water Samples

12.4.1. *Advanced stationary phases*

Although there are still methods that employ traditional reversed phases, the advances in stationary-phase technology has also led to significant improvements in PFAS separation by LC–MS methods. Because PFAS are mainly present as anions, useful characteristics of the new phases are the purity and endcapping of the silica particles. As an example, simultaneous separation of different PFAS classes (PFPA, PFCA, and PFSA) has been achieved on a silica-based C18 column that incorporates a bidentate silane, combined with a double-endcapping process that protects the silica from dissolution up to pH 11.5 and improves peak shape at low or intermediate pH.[35] Trace analysis of PFPA is hampered by analytical challenges, such as poor resolution in HPLC and low detector response in MS. At increasing pH, the di-anionic character of PFPA causes them to be less retained with increasing peak widths. Use of 1-methyl piperidine in HPLC–QTOF analysis of perfluoroalkyl acids (PFAA) resulted in significantly better chromatographic resolution (especially for PFPA) and increased detector response for all PFAA due to improved ionization efficiency.[35]

Three HPLC columns (a Symmetry C18, 50 mm × 2.1 mm, 5 μm; a ZORBAX Extended C18, 50 mm × 2.1 mm, 5 μm; a Fluorosep-RP Octyl, 150 mm × 2.1 mm, 5 μm) and one totally porous packed UHPLC column (ZORBAX Rapid Resolution High Throughput, 30 mm × 2.1 mm, 1.8 μm) were compared for the simultaneous analysis of PFPA and PFOS.[53] Once the mobile phase and the injection solvent has been optimized, the Symmetry C18 column provided wider peaks when compared to the Extended C18 column, although they had the same length and particle size (5 cm, 5 µm). This was probably due to a different endcapping of the columns. No results have been obtained for the Fluorosep-RP Octyl column, because the baseline for the PFOPA transition (*m/z* 499→79) was too high and rose with the progressing gradient. Finally, a gradient has been performed with a UHPLC column providing narrower peaks because of the smaller particle size (1.8 µm). Methanol was replaced by acetonitrile to decrease the back-pressure of the system. With acetonitrile, narrower peaks and fewer tailing PFPA peaks were observed than with methanol, whereas no difference in peak shape was observed for PFOS. The UHPLC column with acetonitrile as modifier was selected as the optimum column.[53]

Alternative stationary phases, such as pentafluorophenyl and perfluorooctyl phases, based on fluorinated materials that exhibit group selectivity for general fluorinated compounds, are commercially available. They exhibit retention ability for fluorous compounds via fluorine–fluorine interaction, which is generally observed between fluorous molecules. This characteristic enhances chromatographic efficiencies for PFAS, reducing coeluting interference from the matrix. Samples analysed with perfluorooctyl phase exhibited a lower signal suppression or enhancement (≤ 10%) compared with traditional C18 phase.[54]

12.4.2. *From conventional HPLC to UHPLC and nano-HPLC*

In order to improve chromatographic efficiency and resolution with a concurrent reduction in analytical time, the effective options are to reduce the column geometry or the particle diameter.

The first approach has been to pass from the fully porous particles to the core shell ones. The latter are produced by sol-gel processing techniques, which allow the growth of durable, homogeneous porous shell on a solid silica core. This particle morphology results in less band broadening compared to fully porous particles and thus delivers higher efficiencies, similar to a sub-2 µm UHPLC column, but with about a half of the column back-pressure. The possibility of using a higher flow rate (0.9 mL/min) than that of a narrow-bore column decreases the chromatographic time without sacrificing separation efficiency. A core shell C18 column (2.1 mm × 50 mm, 2.6 μm) has been applied for the separation of ten perfluorinated chemicals (PFCA from 6 to 12; PFSA: 6 and 8; PFASA) in drinking water.[43] Methanol as the organic mobile phase provided sharper peaks and higher analyte signal intensities than acetonitrile. 10 mM N-methylmorpholine (pH 9.6) as the aqueous mobile phase gave the best peak shapes and signal intensities for the PFCA. The basic aqueous solution increased the dissociation of these acid compounds, thus improving the ionization efficiency at ESI.[43]

Another choice is to reduce particle diameters to <2 μm, which leads to a significant improvement in efficiency and allows the use of higher flow rates; the limiting factor is the increase in back-pressure that has been overcome by the introduction of ad hoc, designed UHPLC pumps. UHPLC methods for PFAS determination in water samples have been rapidly spreading.[27,42,55] UHPLC–MS/MS has been compared to capillary liquid chromatography–mass spectrometry (CLC–MS) for the analysis of 18 PFAS in water samples.[56] CLC offers the possibility of performing highly efficient analyses working at μL/min flow. Due to a miniaturization of the chromatographic system, columns of internal diameter of 500 μm or smaller, lower sample sizes, and flow rates can be used. This results in a decrease of solvent consumption and an improvement of the sensitivity since chromatographic dilution is reduced. UHPLC–MS/MS and CLC–MS analyses were carried out using a Zorbax C-18 column (50 mm × 2.1 mm, 1.8 μm) and a Zorbax SB-C18 column (150 mm × 0.5 mm, 3.5 μm), respectively, in gradient elution mode with a mobile phase of ammonium formate and methanol. Both techniques were compared with conventional LC–MS/MS in terms of speed, sensitivity, selectivity and resolution.[56]

Both UHPLC and CLC showed a marked improvement in analysis time relative to HPLC. UHPLC provides better resolution of PFAS than HPLC (only two pairs of compounds co-elute instead of three) and CLC better resolution than UHPLC (only one pair of compounds co-eluted). The best resolution obtained by UHPLC and

CLC is responsible for some improvement of the detection limits compared to HPLC–MS/MS. UHPLC–MS/MS provides the best linearity, shortest analysis time and for most of the compounds the best sensitivity. However, precision was between that obtained by LC–MS/MS and that obtained by CLC–MS. CLC–MS attains the best precision and sensitivity is clearly comparable to that obtained by UHPLC–MS/MS, providing a viable and economical alternative to determine PFAS in environmental samples.[56]

It is also possible to further reduce the inner column diameter (0.1 mm I.D.); this approach provides substantially improved mass sensitivity with the square of the column radius when compared with capillary LC, in addition to the attractive advantage of coupling to the nanospray MS interface, Wilson *et al.*[49] developed a miniaturized on-line SPE–LC–MS method, with the intent of improving the sensitivity for quantitation of PFOA and PFOS in waters. The mass limits of detection of PFOA and PFOS were 0.5 pg and 1 pg, respectively, corresponding to LODs of 500 pg/L and 1 ng/L, respectively. The total time spent on sample preparation, chromatography, and detection was approximately 12 min per sample. The authors justified the use of nano-bore columns and nanospray-MS instrumentation by sensitivity comparisons with columns of larger diameter and a conventional electrospray interface, but with the same single-MS. With the nano column (0.1 mm I.D., pore size 300 Å) connected to the nanospray interface, the PFOS peak intensity increased 7 times compared with that of a capillary column (0.3 mm I.D., 100 Å) coupled to a regular ESI interface injecting 100 pg in both cases. Theoretically, the peak intensity should increase by a factor of approximately 9 when switching from a 0.3 mm column to a 0.1 mm column, but the band-dilution contribution of the tubing between the column and the spray source is higher in the case of nano-LC system.

12.5. Advances in the Mass Spectrometric Detection of Perfluorinated Compounds

In the quantitative monitoring of PFAS in environmental waters, the most commonly used analyser is still the triple quadrupole (QqQ) and the hybrid triple quadrupole with linear trap (QqLIT), which assures adequate sensitivity and productivity. As a drawback, resolution of quadrupole detection may not be sufficient to avoid the occurrence of a false positive due to the co-eluting of isobaric interferences. The most obvious alternative may be the use of high-resolution mass spectrometry (HRMS), provided by instrumentation such as quadrupole time-of-flight (QTOF) and LTQ–Orbitrap, which are fit for fluorinated chemicals because of their significant mass defects. TOF-MS has already been used in routine monitoring of North Sea and Scheldt estuary samples.[57] The use of very narrow mass tolerance windows (< 10 ppm) resulted in a highly selective MS technique for the detection of 14 PFAS in complex aqueous matrices, such as surface-, sewage- and seawater PFAS. Orbitrap, the alternative high-resolution mass spectrometer, has not yet been explored as a detection method in water monitoring, although it has found limited application in determining PFAS in fish[58] and in the identification of potential transformation products of PFOA in biodegradability studies.[59] In our laboratory an exercise of comparison between Orbitrap and QqQ has been carried out on river water samples, showing good correlations between the two techniques (Valsecchi *et al.*, preliminary and unpublished results).

HRMS also has a fundamental role in unknown-screening applications and in the identification of newly introduced compounds, which should substitute already regulated compounds such as PFOS and PFOA in the industrial processes and formulates. Dagnino *et al.*[60] developed a workflow method on an LC–MS–TOF for discovery of novel fluorinated compounds in biological and environmental matrices. A newly produced and never reported polyfluorinated compound (molecular mass 329.0453; formula $CF_3CF_2CF_2OCF(CF_3)COOH$) was detected in surface water downstream of a chemical plant allowing its environmental monitoring.

UHPLC–ESI–QTOF has also been used in a tiered approach for the screening of PFAS mixtures.[61] To distinguish PFAS from other chemicals, characteristic negative mass defects of perfluorinated compounds, their specific losses of 20 Da (HF), and the presence of series of chromatographic peaks belonging to a homologue series with m/z of n × 50 Da (CF_2) or n × 100 Da (CF_2CF_2) were used. MS/MS methods can benefit from using diagnostic fluorine-containing ions, common molecular ions of homologous series, or neutral losses for screenings. When structural isomers are present (e.g. of di- and trialkylated FTOH-derived PFAS), the MS/MS methods must be set up carefully to avoid systematic underestimations. However, to verify the presence of substances, several requirements must be fulfilled to obtain enough identification points according to the EU Directive 2002/657/EC.[62] These include the significance of the diagnostic product ions (i.e. the amount of structural information that is contained in the ion), the number of required precursor-to-product-ion transitions (two transitions for low resolution MS, and one for high resolution MS) and the relative intensity of the product-ion-to-precursor-ion signal (min 10%). Therefore the quantitative method should contain at least one significant product ion that can be used for confirmation.

Sensitivity and specificity of new mass spectrometers also made it possible to explore the possibility of 'jumping' the chromatographic step in the analysis of PFAS in water samples. A first study demonstrated that the desorption/ionization on porous silicon mass spectrometry (DIOS-MS) technique was able to determine PFOS at ppb level in tap water but the method showed a low sensitivity for the less hydrophobic PFOA.[63] DIOS-MS was introduced as a matrix-free LDI–TOF-MS, which succeeded at almost eliminating the background ion interference and offered a new technology for high-speed analysis of low mass compounds.

Several acidic PFAS, such as PFOS and PFOA, were selected as model analytes for demonstrating the feasibility of a novel combined strategy of SPE followed by matrix-assisted laser desorption/ionization time-of-flight mass spectrometry (MALDI–TOF-MS) detection developed for quantifying PFAS in environmental water samples[64]. 1,8-bis (tetramethylguanidino)-naphthalene (TMGN), a superbasic proton sponge, was firstly employed as the matrix for MALDI-TOF-MS. The results showed that deprotonated ions of these PFAS were detected without any other matrix ions' interference. The calibration curves with a wide linear dynamic range

(0.1–10 ng/L for PFOS, PFHxS, and PFBS, and 0.5–50 ng/L for PFOA, PFNA, and PFDA) were obtained. The LOD for PFOS of this method was 0.015 ng/L. The method was compared with a LC–MS/MS method in the analysis of river and wastewaters and shown to be reliable as an alternative method to detect trace PFAS in environmental water samples.

12.6. Conclusions

The field of environmental analysis of perfluorinated compounds is constantly progressing. New compounds have been introduced as substitutes to the classical PFAS, which are subjected to regulation, and there is a need to monitor new classes of PFAS, such as perfluoro-phosphonic acids, phosphate esters, fluorotelomers and perfluorosulfonamides, in order to better describe their fate and to assess the risks connected with the diffusion of these compounds. For most of the substances the main gap in the environmental assessment is the lack of monitoring and toxicological data, which ought to allow the establishment of reliable and protective environmental standards. The need to increase the spatial resolution of monitoring, with the consequent increase of sample numbers, encourages the use of high-throughput methods which make it possible to reduce time without a concomitant loss in chromatographic resolution and sensitivity. The chromatographic technique of choice is UHPLC because capillary and nano-HPLC, though offering adequate sensitivity and productivity, need suitable instrumentations and settings. UHPLC has been successfully integrated in on-line SPE–UHPLC–MS systems, which ensures productivity and allows the handling of many samples without manipulation, significantly reducing the risk of contamination.

The introduction of high-resolution mass spectrometers, which are fit for fluorinated chemicals because of their significant mass defects, allows the possibility of performing untargeted screening and identification. The efficiency of the use of high-resolution MS analysers in routine monitoring for PFAS determination in environmental samples has yet to be tested in field studies.

References

1. Giesy, J.P. and Kannan, K. (2001). Global distribution of perfluorooctane sulfonate in wildlife, *Environ. Sci. Technol.*, **35**, 1339–1342.
2. Giesy, J.P. and Kannan, K. (2002). Perfluorochemical surfactants in the environment, *Environ. Sci. Technol.*, **36**, 146A–152A.
3. Kannan, K. (2011). Perfluoroalkyl and polyfluoroalkyl substances: current and future perspectives, *Environ. Chem.*, **8**, 333–338.
4. Lindstrom, A.B., Strynar, M.J. and Libelo, E.L. (2011). Polyfluorinated compounds: Past, present, and future. *Environ. Sci. Technol.*, **45**, 7954–7961.
5. Prevedouros, K., Cousins, I.T., Buck, R.C. *et al.* (2006). Sources, fate and transport of perfluorocarboxylates, *Environ. Sci. Technol.*, **40**, 32–44.
6. Buck, R.C., Franklin, J., Berger, U. *et al.* (2011). Perfluoroalkyl and polyfluoroalkyl substances in the environment: terminology, classification, and origins, *Integrated Environ. Assess. Manag.*, 7, 513–541.
7. European Commission, (2013). Directive 2013/39/EU of the European Parliament and of the Council of 12 August 2013 amending Directives 2000/60/EC and 2008/105/EC as regards priority substances in the field of water policy, *Official J. Eur. Union*, 24.8.2013 L 226/1.
8. US Environmental Protection Agency, (2009). *Provisional Health Advisories for Perfluorooctanoic Acid (PFOA) and Perfluorooctane Sulfonate (PFOS).* [Internet] US EPA. Available at: http://water.epa.gov/action/advisories/drinking/upload/2009_01_15_criteria_drinking_pha-PFOA_PFOS.pdf [Accessed January 2014].
9. European Commission, (2006). Directive 2006/122/EC of the European Parliament and of the Council of 12 December 2006 amending for the 30th time Council Directive 76/769/EEC on the approximation of the laws, regulations and administrative provisions of the Member States relating to restrictions on the marketing and use of certain dangerous substances and preparations (perfluorooctane sulfonates), *Official J. Eur. Union*, 27.12.2006 L 372/32.
10. US Environmental Protection Agency, (2006). *PFOA Stewardship Program.* [Internet] US EPA. Available at: http://www.epa.gov/opptintr/pfoa/pubs/stewardship/pfoastewardshipbasics.html [Accessed January 2014].
11. de Voogt, P. and Saez, M. (2006). Analytical chemistry of perfluoroalkylated substances, *TRAC, Trends Anal. Chem.*, **25**, 326–342.

12. Martin, J.W., Kannan, K., Berger, U. *et al.* (2004). Analytical challenges hamper perfluoroalky research, *Environ. Sci. Technol.*, **38**, 248A–255A.
13. Valsecchi, S., Rusconi, M. and Polesello, S. (2013). Determination of perfluorinated compounds in aquatic organisms: A review, *Anal. Bioanal. Chem.*, **405**, 143–157.
14. Naile, J.E., Khim, J.S., House, J.N. *et al.* (2010). Standard purity and response factors of perfluorinated compounds, *Toxicol. Environ. Chem.*, **92**, 1219–1232.
15. Jensen, A.A., Poulsen, P.B. and Bossi, R. (2008). Survey and environmental/health assessment of fluorinated substances in impregnated consumer products and impregnating agents, n. 99, *Danish Environmental Protection Agency.*
16. US Environmental Protection Agency, (2003). Environmental and Health Assessment of Perfluoro Sulfonic Acid and its Salts, AR:226–602.
17. Webster, E.M. and Ellis, D.A. (2011). Equilibrium modeling: A pathway to understanding observed perfluorocarboxylic and perfluorosulfonic acid behavior, *Environ. Toxicol. Chem.*, **30**, 2229–2236.
18. Furdui, V.I., Crozier, P.W., Reiner, E.J. *et al.* (2008). Trace level determination of perfluorinated compounds in water by direct injection, *Chemosphere*, **73**, S24–S30.
19. Schultz, M.M., Barofsky, D.F. and Field, J.A. (2006). Quantitative determination of fluorinated alkyl substances by large-volume-injection liquid chromatography–tandem mass spectrometry: Characterization of municipal wastewaters, *Environ. Sci. Technol.*, **40**, 289–295.
20. Backe, W.J. and Field, J.A. (2012). Is SPE necessary for environmental analysis? A quantitative comparison of matrix effects from large-volume injection and solid-phase extraction based methods, *Environ. Sci. Technol.*, **46**, 6750–6758.
21. Flores, C., Ventura, F., Martin-Alonso, J. *et al.* (2013). Occurrence of perfluorooctane sulfonate (PFOS) and perfluorooctanoate (PFOA) in NE Spanish surface waters and their removal in a drinking water treatment plant that combines conventional and advanced treatments in parallel lines, *Sci. Total Environ.*, **461–462**, 618–326.
22. Villagrasa, M., López de Alda, M. and Barceló, D. (2006). Environmental analysis of fluorinated alkyl substances by liquid chromatography–(tandem) mass spectrometry: A review, *Anal. Bioanal. Chem.*, **386**, 953–972.

23. Petrovic, M., Farré, M., Lopez de Alda, M. *et al.* (2010). Recent trends in the liquid chromatography–mass spectrometry analysis of organic contaminants in environmental samples, *J. Chromatogr. A*, **1217**, 4004–4017.
24. Papadopoulou, A., Román, I.P., Canals, A. *et al.* (2011). Fast screening of perfluorooctane sulfonate in water using vortex-assisted liquid–liquid microextraction coupled to liquid chromatography-mass spectrometry, *Anal. Chim. Acta.*, **691**, 56–61.
25. Jahnke, A. and Berger, U. (2009). Trace analysis of per- and polyfluorinated alkyl substances in various matrices: How do current methods perform?, *J. Chromatogr. A*, **1216**, 410–421.
26. Tseng, C.L., Liu, L.L., Chen, C.M. *et al.* (2006). Analysis of perfluorooctanesulfonate and related fluorochemicals in water and biological tissue samples by liquid chromatography–ion trap mass spectrometry, *J. Chromatogr. A*, **1105**, 119–126.
27. Loos, R., Gawlik, B.M., Locoro, G. *et al.* (2009). EU-wide survey of polar organic persistent pollutants in European river waters *Environ. Pollut.*, **157**, 561–568.
28. Anumol, T., Merel, S., Clarke, B.O. *et al.* (2013). Ultrahigh-performance liquid chromatography–tandem mass spectrometry for rapid analysis of trace organic contaminants in water, *Chem. Cent. J.*, **7**, 104.
29. Taniyasu, S., Kannan, K., Yeung, L.W.Y. *et al.* (2008). Analysis of trifluoroacetic acid and other short-chain perfluorinated acids (C2–C4) in precipitation by liquid chromatography-tandem mass spectrometry: comparison to patterns of long-chain perfluorinated acids (C5–C18), *Anal. Chim. Acta.*, **619**, 221–230.
30. Moeller, A., Ahrens, L., Surm, R. *et al.* (2010). Distribution and sources of polyfluoroalkyl substances (PFAS) in the River Rhine watershed, *Environ. Pollut.*, **158**, 3243–3250.
31. Taniyasu, S., Kannan, K., So, M.K. *et al.* (2005). Analysis of fluorotelomer alcohols, fluorotelomer acids, and short- and long-chain perfluorinated acids in water and biota, *J. Chromatogr. A*, **1093**, 89–97.
32. International Standard ISO 25101 (2009). Water quality determination of perfluorooctanesulfonate (PFOS) and perfluorooctanoate (PFOA) method for unfiltered samples using solid-phase extraction and liquid chromatography/mass spectrometry, Geneva, CH.
33. Zhao, X.L., Li, J.D., Shi, Y.L. *et al.* (2007). Determination of perfluorinated compounds in wastewater and river water samples by mixed

hemimicelle-based solid-phase extraction before liquid chromatography–electrospray–tandem mass spectrometry detection, *J. Chromatogr. A*, **1154**, 52–59.

34. Zhao, R.S., Wang, X., Lin, J.M. *et al.* (2008). Using bamboo charcoal as solid-phase extraction adsorbent for the ultratrace-level determination of perfluorooctanoic acid in water samples by high-performance liquid chromatography–mass spectrometry, *Anal. Bioanal. Chem.*, **390**, 1671–1676.
35. Ullah, S., Alsberg, T. and Berger, U. (2011). Simultaneous determination of perfluoroalkyl phosphonates, carboxylates, and sulfonates in drinking water, *J. Chromatogr. A*, **1218**, 6388–6395.
36. Chen, C., Wang, J., Yang, S. *et al.* (2013). Analysis of perfluorooctane sulfonate and perfluorooctanoic acid with a mixed-mode coating-based solid-phase microextraction fiber, *Talanta*, **114**, 11–16.
37. Takayose, M., Nishimoto, K. and Matsui, J. (2012). A fluorous synthetic receptor that recognizes perfluorooctanoic acid (PFOA) via fluorous interaction obtained by molecular imprinting, *Analyst*, **137**, 2762–2765.
38. Karoyo, A.H. and Wilson, L.D. (2013). Tunable macromolecular-based materials for the adsorption of perfluorooctanoic and octanoic acid anions, *J. Colloid Interf. Sci.*, **402**,196–203.
39. Kawano, S., Kida, T., Takemine, S. *et al.* (2013). Efficient removal and recovery of perfluorinated compounds from water by surface-tethered beta-cyclodextrins on polystyrene particles. *Chem. Lett.*, **42**, 392–394.
40. Xiong, X.Y., Yang, Z.H., Huang, Y.B. *et al.* (2013). Organic–inorganic hybrid fluorous monolithic capillary column for selective solid-phase microextraction of perfluorinated persistent organic pollutants, *J. Sep. Sci.*, **36**, 923–931.
41. Yang, L., Yu, W.J., Yan, X.M. *et al.* (2012). Decyl-perfluorinated magnetic mesoporous microspheres for extraction and analysis perfluorinated compounds in water using ultrahigh-performance liquid chromatography–mass spectrometry, *J. Sep. Sci.*, 35, 2629–2636.
42. Loos, R., Tavazzi, S., Paracchini, B. *et al.* (2013). Analysis of polar organic contaminants in surface water of the northern Adriatic Sea by solid-phase extraction followed by ultrahigh-pressure liquid

chromatography–QTRAP® MS using a hybrid triple-quadrupole linear ion trap instrument, *Anal. Bioanal. Chem.*, **405**, 5875–5885.

43. Chang, Y.C., Chen, W.L., Bai, F.Y. *et al.* (2012). Determination of perfluorinated chemicals in food and drinking water using high-flow solid-phase extraction and ultrahigh-performance liquid chromatography–tandem mass spectrometry, *Anal. Bioanal. Chem.*, **402**, 1315–1325.
44. Nunez, O., Gallart-Ayala, H., Martins, C.P.B. *et al.* (2012). New trends in fast liquid chromatography for food and environmental analysis, *J. Chromatogr. A*, **1228**, 298–323.
45. Takino, M., Daishima, S. and Nakahara, T. (2003). Determination of perfluorooctane sulfonate in river water by liquid chromatography/atmospheric pressure photoionization mass spectrometry by automated on-line extraction using turbulent flow chromatography, *Rapid. Comm. Mass. Spectrom.*, **17**, 383–390.
46. Llorca, M., Farré, M., Pico, Y. *et al.* (2012). Analysis of perfluoroalkyl substances in waters from Germany and Spain, *Sci. Total Environ.*, **431**, 139–150.
47. Enevoldsen, R. and Juhler, R.K. (2010). Perfluorinated compounds (PFASs) in groundwater and aqueous soil extracts: Using on-line SPE–LC–MS/MS for screening and sorption characterisation of perfluorooctane sulphonate and related compounds, *Anal. Bioanal. Chem.*, **398**, 1161–1172.
48. Gosetti, F., Chiuminatto, U., Zampieri, D. *et al.* (2010). Determination of perfluorochemicals in biological, environmental and food samples by an automated on-line solid-phase extraction ultrahigh-performance liquid chromatography–tandem mass spectrometry method, *J. Chromatogr. A*, **1217**, 7864–7872.
49. Wilson, S.R., Malerød, H., Holm, A. *et al.* (2007). On-line SPE–nano-LC–nanospray-MS for rapid and sensitive determination of perfluorooctanoic acid and perfluorooctane sulfonate in river water, *J. Chromatogr. Sci.*, **45**, 146–152.
50. Gode, D., Martin, M.M., Steiner, F. *et al.* (2012). Rapid narrow band elution for on-line SPE using a novel solvent plug injection technique, *Anal. Bioanal. Chem.*, **404**, 433–445.

51. Rodier, D.R. and Birks, J.W. (1994). Dual-injector solvent elution and focusing technique for the on-line analysis of solid-phase extraction cartridges in HPLC, *Chromatographia*, **39**, 45–50.
52. Saito, K., Uemura, E., Ishizaki, E. *et al.* (2010). Determination of perfluorooctanoic acid and perfluorooctane sulfonate by automated in-tube solid-phase microextraction coupled with liquid chromatography–mass spectrometry. *Anal. Chim. Acta.*, **658**, 141–146.
53. Esparza, X., Moyano, E., de Boer, J. *et al.* (2011). Analysis of perfluorinated phosphonic acids and perfluorooctane sulfonic acid in water, sludge and sediment by LC–MS/MS, *Talanta*, **86**, 329–336.
54. Ballesteros-Gómez, A., Rubio, S. and van Leeuwen, S. (2010). Analysis of perfluorinated compounds in biota by microextraction with tetrahydrofuran and liquid chromatography–ion isolation-based ion-trap mass spectrometry, *J. Chromatogr. A*, **1217**, 5913–5921.
55. Valsecchi, S., Mazzoni, M. and Polesello, S. (2014). Analisi multiresiduale LC–MS mediante arricchimento in linea del campione (on-line SPE/UHPLC–ESI–MS/MS) per la determinazione di acidi perfluoroalchilcarbossilati e perfluoroalchilsolfonati nelle acque dolci naturali, *Notiziario dei Metodi Analitici*, 1, 2–12. Downloadable from http://www.irsa.cnr.it/Docs/Notiz/notiz2013_%2812%29.pdf.
56. Onghena, M., Moliner-Martinez, Y., Pico, Y. *et al.* (2012). Analysis of 18 perfluorinated compounds in river waters: Comparison of high-performance liquid chromatography–tandem mass spectrometry, ultra-high-performance liquid chromatography–tandem mass spectrometry and capillary liquid chromatography–mass spectrometry, *J. Chromatogr. A*, **1244**, 88–97.
57. Wille, K., Vanden Bussche, J., Noppe, H. *et al.* (2010). A validated analytical method for the determination of perfluorinated compounds in surface-, sea- and sewage water using liquid chromatography coupled to time-of-flight mass spectrometry, *J. Chromatogr. A*, **1217**, 6616–6622.
58. Kadar, H., Veyrand, B., Antignac, J.P. *et al.* (2011). Comparative study of low- versus high-resolution liquid chromatography–mass spectrometric strategies for measuring perfluorinated contaminants in fish, *Food Addit. Contam. A*, **28**, 1261–1273.

59. Liou, J.S.C., Szostek, B., DeRito, C.M. *et al.* (2010). Investigating the biodegradability of perfluorooctanoic acid, *Chemosphere*, **80**, 176–183.
60. Dagnino, S., Lindstrom, A.B., McMahen, R. *et al.* (2013). 'Identification of novel per- and poly-fluorinated compounds in environmental and biological matrices using LC–MS–TOF', 23rd SETAC Europe Annual meeting, 12-16 May 2013.
61. Trier, X., Granby, K. and Christensen J.H. (2011). Tools to discover anionic and nonionic polyfluorinated alkyl surfactants by liquid chromatography–electrospray ionization–mass spectrometry, *J. Chromatogr. A*, **1218**, 7094–7104.
62. European Commission, (2002). Decision 2002/657/EC implementing Council Directive 96/23/EC concerning the performance of analytical methods and the interpretation of results, *Off. J. Eur. Communities*, 17.08.2002, L 221/8.
63. Kawasaki, H., Shimomae, Y., Watanabe, T. *et al.* (2009). Desorption/ionization on porous silicon mass spectrometry (DIOS-MS) of perfluorooctane sulfonate (PFOS), *Colloids Surf. A*, **347**, 220–224.
64. Cao, D., Wang, Z.D., Han, C.G. *et al.* (2011). Quantitative detection of trace perfluorinated compounds in environmental water samples by matrix-assisted laser desorption/ionization–time-of-flight mass spectrometry with 1,8-bis(tetramethylguanidino)-naphthalene as matrix, *Talanta*, **85**, 345–352.

Chapter 13

Determination of Phenolic Compounds in Food Matrices: Application to Characterization and Authentication

Javier Saurina and Sonia Sentellas

Department of Analytical Chemistry, University of Barcelona, Barcelona, Spain

13.1. Introduction

The quality of food products is an issue of great interest in our society. For this reason, in recent years, the development of new methods focused on the analysis and characterization of food products such as meat, fish, fruits and vegetables has increased dramatically.[1–3] In many cases, thorough assays are required to assess some food aspects dealing with variety, geographical origin, manufacturing practices, etc.[3,4] Consumer preferences regarding food products are often influenced by complex combinations of organoleptic (e.g., color, taste and aroma) and socioeconomic (e.g., ecological production, guaranteed origin and quality) factors.

Tests to estimate product features and quality sometimes rely on sensorial assays by expert panelists. In general, such assays are expensive, time-consuming and difficult to be implemented for routine control, so alternative strategies for solving these drawbacks are more and more in demand. In some cases, instrumental (analytical) approaches have been proposed as a way to gain information on food features.[5,6] It should be pointed out that instrumental methods allow the analysis of large series of samples in a very simple, sensitive and reproducible way, while avoiding the subjectivity of human sensory tests.

Data resulting from instrumental methods needs to be analyzed using appropriate chemometric methods in order to extract the underlying chemical information. Exploratory and interpretative studies are carried out to try to find (bio)chemical markers with food product features.[7–9] In the scientific literature, it has been demonstrated that contents of certain organic and inorganic food constituents are excellent descriptors of food properties. For instance, amino acids, biogenic amines, alcohols, aldehydes, esters, acids, terpenes, and, of course, polyphenols have been exploited for the characterization, classification and authentication of food products.[10]

In this chapter, we discuss the possibilities of the compositional profiles of polyphenols, determined by liquid chromatography with mass spectrometry (LC–MS), as a source of data to be used in food analysis. Steps comprising the overall procedure according to a metabolomic approach will be described, including sample treatment, chromatography and MS detection, data processing, and identification of relevant biomarkers. Illustrative examples dealing with applications of polyphenols to food analysis will be given.

13.2. Polyphenols

Polyphenols comprise a large family of naturally occurring secondary metabolites of plants.[11] Food products such as berries, chocolate, tea, wine and fresh fruits have been recognized as some of the principal dietary sources of polyphenols for humans, with concentrations ranging from 1 mg/kg to hundreds of mg/kg. Polyphenols can also be found in high quantities in transformed products, dietetic supplements and pharmaceuticals.

Chemically, polyphenols are molecules containing, at least, an aromatic ring with one or several –OH groups. Polyphenols can be classified into four main families according to the number of phenol rings that they contain as well as the structural elements that bind these rings together (see Table 13.1):[11,12]

Table 13.1. Chemical structures of polyphenols.

Family	Subfamily	General structure	Substitutions
Phenolic acids	Benzoic acids	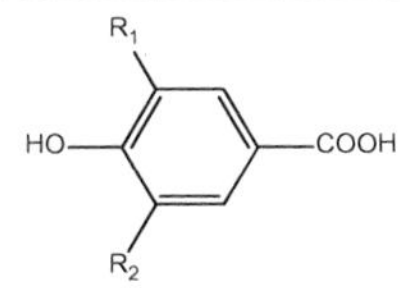	R1: H, OH, OCH_3 R2: H, OH, OCH_3
	Cinnamic acids	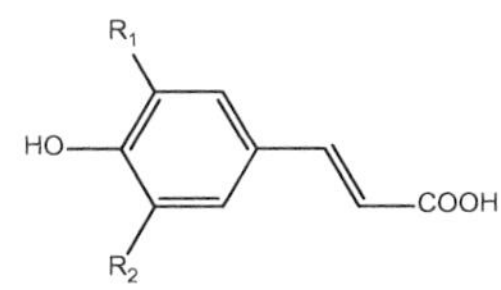	R1: H, OH, OCH_3 R2: H, OCH_3
Flavonoids	Flavones Flavonols	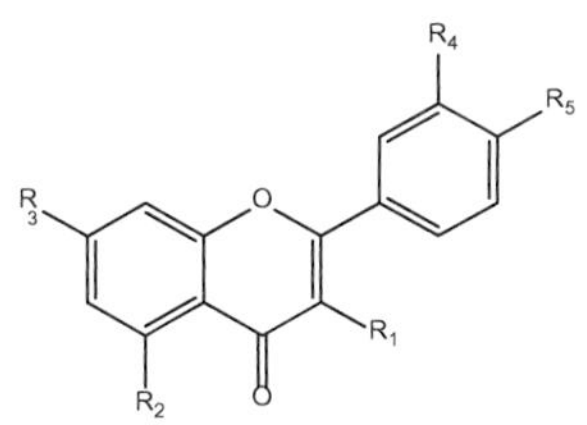	R1: H, OH R2: H, OH R3: OH R4: H, OH
	Flavonones Flavononols	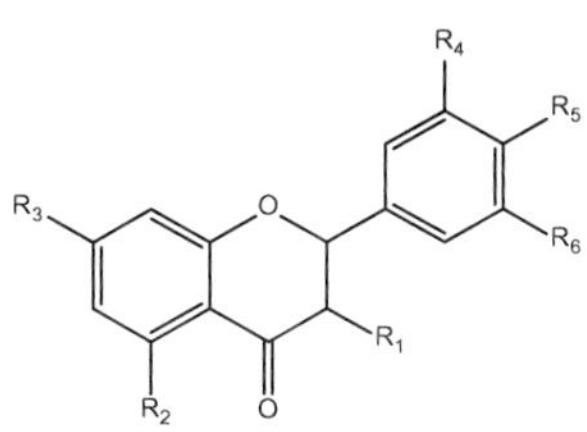	R1: H, OH R2: OH R3: H, OH R4: H, OH R5: OH, OCH_3 R6: H, OH
	Catechins	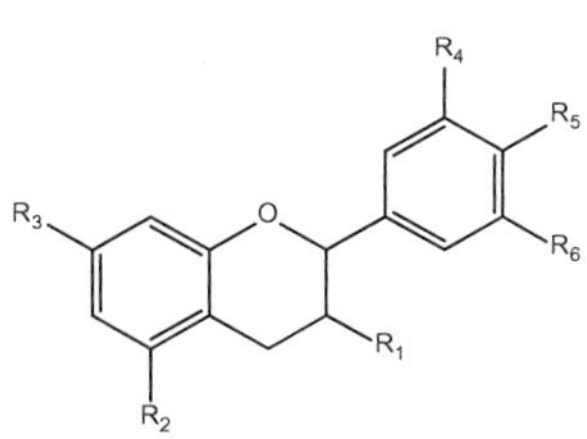	R1 to R3: OH R4: H, OH R5: OH R6: H, OH
	Anthocyanidins	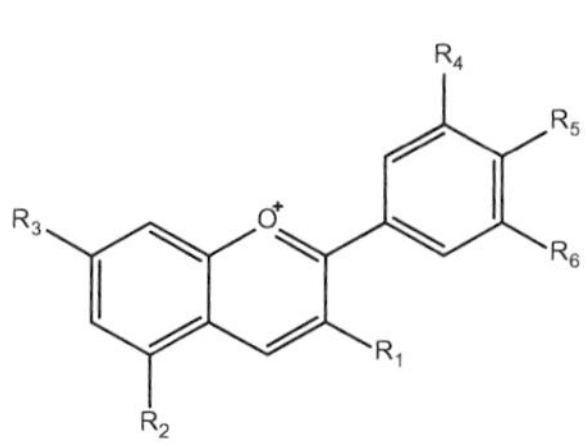	R1: H, OH R2: OH, OCH_3 R3: OH R4: H, OH R5: OH R6: H, OH

(*Continued*)

Table 13.1. *(Continued)*

Family	Subfamily	General structure	Substitutions
	Isoflavone	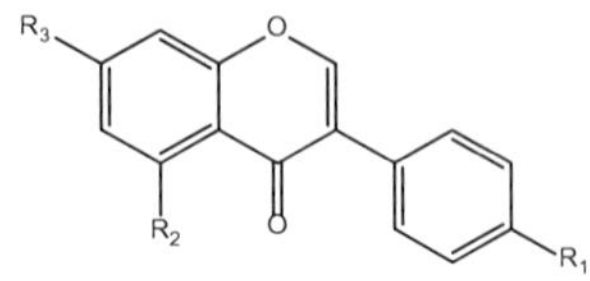	R1: OH R2 to R3: H, OH
Stilbenes		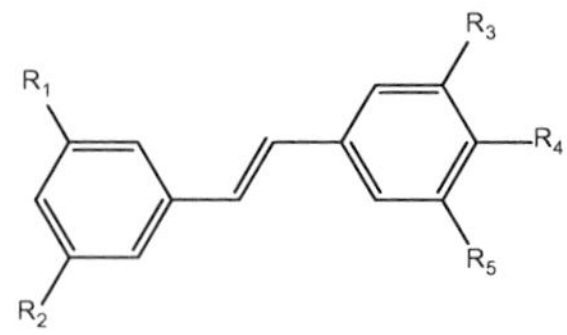	R1 to R4: H, OH, OCH_3 R5: H, OH
Lignans		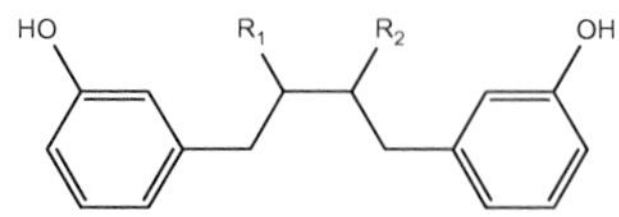	R1, R2: H, OH, others
Others	Chalcones	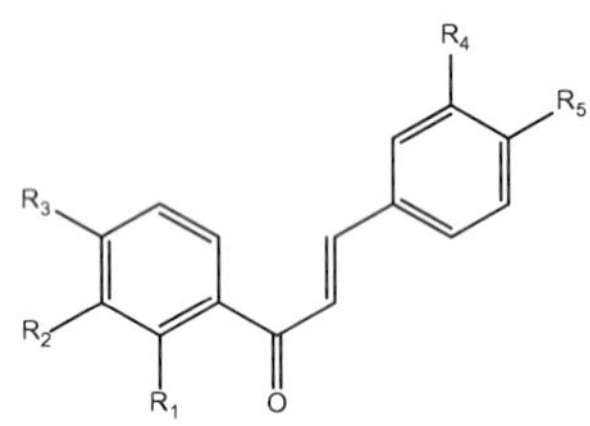	R1 to R5: H, OH

(i) Phenolic acids, comprised of the two subclasses of hydroxybenzoic and hydroxycinnamic acids. They account for 30% of total dietary polyphenols and, in general, cinnamic derivatives (e.g. caffeic and coumaric acids) are more abundant in fruits than the benzoic ones. Additionally, as shown in Fig. 13.1A, phenolic acids are the basis of some complex polyphenols referred to as 'hydrolysable tannins' in which gallic or ellagic acid esterifies one or several the hydroxyl groups of a sugar residue.

(ii) Flavonoids, consisting of two aromatic rings linked by three carbon atoms that form an oxygen heterocycle. Flavonoids account for 60% of total dietary polyphenols. Simple flavonoids can be divided into six subclasses, namely: flavonols, flavones, isoflavones, flavanones, anthocyanidins and flavanols. Additionally,

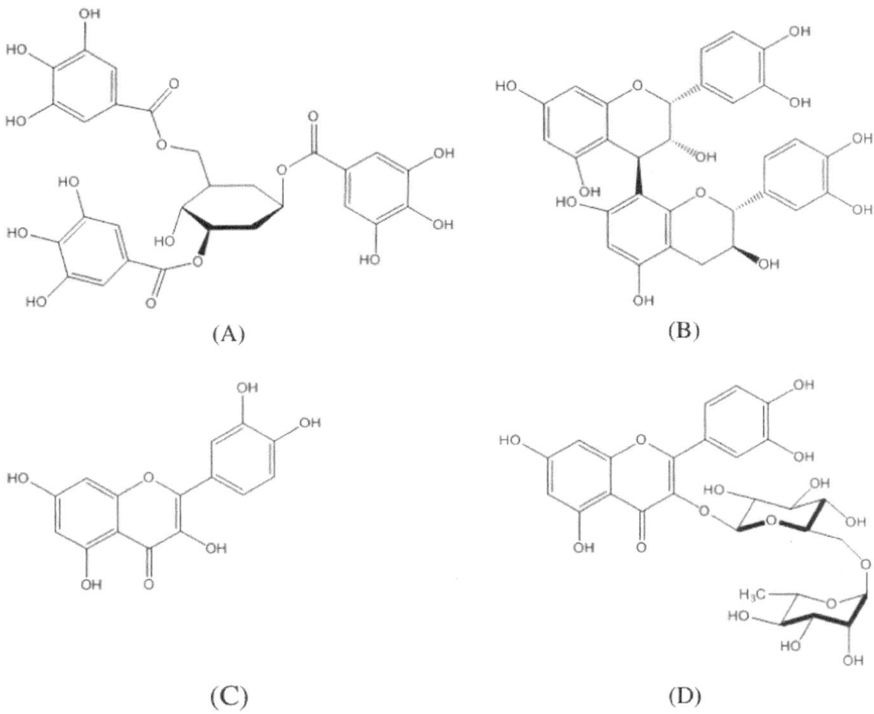

Figure 13.1. Examples of structures of some complex polyphenols. (A) hydrolysable tannin; (B) condensed tannin (procyanidin B2); (C) flavonoid aglycone (quercetin); (D) flavonoid glycoside (rutin).

there are complex flavonoids, commonly included in the generic group of tannins, that consist of one or several flavonoid moieties attached in different ways. On one hand, condensed tannins, such as the so-called proanthocyanidins, contain two or more flavanols linked through C–C or C–O interflavan bonds (see Fig. 13.1B). On the other hand, complex tannins are molecules in which a hydroxyl group (or several) of the flavonoid moiety, often catechin, is bounded glycosidically to a gallotannin or ellagitannin unit.

(iii) Stilbenes, characterized by a double bond connecting two aromatic rings. Despite being found in low quantities in the human diet, the nutritional significance of compounds such as *trans*-resveratrol is outstanding.

(iv) Lignans, a minor class of polyphenols consisting of two phenylpropane units. The main food source of lignans is linseed although they are also found at lower concentrations in cereals, fruits and vegetables. Other groups of compounds structurally related to polyphenols comprise the chalcones, humulones, alcohols, etc. For all these families of polyphenols, compounds occur as single molecules, the so-called aglycones (Fig. 13.1C), or conjugated with one or more sugar residues thus resulting in the corresponding glycosides (Fig. 13.1D).[11,12]

The great relevance of polyphenols in food products comes from their contribution to sensorial and functional properties.[13] First, regarding organoleptic concerns, it has been pointed out in various recent publications that contents of compounds such as anthocyanins and proanthocyanidins have a strong influence on color attributes.[14] For instance, glycosides of anthocyanins (such as malvidin, petunidin and peonidin) have been identified as specific descriptors of pigmentation of wines.[15] Also, other compounds including phenolic acids, catechins and some flavonoids play an important role in food quality, as they affect flavor and color properties.[15] Other sensorial characteristics such as bitterness and astringency have been found to be dependent on tannin compounds.[16]

Several health-promoting properties of food products have partly been attributed to the presence of polyphenols.[17,18] Polyphenols are the principal source of antioxidants in the human diet and products such as green tea or wines are well known because of this activity. Also, anti-bacterial, anti-inflammatory, anti-allergic and anti-thrombotic activities have been related to levels of some polyphenols. For instance, the best known bioactivity of A-type proanthocyanidins (PACs) is related to their capacity to inhibit the adhesion of pathogenic bacteria to cells of the urinary tract, thus preventing bacterial colonization and progression of urinary tract infections.[19,20] Additional antibacterial activity against pathogens of the digestive tract has also been recognized.[21]

Apart from the sensory and functional attributes noted above, the impact of these compounds in characterization, classification and authentication studies cannot be underestimated.[10,22] Polyphenol contents seem to be related to food features such as geographical areas, variety and manufacturing practices. As a result, contents of polyphenols can be exploited as a source of analytical data to establish product classifications.[23–25] The evaluation of food quality and the detection of adulterations can thus be based on polyphenol profiling.

13.3. Metabolomic Approach

Metabolomics deals with the comprehensive study of the small organic molecules involved in the metabolism of a given organism. The metabolome, defined as the overall set of metabolites of a living being, is recognized as an exceptional source of information of interest in scientific fields such as drug research, medical diagnosis and food control and characterization.[26,27] In this approach, samples belonging to different classes are compared to establish patterns or characteristics of each category. The overall study of a metabolome certainly entails great complexity as thousands of metabolites are involved.[28] For this reason, metabolomics is often tackled on a reduced set or family of related compounds, such as acids, alcohols, esters, amino acids, biogenic amines, inorganic species and, of course, polyphenols.[29]

Metabolomics can be carried out in two different ways depending on the type of data analyzed, namely 'profiling' and 'fingerprinting'. Profiling studies are based on targeted assays in which concentrations of relevant polyphenols in the different samples are exploited as food descriptors.[10] This approach obviously requires a previous quantification step using appropriate standards for each analyte. However, when dealing with complex food products, the quantification of all polyphenolic compounds results in a difficult task due to the occurrence of dozens of compounds, many of them unknown, which may interfere with the determination.

Fingerprinting strategies perform untargeted analysis of instrumental responses without assuming any previous knowledge on relevant or irrelevant species. Sample fingerprints can be obtained in different ways by infrared spectroscopy, nuclear magnetic resonance (NMR), chromatography and, of course, MS and LC–MS. In the analysis of food samples by LC–MS, fingerprints contain peaks consisting of intensity values recorded as a function of *m/z* and retention time. These signals are commonly referred to as (peak or instrumental) features so that this concept is extensively used in this chapter. As an example of fingerprint, Fig. 13.2 depicts a chromatogram corresponding to the analysis of a beer sample. In the subplot (A), the total ion chromatogram (TIC) is shown. MS or MS/MS spectra and extracted-ion chromatograms can be disclosed (Fig. 13.2B and 13.2C). Instrumental profiles reflect the complexity of the sample and, although amounts of components are not explicitly known, the intensities of the signals obviously depend on the concentrations of the constituents. Sample fingerprint analysis is gaining popularity because of its simplicity. However, from the point of view of food science, the identification of specific markers responsible for certain product features is greatly welcome.

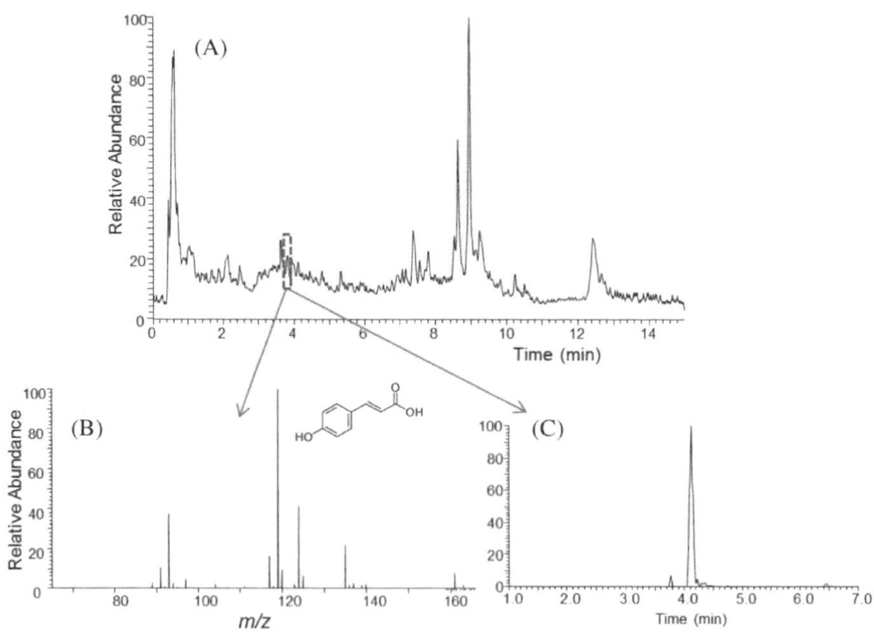

Figure 13.2. LC–MS fingerprint of a beer sample. (A) Total ion chromatogram; (B) MS(/MS) spectrum of a given peak; (C) extracted-ion chromatogram of $[M\text{-}H]^-$. The chemical structure of the unknown compound can be deduced from the exact mass of $[M\text{-}H]^-$ and fragmentations.

13.3.1. *Flowchart of metabolomics*

The scheme of the overall strategy followed in the metabolomic studies for food characterization is given in Fig. 13.3. The first step in this flowchart is the sample preparation. Here, simple, rapid and cheap sample treatment procedures are needed to deal with large series of samples that typically are involved in these kinds of studies.

LC–MS measurements often require data arrangements to make compatible raw instrumental data, stored in different archive formats that depend on the vendor, the instrument, etc., with standardized files such as .mzXML and .mzData.[30] Then, the transformed data files are treated with regular mathematical software such as Excel, R, Matlab, Unscrambler and others.

After format arrangement, data preprocessing tools (including signal alignment, baseline correction, peak feature filtering and noise reduction) are applied to improve the data quality. In the LC–MS methods, peak alignment on the time domain is basic to avoid the variability in the retention time. Analogous alignments are performed on the MS domain. In addition, unwanted instrumental signals can be removed before proceeding with the data analysis. For instance, MS and time windows that mainly contain interfering peaks, unrealistic features, etc. can be suppressed from the data set.

In the following step, chemometric methods are used to try to extract (bio)chemical information from the data sets. First, features are ranked according to their capacity to discriminate among classes by using univariate statistics. The ranking criterion relies on p-values from t-tests, feature intensity, the number of peaks in each class with respect to one another, etc. Subsequently, multivariate methods such as principal component analysis (PCA) and partial least squares-discriminant analysis (PLS-DA) are applied for exploratory and classification purposes.[31] This part deserves more attention so it will be commented on in detail in the following section.

The chemometric results require a through interpretation to gain meaningful information. At this stage, it is very important to determine which features comes from chemical compounds and which are just artifacts or random contributions. The identification of the chemical descriptors responsible for sample discrimination is, by far,

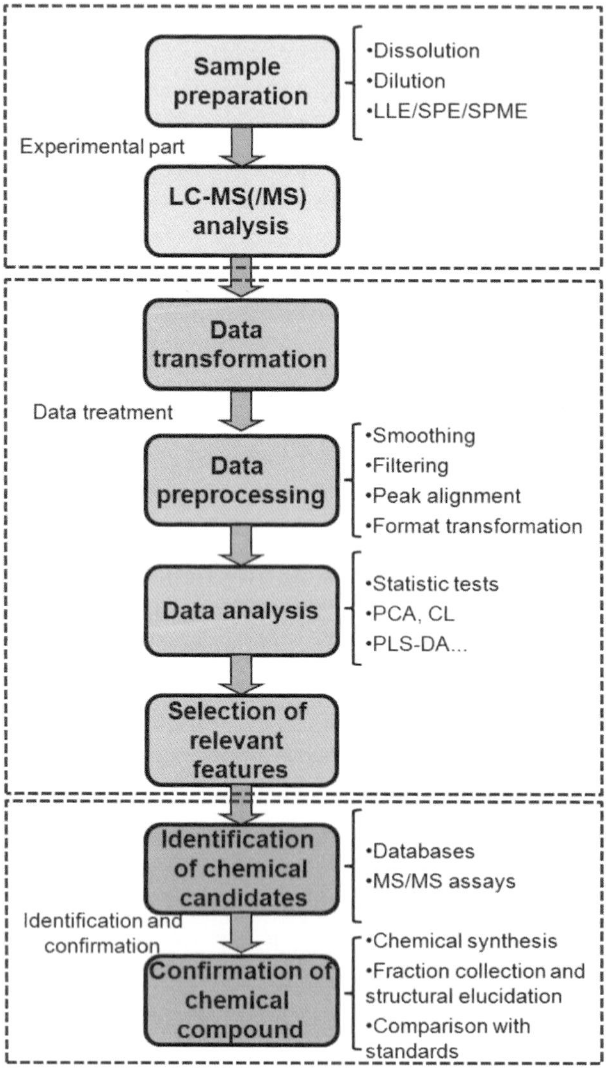

Figure 13.3. Scheme of the strategy for the characterization of food samples based on metabolomics.

the most difficult task in this flowchart. Relevant *m/z* features can be assigned to some molecule candidates. The structure of actual discriminating chemical compounds can be further elucidated by additional MS, MS/MS or NMR assays.[32–34] Hence, databases contain exact *m/z* values, fragmentations and NMR spectra to be compared with our candidates to help to identify the components.

Finally, those features that have tentatively been assigned can be confirmed by comparison, if available, with the corresponding chemical standards. When compounds are not commercial, the alternative procedure consists of the collection and purification of LC-eluting fractions corresponding to the target analytes. After that, the isolated product(s) can be used in complementary assays to confirm the identity(ies).

13.4. Data Analysis

Polyphenolic data handled in food characterization by LC–MS is of a multivariate nature. In both profiling and fingerprinting strategies, sample measurements consist of a list or array of values (i.e. the so-called first order data, according to the chemometric terminology).[35] Hence, when several samples are analyzed simultaneously, the corresponding data is arranged in a data matrix. In the data matrix, each row corresponds to a given sample and each column to concentrations of a given chemical species (profiling) or intensity features (fingerprinting) (Fig. 13.4). In this arrangement, matrix dimensions are m x n, m being the number of samples and n the number of compounds/features.

PCA is the most widely used method for exploratory study of food properties. PCA relies on the concentration of the relevant variance into a small number of new mathematical variables, the so-called principal components (PC).[31,36] The matrix of responses is decomposed into two small matrices of scores (coordinates of the samples) and loadings (eigenvalues), providing information on samples and variables, respectively. The first principal component (PC1) is calculated to capture the maximum amount of data variance; the following PC, PC2, is extracted to retain the maximum amount of the residual variance and is orthogonal to PC1. The rest of the PCs are calculated in the same way. As the first PCs explain a great amount

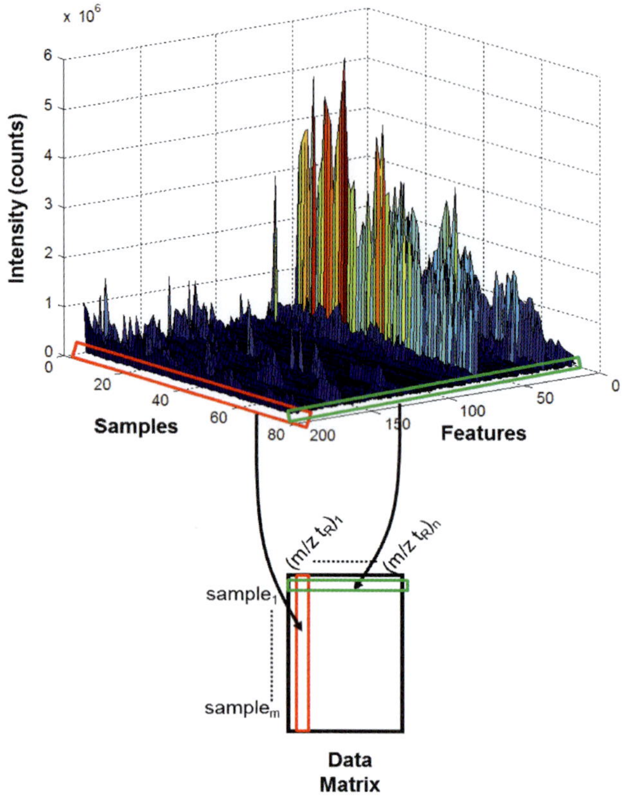

Figure 13.4. Obtaining the data matrix arrangement to be used for principal component analysis.

of the data variance, the chemical information can be displayed efficiently from scores' and loadings' plots.

The scatter plot of the scores of the PCs shows the distribution of samples, with patterns, similarities and differences that might be attributed to features such as origin, manufacturing, product varieties and so on. In the same way, scatter plots of loadings explain the behavior of variables (features) and their correlations, so those that are highly descriptive can be identified and studied. Besides, relationships between samples and variables can also be investigated from the simultaneous study of scores and loadings via the so-called 'bi-plots'. A practical case on the interpretation of PCA is

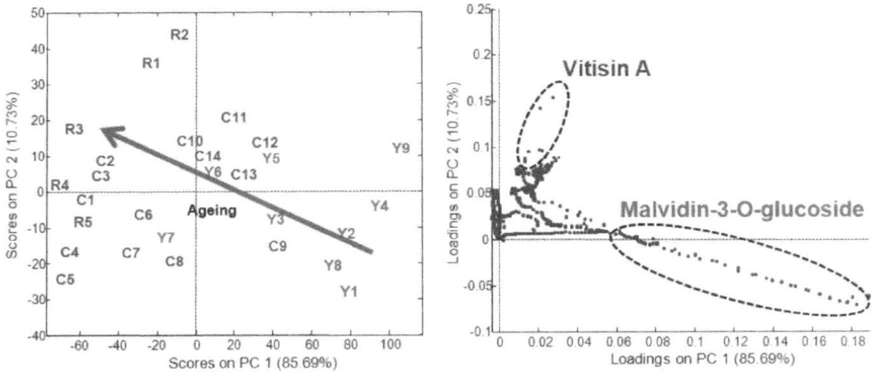

Figure 13.5. Results of wine characterization by PCA using chromatographic fingerprints of colored polyphenols as analytical data. (A) Plot of scores; (B) plot of loadings. Sample assignation: Y, young wine; C, *crianza* wine; R, *reserva* wine. Adapted from Serrano-Lourido, D., Saurina, J., Hernandez-Cassou, S. *et al.* (2012). Classification and characterisation of Spanish red wines according to their appellation of origin based on chromatographic profiles and chemometric data analysis, *Food Chem.*, **135**, 1425–1431.

shown in Fig. 13.5, which corresponds to the fingerprint analysis of a colored polyphenolic fraction of red wines. From the scores' plot, a reasonable discrimination of wines according to their aging period was observed (Fig. 13.5A). Younger wines were mainly located on the right part of the graph while aged wines (*reserva*, *crianza*) were found on the left. The interpretation of the plot of loadings (Fig. 13.5B) suggested that classification was made mainly according to visitin A and malvidin-3-O-glucoside, which appeared in opposite areas of PC1. It was concluded that young wines were characterized by higher levels of malvidin-3-O-glucoside while the amount decreased with aging to form visitin A, a condensed derivative that provides a brownish color aspect.

The classification of food products into pre-established categories can be carried out by discriminant analysis (DA), often combined with partial least squares regression (PLS) and soft independent modeling of class analogy (SIMCA) methods.[31,36] In classification and authentication, two (or several) sets of well-defined samples belonging to the classes of interest (e.g., variety 1 and variety 2, authentic and fake, etc.) are used to create models to be further applied to assign unknown samples to each class. The classification performance is evaluated by using an external test of samples and the ability to correctly assign the samples to their classes is quantified. Additionally, PLS-DA models are interpreted, in a similar way as indicated for PCA, to try to find markers of each class. PLS is also used for most specific modeling and correlation purposes such as in the assessment of relationships of physicochemical variables with agricultural, manufacturing or sensorial attributes.

13.5. Determination of Polyphenols by LC–MS

The determination of polyphenols is one of the priorities of food analysis because of their great implications in taste, health and descriptive issues. Hence, rapid, robust, feasible and accurate analytical methods are required to quantify these compounds in diverse vegetal matrices such as cereals, seeds, nuts, fruits and related products, wine, beer and other beverages, oil, honey, tea and medicinal plants.[37,38]

In the last decade, dozens of new methods for the determination of polyphenols in food products have been developed. In general, such methods combine appropriate sample treatments with liquid chromatography (LC). Apart from LC, other analytical techniques such as cyclic voltammetry,[39,40] amperometic sensors,[41,42] gas chromatography (GC),[43] and capillary electrophoresis (CE)[44–47] have been recently reported for the determination of these compounds.

13.5.1. *Sample treatment*

The complexity of the sample treatment greatly depends on the characteristics of the food matrices.[48] For instance, for alcoholic and non-alcoholic beverages including cold drinks, juices, beer, wine and spirits, only filtration prior to analysis is needed. In some cases, sample dilution may be recommended to reduce the matrix effects and noise of chromatograms. Water or organic solvents (e.g. MeOH,

ACN or DMSO) are typically used in sample : solvent ratios ranging from 10:1 to 1:10 (v/v).

When dealing with solid samples, solvent extraction is carried out to improve the detectability of trace species while minimizing interferences. In the case of fruit matrices, polyphenols can be extracted using organic solvents and aqueous-organic mixtures. Apart from the solvent composition, other chemical variables such as pH, solvent volume, time or temperature may be also relevant to improve the recovery yield. The wide range of physicochemical characteristics (e.g. molecular mass, solubility, polarity and acid-base properties) of compounds belonging to the diverse families of polyphenols requires important differences in the extraction procedures. Some fractions are better extracted as neutral species with organic solvents while others need acidic or basic media. For instance, less polar flavonoids are recovered in high yields as uncharged forms using DMSO.[49] Conversely, basic aqueous solutions are more appropriate for phenolic acids. Hydro-organic solutions at pH ≈ 7 are preferred for reaching an overall compromise to deal with a wide variety of components. As an example, Fig. 13.6 shows the overall extraction percentages of various phenolic acids and flavonoids using different solvent compositions. In this case, the best overall recovery is obtained with DMSO.

Sample sonication and centrifugation are useful to speed up the process kinetics and to obtain cleaner supernatant solutions. Additionally, extracts can be subjected to solvent evaporation under nitrogen current and dry residues then reconstituted in small exact volumes of a proper solvent. Prior to injection, extracts are often filtered through 0.45 μm membranes to avoid sample particles.

In some cases, solvent extraction is combined with an additional purification and preconcentration step based on solid-phase extraction (SPE).[48] Commercial reversed-phase C18 cartridges are very efficient for retaining neutral forms of polyphenols. Analytes are further eluted with a low volume of MeOH or ACN. Ion-exchange SPE can also be exploited for sample cleanup as most polyphenols are anions in basic media. In the case of anthocyanidins, however, cation-exchange cartridges could be considered for

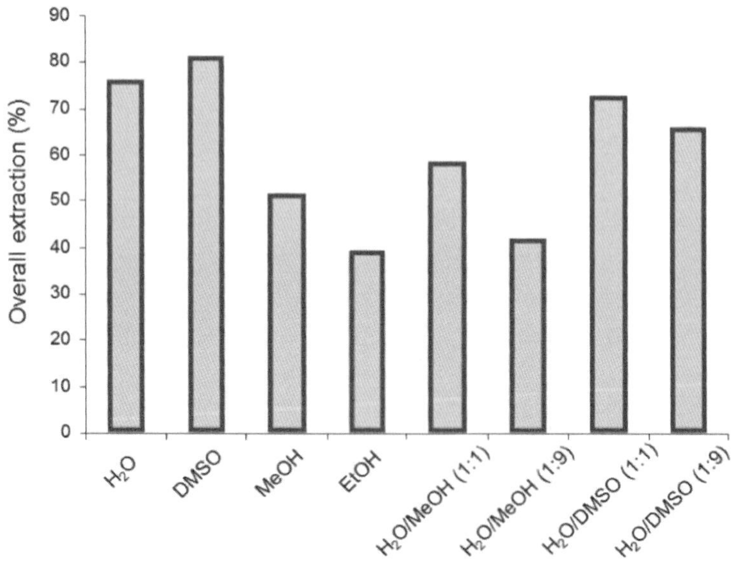

Figure 13.6. Overall extraction of phenolic acids and flavonoids from fruit matrices using different organic solvents and hydro-organic mixtures. Adapted from Raja, M., Hernández-Revelles, J., Hernández-Cassou S. *et al.* (2014). Determination of extractable polyphenols in fruit pulp by solvent extraction and liquid chromatography with UV-Vis detection, *J. Agric. Food Chem*, Submitted.

their purification. Molecularly imprinted polymers (MIP) have been designed for selective SPE of some target compounds such as resveratrol and structurally related compounds.[50] SPME has been optimized for the extraction of phenolic compounds in wine and grape samples. Stationary phases of different natures have been evaluated, that of polystyrene-divinylbenzene-polyacrylonitrile being the most efficient for the preconcentration of polyphenols. This technique has demonstrated a great ability to achieve quantitative extractions of most of the analytes.[51]

Although scarcely applied to food analysis, accelerated solvent extraction, microwave-assisted extraction and supercritical fluid extraction may be of interest.[48,53]

13.5.2. *Chromatographic methods*

Pioneering methods for polyphenol determination were developed in the 1970s, combining liquid chromatography with UV-vis spectroscopic detection. Since then, hundreds of LC methods have been proposed to deal with the high diversity of compounds and the great variety and complexity of some food matrices.[37,38,44]

In general, analytes are separated by reversed-phase mode using C18 (or C8) analytical columns. In the first publications, conventional HPLC columns packed with 5 μm particles were used. Continual advances in LC technology led to columns packed with sub-5 μm particles, which allowed more complex mixtures to be resolved. Mobile phases consisted of hydro-organic mixtures with MeOH or ACN as organic solvent and 10–100 mmol/L formic or acetic acid aqueous solutions. The high complexity of polyphenol samples often involves multi-step elution gradients. Recently, the introduction of hydrophilic interaction liquid chromatography (HILIC) for the separation of polyphenols has opened up great possibilities, especially for dealing with the most polar analytes, which are weakly retained in reversed-phase columns. Furthermore, in recent years the implementation of novel UHPLC methods in analytical laboratories has led to revisiting and re-optimizing the old HPLC methods. As a result, these improved methods exhibit excellent chromatographic performance and selectivity, and reduced analysis time.

Regarding detection, apart from the widespread use of UV-vis spectroscopy, fluorescence[52] and mass spectrometry[53] have been utilized to increase the sensitivity and selectivity of the methods.

13.5.3. *LC–MS methods*

Because of the great versatility and sensitivity of MS, it is not surprising that this detection technique is increasingly applied to food analysis.[53] When using MS alone, the overall *m/z* spectrum containing features of all analytes occurring in the sample is obtained by direct infusion into the MS instrument. MS spectra can be used for target analysis and fingerprinting in different ways, namely quantification, structural elucidation by MS^n experiments, or analyte screening in complex food samples. In these cases, however, as all compounds are simultaneously ionized, problems dealing with ion suppression, isobaric species and spectral interpretation may arise. In general, most of the analytes easily form anions so they are monitored satisfactorily in negative mode as $[M - H]^-$ ions. Only a few of the compounds, including some glycoside derivatives and anthocyanins, display higher sensitivities in positive mode.

As an example, an MS instrument with an electrospray ion (ESI) source and an ion trap (IT) analyzer was used for the analysis of hydroalcoholic extracts of plants containing the polyphenolic fraction.[54] As a different approach, Cajka and coworkers recorded MS profiles obtained in positive and negative ionization modes using a direct analysis in real time (DART) ion source as a way to evaluate special Trappist beers. The objective of this study was the assessment of a strategy able to discriminate these special Trappist beers from other-samples.[55] Matrix-assisted laser desorption/ionization (MALDI)–time-of-flight (TOF)-MS has proved to be an excellent technique for deeper characterization of polyphenols, especially for dealing with polymeric compounds. This topic has been discussed extensively in an excellent review by Monagas *et al.*[56] For instance, the structural diversity and complexity of proanthocyanidins has been assessed in this way. However, sample preparation procedures for MALDI–TOF assays are time-consuming as, apart from the sample pretreatment described above, extracts need to be mixed with matrix and cationizing agent before analysis. From these studies, the contents and structures of proanthocyanidin dimers, trimers, tetramers, hexamers, etc. were investigated, with special focus on the positions of the intermolecular links. The occurrence of proanthocyanidins in different plants and food products was established.

Although the applications detailed in the previous paragraph indicate that MS detection can be carried out with no separation step, coupling LC–MS is very fruitful for enhancing selectivity and decreasing matrix effects. LC–MS provides signals consisting of ion abundances as a function of *m/z* and retention-time values throughout both spectral and chromatographic domains. Hence, the principal shortcomings of direct MS assays (i.e., ionic suppression and isobaric compounds) can be partly overcome chromatographically from the progressive separation of components as a function of time. Applications concerning LC–MS to polyphenol quantification have been focused on the most characteristic compounds. Hence, the resulting compositional profiles can be exploited to establish comparisons among different types of samples. Furthermore, the use of high-resolution mass spectrometers (HRMS) coupled to the LC systems results in an exceptional tool for food analysis. Obtaining accurate mass measurements of the compounds of interest is the major advantage of HRMS in order to provide unambiguous identifications and quantifications.

By far, ESI is the most generalized ionization source, although atmospheric pressure chemical ionization (APCI) has occasionally been used. For instance, amounts of monomeric and polymeric catechins were determined in chocolate by LC–APCI–MS using a single quadrupole as the analyzer.[57] Concentrations of (+)-catechin, (−)-epicatechin and some derivatives were quantified from specific monitoring of protonated molecular ions $[M + H]^+$.

To date, dozens of different applications of (U)HPLC–ESI–MS(/MS) have been published. Some representative examples dealing with instrumentation possibilities, quantification approaches, and characterization and identification of analytes in a broad variety of food samples are described in the following paragraphs.

Rzeppa *et al.* carried out an exhaustive characterization of dimers and trimers of catequin and epicatechin with HPLC–MS using a triple quadrupole (QqQ) analyzer.[58] The lack of commercial standards of some of these products was overcome by chemical synthesis or extraction and purification of the analytes from natural sources. HPLC–ESI–QqQ–MS was also applied to investigate berry samples.[59] In general, the quantification with QqQ detectors was based on multiple reaction monitoring (MRM) of selected transitions established for each compound. IT analyzers were also combined with HPLC to tackle the identification and quantification of several monomeric polyphenols and polymeric tannins. For example, a HPLC–DAD–ESI–IT–MS^n method with both positive and negative ionization modes was established for a thorough evaluation of polyphenolic components of pomegranate samples.[60] About 151 compounds were identified (65 of them not previously reported in these samples) by MS^2 to MS^4 experiments. Other HPLC–DAD–ESI(+/−)–IT–MS^n methods were developed to analyze palm fruits,[61] figs,[62] teas[63] and South American berries.[64,65] TOF analyzers were exploited to identify and determine phenolic acids and flavonoids in almond skins. Accurate *m/z* measurements and isotopic patterns allowed some relevant compounds to be identified.[66] In another study, more than 70 polyphenols were found in cucumber samples using HPLC–ESI–Q–TOF-MS/MS.[67] The study of fragmentations was the basis of a tentative assignation of components.

Regarding UHPLC, Ortega and coworkers reported an UHPLC–ESI–QqQ–MS/MS method for the determination of some polyphenols and alkaloids in carob flour.[68] Non-commercial flavonoid glycosides were identified by full-scan MS and product-ion scan MS/MS experiments. The quantification was based on MRM. Concentrations found ranged from µg/g to mg/g, gallic acid and myricetin rhamnoside being the most abundant compounds. Another UHPLC–ESI–QqQ–MS (and MS/MS) method was applied to determine multiclass polyphenols in vegetables such as tomato, broccoli, eggplant, etc.[69] Most of the analytes were detected by negative ionization and only a few of them (e.g. genistein, kaempherol and various flavone glycosides) displayed higher sensitivities in positive ionization mode. MRM transitions were found out for both quantification and confirmation purposes. Similar instrumental equipment was considered for the study of major phenolic constituents of *Salvia officinalis*,[70] tomato and tomato-based products,[71] tea, and dietary supplements.[72] Owing to the great performance of these techniques, in some cases LC–MS runs take less than 3 min.

Some applications concerning HRMS have been reported in the literature. In one example, ellagic acid and its tannin derivatives were determined from blackberry and raspberry cultivars using an UHPLC–Q–TOF-HRMS method furnished with an ESI source.[73] This study allowed the identification of several new ellagic acid derivatives. Besides, UHPLC–HRMS using negative and positive electrospray ionization and Orbitrap analyzer was used to optimize an analytical method for the analysis of blueberry and red radish.[74] MS^n characterization of compounds allowed complex anthocyanins to be identified. Compositional profiles of the different samples were compared to investigate specific components of each food product.

Alternative LC techniques have been applied to implement highly powerful analytical methods. For instance, Hashim and coworkers developed a micro-LC–ESI–IT-MS/MS method for the determination of resveratrol and related compounds in red wines.[75] Cifuentes group established a new method relying on comprehensive two-dimensional LC (LC × LC) with MS detection.[76] HILIC and reversed-phase modes were used in the first and second chromatographic dimensions, respectively. MS consisted of an ESI source and an IT analyzer. The overall polyphenol profiling, containing peaks of single and condensed compounds, was obtained in 50 min.

Beyond the quantitative determination of polyphenols, an emerging trend with high scientific impact relies on the application of polyphenol profiling and fingerprinting to classification, authentication and related studies via chemometic analysis.[26,77–78] In the publication by Biasoto *et al.* spectra acquired by ESI–Q–TOF-MS were used for flavor characterization of wines.[72] Data were explored by PCA to assess the influence of grape varieties on the compositional profiles. Sensory attributes given by expert panelists were correlated with contents of some constituents in order to try to identify chemical descriptors of taste properties. Relevant peaks in MS spectra were characterized structurally by MS/MS and compounds were tentatively identified by comparison with information given in the literature. Various organic acids and polyphenols occurring in the wine fingerprints were found to be significant from the sensorial point of view.

Concerning LC–MS, Spanish wines from three protected designations of origin (PDO) were analyzed by HPLC–DAD-F and HPLC–ESI–QqQ–MS.[25] Gross UV-vis data recorded at several working wavelength was characterized preliminarily by PCA in order to find the most discriminant peaks. For classification purposes, 13 peaks were selected and their areas were used as a source

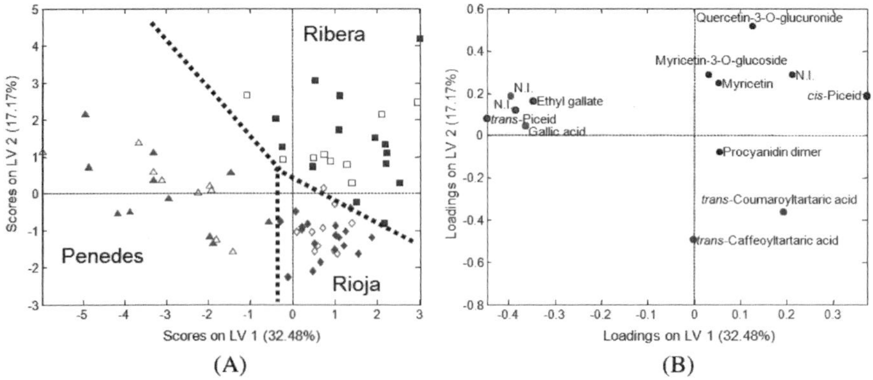

Figure 13.7. Classification of Spanish wines from three producing areas (Penedes, Rioja and Ribera del Duero) by PLS-DA. (A) Scatter plot of scores of latent variables LV1 vs LV2. Solid symbols = calibration samples; empty symbols = test samples; triangle = PDO Penedes; rhombus = PDO Rioja; square = PDO Ribera. (B) Scatter plot of loadings of LV1 vs LV2: N.I. = non-identified compound. Adapted from Serrano-Lourido, D., Saurina, J., Hernandez-Cassou, S. *et al.* (2012). Classification and characterisation of Spanish red wines according to their appellation of origin based on chromatographic profiles and chemometric data analysis, *Food Chem.*, **135**, 1425–1431.

of analytical information. Wine samples belonging to the different PDO were assigned into each class using PLS-DA. As shown in Fig.13.7, all samples were correctly identified. Components contributing to wine discrimination were deduced from the inspection of the plot of loadings (Fig. 13.7B). The identification of the chemical compound behind each of these contributions was based on MS and MS/MS analysis. For instance, tentative assignations of malvidin-3-O-glucoside, visitin, etc. were confirmed by the presence of characteristic positive *m/z* peaks. The same approach was followed for other components. Conclusions on characteristic components of each PDO were extracted. For instance, gallic acid and *trans*-piceid were present in Penedes wines in higher concentrations, *trans*-caffeoyltartaric and *trans*-coumaroyltartaric were representative of Rioja, while quercetin-3-O-glucuronide and *cis*-piceid were descriptive of Ribera del Duero wines.

In other examples, Mattarucchi and coworkers reported a method for food authentication purposes.[79] The profiles generated by this technique allowed the discrimination of Rochefort Trappist beers from other varieties. Boselli and coworkers studied the influence of specific polyphenols on color attributes of wines.[80] Concentrations of colored components, quantified by HPLC–MS/MS, were related to pigmentation and taste. Malvidin, petunidin and peonidin (di)glucosides were recognized as characteristic descriptors of a given Italian controlled designation of origin. A method for the classification of fruits according to the polyphenolic contents was developed by Vrhovsek *et al.*[81] Concentrations of about 90 compounds determined were treated by cluster analysis and analogies and differences among different types of fruits were established.

13.6. Concluding Remarks

LC–MS has demonstrated great analytical potential for the determination and characterization of polyphenols in food matrices. Nowadays, LC–MS is the technique of choice in these kinds of applications since it provides excellent sensitivity and selectivity, especially when using high-resolution instruments in both chromatography and detection. The broad variety of phenolic acids and flavonoids, including monomeric compounds and polymeric structures, complicates the studies. Hence, MS^n detection modes, sometimes complemented with searches in databases and additional assays by NMR, may be needed for structural elucidation. This is an important issue that cannot be underestimated as the molecular structure strongly affects the biological activity of the analytes.

Following metabolomic approaches, the most outstanding instrumental features, consisting of intensity counts as a function of *m/z* and retention time, can be detected. Chemometric methods such as PCA and PLS-DA have proved to be very efficient in facilitating the extraction of relevant information regarding functional and descriptive characteristics of food products. Hence, analogies and differences among food varieties and chemical descriptors of the different classes can be established. In this field, the number of applications concerning statistical and mathematical data analysis has increased dramatically in recent years, especially for dealing with classification and authentication purposes.

References

1. Luykx, D.M.A.M. and Van Ruth, S.M. (2008). An overview of analytical methods for determining the geographical origin of food products, *Food Chem.*, **107**, 897–911.
2. Reid, L.M., O'Donnell, C.P. and Downey, G. (2006). Recent technological advances for the determination of food authenticity, *Trends Food Sci. Technol.*, **17**, 344–353.
3. Arvanitoyannis, I.S., Katsota, M.N., Psarra, E.P. *et al.* (1999). Application of quality control methods for assessing wine authenticity: use of multivariate analysis (chemometrics), *Trends Food Sci. Technol.*, **10**, 321–336.
4. Cayot, N. (2007). Sensory quality of traditional foods, *Food Chem.*, **101**, 154–162.
5. Lopez-Feria, S., Cárdenas, S. and Valcárcel, M. (2008). Simplifying chromatographic analysis of the volatile fraction of foods, *Trends Anal. Chem.*, **27**, 794–803.
6. Plutowska, B. and Wardencki, W. (2007). Aromagrams: aromatic profiles in the appreciation of food quality, *Food Chem.*, **101**, 845–872.
7. Kozak, M. and Scaman, C.H. (2008). Unsupervised classification methods in food sciences: discussion and outlook, *J. Sci. Food Agric.*, **88**, 1115–1127.
8. Gishen, M., Dambergs, R.G. and Cozzolino, D. (2005). Grape and wine analysis: enhancing the power of spectroscopy with chemometrics. A review of some applications in the Australian wine industry, *Aust. J. Grape Wine Res.*, **11**, 296–305.
9. Ogrinc, N., Kosir, I.J., Spangenberg, J.E. *et al.* (2003). The application of NMR and MS methods for detection of adulteration of wine, fruit juices, and olive oil. A review, *Anal. Bioanal. Chem.*, **376**, 424–430.
10. Saurina, J. (2010). Characterization of wines using compositional profiles and chemometrics, *Trends Anal. Chem.*, **29**, 234–245.
11. Valentao, P., Guerra, L., Pereira, D.M. *et al.* (2009). 'Red wine phenolics: reasons for their variation', in O'Byrne, P. (ed.), *Red Wine and Health*, Nova Science Publishers Inc., New York, 53–90.
12. Manach, C., Scalbert, A., Morand, C. *et al.* (2004). Polyphenols: food sources and bioavailability, *Am. J. Clin. Nut.*, **79**, 727–747.

13. Bueno, J.M., Ramos-Escudero, F., Saez-Plaza, P. *et al.* (2012). Analysis and antioxidant capacity of anthocyanin pigments. Part I: General considerations concerning polyphenols and flavonoids, *Crit. Rev. Anal. Chem.*, **42**, 102–125.
14. Bueno, J.M., Saez-Plaza, P., Ramos-Escudero, F. *et al.* (2012). Analysis and antioxidant capacity of anthocyanin pigments. Part II: Chemical structure, color, and intake of anthocyanins, *Crit. Rev. Anal. Chem.*, **42**, 126–151.
15. Boulton, R. (2001). The copigmentation of anthocyanins and its role in the color of red wine: a critical review, *Am. J. Enol. Vitic.*, **52**, 67–87.
16. Lesschaeve, I. and Noble, A.C. (2005). Polyphenols: factors influencing their sensory properties and their effects on food and beverage preferences, *Am. J. Clin. Nut.*, **81**, 330S–335S.
17. Nijveldt, R.J., van Nood, E., van Hoorn, D.E.C. *et al.* (2001). Flavonoids: a review of probable mechanisms of action and potential applications, *Am. J. Clin. Nutr.*, **74**, 418–425.
18. Cabrera, C., Artacho, R. and Gimenez, R. (2006). Beneficial effects of green tea: a review, *J. Am. Coll. Nut.*, **25**, 79–99.
19. Foo, L.Y., Lu, Y., Howell, A.B. *et al.* (2000). The structure of cranberry proanthocyanidins which inhibit adherence of uropathogenic P-fimbriated *Escherichia coli* in vitro, *Phytochem.*, **54**, 173–181.
20. Howell, A.B., Reed, J.D., Krueger, C.G. *et al.* (2005). A-type cranberry proanthocyanidins and uropathogenic bacterial anti-adhesion activity, *Phytochem.*, **66**, 2281–2291.
21. Weiss, E.I., Kozlovsky, A., Steinberg, D. *et al.* (2004). A high molecular mass cranberry constituent reduces mutants streptococci level in saliva and inhibits in vitro adhesion to hydroxyapatite, *FEMS Microbiol. Lett.*, **232**, 89–92.
22. Makris, D.P., Kallithraka, S. and Kefalas, P. (2006). Flavonols in grapes, grape products and wines: burden, profile and influential parameters, *J. Food Comp. Anal.*, **19**, 396–404.
23. Makris, D.P., Kallithraka, S. and Mamalos, A. (2006). Differentiation of young red wines based on cultivar and geographical origin with application of chemometrics of principal polyphenolic constituents, *Talanta*, **70**, 1143–1152.

24. Kallithraka, S., Mamalos, A. and Makris, D.P. (2007). Differentiation of young red wines based on chemometrics of minor polyphenolic constituents, *J. Agric. Food Chem.*, **55**, 3233–3239.
25. Serrano-Lourido, D., Saurina, J., Hernandez-Cassou, S. *et al.* (2012). Classification and characterisation of Spanish red wines according to their appellation of origin based on chromatographic profiles and chemometric data analysis, *Food Chem.*, **135**, 1425–1431.
26. Antignac, J.P., Courant, F., Pinel, G. *et al.* (2011). Mass spectrometry-based metabolomics applied to the chemical safety of food, *Trends Anal. Chem.*, **30**, 292–301.
27. Ellis, D.I., Dunn, W.B., Griffin, J.L. *et al.* (2007). Metabolic fingerprinting as a diagnostic tool, *Pharmacogenomics*, **8**, 1243–1266.
28. Nicholson, J.K. and Wilson, I.D. (2003). Understanding 'global' systems biology: metabonomics and the continuum of metabolism, *Nat. Rev. Drug Discov.*, **2**, 668–676.
29. Wishart, D.S. (2008). Metabolomics: applications to food science and nutrition research, *Trends Food Sci. Technol.*, **19**, 482–493.
30. Christin, C., Bischoff, R. and Horvatovich, P. (2011). Data processing pipelines for comprehensive profiling of proteomics samples by label-free LC–MS for biomarker discovery, *Talanta*, **83**, 1209–1224.
31. Brown, S.D., Tauler, R. and Walczak, B. (2009). *Comprehensive Chemometrics. Chemical and Biochemical Data Analysis, Volume 3*, Elsevier, Amsterdam.
32. Hill, D.W., Kertesz, T.M., Fontaine, D. *et al.* (2008). Mass spectral metabonomics beyond elemental formula: chemical database querying by matching experimental with computational fragmentation spectra, *Anal. Chem.*, **80**, 5574–5582.
33. Katajamaa, M. and Oresic, M. (2005). Processing methods for differential analysis of LC–MS profile data, *BMC Bioinformatics*, **6**, 179.
34. Wishart, D.S. (2008). Quantitative metabolomics using NMR, *Trends Anal. Chem.*, **27**, 228–237.
35. Sentellas, S. and Saurina, J. (2003). Chemometrics in capillary electrophoresis. Part B: Methods for data analysis, *J. Sep. Sci.*, **26**, 1395–1402.
36. Massart, D.L., Vandeginste, B.G.M., Buydens, L.M.C. *et al.* (1997). *Handbook of Chemometrics and Qualimetrics*, Elsevier, Amsterdam.

37. Ignat, I., Volf, I. and Popa, V.I. (2011). A critical review of methods for characterisation of polyphenolic compounds in fruits and vegetables, *Food Chem.*, **126**, 1821–1835.
38. Escarpa, A. and Gonzalez, M.C. (2001). An overview of analytical chemistry of phenolic compounds in foods, *Crit. Rev. Anal. Chem.*, **31**, 57–139.
39. Sanchez Arribas, A., Martinez-Fernandez, M. and Chicharro, M. (2012). The role of electroanalytical techniques in analysis of polyphenols in wine, *Trends Anal. Chem.*, **34**, 78–96.
40. Dobes, J., Zitka, O., Sochor, J. *et al.* (2013). Electrochemical tools for determination of phenolic compounds in plants. A Review, *Int. J. Electrochem. Sci.*, **8**, 4520–4542.
41. Fernandes, C.I.S. and Rebelo, M.J.F. (2009). Polyphenolic biosensors: application in red wines, *Portugaliae Electrochim. Acta*, **27**, 457–462.
42. Photinon, K., Chalermchart, Y., Khanongnuch, C. *et al.* (2010). A thick-film sensor as novel device for determination of polyphenols an their antioxidant capacity in white wine, *Sensors*, **10**, 1670–1678.
43. Viñas, P., Campillo, N., Martinez-Castillo, N. *et al.* (2009). Solid-phase microextraction on-fiber derivatization for the analysis of some polyphenols in wine and grapes using gas chromatography-mass spectrometry, *J. Chromatogr. A*, **1216**, 1279–1284.
44. Kartsova, L.A. and Alekseeva, A.V. (2008). Chromatographic and electrophoretic methods for determining polyphenol compounds, *J. Anal. Chem.*, **63**, 1024–1033.
45. Jac, P., Polasek, M. and Pospisilova, M. (2006). Recent trends in the determination of polyphenols by electromigration methods, *J. Pharm. Biomed. Anal.*, **40**, 805–814.
46. García-Cañas, V. and Cifuentes, A. (2008). Recent advances in the application of capillary electromigration methods in food analysis, *Electrophoresis*, **29**, 294–309.
47. Gu, X., Chu, Q., O'Dwyer, M. *et al.* (2000). Analysis of resveratrol in wine by capillary zone electrophoresis, *J. Chromatogr. A*, **881**, 471–481.
48. Naczk, M. and Shahidi, F. (2004). Extraction and analysis of phenolics in food, *J. Chromatogr. A*, **1054**, 95–111.

49. Raja, M., Hernández-Revelles, J., Hernández-Cassou S. *et al.* (2014). Determination of extractable polyphenols in fruit pulp by solvent extraction and liquid chromatography with UV-Vis detection, *J. Agric. Food Chem.*, Submitted.
50. Hashim, S.N.N.S., Schwarz, L.J., Boysen, R.I. *et al.* (2013). Rapid solid-phase extraction and analysis of resveratrol and other polyphenols in red wine, *J. Chromatogr A*, **1313**, 284–290.
51. Mirnaghi, F.S., Mousavi, F., Rocha, S.M. *et al.* (2013). Automated determination of phenolic compounds in wine, berry, and grape samples using 96-blade solid phase microextraction system coupled with liquid chromatography-tandem mass spectrometry, *J. Chromatogr A*, **1276**, 12–19.
52. Aznar, O., Checa, A., Oliver, R. *et al.* (2011). Determination of polyphenols in wines by liquid chromatography with UV spectrophotometric detection, *J. Sep. Sci.*, **34**, 527–535.
53. Motilva, M.J., Serra, A. and Macià, A. (2010). Analysis of food polyphenols by ultrahigh-performance liquid chromatography coupled to mass spectrometry: an overview, *J. Chromatogr A*, **1292**, 66–82.
54. Mari, A., Eletto, D., Pizza, C. *et al.* (2013). Integrated mass spectrometry approach to profile proanthocyanidins occurring in food supplements: analysis of *Potentilla erecta L.* rhizome, *Food Chem.*, **141**, 4171–4178.
55. Cajka, T., Riddellova, K., Tomaniova, M. *et al.* (2010). Ambient mass spectrometry employing a DART ion source for metabolomic fingerprinting/profiling: a powerful tool for beer origin recognition, *Metabolomics*, 7, 500–508.
56. Monagas, M., Quintanilla-Lopez, J.E., Gomez-Cordoves, C. *et al.* (2010). MALDI–TOF-MS analysis of plant proanthocyanidins, *J. Pharm. Biomed. Anal.*, **51**, 358–372.
57. Nelson, B.C. and Sharpless, K.E. (2003). Quantification of the predominant monomeric catechins in baking chocolate standard reference material by LC–APCI-MS, *J. Agric. Food Chem.*, **51**, 531–537.
58. Rzeppa, S., Von Bargen, C., Bittner, K. *et al.* (2011). Analysis of flavan-3-ols and procyanidins in food samples by reversed-phase high-performance liquid chromatography coupled to electrospray ionization tandem mass spectrometry (RP-HPLC–ESI–MS/MS), *J. Agric. Food Chem.*, **59**, 10594–10603.

59. Gordon, A., Schadow, B., Quijano, C.E. *et al.* (2011). Chemical characterization and antioxidant capacity of berries from *Clidemia rubra* (Aubl.) Mart. (Melastomataceae), *Food Res. Int.*, **44**, 2120–2127.
60. Sentandreu, E., Cerdan-Calero, M. and Sendra, J.M. (2013). Phenolic profile characterization of pomegranate (*Punica granatum*) juice by high-performance liquid chromatography with diode array detection coupled to an electrospray ion trap mass analyzer, *J. Food Comp. Anal.*, **30**, 32–40.
61. Finco, F.D.B.A., Kammerer, D.R., Carle, R. *et al.* (2012). Antioxidant activity and characterization of phenolic compounds from bacaba (*Oenocarpus bacaba Mart.*) fruit by HPLC–DAD-MS^n, *J. Agric. Food Chem.*, **60**, 7665–7673.
62. Vallejo, F., Marin, J.G. and Tomas-Barberan, F.A. (2012). Phenolic compound content of fresh and dried figs (*Ficus carica L.*), *Food Chem.*, **130**, 485–492.
63. Fraser, K., Harrison, S.J., Lane, G.A. *et al.* (2012). HPLC–MS/MS profiling of proanthocyanidins in teas: a comparative study, *J. Food Comp. Anal.*, **26**, 43–51.
64. Simirgiotis, M.J., Borquez, J. and Schmeda-Hirschmann, G. (2013). Antioxidant capacity, polyphenolic content and tandem HPLC–DAD–ESI-MS profiling of phenolic compounds from the South American berries *Luma apiculata* and *L. chequen*, *Food Chem.*, **139**, 289–299.
65. Ruiz, A., Hermosin-Gutierrez, I., Vergara, C. *et al.* (2013). Anthocyanin profiles in south Patagonian wild berries by HPLC–DAD–ESI–MS/MS, *Res. Int.*, **51**, 706–713.
66. Arraez-Roman, D., Fu, S.P., Sawalha, S.M.S. *et al.* (2010). HPLC/CE–ESI–TOF-MS methods for the characterization of polyphenols in almond-skin extracts, *Electrophoresis*, **31**, 2289–2296.
67. Abu-Reidah, I.M., Arraez-Roman, D., Quirantes-Pine, R. *et al.* (2012). HPLC–ESI–Q–TOF-MS for a comprehensive characterization of bioactive phenolic compounds in cucumber whole fruit extract, *Food Res. Int.*, **46**, 108–117.
68. Ortega, N., Macia, A., Romero, M.P. *et al.* (2009). Rapid determination of phenolic compounds and alkaloids of carob flour by improved liquid chromatography–tandem mass spectrometry, *J. Agric. Food Chem.*, **57**, 7239–7244.

69. Alarcon-Flores, M.I., Romero-Gonzalez, R., Vidal, J.L.M. *et al.* (2013). Multiclass determination of phytochemicals in vegetables and fruits by ultrahigh-performance liquid chromatography coupled to tandem mass spectrometry, *Food Chem.*, **141**, 1120–1129.
70. Zimmermann, B.F., Walch, S.G., Tinzoh, L.N. *et al.* (2011). Rapid UHPLC determination of polyphenols in aqueous infusions of *Salvia officinalis L.* (sage tea), *J. Chromatogr. B*, **879**, 2459–2464.
71. Di Lecce, G., Martinez-Huelamo, M., Tuipani, S. *et al.* (2013). Setup of a UHPLC–QqQ–MS method for the analysis of phenolic compounds in cherry tomatoes, tomato sauce, and tomato juice, *J. Agric. Food Chem.*, **61**, 8373–8380.
72. Biasoto, A.C.T., Catharino, R.R., Sanvido, G.B. *et al.* (2010). Flavour characterization of red wines by descriptive analysis and ESI mass spectrometry, *Food Qual. Prefer.*, **21**, 755–762.
73. Gasperotti, M., Masuero, D., Vrhovsek, U. *et al.* (2010). Profiling and accurate quantification of Rubus ellagitannins and ellagic acid conjugates using direct UPLC–Q–TOF-HDMS and HPLC–DAD analysis, *J. Agric. Food Chem.*, **58**, 4602–4616.
74. Sun, J.H., Lin, L.Z. and Chen, P. (2012). Study of the mass spectrometric behaviors of anthocyanins in negative ionization mode and its applications for characterization of anthocyanins and non-anthocyanin polyphenols, *Rapid Commun. Mass Spectrom.*, **26**, 1123–1133.
75. Hashim, S.N.N.S., Schwarz, L.J. and Yang, Y.Z. (2013). Rapid solid-phase extraction and analysis of resveratrol and other polyphenols in red wine, *J. Chromatogr. A*, **1313**, 284–290.
76. Montero, L., Herrero, M., Ibañez, E. *et al.* (2013). Profiling of phenolic compounds from different apple varieties using comprehensive two-dimensional liquid chromatography, *J. Chromatogr. A*, **1313**, 275–283.
77. Ibañez, C., Garcia-Canas, V., Valdes, A. *et al.* (2013). Novel MS-based approaches and applications in food metabolomics, *Trends Anal. Chem.*, **52**, 100–111.
78. Hu, C. and Xu, G. (2013). Mass-spectrometry-based metabolomics analysis for foodomics, *Trends Anal. Chem.*, **52**, 36–46.
79. Mattarucchi, E., Stocchero, M., Moreno-Rojas, J.M. *et al.* (2010). Authentication of Trappist beers by LC–MS fingerprints and multivariate data analysis, *J. Agric. Food Chem.*, **58**, 12089–12095.

80. Boselli, E., Giomo, A., Minardi, M. *et al.* (2008). Characterization of phenolics in Lacrima di Morro d'Alba wine and role on its sensory attributes, *Eur. Food. Res. Technol*, **227**, 709–720.
81. Vrhovsek, U., Masuero, D., Gasperotti, M. *et al.* (2012). A versatile targeted metabolomics method for the rapid quantification of multiple classes of phenolics in fruits and beverages, *J. Agric. Food Chem.*, **60**, 8831–8840.

Chapter 14

Liquid Chromatography–Mass Spectrometric Analysis of Mycotoxins in Food

Veronica M.T. Lattanzio and Angelo Visconti

National Research Council,
Institute of Sciences of Food Production CNR – ISPA, Italy

14.1. Introduction

Mycotoxins are naturally occurring toxic metabolites that can be produced by fungi infecting agricultural crops during their growth, drying, and subsequent storage. The natural fungal flora associated with foods is dominated by the genera *Aspergillus*, *Fusarium*, *Penicillium,* and *Alternaria.*[1] Especially, environmental and biological factors such as water activity, temperature, humidity, and insect damage can have a great influence on growth of certain fungi and, therefore, on the spectrum of produced secondary metabolites. The range of foods susceptible to fungal growth and subsequent mycotoxin contamination is large and represents many of the staple food crops worldwide.

Hundreds of fungal secondary metabolites are known, but agriculturally important toxins can be related to five major chemical families: aflatoxins, fumonisins, ochratoxin A (OTA), trichothecenes and zearalenone (ZEA), whose known or suspected effects on human and animal health is of a nature to deserve significant attention.[2] Table 14.1 lists major mycotoxins with relevant producing fungi and commodities most at risk of contamination. For the aflatoxins, fumonisins and trichothecenes, each group contains a number of structurally related analogs.

Table 14.1. Overview of major mycotoxins: molecular structures, main producer fungi and main crops affected.

Aflatoxin B_1	**Aflatoxin B_2**	**Aflatoxin G_1**	**Aflatoxin G_2**
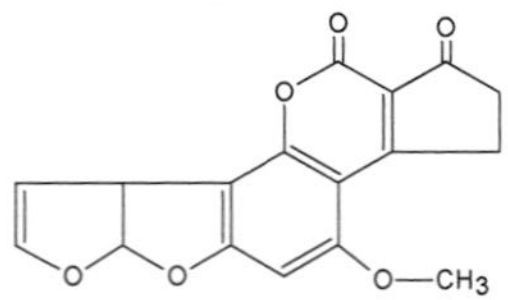	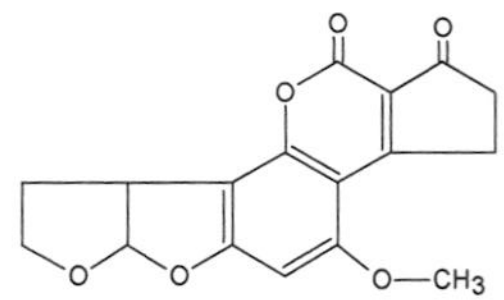	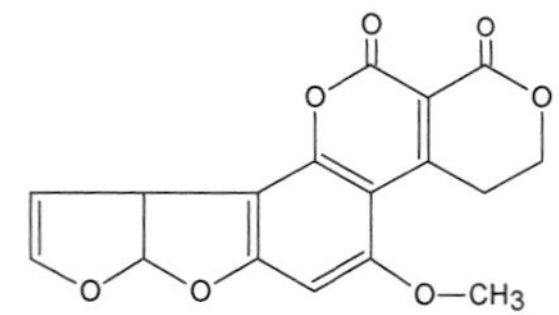	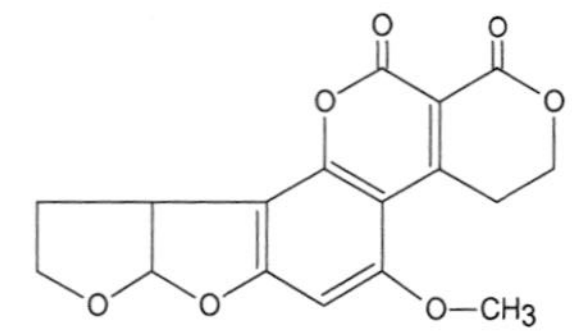

Main producers: *Aspergillus flavus, A. parasiticus.*

Main crops affected: maize and other cereal grains, nuts, fresh and dries fruits (dried figs), spices and herbs, cassava and other roots and tubes.

Ochratoxin A

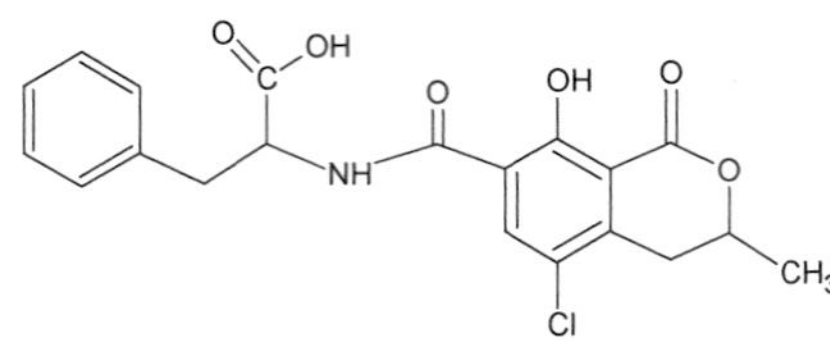

Main producers: *Aspergillus ochraceous, A. alliaceus, A. niger, Penicillium verrucosum.*

Main crops affected: grapes and dried wine fruits, wheat, maize, and other cereals, coffee, wine and beer, spices, sunflower seeds.

(Continued)

Table 14.1. *(Continued)*

Fumonisin B_1

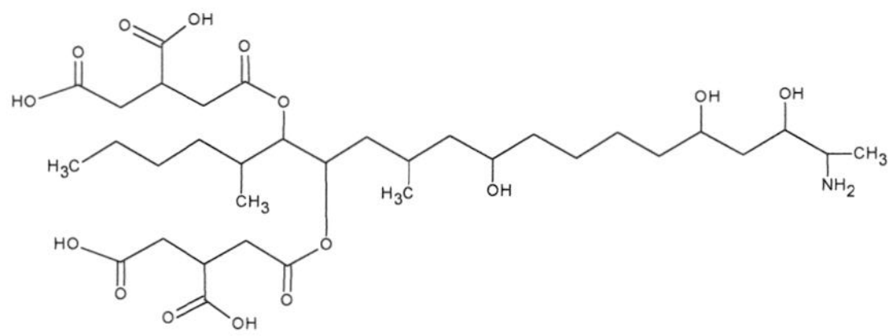

Fumonisin B_2

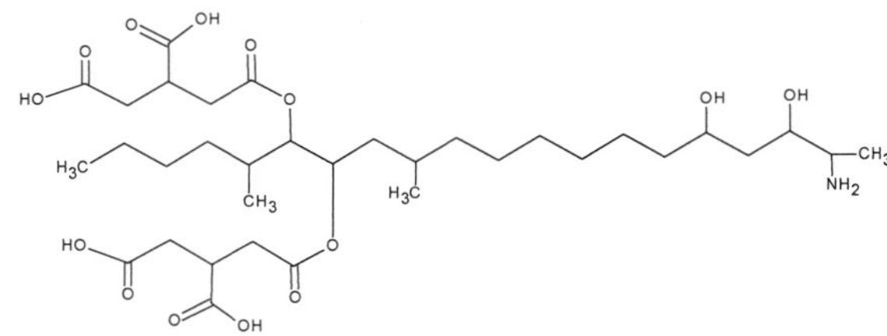

Main producers: *Fusarium verticillioides* (syn. *moniliforme*), *F. proliferatum.*

Main crops affected: maize.

Zearalenone

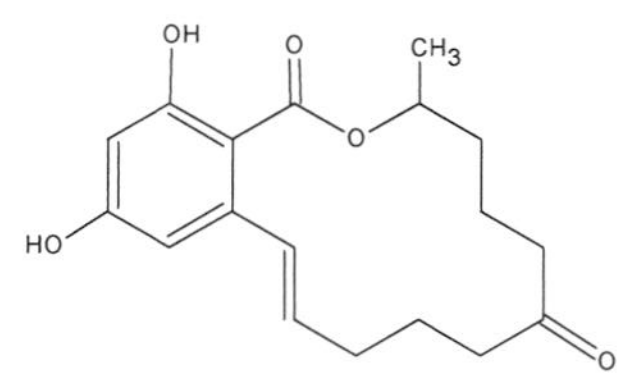

Main producers: *F. graminearum, F. culmorum, F. cerealis, F. equiseti, F. semitectum.*

Main crops affected: maize, wheat.

(Continued)

Table 14.1. (*Continued*)

Type-A trichotecenes	Type-B trichothecenes
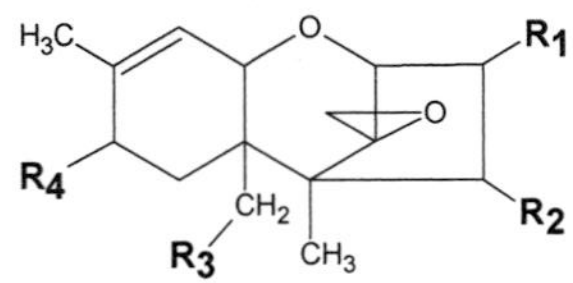	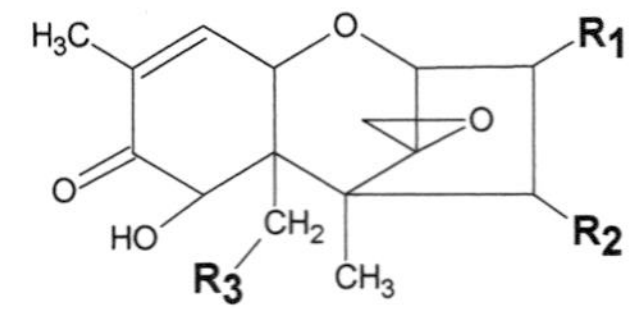
HT-2 toxin R1=R2=OH, R3=OAc, R4=$OCOCH_2CH(CH_3)_2$	Deoxynivalenol R1=R3=OH, R2=H
T-2 toxin R1=OH, R2=R3=OAc, R4=$OCOCH_2CH(CH_3)_2$	Nivalenol R1=R2=R3=OH
Nesosolaniol R1=OH, R2=R3=OAc, R4=OH	3-Acetyldeoxynivalenol R1=OAc, R2=H, R3=OH
Diacetoxyscirpenol R1=OH, R2=R3=OAc, R4=H	15-Acetyldeoxynivalenol R1=OH, R2=H, R3=OAc
Main producers: *F. sporotrichioides, F. poae, F. langsethiae, F. acuminatum, F. equiseti, F. sambucunum.*	**Main producers:** *F.graminearum, F.culmorum, F.cerealis* (syn. *Crookwellense*).
Main crops affected: maize, wheat, oats.	**Main crops affected:** maize, wheat.

Patulin

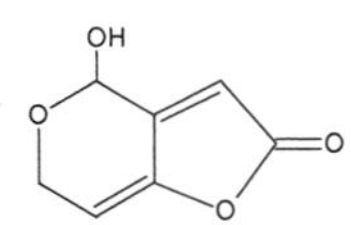

Main producers: *Penicillium expansum, Aspergillus* and *Byssochlamys spp.*

Main crops affected: apple, pear and other fruits.

(*Continued*)

Table 14.1. (*Continued*)

Beauvericin

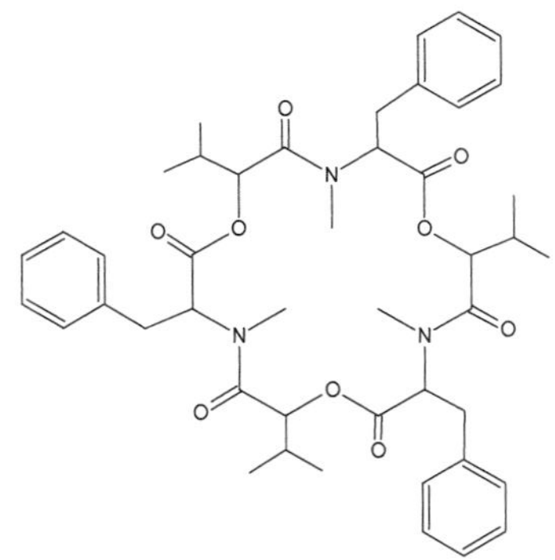

Main producers: *F. semitectum, F. subglutinans, F. proliferatum, F. avenaceum.*

Main crops affected: maize, wheat, barley, oats, rice.

(*Continued*)

Table 14.1. *(Continued)*

Enniatins

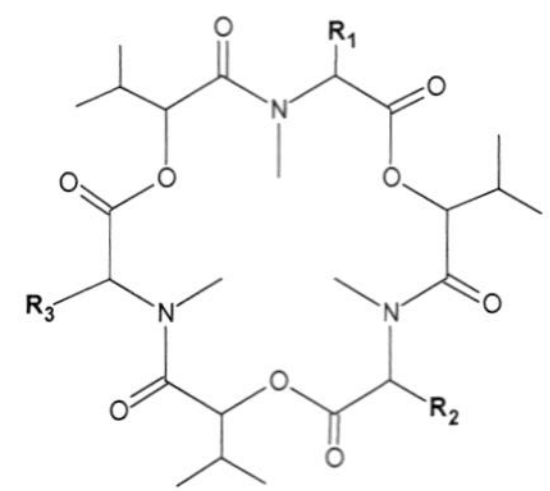

Enniatin A	R1=$CH(CH_3)CH_2CH_3$	R2=$CH(CH_3)CH_2CH_3$	R3=$CH(CH_3)CH_2CH_3$
Enniatin A1	R1=$CH(CH_3)CH_2CH_3$	R2=$CH(CH_3)CH_2CH_3$	R3=$CH(CH_3)_2$
Enniatin B	R1=$CH(CH_3)_2$	R2=$CH(CH_3)_2$	R3=$CH(CH_3)_2$
Enniatin B1	R1=$CH(CH_3)_2$	R2=$CH(CH_3)_2$	R3=$CH(CH_3)CH_2CH_3$

Main producers: *F. avenaceum, F. oxysporum.*

Main crops affected: wheat, maize, barley, oats, rice.

(Continued)

Table 14.1. *(Continued)*

Ergot alkaloids

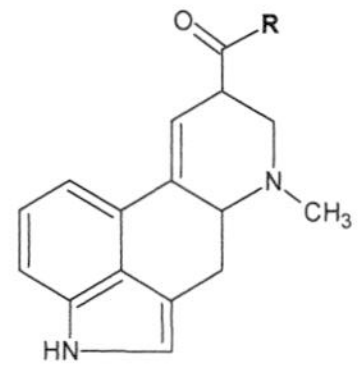

Ergometrin: R=NHCH(CH_3)CH_2OH

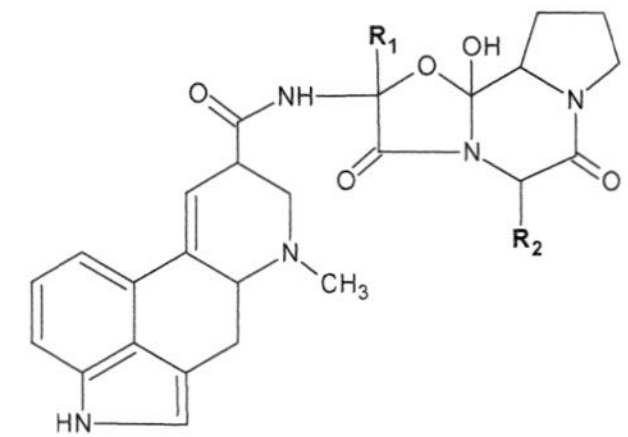

Ergotamine: R1=CH_3, R2=$CH_2C_6H_5$
Ergocristin: R1=CH$(CH_3)_2$, R2=$CH_2C_6H_5$
α-Ergosin: R1=CH_3, R2=CH_2CH$(CH_3)_2$
Ergocornin: R1=CH$(CH_3)_2$, R2=CH$(CH_3)_2$
α-Ergocryptin: R1=CH$(CH_3)_2$, R2=CH_2CH$(CH_3)_2$

Main producers: *Claviceps purpurea,* other *Claviceps spp.*

Main crops affected: wheat, rye, barley, millet oats, sorghum, triticale.

Altenuene

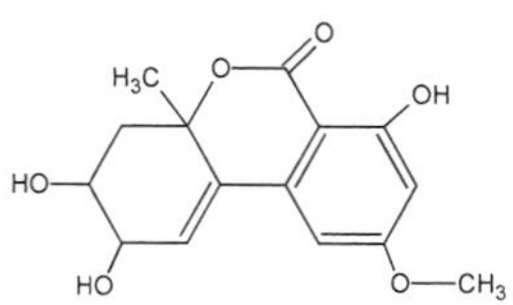

Alternariol

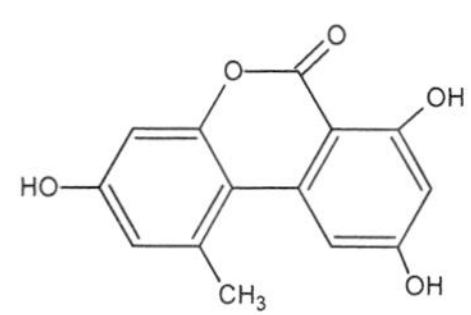

Alternariol methylether

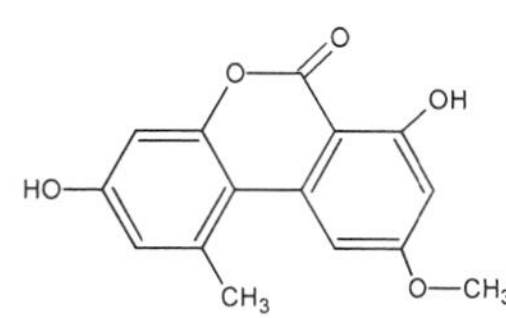

Tenuazonic Acid

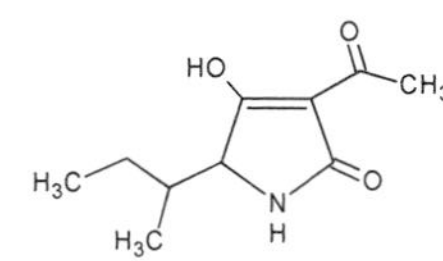

Main producers: *Alternaria alternata* and other *Alternaria spp.*

Main crops affected: tomatoes, olives, citrus fruits, small-grain cereals.

Mycotoxins are small molecules with various chemical structures and, therefore, various biological effects. At the farm level mycotoxin contamination can result in reduced crop yields, as well as reduced livestock productivity stemming from health problems due to consumption of contaminated feed.[3] The human exposure is possible through intake of contaminated agricultural products like grains, dried fruits, grapes, etc. Exposure can also occur by mycotoxin residues due to carry-over and/or metabolization products occurring in foods of animal origin such as animal tissue, milk, and eggs. When present in food in sufficiently high amounts, these fungal metabolites can have toxic effects that range from acute to chronic symptoms. Some mycotoxins have been shown to be mutagenic, teratogenic, or/and carcinogenic. Symptoms of intoxications range from skin irritation to immunosuppression, hepatotoxicity, and nephrotoxicity. A summary of available data and knowledge relevant to the toxicity of major mycotoxins can be found in evaluations by The Joint Expert Committee on Food Additives (JECFA), a scientific advisory board of the World Health Organization (WHO) and Food and Agriculture Organization (FAO) (available at: http://www.inchem.org/documents/jecfa/jecmono/v47je01.htm [Accessed 12 December 2013]), and in Europen Food Safety Association (EFSA) Opinions (available at http://www.efsa.europa.eu/en/topics/topic/mycotoxins.htm [Accessed 12 December 2013]). Health risks and nutrition issues related to mycotoxin contamination have been reviewed by Shephard.[4]

The knowledge of the serious effects of mycotoxins in humans and animals has led many countries to establish regulations for mycotoxins in food and feed to safeguard human and animal health as well as the economic interests of producers and traders. Setting mycotoxin regulations is a complex activity, which involves many factors and interested parties. Detailed information about all the regulatory limits in force for various commodities have been published in the FAO Food and Nutrition Paper 81.[5] The paper reports information relevant to the legal basis, the responsible authorities, the methods used for sampling and analysis, and a series of graphical presentations of the regulatory situations at global level updated to the year 2003. In Europe, harmonized maximum levels for mycotoxins in foodstuffs have been specified in the Commission Regulation 2006/1881/EC,[6] which has been further amended by the Regulations 2007/1126/EC[7] regarding *Fusarium* toxins in maize and maize products, 2012/594/EC[8] regarding ochratoxin A in foodstuffs, and 2012/1058/EC[9] regarding aflatoxins in dried figs. Very recently, the Recommendation 2013/165/EC[10] has been issued setting maximum recommended levels for the sum of T-2 (T-2) and HT-2 (HT-2) toxins in cereals and cereal products.

Reliable analytical methods must be available to enable enforcement of the regulations in daily practice. In addition to reliability, simplicity is desired, as it will affect the amount of data generated and the feasibility of the ultimate measures to be undertaken. The reliability of mycotoxin analysis data can be improved by means of interlaboratory-validated methods of analysis (e.g. the official methods of AOAC International and the methods standardized by European Standardization Committee (CEN)). To date none of the AOAC or CEN methods, which refer mainly to single or closely related mycotoxins in different food matrices, is based on LC–MS. However, several LC–MS methods are currently available for the determination of single and multi-mycotoxins in foods.[11–14] Control laboratories are not forced to use official or standard methods as published by AOAC International or CEN. However, for each mycotoxin, the values of recovery, repeatability and reproducibility of the analytical method selected by each laboratory must fall within the range of acceptability as prescribed in the Commission Regulation 401/2006.[15] A survey on the use and application of methods for mycotoxin determination in food and feed revealed that 42% of participant laboratories routinely used LC–MS(/MS) methodologies for their single or simultaneous multiple determination.[16]

The need for LC–MS methods for mycotoxin determination in control laboratories is highlighted by a recent mandate by the European Commission (EC) for standardization of methods of analysis for mycotoxins in food (M/520 EN) by which the Commission invites CEN to establish European Standards/Technical Specifications that provide standardized methods of analysis for mycotoxins in food (ftp://ftp.cen.eu/CEN/Sectors/List/Food/M520_EN.pdf [Accessed 12 December 2013]). Six of the 11 methods of analysis listed in this mandate are specifically requested to be based on LC–MS/MS.

While relatively extensive information is available on the occurrence and toxicity of the regulated mycotoxins, the requirement for more comprehensive information on food crop contamination by 'emerging' toxins such as nivalenol (NIV), ergot alkaloids, beauvericin or enniatins have been raised only recently. As an example, in 2012–2013 the EFSA's Panel on Contaminants in the Food Chain (CONTAM) issued Opinions on nivalenol, sterigmatocystin, ergot alkaloids, beauvericin and enniatins (available at http://www.efsa.europa.eu/en/topics/topic/mycotoxins.htm [Accessed 12 December 2013]). The overall conclusion was that more occurrence data are needed for these toxins alone or in combination with related known mycotoxins.

An additional emerging issue in the area of mycotoxins is represented by the so-called 'masked' mycotoxins that can be produced by fungal or plant metabolism, or during food processing. These derivatives are mainly produced by plants via enzymatic transformations related to detoxification processes (phase II metabolites) and have been related to a resistance mechanism exerted by plants to counteract pathogen invasion. In addition, chemical transformations of mycotoxins may also occur during food processing and/or fermentations.[17] Masked mycotoxins may occur in conjugated forms, usually formed via reaction of the parent compounds with sugars or amino acids. Modification of parent compounds may also take place via covalent binding or non-covalent association to macromolecules, such as starch and proteins, within the food matrix. Although data on occurrence and toxicity of such compounds are yet limited, human and animals consuming contaminated foods are potentially exposed not only to the native mycotoxins but also to their metabolites that normally escape routine analytical methods.

In recent decades, high- and ultrahigh-performance liquid chromatography have become the most important methods for the analysis of the known mycotoxins in food and feed. However LC coupled with UV or fluorescence (FL) detection cannot fully address emerging issues such as the need for multiple-mycotoxin determination, or for the identification and determination of masked (conjugated) mycotoxins. LC–MS techniques offer a powerful tool to achieve these purposes. In particular, tandem mass spectrometry has opened new perspectives in the determination and identification of mycotoxins in food and feed as well as their metabolic products in biological samples (e.g. urine, blood, or feces). However mycotoxins present largely differing chemical structures, and consequently their physicochemical properties vary in a wide range. Therefore appropriate sample preparation and chromatographic separation of the target mycotoxins from matrix compounds are necessary to ensure accurate quantification and unambiguous identification.

Advances in LC–MS(/MS) methods for the determination of single mycotoxins or closely related mycotoxins have been extensively reviewed and discussed.[11,12,18] This chapter aims to give a critical overview of the application of modern LC–MS(/MS) techniques for the simultaneous determination of multiple-mycotoxins in foods. Critical aspects such as sample preparation, influence of the matrix on quantification results, and evaluation of method performances in relation to legislation requirements are discussed together with emerging issues and future perspectives such as discovery and characterization of masked mycotoxins, and multi-contaminant screening methods.

14.2. LC–MS Analysis of Multiple Mycotoxins: Sample Preparation Aspects

For enforcement purposes, the availability of precise and reliable analytical methods applicable at the regulatory levels for the relevant mycotoxins and commodities is essential. The development of methods for multiple-mycotoxin analysis with one common sample preparation and a single final determination is highly desirable. Due to increasing instrument availability with technical developments enabling high sensitivity and selectivity, LC–MS plays an important role in this field, and sample preparation can become the most challenging task. Several strategies based on either traditional clean up devices or innovative systems have been proposed and evaluated for their contribution to the achievement of satisfactory method performances, including acceptable matrix effects. The higher the selectivity of the sample preparation protocol is, the lower the number of analytes that can be included in the method. But sometimes this is compensated for by the improvement of method performances. On the contrary, less selective clean up strategies, allowing the analysis of a larger number of mycotoxins and other metabolites, require higher instrumental technical specifications (selectivity and sensitivity) to achieve satisfactory method performances.

Immunoaffinity columns (IACs) consist of anti-mycotoxin antibodies coupled covalently to an appropriate carrier and stored normally in phosphate-buffered saline. The mycotoxins bind to specific antibodies when the sample extract is loaded into the column, whereas the impurities are removed without retention during the loading and washing steps. Afterwards the analyte can be eluted by denaturation of the antibody with appropriate organic solvents such as methanol or acetonitrile. The main advantage offered by this technique is the enrichment of the analyte coupled with a highly efficient removal of matrix-interfering components, resulting in better detection and quantification limits, and providing interference-free chromatograms.[19] This makes IACs particularly suitable for LC with UV or FL detection. The most recent standard methods adopted by international bodies such as CEN or AOAC International use immunoaffinity column clean up coupled to LC[20] for the analysis of single mycotoxins or groups of closely related mycotoxins. The main disadvantage of using IACs is the high costs of the columns due to the large amount of antibody needed per single analysis, considering that the columns can be used only once due to antibody denaturation. Another important restriction of IACs is that specific antibodies are needed, and for some mycotoxins no antibodies have been developed yet. The newest developments in immunoaffinity clean up are represented by multi-mycotoxin IACs dedicated to LC–MS detection. Commercial multi-antibody immunoaffinity columns have been successfully used for the simultaneous determination of mycotoxins belonging to different chemical families by LC–MS/MS.[19,21] The latest advance in this field is represented by a multi-analyte column (Myco6in1™, Vicam) containing six different antibodies, for aflatoxins (AFB_1, AFB_2, AFG_1, AFG_2), OTA, fumonisins (FB_1, FB_2), deoxynivalenol (DON), ZEA, T-2 and HT-2 respectively, that has been used for clean up of extracts from maize[22] and other cereals.[23,24] Recently this column has been upgraded by introducing a new antibody able to retain DON and NIV, thus further increasing the number of analytes included in the method.[25]

Thanks to the high selectivity of tandem MS or high-resolution MS techniques, solid-phase extraction (SPE) columns have often been used as a valuable tool for multiple-mycotoxin clean up prior to LC–MS determination. SPE can be used either for analyte enrichment or for matrix-interfering-compounds elimination. It can be applied as a reversed-phase or a normal-phase separation. Retention in the column takes place owing to interactions between the functional groups of the analytes and the surface of the sorbent. When performing multiple-mycotoxin clean up, one of the challenges is to find a suitable SPE sorbent allowing the simultaneous purification of different mycotoxins, which may vary considerably in polarity. As an example, the application of polymeric reversed-phase columns (*N*-vinylpyrrolidone/divinylbenzene columns, Oasis HLB®, Phenomenex) has been reported for the determination of multiple mycotoxins in a wide range of food matrices such as corn silage,[26] beer,[27] and bovine milk.[28] This sample preparation strategy coupled with LC–MS/MS determination gave satisfactory method performances in terms of recoveries, repeatability and detection limits for the determination of most of regulated mycotoxins in cereals and cereal-based foods.[29,30] Among commercially available SPE columns, multifunctional cartridges are also frequently used. These columns, available under the trade name of Mycosep® and/or Multisep® (Romer Labs), contain charcoal, celite, alumina, polymers, and ion-exchange resins in a specially designed column. The multifunctional clean up strategy, first tested for trichothecene analysis and then extended to other major mycotoxins, provides a quick sample purification method, able to remove impurities, but less selective if compared to IACs.[31,32] As for IACs, the most recent advance in this field is represented by a multi-analyte column. MycoSpin™ 400 columns are currently proposed by the supplier (Romer Labs) for multiple-mycotoxin enrichment prior to LC–MS detection, although no literature data are available showing performances in the analysis of real samples.

Some studies describe the potential use of molecularly imprinted polymers (MIPs) as adsorbents for SPE of mycotoxins.[33,34] MIPs are cross-linked polymers synthesized by reaction of a monomer and a cross-linker in presence of the analyte (or mimic compounds) used as a template. After polymerization the analyte is removed, leaving specific recognition sites inside the polymer. MIPs are cheap, easy to obtain, and have high chemical stability and long shelf life. Again, the development of cartridges for multiple-mycotoxin detection represents the latest advance in this field. A column containing a mix of MIPs for clean up of aflatoxins, ZEA, OTA, fumonisins, T-2 and HT-2 has been very recently introduced onto the market (AFFINIMIP® SPE Multimyco10, POLYNTELL). These columns have been developed for multiple-mycotoxin clean up prior to LC–MS(/MS) analysis, however, demonstration of their applicability in routine analysis as well as comparison with validated methods are not yet available.

Important advantages of some SPE columns are reusability and ease of on-line coupling with the possibility of automation for high-throughput applications. As an example, the use of TurboFlow™ technology (TLX), a new automated on-line sample clean up system directly coupled to LC–HRMS equipment, has been recently evaluated for the simultaneous determination of *Fusarium* toxins (DON, T-2, HT-2, ZEA, FB_1 and FB_2) in maize, wheat and animal feed.[35] The principle of TLX-LC chromatography is that target compounds are injected and loaded onto a suitable SPE column whilst matrix interferences with higher molecular weight and different chemical properties are sent to waste. Then, target compounds are eluted by the loop (which is filled in advance with a stronger eluent) and transferred onto the analytical column for compound separation and mass spectrometric detection. After testing different stationary phases, a mixed-mode stationary phase with both strong anion exchange and reversed phase material was found to be able to retain all target compounds. In this study, HRMS detection in full-scan mode was chosen, having the advantage of retrospective data analysis for the re-evaluation of measured samples for additional targeted or unknown compounds.

The advantages of retrospective analysis can be fully exploited only if adopting a generic (i.e. poorly selective) sample preparation strategy. An example of a very fast and generic extraction/purification strategy is based on the use of QuEChERS (Quick, Easy, Cheap, Effective, Rugged and Safe), currently widely used in multi-pesticide analysis. The use of QuEChERS as generic sample pre-treatment enabling the simultaneous analysis of a wide range of mycotoxins is becoming a popular alternative to the direct injection of crude extracts, and deserves further investigation. The key principle of the QuEChERS approach is analyte partitioning in an acetonitrile/water mixture induced by addition of inorganic salts. While the analytes are transferred in the organic phase, the more polar matrix impurities are left in an aqueous layer. The residual impurities in acetonitrile can be removed by dispersive SPE by the addition of primary secondary amine (PSA) sorbent. The QuEChERS approach has been adapted, applied and evaluated for the determination of multiple mycotoxins in cereals,[36] but also in more complex matrices such as silage[37] or highly pigmented spices.[38] When applying this approach significant matrix effects are generally observed and these need to be compensated for by matrix-matched calibration curves or the use of internal standards. It is worth noting that the poor selectivity of this sample preparation strategy is often compensated for by coupling it with highly selective detection by HRMS.

In parallel with the increasing availability of bench-top highly sensitive LC–MS instrumentation, the direct injection of crude extracts is becoming more and more popular for the development of relatively rapid methods. The so called 'dilute-and-shoot' approach, omitting any sample clean-up, is generally preferred for the screening of a large number of contaminants, including plant and/or fungal metabolites.[39–42] Matrix effects are generally managed by extract dilution that, on the other hand, can result in high detection limits, unless highly sensitive and selective LC–MS instrumentation is used. In some cases two chromatographic runs are set to guarantee optimal MS conditions (i.e. dwell time, ion source polarity, etc.) for all analytes.[39,40] A drawback of most multi-target methods is that they require extensive validation which is time- and cost-consuming and, hence, often reduced to a minimum. Therefore, these methods are mainly proposed for semi-quantitative screening.

Methodologies based on QuEChERS extraction or direct injection of crude extracts are generally adopted to screen for a large number of contaminants, and provide data about the co-occurrence of multiple mycotoxins in the same sample, including a wide array of less known or 'emerging' mycotoxins and other metabolites.[41,42] For example, LC–MS/MS after QuEChERS extraction has been used to investigate on the co-occurrence of 'traditional' mycotoxins, such as type-A and -B trichothecenes and ZEA, together with less routinely determined *Alternaria* toxins, ergot alkaloids and 'emerging' mycotoxins (enniatins and beauvericin) in cereal-based food samples.[43] The study highlighted a high incidence of enniatins and of the conjugated form of DON, DON-3-glucoside (DON-3-G) occurring in 100% and 80%, respectively, of the analyzed samples from the Czech retail market. The dilute-and-shoot approach has been used for LC–MS/MS screening for a broader range (up to 139) of mycotoxins and other fungal metabolites in food and feed samples, revealing the presence of up to 69 metabolites in a single sample.[44,45] Although the observed concentrations of the individual analytes were generally in the low μg/kg range, these studies emphasize the great variety of potential mycotoxin co-exposure.

14.3. The Potential of High-Resolution Mass Spectrometry in Mycotoxin Analysis

While the potential of tandem mass spectrometry for quantitative determination of multiple mycotoxins has been largely documented, there is actually an increasing interest in evaluating new MS detection approaches mainly based on high-resolution mass spectrometry. In the field of mycotoxin analysis this trend is very well reflected by recent studies[35,36,46-49] exploring the potential of application of HRMS, mostly based on Orbitrap™ technology, as a tool for obtaining quantitative determination and full spectral information in a unique analysis.

The process of obtaining mycotoxin fragmentation patterns by high-energy collision-induced dissociation (HCD) has been investigated to obtain quantitative and confirmatory information (two characteristic masses per mycotoxin) using Orbitrap® based HRMS. The main perspective of HRMS technology, coupled with full-scan analysis, is the inclusion of masked mycotoxins and/or the identification of other metabolites by retrospective analysis. The comparison between full-scan HRMS and triple quadrupole detection gave comparable results for the quantitative determination of regulated mycotoxins in cereals and derived products,[47] provided that effective extract clean up and proper chromatographic separation was applied. On the other hand, difficulties when using single-stage MS are encountered in obtaining confirmatory ions or detecting them with adequate mass accuracy at low concentrations.[36] High detection limits, unsuitable for assessing mycotoxin contamination at regulatory levels, are obtained when poor sample preparation is applied to recover a wide range of different analytes.[48]

Another emerging development in the field of HRMS analysis of multiple mycotoxins is associated with the introduction of novel ambient ionization techniques, represented mainly by desorption ionization (DESI) and direct analysis in real time (DART). DESI utilizes the impact of electrosprayed solvent droplets upon sample surfaces to generate analyte ions, whereas DART ionization is based on bombardment of sample with electrical discharged metastable and atmospheric gases. As in other ambient MS techniques, chromatographic separation is omitted, therefore requiring highly selective MS detection such as HRMS to identify target analytes. The applicability of DESI has only been shown for the analysis of fumonisins in intact maize kernels,[50] whereas the potential of

DART coupled to HRMS has been more deeply investigated for the (semi-)quantitative analysis of multiple mycotoxins in cereals. In particular, Zachariasova *et al.* used positive and negative DART–Orbitrap MS profiles of beer samples to rapidly assess the efficiency of the developed clean up strategy based on the partitioning in acetonitrile.[46] Vaclavik *et al.* evaluated the application of DART–Orbitrap MS to the analysis of multiple mycotoxins in wheat and maize after QuEChERS extraction.[51] Under the applied experimental conditions, only 11 of the 24 tested mycotoxins could be analyzed, since aflatoxins and T-2/H-2 showed poor ionization, whereas OTA, ergot alkaloids and fumonisins could not be ionized. The DART–MS based method was shown to be applicable for high-throughput control of maximum limits of ZEA and DON established by EC regulation for unprocessed wheat/maize.

The few available studies demonstrate that the direct analysis of mycotoxins on food surfaces is possible, although the real applicability in routine food control at maximum permitted levels needs to be further investigated. Reliability and accuracy of quantitative measurements is guaranteed only by using suitable internal standards to compensate for matrix effects and the relatively high signal fluctuation of ions intensities obtained by repeated measurements.

14.4. Matrix Effects in LC–MS Determination of Mycotoxins

Despite the high sensitivity and selectivity, the influence of matrix components on the analyte ionization can represent a limit to the accuracy of the LC–MS(/MS) methods. The co-elution of matrix compounds can result in enhancement or suppression of the analyte chromatographic signal. A general review dealing with matrix effects in LC–MS has been published by Gossetti *et al.*[52] Prediction of matrix effects is difficult because they are influenced by several factors, like target compound (chemical structure, polarity), matrix type, and the relative concentrations of the substances competing for the limited number of charges. Additionally, sample preparation (extraction, clean up), chromatographic conditions, mass spectrometric instrumentation (e.g. design of ion source) and ionization conditions influence the extent of matrix effects.[52] Therefore a universal approach to managing matrix effects by introducing a general type of corrective action does not work, since it has been extensively experienced that every mycotoxin/matrix combination can show different matrix effects and of unpredictable magnitudes.

The magnitude of matrix effects is generally estimated by comparing the slopes of standard and matrix-assisted calibrations. A quantitative estimation can be achieved by calculating the so-called 'signal suppression/enhancement' (SSE) ratio using the formula:

$$\mathrm{SSE}(\%) = 100 \times \frac{\mathrm{slope}_{\text{matrix calibration}}}{\mathrm{slope}_{\text{standard calibration}}}$$

according to Matuszewski *et al.*[53] or by statistically evaluating slope differences using the Student's *t* test.[29]

Two basic approaches can be adopted and integrated to manage matrix effects: the reduction of the absolute amount of matrix components in the injected sample and the selection of an optimal calibration strategy. The first goal can be achieved by extract clean up or extract dilution. It has been demonstrated that increasing the selectivity of sample clean up allows the minimization of matrix effects,[29,30,54] but does not completely eliminate them even if based on immunoaffinity columns.[22] On the other hand, the drastic dilution of the sample extract combined with the use of highly sensitive MS/MS instruments is sometimes proposed to minimize or eliminate matrix effects. Also, in this case significant signal enhancement/suppression can be still observed, requiring appropriate correction/compensation.[40] Minimization of matrix effects is of utmost importance since they can also affect the stability of the MS signal, due to progressive dirtiness of the MS interface and the shelf life of the LC column, and therefore the repeatability and robustness of analytical determinations in routine analysis.

The most common approach used for matrix effect compensation in multi-mycotoxin analysis is the matrix-matched calibration (i.e. the preparation of calibrant solutions in blank extracts of the respective matrix).[29,55–58] The main drawback of this approach is the need to use a blank matrix, which is impractical under routine conditions where control laboratories are faced daily with a variety of matrices to be analyzed.

The standard addition approach is also often used in routine analysis, but it at least doubles the number of LC runs per sample.[59] The most suitable approach to control the matrix effect is the use of an internal standard with molecular structure and physicochemical properties as close as possible to those of the target analyte. The addition of the internal standard to the sample to be analyzed allows calculatation and correction for recovery losses during the sample preparation process and ionization-suppression effects in the MS ion source. This approach overcomes the problem of finding blank samples for external matrix-assisted calibration and decreases the total time of analysis. Structurally related compounds can be used as internal standards for mycotoxin determination. Zearalanone (ZAN), differing from zearalenone for the absence of one double bond, was used to compensate the matrix effects for ZEA and trichothecene analysis.[60,61] Verrucarol[62] and deepoxy-deoxynivalenol[63] were used as internal standards for A- and B-trichothecenes. A main drawback of using structurally related mycotoxins is that they do not co-elute and may show different ionization properties, thus not ensuring a proper compensation of matrix effects.

The best way to correct for the analyte signal suppression/enhancement is to use a stable isotope-labeled analog of the analyte co-eluting with the analyte itself. Several examples of the use of isotope-labeled standards for the determination of single or multiple mycotoxins in foods can be found in the literature. For instance, deuterium-labeled mycotoxins such as [$^2H_6$]-FB_1,[64] 15-[$^2H_1$]-DON[65] and 3-[$^2H_3$]-acetyldeoxynivalenol,[65] [3,5-$^2H_2$]-ZEN,[66] [$^2H_5$]-OTA,[67] [$^2H_2$]-AFB_2 and [$^2H_2$-4]-AFG_2[68] have been used for the determination of the relevant naturally occurring mycotoxins in foodstuffs. Applications of stable isotope-labeled standards in mycotoxin analysis, including prerequisites and limitations, have been reviewed by Rychlik and Asam.[69]

Fully ^{13}C isotope-labeled mycotoxins are now commercially available for all regulated mycotoxins. The physicochemical properties of such substances, together with their chromatographic behavior and their ionization potential, are very similar (almost the same) to those of the relevant naturally occurring target mycotoxins. This results in comparable ionization properties in the presence of co-eluting components, making them suitable for compensation for matrix effects. Target mycotoxins and their isotope-labeled analogs show the same fragmentation pattern (as shown by the example of OTA reported in Fig. 14.1), but can be separated and distinguished owing to their different molecular weight.

Most of the more recently published methods are based on fully ^{13}C-labeled mycotoxins, mirroring the growing supply of commercially available standards. The first reports deal with single trichothecene analysis and showed the use of $[^{13}C_{15}]$-labeled DON and $[^{13}C_{24}]$-labeled T-2 for the determination of DON and T-2, respectively, in cereal grains.[70,71] Since then, several applications to multiple-trichothecene and multiple-mycotoxin LC–MS analysis have demonstrated the reliability of this approach in compensating for matrix effects and then enhancing the overall method accuracy.[30,39,72] Two main drawbacks for the use of isotope-labeled internal standards are their cost and their commercial availability being limited to a restricted number of mycotoxins.

Finally, within this context another aspect that deserves some consideration is the influence of matrix components on the accuracy of high-resolution MS measurements. In high-resolution full-scan MS measurements, selectivity is obtained by the creation of extracted ion chromatograms (XICs) of diagnostic ions of the compound of interest. Correct mass assignment over the entire chromatographic elution profile of the analyte is an essential parameter for peak quantification and identification/confirmation. Matrix compounds overlapping with nearby masses can affect mass accuracy of the ions selected for target mycotoxin monitoring. The influence of matrix compounds on the analyte identification can be evaluated, for instance, by comparing mass accuracy values for diagnostic ions of each mycotoxin, measured in standard solution and food extracts.[47,73] Besides instrumental

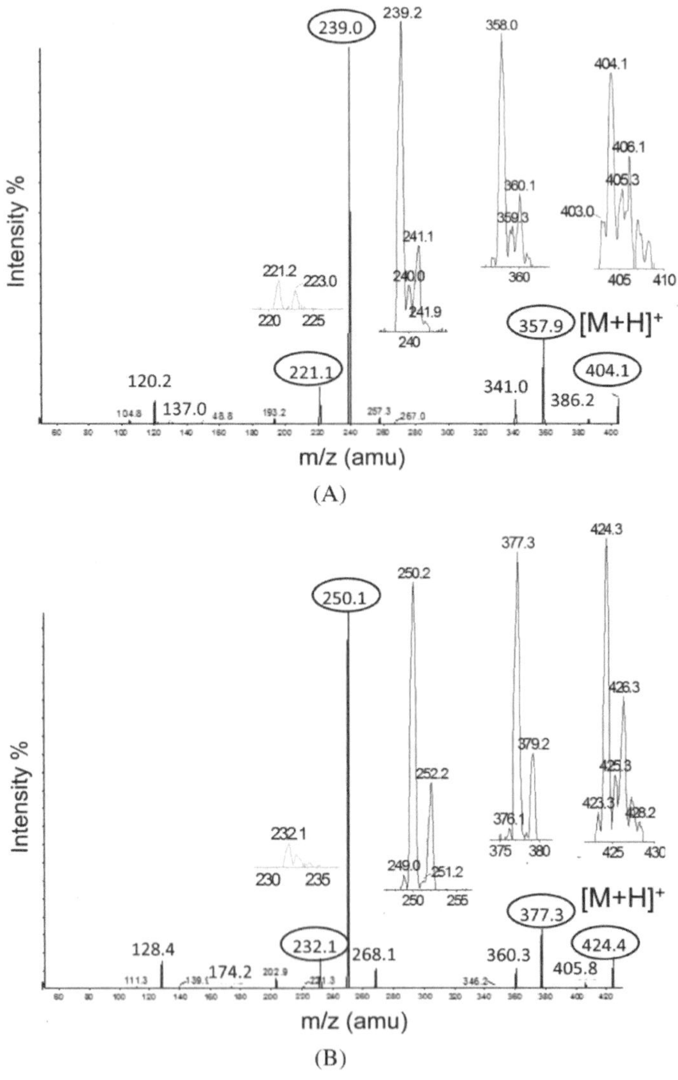

Figure 14.1. Comparison of fragmentation patterns (MS/MS spectra) in positive-ion mode of OTA and ^{13}C-labelled OTA. Isotopic patterns of main fragments. Reproduced with permission from Lattanzio, V.M.T., Della Gatta, S., Suman, M *et al.* (2011). Development and in house validation of a robust and sensitive solid phase extraction: LC–MS/MS method for the quantitative determination of aflatoxins B_1, B_2, G_1, G_2, ochratoxin A, deoxynivalenol, zearalenone, T-2 and HT-2 toxins in cereal based foods, *Rapid Comm. Mass Spectrom.*, **25**, 1869–1880. Copyright (2011) John Wiley and Sons.

resolving power, the efficiency of separation from matrix compounds, achieved by extract clean up (or dilution) and good chromatography, is an important parameter for unbiased mass assignment in the analysis of small molecules in complex food matrices.

14.5. Performance Evaluation of LC–MS Methods for Multiple-Mycotoxin Determination

Despite the large number of currently available LC–MS(/MS) methods for multiple-mycotoxin determination in foods, most of them proved to be suitable for semi-quantitative or screening purposes only due to the lack of sufficient validation efforts beyond single laboratory studies and low recovery rates or poor detectability for some target toxins.[16,30] Acceptability criteria for analytical methods are set in the Commission Regulation 401/2006[15] for each regulated mycotoxin, in terms of recovery values, repeatability and reproducibility. Requirements for identification by mass spectrometry are specified in official documents such as the Commission Decision 2002/657/EC,[74] which relies on food of animal origin, whereas for food of non-animal origin no such criteria document yet exists. Therefore criteria set for pesticide analysis in the Document SANCO/12495/2011[75] are often applied for mycotoxins too. However until specific criteria are defined, laboratories can employ a 'fit-for-purpose' approach.

Still far from harmonization, efforts are being made for method comparison and deeper understanding of performances of the available LC–MS(/MS) methodologies for multiple-mycotoxin analysis. Within the EU Network of Excellence MoniQA (www.MoniQA.eu) a proficiency test was conducted to benchmark laboratories using LC–MS/(MS) for multi-mycotoxin analysis and to obtain information on currently used methodologies and related method performances.[76,77] The study involved 41 laboratories from 14 countries and was conducted for the simultaneous determination of up to 11 mycotoxins (aflatoxins, OTA, FB_1, FB_2, ZEA, DON, T-2 and HT-2) in spiked and naturally contaminated maize. A robust and reliable method for simultaneous determination of 11 mycotoxins in maize could not be identified from this study, highlighting the need for more experimental work to set up a method suitable for inter-laboratory validation; only one laboratory obtained acceptable z-scores for all mycotoxins. In general, extraction mixtures of water with acetonitrile, methanol, or both provided good results for quantitative extraction of mycotoxins from maize. Laboratories using extract clean up reported acceptable results for the majority of mycotoxins. Good results were also obtained by laboratories that analyzed crude extracts, although a high variability of results was observed for all tested mycotoxins. Matrix-matched calibration or isotope-labeled internal standards efficiently compensated for matrix effects, whereas external calibration gave reliable results only when injecting <10 mg of matrix equivalent amounts. It is worth mentioning that unacceptably high recovery and variability of the fumonisin results were obtained by the majority of laboratories for spiked maize.

Results of an inter-laboratory study concerning relative and absolute matrix effects in multiple-mycotoxin determination have been reported by Malachova *et al.*[43] The applicability of commonly used strategies in matrix-effect reduction was tested in the quantitative determination of NIV, DON, FB_1, FB_2, and ZEA in complex feed matrices. The study showed that the use of any purification technique helped to improve absolute matrix effects for some analytes, whereas other factors such as changes in LC conditions or switching ion source polarity had a minor impact.

An interesting critical comparison between the use of IAC and direct analysis of crude extracts by LC–MS/MS with respect to compliance with EC acceptability criteria for MS detection[74] (Regulation 657/2002/EC) has been carried out by Senyuva *et al.*[78] LC–MS/MS ion ratios and peak profiles, repeatability, and quantification limits (LOQs) were used as the basis for a detailed comparison of the two approaches. On the basis of visual inspection, ion chromatograms of sample extracts not submitted to clean up showed frequent background interference, although MS/MS product ions could be detected at the correct retention times. Ions were visible, but showed poor peak shape (e.g. shoulders and lack of symmetry) and resolution from the background ions. After IAC clean up, ions had Gaussian peak shapes and were essentially indistinguishable from standards. Both tested approaches gave satisfactory results in the analysis of FAPAS materials. However the overall study, which included more complex matrices such as feed samples, indicated that with no sample clean up it was generally not possible to meet identification criteria; therefore, any further data processing was unreliable. The authors concluded that for DON, ZEA, HT-2, and T-2 determination in animal feed samples, methods with no clean up can only be regarded as screening tools. Where definitive identification is an essential requirement prior to quantification (e.g. for food regulatory control purposes), sample clean up prior to LC–MS/MS quantification is essential.

These studies provide a great deal of information on current methodologies, enabling a deeper understanding of the performances of different LC–MS-based approaches for multiple-mycotoxin analysis in real food matrices, and would be helpful in deriving performance characteristics of a method for simultaneous determination of the EC-regulated mycotoxins in maize.

14.6. LC–MS Identification and Determination of Masked Mycotoxins

The increasing instrumental availability and technological advances in LC–MS have had a strong impact on the knowhow about formation and determination of the so called ‘masked’ or ‘hidden’ mycotoxins, which are mycotoxin derivatives that usually escape routine analysis due to their different chemical behavior with respect to the parent mycotoxins. Detection and characterization of known or ‘novel’ compounds in fungal cultures, plant material, processed food or biological fluids can provide insights in mycotoxin detoxification and metabolism. Furthermore, the direct determination of free and bound mycotoxins would enable a more accurate evaluation of the risk of exposure. LC–MS is the technique of choice for the detection and characterization of mycotoxins metabolites such as masked and other conjugated mycotoxins in processed and unprocessed foods and feeds.[17]

High-resolution full scans over a given mass range, acquired by time-of-flight (TOF) or Fourier transform (e.g. Orbitrap) mass analyzers, are often used to screen for accurate masses corresponding to molecular ions of the hypothesized metabolites. Signal-to-noise ratios are usually good due to the high resolution of such instruments, which limits the occurrence of interfering background substances with *m/z* values similar to those of the analyte. Additional mathematical operations, such as the application of mass defect filters, or the use of specific software, can greatly help to identify unknown metabolites. Examples of structure elucidation by LC–HRMS were given for glucoside derivatives of T-2, HT-2, and neosolaniol (NEO)[79–81] in naturally contaminated cereals and fungal cultures and hydrolyzed derivatives of fumonisins in maize and processed products.[82]

For the characterization of conjugated mycotoxins, tandem mass spectrometers offer the possibility of integrating information derived from different scan modes. Precursor ion scans with triple quadrupole (QqQ) MS may find higher-mass conjugates when the mass of the mycotoxin (or a specific highly abundant fragment) is fixed in the third quadrupole. Neutral-loss scans can identify specific derivatives of sample compounds (e.g. glucosides (loss of 162 Da), glucuronides (loss of 176 Da), etc.). Unfortunately, these two scan types are not very sensitive. High sensitivity of tandem mass spectrometers (such as QqQ instruments, ion traps, quadrupole TOF instruments, quadrupole traps, TOF–TOF instruments, etc.) can be achieved through product ion scans. A screening for calculated monoisotopic masses of presumed ions (e.g. $[M + H]^+$, $[M - H]^-$, $[M + NH_4]^+$) from predicted mycotoxin conjugates can be easily performed by full-scan chromatograms. Then acquisition of the fragmentation pattern (product ion spectrum) and comparison with that of native toxin can confirm the existence of the hypothesized metabolite. Such approaches have been used to identify and characterize glucoside derivatives of diacetoxyscirpenol (DAS), NEO in fungal cultures,[83] and glucoronide derivatives of DON in human and rat urine.[84] A very exhaustive example of an ion trap MS/MS application to characterize fumonisin-like compounds in rice cultures infected by *Fusarium verticillioides* has been reported by Bartók *et al.*[85] In a subsequent study the same authors reported on the first identification of fumonisins esterified with three acyl groups (palmitoyl, linoleoyl, and oleoyl esterified fumonisin B_1), highlighting the importance of using two combined approaches based on low-resolution (ion trap) and high-resolution (TOF) MS for a complete characterization of the new compounds.[86]

All available studies dealing with the characterization of new mycotoxin derivatives show the power of MS techniques for the characterization of new compounds, even if present at trace levels, without the need for isolating them. Of course, neither advanced nor combined MS techniques enable the determination of the relative and absolute configuration of these toxins; this represents the main limitation of these approaches. The combination of MS with ion mobility spectrometry (IMS) seems to be a promising tool for increasing the potential of this technique to fully resolve molecular structures. In ion mobility spectrometry, an ionized sample enters a drift tube that contains an electric field and a carrier buffer gas. As ions move through the matrix, they are separated according to their size, shape, and charge. Differential migration time through the tube can be used, for example, to discriminate between analytes having the same molecular weight (and molecular formula) but different structural configuration (e.g. isomers having different anomeric or stereochemical configurations). In the near future, IMS–MS is expected to play a significant role in the characterization of conjugated mycotoxins.

14.7. LC–MS-Based Multi-Class Methods

In the previous paragraphs it has been shown that combined determination of mycotoxins with a wide variety of physicochemical properties is feasible. Finally, it is worth mentioning the strong trend towards the use of LC–MS(/MS) for the analysis of mycotoxins in multi-class methods. An emerging issue in chemical food safety control is the effort to integrate analyses of various groups of food contaminants/toxicants into a single, high-throughput method.[13,14] Again, the power of modern LC–MS(/MS) instrumentation to screen for a large number of contaminants being widely documented, increasing attention has been given to the choice of optimal sample-preparation procedure, which remains one of the key steps to achieving satisfactory performance characteristics. With respect to comprehensive screening for residues and contaminants, full-scan techniques based on HRMS are generally preferred to techniques using targeted acquisition like MS/MS detection. Due to the acquisition of high-resolution full scan spectra, it is feasible to apply both target and non-target approaches for the rapid qualitative screening of multiple contaminants.

The first work covering this topic has been carried out by Mol *et al.*[87] with the purpose of developing a generic extraction method able to cover a vast number of target analytes (pesticides, veterinary drugs, plant toxins, and mycotoxins) and also applicable to different types of food and feed matrices. To achieve this goal a generic extraction procedure using slightly acidified conditions was optimized. The avoidance of phase separation and the use of acidic conditions were found to be the key factors for the high extraction efficiencies of the wide variety of analytes considered. The sample preparation did not involve any clean up; as a consequence, significant matrix effects were observed, especially for complex samples like compound feeds. This was minimized by injecting small volumes of extracts containing low amounts of matrix equivalent. The suitability of this sample preparation protocol for screening purposes was shown either in combination with target LC–MS/MS or untargeted LC–TOF-MS. This approach was implemented by Herrmann *et al.*[88] expanding the number of matrices and focusing on 108 representative compounds related to emergency situations, including 36 mycotoxins.

Besides the dilute-and-shoot approach, the most investigated sample preparation strategy for multi-class methods is the use of QuEChERS. As an example, Lacina *et al.*[89] reported results of a study dedicated to test different sample preparation procedures: aqueous acetonitrile extraction followed by partition (QuEChERS-like method); aqueous acetonitrile extraction; and pure acetonitrile extraction for the analysis of a list of target analytes, including 288 pesticides together with 38 mycotoxins. The QuEChERS-like method showed the best performances in terms of co-extraction of focused analytes. Indeed the QuEChERS approach is the most commonly used when developing multi-contaminant extraction, either in combination with HRMS based on TOF or Orbitrap mass analyzers, or with last-generation QqQ enabling fast detection of multiple analytes. QuEChERS-like methods have been developed and optimized for the simultaneous determination of 22 carbamate insecticides and 17 mycotoxins in cereals[90] by UPLC–MS/MS, for the simultaneous analysis of veterinary drugs and mycotoxins in hen eggs by LC–MS/MS,[91] and for purifying bakery raw materials and finished products to detect pesticides, mycotoxins and veterinary drugs by LC–HRMS.[92] A method combining generic extraction and analysis protocols, such as QuEChERS extraction, with full scan LC–HRMS was evaluated for its use in screening for 118 contaminants including mycotoxins (DON, AFB_1, T-2), plant alkaloids, carbamate and organophosphate pesticides, and several types of veterinary drugs in whole milk, muscle tissue, liver tissue and maize silage.[93] Accurate mass-to-charge ratios for expected pseudo-molecular ions together with retention times were used to identify analytes. QuEChERS extraction in combination with LC–MS/MS for a total of more than 90 compounds was used for the simultaneous determination of pesticides, biopesticides and mycotoxins in organic products.[94]

On the other hand, some authors proposed commodity-dedicated methods for the determination of a restricted number of mycotoxins together with other contaminants of interest in specific matrices. In particular, Mornar *et al.*[95] developed a method for the simultaneous determination of citrinin and cholesterol-lowering compounds (monacolins) in red fermented rice using connected diode arrays, fluorescence and MS detectors, whereas Song *et al.*[96] reported a LC–MS method for the simultaneous determination of AFB_1, OTA, and patulin together with bisphenol A for regulatory purposes in beverages and food products. A screening method for the detection and identification of undesirable organic compounds, including antibiotics, pesticides, and mycotoxins in aquaculture products based on direct injection of acetonitrile/water extracts in the LC–HRMS system has been reported by Nácher-Mestre *et al.*[97] In the same study, retrospective analysis of accurate-mass full-spectrum acquisition data provided by Q–TOF-MS was exploited to detect and tentatively identify other undesirable organic compounds different from those included in the validated list.

14.8. Conclusions

The aim of this chapter was to give a critical overview of the applications of modern LC–MS techniques to the field of mycotoxin analysis in foods. In the area of regulated mycotoxins, where validated official AOAC and CEN methods based on conventional chromatographic techniques and dedicated to single or closely related groups of mycotoxins already exist, the real competitiveness of LC–MS is represented by the possibility of performing multiple-mycotoxin analysis. In addition, tandem mass spectrometry provides the highest degree of certainty in analyte identification and may be employed in accordance with recent EC guidelines to obtain relevant, unambiguous data. Quantitative and confirmatory information can be obtained from a single chromatographic run.

However, from the reviewed studies it can be concluded that LC–MS is definitely not a trouble-free solution for the analysis of multiple mycotoxins in complex food matrices. Sample clean up still remains an important and necessary step for obtaining reliable analytical results. This is clearly demonstrated by specifically designed comparative studies.

A universally recognized limitation of LC–MS is the matrix effects, which cannot be predicted nor completely eliminated. Validation of newly developed methods or method applications to new matrices must include the evaluation of matrix effects and which additional measures, such as internal standard, standard addition or matrix assisted calibration, have to be taken to ensure accurate and reliable mycotoxin quantification. Particularly suitable in this context is the use of isotope-labeled internal standards that allows the skipping of external calibration and ensures proper compensation for matrix effects. In the near future, the use of a dilute-and-shoot approach coupled with highly sensitive MS instrumentation is expected to eliminate or significantly reduce the problem of matrix effects.

The great potential of LC–HRMS, coupled with generic sample preparation protocols to screen for a very large number of mycotoxins together with their metabolites or other class of contaminants, has been shown by recent studies. The driving force in developing such multi-contaminant methods is probably the increasing availability of bench top highly selective and sensitive instrumentation. However, the real scientific meaning of searching for hundreds of contaminants in a unique food matrix notwithstanding, it might be useful for control and research laboratories to have available a high-throughput method to screen for a large number of mycotoxins and metabolites in a variety of different matrices.

MS-based screening is playing an important role in the discovery of novel mycotoxin conjugates and this trend is expected to continue in the future. However, advances in metabolite discovery need to be integrated with occurrence data as well as toxicity and bioavailability studies to give the real dimension of the risks related to food contamination by masked mycotoxins.

References

1. Miller, J.D. and Trenholm, H.L. (1994). *Mycotoxins in Grain: Compounds Other than Aflatoxins*, Eagan Press, St. Paul, MN, USA.
2. Miller, J.D. (1998). "Global significance of mycotoxins", in Miraglia, M., van Egmong, H.P., Brera, C. *et al.* (eds), *Mycotoxins and Phycotoxins: Developments in Chemistry, Toxicology and Food Safety,* Alaken Inc., Fort Collins, CO, USA, pp. 3–15.
3. Council for Agriculture Science and Technology (2003). Mycotoxins: risks in plant, animal, and human systems, *Task Force Report 139*, Ames, IA, USA.
4. Shephard, G.S. (2006). 'Mycotoxins in the context of food risk and nutrition issues', in *The Mycotoxin Fact Book: Food and Feed Topics*, Wageningen Academic Publishers, Wageningen, The Netherlands, pp. 21–36.

5. Food and Agriculture Organization (FAO) (2004). *Worldwide Regulations for Mycotoxins in Food and Feed in 2003*, Food and Nutrition Paper no. 81, Rome, Italy.
6. Commission Regulation (EC) No 1881/2006 of 19 December 2006 setting maximum levels for certain contaminants in foodstuffs, *Off. J. Eur. Union,* L 364/5.
7. Commission Regulation (EC) No 1126/2007 of 28 September 2007 amending Regulation (EC) No 1881/2006 setting maximum levels for certain contaminants in foodstuffs as regards *Fusarium* toxins in maize and maize products, *Off. J. Eur. Union,* L 255/14.
8. Commission Regulation (EU) No 594/2012 of 5 July 2012 amending Regulation (EC) 1881/2006 as regards the maximum levels of the contaminants ochratoxin A, non dioxin-like PCBs and melamine in foodstuffs, *Off. J. Eur. Union,* L 176/43.
9. Commission Regulation (EU) No 1058/2012 of 12 November 2012 amending Regulation (EC) No 1881/2006 as regards maximum levels for aflatoxins in dried figs, *Off. J. Eur. Union,* L 313/14.
10. Commission Recommendation of 27 March 2013 on the presence of T-2 and HT-2 toxin in cereals and cereal products, *Off. J. Eur. Union,* L 91/12.
11. Li, P., Zhang, Z., Hu, X. *et al.* (2013). Advanced hyphenated chromatographic–mass spectrometry in mycotoxin determination: current status and prospects, *Mass Spectrom. Rev.*, **32**, 420–452.
12. Songsermsakul, P. and Razzazi-Fazeli, E. (2008). A review of recent trends in applications of liquid chromatography–mass spectrometry for determination of mycotoxins, *J. Liq. Chromatogr. Rel. Technol.*, **31**, 1641–1686.
13. Shephard, G.S., Berthiller, F., Burdaspal, P.A. *et al.* (2013). Developments in mycotoxin analysis: an update for 2011–2012, *World Mycotoxin J.*, **6**, 3–30.
14. Berthiller, F., Burdaspal, P.A., Crews, C. *et al.* (2013). Developments in mycotoxin analysis: an update for 2012–2013, *World Mycotoxin J.*, **7**, 3–33.
15. Commission Regulation (EU) No 401/2006 of 23 February 2006 laying down the methods of sampling and analysis for the official control of the levels of mycotoxins in foodstuffs, *Off. J. Eur. Union,* L 70/12.
16. Solfrizzo, M., Alldrick, A.J. and van Egmond, H.P. (2009). The use of mycotoxin methodology in practice: a need for harmonization, *Qual. Assur. Saf. Crop.*, **1**, 121–132.

17. Berthiller, F., Schuhmacher, R., Adam, G. *et al.* (2009). Formation, determination and significance of masked and other conjugated mycotoxins, *Anal. Bioanal. Chem.*, **395**, 1243–1252.
18. Zöllner, P. and Mayer-Helm, B. (2006). Trace mycotoxin analysis in complex biological and food matrices by liquid chromatography–atmospheric pressure ionisation mass spectrometry, *J. Chromatogr. A*, **1136**, 123–169.
19. Pascale, M. and Visconti, A. (2006). 'Immunoaffinity clean up/fluorescence detection methods for mycotoxins', in *The Mycotoxin Fact Book: Food and Feed Topics*, Wageningen Academic Publishers, Wageningen, The Netherlands, pp. 269–283.
20. Senyuva, H.Z. (2011). 'Official methods and performance criteria for determining mycotoxins in food and feed', in *Determining Mycotoxins and Mycotoxigenic Fungi in Food and Feed*, Woodhead Publishing Series in Food Science, Technology and Nutrition, Number 203, pp. 171–193.
21. Tanaka, H., Takino, M., Sugita-Konishi, Y. *et al.* (2010). Determination of *Fusarium* mycotoxins by liquid chromatography–tandem mass spectrometry coupled with immunoaffinity extraction, *Rapid Commn. Mass Spectrom.*, **24**, 2445–2452.
22. Lattanzio, V.M.T., Solfrizzo, M., Powers, S. *et al.* (2007). Simultaneous determination of aflatoxins, ochratoxin A and *Fusarium* toxins by liquid chromatography–tandem mass spectrometry after multitoxin immunoaffinity clean up, *Rapid Commun. Mass Spectrom.*, **21**, 3253–3261.
23. Tang, Y.Y., Lin, H.Y., Chen, Y.C *et al.* (2013). Development of a quantitative multi-mycotoxin method in rice, maize, wheat and peanut using UPLC–MS/MS, *Food Anal. Meth.*, **6**, 727–736.
24. Vaclavikova, M., MacMahon S., Zhang K. *et al.* (2013). Application of single immunoaffinity clean up for simultaneous determination of regulated mycotoxins in cereals and nuts, *Talanta*, **117**, 345–351.
25. Lattanzio, V.M.T., Ciasca, B., Haidukowky, M. *et al.* (2013). 'LC–MS/MS multi-mycotoxin determination after immunoaffinity column clean up: a single laboratory validation design to evaluate contribution of selected factors to method performance characteristics', 6th International Symposium on Recent Advances in Food Analysis, 5–8 November 2013, Prague.

26. Van Pamel, E., Verbeken, A., Vlaemynck, G. *et al.* (2011). Ultrahigh-performance liquid chromatographic tandem mass spectrometric multimycotoxin method for quantitating 26 mycotoxins in maize silage, *J. Agric. Food Chem.*, **59**, 9747–9755.
27. Ventura, M., Guillén, D., Anaya, I. *et al.* (2006). Ultraperformance liquid chromatography–tandem mass spectrometry for the simultaneous analysis of aflatoxins B_1, G_1, B_2, G_2 and ochratoxin A in beer, *Rapid Commun. Mass Spectrom.*, **20**, 3199–3204.
28. Sorensen, L.K. and Elbaek, T.H. (2005). Determination of mycotoxins in bovine milk by liquid chromatography tandem mass spectrometry, *J. Chromatogr. B*, **820**, 183–196.
29. Lattanzio, V.M.T., Solfrizzo, M. and Visconti, A. (2008). Determination of trichothecenes in cereals and derived products by liquid chromatography–tandem mass spectrometry, *Food Addit.Contam.*, **25**, 320–330.
30. Lattanzio, V.M.T., Della Gatta, S., Suman, M. *et al.* (2011). Development and in house validation of a robust and sensitive solid phase extraction: LC-MS/MS method for the quantitative determination of aflatoxins B_1, B_2, G_1, G_2, ochratoxin A, deoxynivalenol, zearalenone, T-2 and HT-2 toxins in cereal based foods, *Rapid Commun. Mass Spectrom.*, **25**, 1869–1880.
31. Weingaertner, J., Krska, R., Praznik, W. *et al.* (1997). Use of Mycosep multifunctional clean-up columns for the determination of trichothecenes in wheat by electron-capture gas chromatography, *Fresenius' J. Anal. Chem.*, **357**, 1206–1210.
32. Krska, R. (1998). Performance of modern sample preparation techniques in the analysis of *Fusarium* mycotoxins in cereals, *J. Chromatogr. A*, **815**, 49–57.
33. Lattanzio, V.M.T., Pascale, M. and Visconti A. (2009). Current analytical methods for trichothecene mycotoxins in cereals, *Trends Anal. Chem.*, **28**, 758–768.
34. Baggiani, C., Anfossi, L. and Giovannoli, C. (2007). Solid phase extraction of food contaminants using molecular imprinted polymers, *Anal. Chim. Acta*, **591**, 29–39.
35. Ates, E., Mittendorf, K., Stroka, J. *et al.* (2013). Determination of *Fusarium* mycotoxins in wheat, maize and animal feed using on-line clean-up with high resolution mass spectrometry, *Food Addit. Contam.*, **30**, 156–165.

36. Zachariasova, M., Lacina, O., Malachova, A. *et al.* (2010). Novel approaches in analysis of *Fusarium* mycotoxins in cereals employing ultraperformance liquid chromatography coupled with high-resolution mass spectrometry, *Anal. Chim. Acta*, **662**, 51–61.
37. Rasmussen, R.R., Storm, I.M.L.D., Rasmussen, P.H. *et al.* (2010) Multimycotoxin analysis in maize silage by LC–MS/MS, *Anal. Bioanal. Chem.*, **397**, 765–776.
38. Yogendrarajah, P., Van Poucke, C., De Meulenaer, B. *et al.* (2013). Development and validation of a QuEChERs-based liquid chromatography–tandem mass spectrometry method for the determination of multiple mycotoxins in spices, *J. Chromatogr. A*, **1297**, 1–11.
39. Varga, E., Glauner, T., Berthiller, F. *et al.* (2013). Development and validation of a (semi-)quantitative UHPLC–MS/MS method for the determination of 191 mycotoxins and other fungal metabolites in almonds, hazelnuts, peanuts and pistachios, *Anal. Bioanal. Chem.*, **405**, 5087–5104.
40. Sulyok, M., Berthiller, F., Krska, R. *et al.* (2006). Development and validation of a liquid chromatography–tandem mass spectrometric method for the determination of 39 mycotoxins in wheat and maize, *Rapid Commun. Mass Spectrom.*, **20**, 2649–2659.
41. Sulyok, M., Krska, R. and Schuhmacher, R. (2007). A liquid chromatography–tandem mass spectrometric multi-mycotoxin method for the quantification of 87 analytes and its application to semi-quantitative screening of moldy food samples, *Anal. Bioanal. Chem.*, **389**, 1505–1523.
42. Spanjer, M., Rensen, P.M. and Sholten, J.M. (2008). LC–MS/MS multi-method for mycotoxins after single extraction, with validation data for peanut, pistachio, wheat, maize, cornflakes, raisins and figs, *Food Addit. Contam.*, **25**, 472–489.
43. Malachova, A., Dzuman, Z., Veprikova, Z. *et al.* (2011). Deoxynivalenol, deoxynivalenol-3-glucoside, and enniatins: the major mycotoxins found in cereal-based products on the Czech market, *J. Agric. Food Chem.*, **59**, 12990–12997.
44. Warth, B., Parich, A., Atehnkeng, J. *et al.* (2012). Quantitation of mycotoxins in food and feed from Burkina Faso and Mozambique using a modern LC–MS/MS multitoxin method, *J. Agric. Food Chem.*, **60**, 9352–9363.

45. Streit, E., Schwab, C., Sulyok, M. *et al.* (2013). Multi-mycotoxin screening reveals the occurrence of 139 different secondary metabolites in feed and feed ingredients, *Toxins*, **5**, 504–523.
46. Zachariasova, M., Cajka, T., Godula, M. *et al.* (2010). Analysis of multiple mycotoxins in beer employing (ultra)high-resolution mass spectrometry, *Rapid Commun. Mass Spectrom.*, **24**, 3335–3367.
47. Lattanzio, V.M.T., Della Gatta, S., Godula, M. *et al.* (2011). Quantitative analysis of mycotoxins in cereal foods by collision cell fragmentation–high-resolution mass spectrometry: performances and comparison with triple stage quadrupole detection, *Food Addit. Contam.*, **28**, 1424–1437.
48. Herebian, D., Zuhlke, S., Lamshoft, M. *et al.* (2009). Multi-mycotoxin analysis in complex biological matrices using LC–ESI/MS: experimental study using triple-stage quadrupole and LTQ-Orbitrap, *J. Sep. Sci.*, **32**, 939–948.
49. Rubert, J., Soler, C., Marín, R. *et al.* (2013). Mass spectrometry strategies for mycotoxins analysis in European beers, *Food Contr.*, **30**, 122–128.
50. Maragos, C.M. and Busman, M. (2010). Rapid and advanced tools for mycotoxin analysis: a review. *Food Addit. Contam.*, **27**, 688–700.
51. Vaclavik, L., Zachariasova, M., Hrbek, V. *et al.* (2010). Analysis of multiple mycotoxins in cereals under ambient conditions using direct analysis in real time (DART) ionization coupled to high-resolution mass spectrometry, *Talanta*, **82**, 1950–1957.
52. Gosetti, F., Mazzucco, E., Zampieri, D. *et al.* (2010). Signal suppression/enhancement in high-performance liquid chromatography–tandem mass spectrometry, *J. Chromatogr. A*, **1217**, 3929–3937.
53. Matuszewski, B.K., Constanzer, M.L. and Chavez-Eng, C.M. (2003). Strategies for the assessment of matrix effect in quantitative bioanalytical methods based on HPLC–MS/MS. *Anal. Chim. Acta*, **75**, 3019–3030.
54. Malachova, A., Sulyok, M., Schuhmacher, R. *et al.* (2013). Collaborative investigation of matrix effects in mycotoxin determination by high-performance liquid chromatography coupled to mass spectrometry, *Qual. Ass. Saf. Crop.*, **5**, 91–103.
55. Laganà, A., Curini, R., D'Ascenzo, G. *et al.* (2003). Liquid chromatography–tandem mass spectrometry for the identification and determination of trichothecenes in maize, *Rapid Commun. Mass Spectrom.*, **17**, 1037–1043.

56. Cavaliere, C., Foglia, P., Pastorini, E. *et al.* (2005). Development of a multiresidue method for analysis of major *Fusarium* mycotoxins in corn meal using liquid chromatography–tandem mass spectrometry, *Rapid Commun. Mass Spectrom.*, **19**, 2085–2093.
57. Dall'Asta, C., Sforza, S., Galaverna, G. *et al.* (2004). Simultaneous detection of type A and type B trichothecenes by LC/ESI-MS by using NaCl as cationization agent, *J. Chromatogr. A*, **1054**, 389–395.
58. Razzazi-Fazeli, E., Böhm, J., Jarukamjorn, K. *et al.* (2003). Simultaneous determination of major B-trichothecenes and de-epoxy-metabolite of deoxynivalenol in pig urine and maize using high-performance liquid chromatography–mass spectrometry, *J. Chromatogr. B*, **796**, 21–33.
59. Di Mavungu, J.D., Monbaliu, S., Scippo, M.L. *et al.* (2009). LC–MS/MS multi-analyte method for mycotoxin determination in food supplements, *Food Addit. Contam.*, **26**, 885–895.
60. Pallaroni, L. and von Holst, C. (2003). Determination of zearalenone from wheat and corn by pressurized liquid extraction and liquid chromatography–electrospray mass spectrometry, *J. Chromatogr. A*, **993**, 39–45.
61. Ren, Y., Zhang, Y., Shao, S. *et al.* (2007). Simultaneous determination of multi-component mycotoxins contaminants in foods and feeds by ultraperformance liquid chromatography–tandem mass spectrometry, *J. Chromatogr. A*, **1143**, 48–64.
62. Berger, U., Oehme, M. and Kuhn, F. (1999). Quantitative determination and structure elucitation of type A- and B-trichothecenes by HPLC/ion trap multiple mass spectrometry, *J. Agric. Food Chem.*, **47**, 4240–4245.
63. Klötzel, M., Gutsche, B., Lauber, U. *et al.* (2005). Determination of 12 type A and B trichothecenes in cereals by liquid chromatography–electrospray ionization–tandem mass spectrometry. *J. Agric. Food Chem.*, **53**, 8904–8910.
64. Hartl, M., Herderich, M. and Humpf, H.U. (1999). Rapid determination of fumonisin FB_1 in corn products by high-performance liquid chromatography–electrospray mass spectrometry, *Eur. Food Res. Technol.*, **209**, 348–351.
65. Bretz, M., Beyer, M., Cramer, B. *et al.* (2006). Stable isotope dilution analysis of the *Fusarium* mycotoxins deoxynivalenol and 3-acetyldeoxynivalenol, *Mol. Nutr. Food Res.*, **50**, 251–260.

66. Cramer, B., Bretz, M. and Humpf, H.U. (2007). Stable isotope dilution analysis of *Fusarium* mycotoxin zearalenone, *J. Agric. Food Chem.*, **55**, 8353–8358.
67. Lindenmeier, M., Schieberle, P. and Rychlik, M. (2004). Quantification of ochratoxin A in foods by a stable isotope dilution assay using high-performance liquid chromatography–tandem mass spectrometry, *J. Chromatogr. A*, **1023**, 57–66.
68. Cervino, C., Asam, S., Knopp, D. *et al.* (2008). Use of isotope-labelled aflatoxins for LC–MS/MS stable isotope dilution analysis of foods, J. *Agric. Food Chem.*, **56**, 1873–1879.
69. Rychlik, M. and Asam, S. (2008). Stable isotope dilution assays in mycotoxin analysis, *Anal. Bioanal. Chem.*, **390**, 617–628.
70. Häubl, G., Berthiller, F., Krska, R. *et al.* (2006). Suitability of a fully 13C isotope-labeled internal standard for the determination of the mycotoxin deoxynivalenol by LC–MS/MS without clean up, *Anal. Bioanal. Chem.*, **384**, 692–696.
71. Häubl, G., Berthiller, F., Hametner, C. *et al.* (2007). Characterization of ($^{13}C_{24}$) T-2 toxin and its use as an internal standard for the quantification of T-2 toxin in cereals with HPLC–MS/MS, *Anal. Bioanal. Chem.*, **389**, 931–940.
72. Asam, S. and Rychlik, M. (2007). Quantitation of type B-trichothecene mycotoxins in foods and feeds by a multiple stable isotope dilution assay, *Eur. Food Res. Technol.*, **224**,769–783.
73. Kellmann, M., Muenster, H., Zomer, P. *et al.* (2009). Full-scan MS in comprehensive qualitative and quantitative residue analysis in food and feed matrices: how much resolving power is required?, *J. Am. Soc. Mass Spectrom.*, **20**, 1464–1476.
74. Commission Decision No 657/2002 of 12 August 2002 implementing Council Directive 96/23/EC concerning the performance of analytical methods and the interpretation of result, *Off. J. Eur. Union*, L 221/8.
75. Document SANCO/12495/2011. Method Validation and Quality Control Procedures for Pesticide Residues Analysis in Food and Feed. EU Reference Laboratories for Residues of Pesticides, European Union. Available at http://eurl-pesticides.eu/library/docs/allcrl/AqcGuidance_Sanco_2011_12495.pdf

76. Solfrizzo, M., De Girolamo, A., Lattanzio, V.M.T. *et al.* (2013). Results of a proficiency test for multi-mycotoxin determination in maize by using methods based on LC–MS/(MS), *Qual. Assur. Saf. Crop.*, **5**, 15–48.
77. De Girolamo, A., Solfrizzo, M., Lattanzio, V.M.T. *et al.* (2013). Critical evaluation of LC–MS-based methods for simultaneous determination of deoxynivalenol, ochratoxin A, zearalenone, aflatoxins, fumonisins and T-2/HT-2 toxins in maize, *World Mycotoxin J.*, **6**, 317–334.
78. Senyuva, H.Z., Gilbert, J., Türköz, G. *et al.* (2012). Analysis of deoxynivalenol, zearalenone, T-2, and HT-2 toxins in animal feed by LC–MS/MS: a critical comparison of immunoaffinity column cleanup with no cleanup, *J. AOAC Int.*, **95**, 1701–1708.
79. Lattanzio, V.M.T., Visconti, A., Haidukowski, M. *et al.* (2012). Identification and characterization of new *Fusarium* masked mycotoxins, T2 and HT2 glycosyl derivatives, in naturally contaminated wheat and oats by liquid chromatography–high-resolution mass spectrometry, *J. Mass Spectrom.*, **47**: 466–475.
80. Veprikova, Z., Vaclavikova, M., Lacina, O. *et al.* (2012). Occurrence of mono- and di-glucosylated conjugates of T-2 and HT-2 toxins in naturally contaminated cereals, *World Mycotoxin J.*, **5**, 231–240.
81. Nakagawa, H., Sakamoto, S., Sago, Y. *et al.* (2013). Detection of masked mycotoxins derived from type-A trichothecenes in corn by high-resolution LC-Orbitrap mass spectrometer, *Food Addit. Contam.*, **30**, 1407–1414.
82. De Girolamo, A., Lattanzio, V.M.T., Schena, R. *et al.* (2013) Use of liquid chromatography–high-resolution mass spectrometry for isolation and characterization of hydrolyzed fumonisins and relevant analysis in maize based products, *J. Mass Spectrom., submitted.*
83. Lattanzio, V.M.T., Ciasca, B., Haidukowski, M. *et al.* (2013). Mycotoxin profile of *Fusarium langsethiae* isolated from wheat in Italy: production of type-A trichothecenes and relevant glucosyl derivatives, *J. Mass Specrom.*, **48**, 1291–1298.
84. Lattanzio, V.M.T., Solfrizzo, M., De Girolamo, A. *et al.* (2011). LC–MS/MS characterization of the urinary excretion profile of the mycotoxin deoxynivalenol in human and rat, *J. Chromatogr. B*, **879**, 707–715.
85. Bartók, T., Szécsi, A., Szekeres, A. *et al.* (2006). Detection of new fumonisin mycotoxins and fumonisin-like compounds by reversed-phase

high-performance liquid chromatography–ion trap mass spectrometry, *Rapid Commun. Mass Spectrom.*, **20**, 2447–2462.

86. Bartók, T., Tölgyesi, L., Mesterházy, A. *et al.* (2010). Identification of the first fumonisin mycotoxins with three acyl groups by ESI–ITMS and ESI–TOF-MS following RP–HPLC separation: palmitoyl, linoleoyl and oleoyl EFB1 fumonisin isomers from a solid culture of *Fusarium verticillioides*, *Food Addit. Contam.*, **27**, 1714–1723.
87. Mol, H.G.J., Plaza-Bolaños, P., Zomer, P. *et al.* (2008). Toward a generic extraction method for simultaneous determination of pesticides, mycotoxins, plant toxins, and veterinary drugs in feed and food matrixes, *Anal. Bioanal. Chem.*, **80**, 9450–9459.
88. Herrmann, A., Rosén, J., Jansson, D. *et al.* (2012). Evaluation of a generic multi-analyte method for detection of >100 representative compounds correlated to emergency events in 19 food types by ultrahigh-pressure liquid chromatography–tandem mass spectrometry, *J. Chromatogr. A*, **1235**, 115–124.
89. Lacina, O., Zachariasova, M., Urbanova, J. *et al.* (2012). Critical assessment of extraction methods for the simultaneous determination of pesticide residues and mycotoxins in fruits, cereals, spices and oil seeds employing ultrahigh- performance liquid chromatography–tandem mass spectrometry, *J. Chromatogr. A*, **1262**, 8–18.
90. Zhang, J.M., Wu, Y.L. and Lu, Y.B. (2013). Simultaneous determination of carbamate insecticides and mycotoxins in cereals by reversed-phase liquid chromatography–tandem mass spectrometry using a quick, easy, cheap, effective, rugged and safe extraction procedure, *J. Chromatogr. B*, **915-916**, 13–20.
91. Capriotti, A.L., Cavaliere, C., Piovesana, S. *et al.* (2012). Multiclass screening method based on solvent extraction and liquid chromatography–tandem mass spectrometry for the determination of antimicrobials and mycotoxins in egg, *J. Chromatogr. A*, **1268**, 84–90.
92. De Dominicis, E., Commissati, I. and Suman, M. (2012). Targeted screening of pesticides, veterinary drugs and mycotoxins in bakery ingredients and food commodities by liquid chromatography–high-resolution single-stage Orbitrap mass spectrometry, *J. Mass Spectrom.*, **47**, 1232–1241.
93. Filigenzi, M.S., Ehrke, N., Aston, L.S. *et al.* (2011). Evaluation of a rapid screening method for chemical contaminants of concern in four

food-related matrices using QuEChERS extraction, UHPLC and high-resolution mass spectrometry, *Food Addit. Contam., Part A*, **28**, 1324–1339.

94. Romero-Gonzalez, R., Garrido Frenich, A., Martinez Vidal, J.L. *et al.* (2011). Simultaneous determination of pesticides, biopesticides and mycotoxins in organic products applying a quick, easy, cheap, effective, rugged and safe extraction procedure and ultrahigh-performance liquid chromatography–tandem mass spectrometry, *J. Chromatogr. A*, **1218**, 1477–1485.
95. Mornar, A., Sertic, M. and Nigovic, B. (2013). Development of a Rapid LC/DAD/FLD/MSn Method for the simultaneous determination of monacolins and citrinin in red fermented rice products, *J. Agric. Food Chem.*, **61**, 1072–1080.
96. Song, S., Ediage, E.N., Wu, A. *et al.* (2013). Development and application of salting-out assisted liquid–liquid extraction for multi-mycotoxin biomarkers analysis in pig urine with high-performance liquid chromatography–tandem mass spectrometry, *J. Chromatogr. A*, **1292**, 111–120.
97. Nácher-Mestre, J., Ibáñez, M., Serrano, R. *et al.* (2013). Qualitative screening of undesirable compounds from feeds to fish by liquid chromatography coupled to mass spectrometry, *J. Agric. Food Chem.*, **61**, 2077–2087.

Index